Recent Advances in Isotope Hydrology

Recent Advances in Isotope Hydrology

Edited by Teddy Reynolds

MURPHY & MOORE
www.murphy-moorepublishing.com

Published by Murphy & Moore Publishing,
1 Rockefeller Plaza,
New York City, NY 10020, USA
www.murphy-moorepublishing.com

Recent Advances in Isotope Hydrology
Edited by Teddy Reynolds

International Standard Book Number: 978-1-63987-477-4 (Hardback)

Cataloging-in-Publication Data

Recent advances in isotope hydrology / edited by Teddy Reynolds.
p. cm.
Includes bibliographical references and index.
ISBN 978-1-63987-477-4
1. Radioisotopes in hydrology. 2. Hydrologic cycle. 3. Stable isotopes. I. Reynolds, Teddy.
GB656.2.R34 R43 2022
551.480 28--dc23

TABLE OF CONTENTS

PREFACE

The interdisciplinary field of geochemistry and hydrology, which deals with isotope utilization to estimate the age and origin of surface and groundwater is known as isotope hydrology. It also studies the circulation and movement of water with the hydrologic cycle. Water molecules carry unique isotope fingerprints located in parts of different proportions of oxygen and hydrogen isotopes which constitute the water molecule. It also helps in determining the age of ice or snow using stable isotopes. The techniques used are beneficial for conserving water supplies, controlling pollution and mapping aquifers. This field also has applications in ecohydrology. It is used to understand the impact of humans on the varied dimensions of the hydrological cycle and other ecosystem services. This book unravels the recent studies in the field of isotope hydrology. The ever-growing need of advanced technology is the reason that has fueled the research in this field in recent times. For all those who are interested in isotope hydrology, this book can prove to be an essential guide.

After months of intensive research and writing, this book is the end result of all who devoted their time and efforts in the initiation and progress of this book. It will surely be a source of reference in enhancing the required knowledge of the new developments in the area. During the course of developing this book, certain measures such as accuracy, authenticity and research focused analytical studies were given preference in order to produce a comprehensive book in the area of study.

This book would not have been possible without the efforts of the authors and the publisher. I extend my sincere thanks to them. Secondly, I express my gratitude to my family and well-wishers. And most importantly, I thank my students for constantly expressing their willingness and curiosity in enhancing their knowledge in the field, which encourages me to take up further research projects for the advancement of the area.

Editor

1

Continual in situ monitoring of pore water stable isotopes in the subsurface

T. H. M. Volkmann and M. Weiler

Chair of Hydrology, Faculty of Environment and Natural Resources, University of Freiburg, Fahnenbergplatz, 79098 Freiburg, Germany

Correspondence to: T. H. M. Volkmann (till.volkmann@hydrology.uni-freiburg.de)

Abstract. Stable isotope signatures provide an integral fingerprint of origin, flow paths, transport processes, and residence times of water in the environment. However, the full potential of stable isotopes to quantitatively characterize subsurface water dynamics is yet unfolded due to the difficulty in obtaining extensive, detailed, and repeated measurements of pore water in the unsaturated and saturated zone. This paper presents a functional and cost-efficient system for non-destructive continual in situ monitoring of pore water stable isotope signatures with high resolution. Automatic controllable valve arrays are used to continuously extract diluted water vapor in soil air via a branching network of small microporous probes into a commercial laser-based isotope analyzer. Normalized liquid-phase isotope signatures are then obtained based on a specific on-site calibration approach along with basic corrections for instrument bias and temperature dependent isotopic fractionation. The system was applied to sample depth profiles on three experimental plots with varied vegetation cover in southwest Germany. Two methods (i.e., based on advective versus diffusive vapor extraction) and two modes of sampling (i.e., using multiple permanently installed probes versus a single repeatedly inserted probe) were tested and compared. The results show that the isotope distribution along natural profiles could be resolved with sufficiently high accuracy and precision at sampling intervals of less than four minutes. The presented in situ approaches may thereby be used interchangeably with each other and with concurrent laboratory-based direct equilibration measurements of destructively collected samples. It is thus found that the introduced sampling techniques provide powerful tools towards a detailed quantitative understanding of dynamic and heterogeneous shallow subsurface and vadose zone processes.

1 Introduction

Stable isotope ratios of hydrogen ($^{2}H/^{1}H$) and oxygen ($^{18}O/^{16}O$) provide powerful tracers of water in the hydrologic cycle (Gat, 1996). This is due to the conservative nature of the stable isotopes as an intrinsic part of the water molecule and the measurable variations of their ratios in precipitated and other natural waters imposed by large and variable fractionations during physicochemical processes and reactions such as, and most importantly, phase changes (Craig, 1961; Dansgaard, 1964; Friedman et al., 1964; Ehhalt and Knott, 1965; Majoube, 1971). The isotopic distinctness of input waters to a system in space and time provides for a continuously and naturally imprinted signature that can be traced to reveal information on transit times, origin, flow pathways, and processes such as phase transitions, transport, and exchange.

As such, the analysis of stable water isotopes is being widely employed in hydrology and related disciplines and has brought major advances to our understanding of natural systems (see reviews by Ehleringer and Dawson, 1992; Kendall and McDonnell, 1998; Yakir and Sternberg, 2000; Vitvar et al., 2005; Aggarwal et al., 2007; West et al., 2010b; Soderberg et al., 2012). In particular, stable isotope signatures of pore water have provided insight into soil evaporation (Zimmermann et al., 1967; Allison, 1982; Allison et al., 1983; Barnes and Allison, 1988; Walker et al., 1988); recharge rates (Dincer et al., 1974; Saxena, 1984; Darling and Bath, 1988); unsaturated and saturated subsurface flow processes, mixing, and residence times (Sklash and Farvolden, 1979; Buttle and Sami, 1990; McDonnell, 1990; Stewart and McDonnell, 1991; Gazis and Feng, 2004; Laudon et al., 2004; Garvelmann et al., 2012);

evapotranspiration partitioning (Hsieh et al., 1998a; Wang and Yakir, 2000; Yepez et al., 2005; Rothfuss et al., 2010); hydraulic lift (Dawson, 1993, 1996; Schulze et al., 1998); and sources of plant water uptake (White et al., 1985; Dawson and Ehleringer, 1991; Walker and Richardson, 1991; Thorburn et al., 1992).

However, the full potential of stable water isotopes to quantitatively characterize water dynamics is yet to be unfolded. This particularly holds with regard to complex and heterogeneous processes, such as infiltration and shallow subsurface flow in the vadose zone subject to strong gradients and fluxes of energy and matter, demanding stable isotope data with high spatial and temporal resolution to complement traditional observations. Instead, a major limitation to the extent and scope of stable isotope applications was imposed by the available techniques for sampling and analysis (e.g., Kerstel and Meijer, 2005; Helliker and Noone, 2010; Munksgaard et al., 2011). Conventionally, measurement of water stable isotopic composition was relatively labor-, time-, and cost-intensive and constrained to laboratory-bound analysis of previously collected and processed discrete samples based on gas source isotope ratio mass spectrometry (IRMS) (Horita and Kendall, 2004). Pore water samples were most commonly obtained through destructive collection of geologic material and subsequent extraction using various methods (e.g., Buttle and Sami, 1990; Ingraham and Shadel, 1992; Walker et al., 1994; West et al., 2006) or, less frequently, using suction or gravity lysimeters (e.g., Stewart and McDonnell, 1991; Wels et al., 1991; Landon et al., 1999; Figueroa-Johnson et al., 2007). Each method is, however, associated with considerable uncertainty and unclear implications for the representativeness of obtained pore water isotopic signatures.

In recent years, major advances in analyzing techniques have come with the advent of commercially available, portable, and field-deployable laser spectroscopic instruments for isotopic analysis of water vapor (see reviews by Kerstel and Meijer, 2005; Kerstel and Gianfrani, 2008; Helliker and Noone, 2010). Available instruments based on, for instance, wavelength-scanned cavity ring-down spectroscopy (WS-CRDS) allow for direct simultaneous high-frequency measurements of oxygen and hydrogen stable isotopic composition of water vapor at a precision and accuracy approaching that of IRMS (Berden et al., 2000; Crosson, 2008; Kerstel and Gianfrani, 2008; Gupta et al., 2009). Such isotope analyzers promote broader usage of stable isotope analysis by allowing measurement of collected liquid and vapor phase samples with minimal or no preparation, higher throughput rates, and lower cost (Gupta et al., 2009).

In addition, current instruments facilitate direct continuous in situ monitoring of stable isotopes in, for example, atmospheric moisture (e.g., Gupta et al., 2009; Iannone et al., 2010; Sturm and Knohl, 2010; Johnson et al., 2011; Aemisegger et al., 2012). Other recent studies have used WS-CRDS for continuous determination of liquid water isotopic composition in laboratory experiments (Munksgaard et al., 2011; Herbstritt et al., 2012) and field applications (Munksgaard et al., 2012a, b) by direct intake and analysis of vapor diffused into a dry gas stream via a submerged semipermeable membrane body. A similar direct equilibration methodology can also be applied to determine the isotope signature of liquid pore water in situ as recently shown by Rothfuss et al. (2013) in a sand beaker laboratory experiment. In fact, the feasibility of inferring the stable isotopic composition of soil water by in situ extraction of water vapor or carbon dioxide (CO_2) from soil air has already been indicated by Thoma et al. (1979) and Allison et al. (1987), while more recently IRMS and laser spectroscopic instruments have been applied to analyze pore water in destructively collected geologic material by headspace sampling of equilibrated water vapor and CO_2 (Hsieh et al., 1998b; Koehler et al., 2000; Wassenaar et al., 2008; Garvelmann et al., 2012).

Building upon the previous work, this paper presents the development and application of a novel field system for continual unattended in situ monitoring of liquid pore water stable isotopes by delivering diluted soil water vapor via small microporous soil water isotope probes (SWIPs) from an arbitrary number of subsurface sources into a commercial WS-CRDS instrument. Representing a trade-off between expenditure and universality in application, two different implementations of the monitoring system are investigated. These are henceforth referred to as the advection dilution sampling (ADS) method and the diffusion dilution sampling (DDS) method, according to the respective dominant vapor transport mechanism across the microporous probing tube (as in Munksgaard et al., 2011). The aim was to develop a functional system suitable for monitoring dynamic and heterogeneous subsurface processes at the relevant temporal and spatial scales. The system shall therefore facilitate accurate and precise in situ measurements while meeting the following desirable attributes:

- inexpensive in terms of both acquisition and maintenance and expandable to a large number of probing locations as prerequisite to broad and locally dense application and thus, high spatial resolution;
- short sampling interval to facilitate frequent measurements and thus, high temporal resolution;
- unattended applicability under most conditions encountered in the subsurface throughout the day and year as a prerequisite for continuous observations;
- compact (subsurface components), easy and quick to install imposing little disturbance and obstruction to the natural subsurface system and processes under investigation;

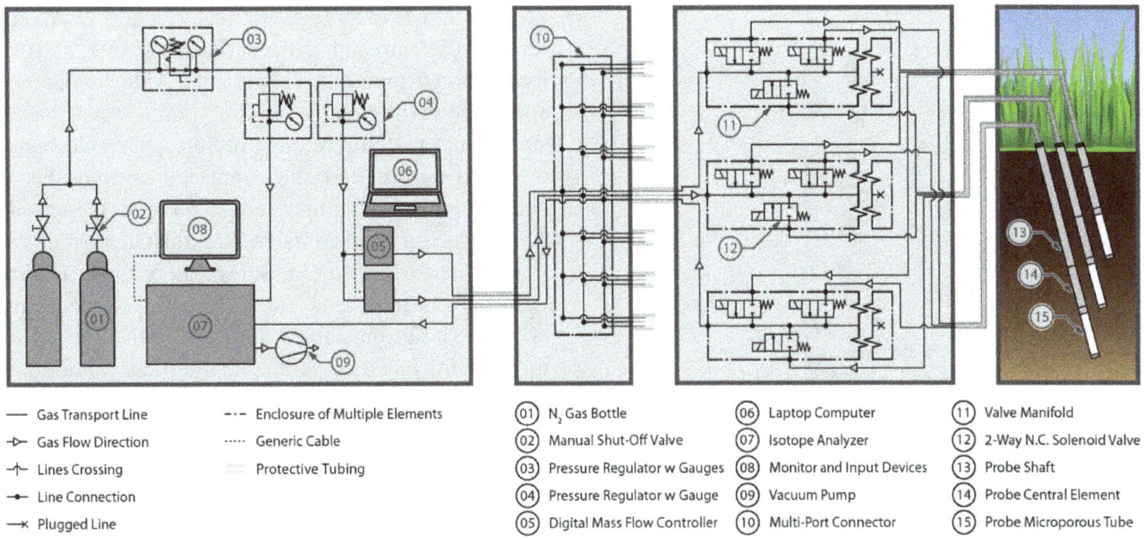

Figure 1. Schematic representation of the in situ pore water stable isotope monitoring system setup according to the DDS method.

- mechanically robust and largely chemically inert, weathering resistant components to facilitate safe insertion and prolonged duration in natural subsurface environments.

2 Methods

In this section, the basic setup, working principle, operation, and data processing of the monitoring system are first described, followed by the field evaluation approach. Throughout this paper, the δ-notation (e.g., δ^{18}O or δ^{2}H) is used to express the isotopic composition of a sample in units of parts per thousand (‰). For a given element (e.g., O or H), this is defined as

$$\delta = \left(\frac{R_{\text{sample}}}{R_{\text{reference}}} - 1 \right) \cdot 1000\text{‰}, \tag{1}$$

where R_{sample} denotes the atomic ratio of a rare to the most abundant isotope of this element (e.g., $^{18}\text{O}/^{16}\text{O}$ or $^{2}\text{H}/^{1}\text{H}$) in a sample, and $R_{\text{reference}}$ is the same ratio of the employed international reference standard, Vienna Standard Mean Ocean Water (VSMOW). Subscripts "l" and "v" are used to indicate reference to liquid and vapor phase water isotopic composition, respectively.

2.1 In situ monitoring system

2.1.1 Monitoring system setup

A schematic representation of the monitoring system setup is provided in Fig. 1. All components were selected in compliance with the above stated desirable attributes. The system is builtup modularly and can simply be modified or augmented to accommodate a certain number of probes with relatively little extra cost. A single set of central analyzing and supply devices is sufficient to operate a multitude of SWIP sampling locations via a branching tubing network and valve control system. A commercial laser spectroscopic instrument (Picarro L2120-i WS-CRDS, Santa Clara, CA, USA) is used, analyzing an air stream drawn through the attached main sample line by a downstream vacuum pump for $\delta^{18}\text{O}_\text{v}$, $\delta^{2}\text{H}_\text{v}$, and vapor concentration at a frequency of 0.5 Hz. Further, a nitrogen gas (N_2) supply installation is employed, comprising a regulated pressurized source to the main dilution line with a flow rate adjustable to a target value sequence imposed by a programmable digital mass flow controller (Analyt-MTC, Müllheim, Germany). In case of the diffusion dilution sampling (DDS) method, the N_2 supply is split upstream of the first mass flow controller and a second mass flow controller is used to regulate the flow rate through the main throughflow line. The CRDS instrument and N_2 supply installations are housed in a customized trailer for field deployment. The trailer comprises a voltage transformer and high capacity accumulators to enable stable and gap-free operation using an electric generator power source in remote field applications and is vented to prevent instrument overheating.

Each of the main gas transport lines is branched in two steps to facilitate operation of a multitude of probes at varying locations with minimum total tubing length. The main gas transport lines are guided from the trailer to a central distribution box where radial multi-port connectors (Upchurch Scientific, Oak Harbor, WA, USA) are used to distribute the main lines towards local control boxes associated with a cluster of probes installed in the subsurface (e.g., at varying depths beneath a specific surface location). Arrays of two-way normally closed electromechanically actuated solenoid valves (Clippard Minimatic, Cincinnati, OH, USA) mounted on manifold blocks (custom manufactured from polyvinylidene fluoride (PVDF) by Fischer Plastics,

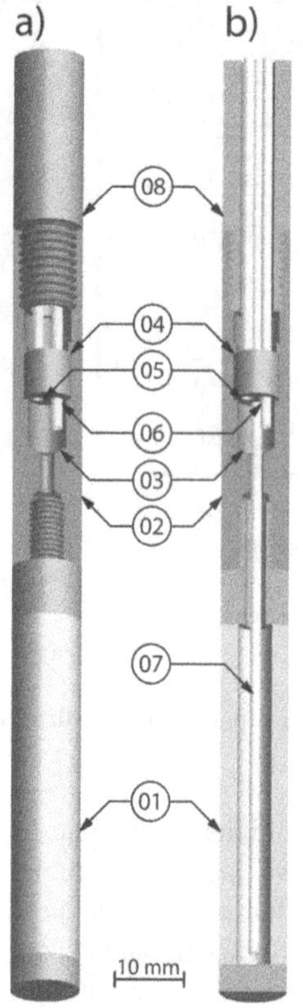

Figure 2. Schematic three-dimensional representation of the in situ soil water isotope probes according to **(a)** the ADS and **(b)** the DDS method. Labels indicate (1) microporous tube, (2) central element, (3) mixing chamber, (4) line retainer, (5) sample line, (6) dilution line, (7) throughflow line, and (8) insertion and protection shaft.

Gundelfingen, Germany) are used to further split and control the incoming gas transport lines towards individual probes. The valves are actuated either manually or automatically by a microcontroller switching unit. Valves controlling the ports receiving gas transport lines of a specific probe on each manifold are coupled and operated simultaneously.

All tubing used for the monitoring system gas transport lines consists of fluorinated ethylene propylene (FEP) and is protected by mesh hoses. All sample lines and all lines immediately attached to the probes have an inner diameter (ID) of 1 mm, while the remaining N_2 supply lines have an ID of 1/16 in. The total effective length of each gas transport line (from each probe to each end device) depends on the study design and field situation and amounted to ca. 10 m in the present study. Outside the trailer, all tubing connections can quickly and without tools be finger tightened using removable and reusable fittings (Upchurch Scientific), while stainless steel fittings (Swagelok, Solon, OH, USA) are used inside the trailer.

Each soil water isotope probe (Fig. 2) is composed of three major custom parts connectable via threads. The core of the probe is a rigid hydrophobic microporous polyethylene (PE) probing tube (Porex Technologies, Aachen, Germany) with a length of 50 mm and an outer diameter (OD) of 10 mm. The hydrophobic material allows water vapor contained in the soil air to pass while prohibiting liquid water penetration at sufficiently low differential pressures. The microporous tube is sealed on one end and connected on the other end to a central element manufactured from PVDF (Fischer Plastics). The central element contains a mixing chamber and accepts a line retainer with two (ADS; Fig. 2a) or three (DDS; Fig. 2b) gas transport lines attached. The central element can be connected to a rigid insertion shaft of variable length (here, 100 to 700 mm) made of anodized aluminum, guiding and protecting the gas transport lines in the subsurface. This design facilitates convenient probe installation from the soil surface with little disturbance of the subsurface environment. The probes are installed into pre-drilled holes using a steel tube with an OD of 10 mm, thus favoring immediate close contact of the subsequently inserted probes with and preventing compaction of the surrounding soil matrix.

The system can be used to sample the distribution of pore water stable isotopic composition in two different modes. On the one hand, a multitude of soil probes can be permanently installed in clusters to continually monitor, for example, depth profiles below various surface locations (i.e., continual mode). On the other hand, a single probe can be used to subsequently sample spatial distributions by repeated insertion at various locations and depths (i.e., push-in mode). Of course, continuous sampling using a single permanently installed probe is also possible; however, this will seldom present a relevant mode of field application.

A specific calibration approach following the principle of identical treatment is used to facilitate normalization of measured isotope signatures to the VSMOW reference scale. Therefore, four polyvinyl chloride (PVC) boxes of size 20 cm × 30 cm × 30 cm were equipped with a centrally placed SWIP, closely neighbored by two 5TE probes and one MPS-1 probe (both Decagon Devices, Pullman, WA, USA) sensing soil water content, temperature, electrical conductivity, and matric potential. The boxes were then filled with oven-dried soil material from the upper 60 cm of the study site. Subsequently, water of known isotopic composition was added to generate a soil water content of ca. 20 vol% at the depth of the probes and the boxes were immediately sealed. The waters used included an isotopically light and a heavy laboratory standard serving for normalization during calibration and an intermediate laboratory standard used twice (i.e., for two boxes) and serving for quality control. The respective boxes are henceforth referred to as calibration and control standard boxes. During field sampling, three of the boxes were placed into a thermo-insulated housing to reduce temperature variations, while one control standard box was placed outside the housing, thus being subject to stronger temperature variations. The initial setup of the boxes requires approximately two hours of work and less than USD 100 of monetary effort (excluding soil physical and soil water isotope probes). Preparation for a specific application requires approximately three hours of work (excluding collection of soil material).

2.1.2 Monitoring system operation

An individual probe is operated by simultaneously opening the associated valve on each of the manifolds, while all other valves are closed. Once the valves are open, sample air is drawn through the sample line at a rate of 30–35 mL min^{-1} imposed by the vacuum pump of the CRDS instrument. At the same time, N_2 is supplied via the dilution line into the mixing chamber with a controllable rate. If the N_2 supply rate is lower than the sample intake rate, soil air saturated with water vapor at the local temperature advects through the walls of the microporous probing tube in response to the applied absolute pressure gradient. The soil air flows into the mixing chamber, where it is diluted to a lower vapor concentration as determined by the ratio of N_2 supply rate to sample intake rate. This diluted vapor-sample stream enters the sample line and flows to the CRDS instrument (ADS method). By varying the N_2 supply rate, the vapor mixing ratio can be controlled from close to zero to saturation at soil source temperature. The primary purpose of the vapor dilution is to preclude any condensation of water due to varying temperature along the sample line while avoiding a sample line heating system.

In case of the DDS method, an additional third gas transport line (throughflow line) supplies N_2 to the lower end of the probe. The rate is regulated by a second mass flow controller to equal the difference between sample intake rate and dilution rate, such that no absolute pressure gradients result across the microporous wall. Instead, driven by a partial pressure gradient, vapor diffuses into the probing tube to saturate the through flowing N_2 carrier gas. As for the ADS method, the sample stream is transported to the analyzer and diluted in the process. While this method raises the expenses for monitoring system acquisition and operation, it provides the advantage that it can be applied under saturated conditions as long as the water entry pressure of the employed microporous material is not exceeded.

The monitoring system allows for manual or automated sampling using programmed probe and dilution rate sequences. In a typical probe sequence, all SWIPs are subsequently sampled, moving from one cluster to the next, occasionally intermitted by sampling the standard boxes. A given location is typically probed using a dilution rate sequence starting with a short flushing phase with high dilution, thus clearing the mixing chamber, sample lines, and cavity (i.e., the entire path for the new sample vapor) from vapor of previous samples. This is followed by a sampling phase with lower dilution. In the present application, each in situ measurement was conducted using a sequence of 60 s flushing and 150 s sampling. Thereby, dilution flow rates of 36 and 18 mL min^{-1} were used, respectively. The former

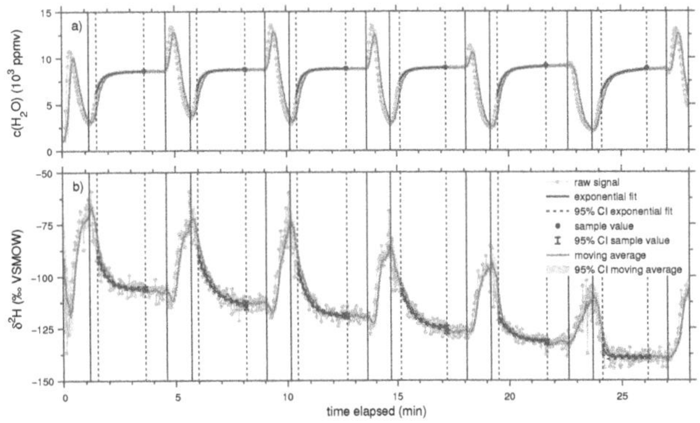

Figure 3. Exemplary time series signal representing the acquisition of a pore water stable isotope profile using the in situ sampling system with permanently installed ADS probes and a dilution sequence of 60 s flushing and 210 s sampling. Solid vertical lines enclose periods of sample arrival. Exponential fits use data from the asymptotic period up to 150 s after start of sample arrival, as enclosed by the dashed vertical lines. A moving average (15 s backward) is provided for illustrative purposes.

slightly exceeds the nominal sample intake rate, while the latter presents the minimal dilution rate necessary to avoid condensation during > 95 % of the time of a day and year in the region of application, as computed from multi-year temperature records.

2.1.3 Signal processing

Processing of the acquired time series signals is fully automated into a MATLAB® (The MathWorks, Inc., Natick, MA, USA) toolbox. Standard processing for the CRDS-based measurements includes derivation of raw sample values from the observed high-frequency signals, subsequent corrections to account for instrument-specific vapor concentration biases, drift, and deviations from the internationally accepted reference scale (e.g., Schmidt et al., 2010; Sturm and Knohl, 2010; Gröning, 2011; Aemisegger et al., 2012), as well as conversion of the corrected vapor to respective liquid water isotopic compositions.

Inspecting the acquired time series of water vapor concentration (Fig. 3a), individual sample phases can be easily identified as step input responses with a vapor concentration asymptotically approaching a level depending on the soil temperature and the imposed dilution rate. The asymptotic behavior can be primarily attributed to advective–diffusive–reactive transport through the soil pore space and the various segments of the probing system as well as storage effects such as cavity reservoir gas exchange. The preceding flush phase is characterized by an initial peak followed by declining vapor concentration, emptying the sample line for the next sample intake. Note also that a short time lag (ca. 20 s) exists between observed responses and switching of valves and dilution rates due to the bulk transport through the sample lines.

To determine the vapor concentration and raw isotopic composition of each sample, the acquired time series of vapor concentration is first scanned to identify characteristic times based on differentiation of the signal (i.e., using a multi-point moving slope to reduce the impact of noise). These include the defined start time of sample arrival, start time of asymptotic behavior, and end time of sample arrival, which are identified as the stationary point, first inflection point (from convex to concave downwards), and second inflection point (from concave to convex downwards) following a declining flush phase, respectively. Second, sample values are obtained as steady-state solution of a simple exponential function fitted to the respective asymptotic period of sample arrival in a least squares sense. Using the root mean squared error (RMSE) as quality criterion, it was found that exponential models of type

$$\delta(t) = \delta_\infty + (\delta_0 - \delta_\infty)e^{(-t/\tau)}, \quad (2)$$

where t indicates the time, τ is a time constant, and subscripts 0 and ∞ indicate the initial and steady-state solution, respectively, fit well (RMSE < 0.4 ‰ for $\delta^{18}O_v$ and < 3.3 ‰ for δ^2H_v) to the observed δ-value response signal. Further, exponential models of type

$$c(t) = c_\infty + (c_0 - c_\infty)e^{(-t/\tau_1)} + \lambda e^{(-t/\tau_2)}, \quad (3)$$

where τ_1 and τ_2 are time constants and λ is a parameter, fit well (RMSE < 80 ppmv) to the observed vapor concentration c response signal. Note that device related corrections (e.g., vapor concentration dependency) may be applied prior to fitting a model in case they are non-linear. The 95 % confidence interval (CI) of the exponential model fits is computed based on the estimated coefficient covariance matrix (on average 0.19 ‰ for $\delta^{18}O_v$, 1.5 ‰ for δ^2H_v, and 11 ppmv for c) and illustrates the uncertainty of sample value estimation from the signal (Fig. 3). While some average value over a specified segment of the asymptotic period could actually be used without significant deterioration of data quality, the asymptotic model makes better use of the data, can reduce the impact of noise and increase the precision, is independent of an averaging interval, less dependent on the selected period, and provides for a statistically more valid approach. These aspects are of particular relevance when short sampling times are targeted. However, we omit fitting a more complex model to the entire response signal at this point, partially to avoid the impact of noisy observations at low vapor concentration.

The obtained raw sample values of $\delta^{18}O_v$ and δ^2H_v are then corrected for analyzer specific vapor concentration dependent bias relative to an arbitrary reference concentration. Least-squares-fitted relations were derived prior to the field sampling based on vaporized discrete laboratory standard water aliquots of three isotopic compositions and varying volume. The aliquots were injected into a vaporizer using a liquid auto sampler and N_2 was used as a carrier gas. The sample liquid water isotopic composition (i.e., $\delta^{18}O_l$ and δ^2H_l) associated with each $\delta^{18}O_v$ and δ^2H_v value is then estimated using soil temperatures measured at the respective depth of vapor extraction along with the equations for free water liquid–vapor equilibrium fractionation established by Majoube (1971). Finally, a combined calibration and drift correction is applied for normalization of $\delta^{18}O_l$ and δ^2H_l values to the VSMOW reference scale and elimination of temporal biases. Therefore, calibration standard measurements obtained approximately every two to three hours and processed identically to in-soil measurements up to this point are linearly interpolated to the time of each other sample to conduct a linear two-point calibration (e.g., Gröning, 2011).

2.2 Field application and evaluation

2.2.1 Site description and instrumentation

The in situ monitoring system was tested under field conditions and evaluated against results from concurrent destructive sampling. Measurements were conducted during one day in September 2012 on an experimental site located at 310 m above sea level on a vineyard terrace of the Kaiserstuhl mountain range (48°05′33.48″ N, 7°42′24.06″ E) in southwest Germany. The mean annual temperature is 11 °C and the mean annual precipitation amounts to 900 mm. A local meteoric water line (LMWL) was obtained based on weekly bulk samples of precipitation collected on the site from June 2011 to August 2012. The slightly clayey silt soils (Calcaric Regosol) of the study site developed from a thick layer of Pleistocene aeolian loess covering the Tertiary volcanic rock forming the Kaiserstuhl (Hädrich and Stahr, 2001). Due to the low slope (< 5 %) and deep permeable soils, water fluxes are predominantly vertical.

Plots of size 3 m^2 were instrumented and planted with various species in 2011 as part of ongoing research. 5TE and MPS-1 or MPS-2 probes installed at depths of 5, 10, 20, ..., 60 cm deliver data at 10 min intervals. Three plots planted with grass, beech, and no vegetation (i.e., bare soil) were selected for this study. The plots had been sheltered from precipitation approximately two weeks prior to sampling such that quasi-static conditions in terms of pore water content and isotopic composition may be assumed.

2.2.2 Destructive pore water isotope sampling

For each plot, three soil cores were retrieved using a Pürckhauer drill. The cores were immediately partitioned into samples representing depth intervals of 0–5, 5–15,..., 75–85 cm and stored into gas-impermeable bags (Weber Packaging, Güglingen, Germany). The bags were subsequently filled with N_2, heat-sealed, and analyzed in the laboratory using the direct equilibration method as detailed by Wassenaar et al. (2008) and Garvelmann et al. (2012). The headspace water vapor isotopic composition of each sample was probed

Table 1. Summary statistics of in situ and destructive laboratory-based pore water stable isotope measurements of quality control standards with and without thermo-insulation (TI) and soil profiles.

		ADS				DDS			Destructive
		Control		Soil		Control		Soil	Control
		w/ TI	w/o TI	Installed	Push-in	w/ TI	w/o TI	Push-in	
$\delta^{18}O_l$ (‰)	RMSE	0.16	0.17	0.38	0.43	0.15	0.18	0.42	0.15
	MAE	0.13	0.14	0.31	0.35	0.12	0.14	0.34	0.11
	BIAS	−0.02	0.02	0.04	−0.02	−0.02	0.03	−0.04	−0.01
	SD	0.16	0.17	0.37	0.43	0.15	0.18	0.42	0.15
	LOA upper	0.29	0.35	0.77	0.82	0.27	0.38	0.78	0.28
	LOA lower	−0.33	−0.31	−0.69	−0.86	−0.31	−0.32	−0.86	−0.30
$\delta^{2}H_l$ (‰)	RMSE	1.23	1.15	2.87	2.92	1.32	1.20	3.09	1.03
	MAE	1.05	0.98	2.38	2.35	1.10	1.02	2.53	0.86
	BIAS	−0.12	0.10	0.32	−0.20	−0.17	0.12	−0.37	−0.10
	SD	1.22	1.15	2.85	2.91	1.32	1.20	3.06	1.03
	LOA upper	2.27	2.35	5.91	5.50	2.42	2.47	5.63	1.92
	LOA lower	−2.51	−2.15	−5.27	−5.90	−2.76	−2.23	−6.37	−2.12
T^a (°C)	Min/max	13/21	11/23	14/19	14/19	14/19	12/22	14/19	21/22
c^b (10^3 ppmv)	Min/max	7/13	6/14	7/10	7/10	7/10	6/13	7/10	26/27
n^c		18	18	36	36	9	9	36	9

[a] Temperature; [b] Vapor concentration; [c] Sample size.

for 390 s at constant temperature using the same WS-CRDS analyzer as above. Raw sample $\delta^{18}O_v$ and $\delta^{2}H_v$ values were obtained as the arithmetic mean over the last 90 s of sample arrival and consequently normalized, drift corrected, and converted to $\delta^{18}O_l$ and $\delta^{2}H_l$ values as described for the in situ sampling. Following a similar identical treatment approach, calibration and quality control were conducted based on reference samples prepared from oven-dried soil material from the study site mixed with three laboratory standard waters. The reference samples were prepared to have the same dry mass as the average soil samples and a water content of 20 vol%, otherwise treated identically to the soil core samples, and alternatingly analyzed between groups of three soil core samples.

2.2.3 In situ pore water isotope sampling and evaluation

Three in situ sampling approaches were selected, each to obtain two pore water isotope profiles for each plot with discrete measurements at depths of 10, 20, ..., 60 cm. In situ measurements were conducted and processed as described above and using (i) a subset of 18 ADS probes permanently installed at the respective locations and depths in July 2012 as part of ongoing research, (ii) an ADS probe in push-in mode (i.e., subsequently inserted at the respective locations and depths), and (iii) a DDS probe in push-in mode.

The methodological accuracy of and the agreement between processed in situ and destructive laboratory-based isotope measurements are assessed in terms of the differences to known control standard values and to each other (i.e., evaluating all measurements from equal depths and similar vegetation cover against each other), respectively. Summary statistics provided include the respective root mean squared differences (RMSE), mean absolute differences (MAE), mean differences (BIAS), and 95 % limits of agreement (LOA) estimated as the BIAS ± 1.96-fold standard deviation (SD) of the differences (Bland and Altman, 1986). For each method, the methodological precision and the reproducibility (or variability) of soil profile measurements are assessed in terms of the (pooled) SD of processed control standard measurements and repeated soil measurements at equal depths, respectively. Control standards placed inside and outside a thermo-insulated housing are evaluated separately to assess the impact of larger temperature and associated vapor concentration variations and differences to calibration conditions. In addition to data obtained during the day of profile sampling, a multi-day record of control standard measurements obtained as part of ongoing research is used to increase the sample size.

3 Results

3.1 Evaluation of sampling methods

The monitoring system facilitates rapid signal stabilization enabling high-quality sample value estimates at short sampling intervals for the employed dilution sequence and under the field conditions encountered (Table 1; Fig. 3). The time required to reach a stable signal depends upon a variety of operational and environmental factors, such as dilution sequence and rates or magnitude of isotopic differences between successive samples. While a systematic assessment is beyond the scope of this paper and sample value estimation was based on exponential models, it is noted that time constants for exponential fits to the asymptotic periods of the acquired signals, starting ca. 18 ± 4 s (arithmetic mean $\pm$ SD) after first sample arrival, were similar for all in situ sampling approaches and averaged to ca. 31 ± 12 s for $\delta^{18}O_v$ and 34 ± 13 s for δ^2H_v. This given, the time required for the signal to approach a steady value to within a range of, e.g., ± 0.1‰ for $\delta^{18}O_v$ and ± 0.5‰ for δ^2H_v, which corresponds to half the analyzer precision specification (Picarro, 2012), can be estimated to amount 108 ± 45 s for $\delta^{18}O_v$ and 128 ± 48 s for δ^2H_v. This indicates that short sampling intervals are feasible also without employing an asymptotic model.

Evaluation of control standard measurements indicates high accuracy and precision for both the ADS and the DDS method over multiple days of field operation. Summary statistics (Table 1) show that systematic errors were small, while random errors were on the order of the precision specified for the employed CRDS instrument (Picarro, 2012) and approximately twice to thrice as high as that currently attainable with IRMS instrumentation under laboratory conditions (Brand, 2004; Horita and Kendall, 2004); however, this is excluding effects related to sample collection and preparation in case of the latter, which often compromise the overall accuracy (Kerstel and Gianfrani, 2008). Thereby, larger temperature variations and deviations from calibration conditions (ca. -3 to $+4$ °C) did not significantly impact the control standard measurement error (two-sample t test, $p > 0.05$). Further, the quality of in situ measurements approached that attained for laboratory-based measurements conducted using the direct equilibration technique under constant temperature conditions and with higher vapor concentrations. Note that in case of the laboratory-based measurements, some variability may be associated with the preparation of individual reference samples for calibration and control of soil core samples in contrast to repeated in situ measurements of identical calibration and control standard boxes.

3.2 Evaluation of soil profile measurements

Depth profiles of pore water isotopic composition obtained based on the various in situ sampling approaches and destructive sampling are physically plausible and generally in good agreement for all three plots under investigation (Fig. 4). This holds with respect to all, the range, distribution, and magnitude of observed isotope signatures and under the conditions encountered, with observed soil water contents and matric potentials at depths below 5 cm generally ranging from ca. 12 to 24 vol% and -150 to -27 kPa, respectively.

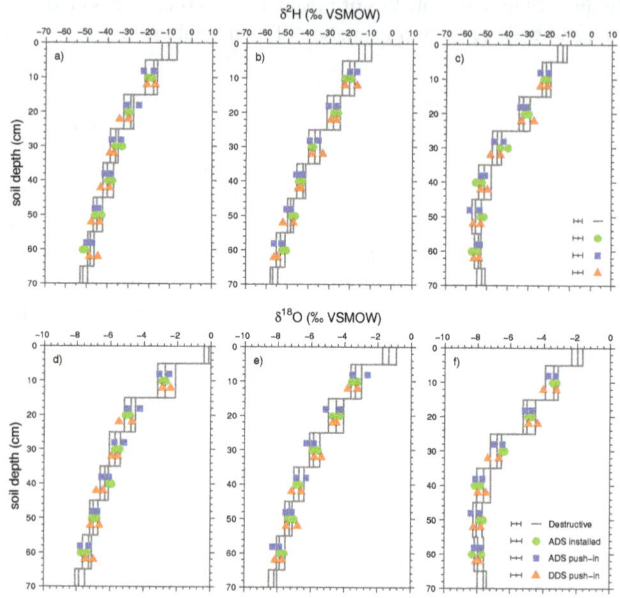

Figure 4. Depth profiles of pore water δ^2H_l and $\delta^{18}O_l$ for a bare soil **(a, d)**, grass **(b, e)**, and beech **(c, f)** plot obtained based on in situ (vertically shifted for visibility) and destructive sampling. The methodological accuracy associated with each symbol is indicated in terms of 95 % limits of agreement.

Repeated in-soil measurements obtained for installed identical ADS probes coincide closely (SD 0.22 ‰ for $\delta^{18}O_l$ and 1.6 ‰ for δ^2H_l), exhibiting a reproducibility approaching that obtained for control standard measurements and indicating the value of continual in situ multi-probe sampling. Stronger inter-profile variability exceeding the respective estimated methodological precision is observed for push-in profile sampling using the ADS (SD 0.38 ‰ for $\delta^{18}O_l$ and 2.6 ‰ for δ^2H_l) or DDS (SD 0.37 ‰ for $\delta^{18}O_l$ and 2.8 ‰ for δ^2H_l) probes and for destructive profile sampling (SD 0.34 ‰ for $\delta^{18}O_l$ and 2.4 ‰ for δ^2H_l). However, this can primarily be attributed to varying target soil water isotopic composition encountered at differing sampling locations.

Comparing in situ and destructive profile sampling results, observed differences are overall similar for all in situ sampling approaches (Table 1). Comparable agreement with destructive sampling results is obtained for spatially more proximate push-in mode sampling using either method compared to continual mode sampling, supporting the assumption of sampling distance as source of intra-method variability. Similar results could be observed when comparing in situ

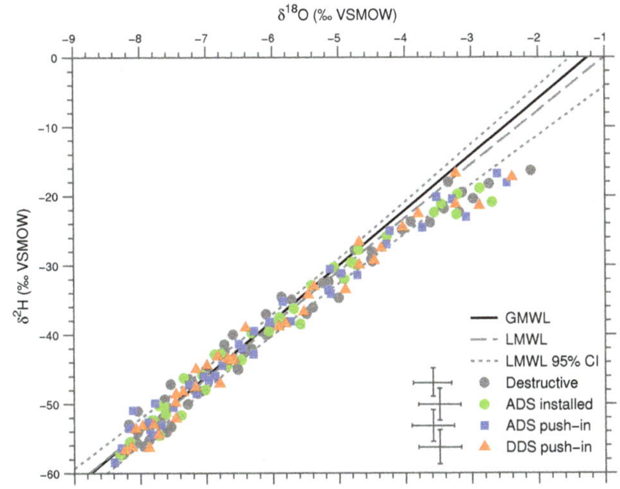

Figure 5. Plot of pore water δ^2H_l versus $\delta^{18}O_l$ obtained based on in situ and destructive sampling of soil profiles (depths from 10 to 60 cm). The methodological accuracy associated with each symbol is indicated in terms of 95 % limits of agreement.

sampling results obtained from different methods and modes of operation with each other (Fig. 4). In all cases, random deviations appear dominant, while biases are small. The inter-method SD and RMSE are on the order of the estimated inter-profile variability for the non-continual sampling methods, indicating that intra-method variability is the major source of inter-method disagreement. Clearly, additional sources of error and uncertainty including spatial variation of targeted pore water signatures along with lack of spatial coincidence of sampling location, imprecise depth assignment, and compression during destructive sampling add to the methodological differences and errors. However, no major bias or obvious systematic dependency can be observed regarding the agreement and relative variability along the soil profiles and thus, with varying soil properties or states and gradients of soil water content, matric potential, and temperature. This indicates that any such effect, over the range of field conditions encountered and if present, does not substantially exceed the reported level of agreement or reproducibility and may thus be deemed negligible, and is at least similar for both sampling techniques, such that the methods may be used interchangeably. Further, the spatial variations in isotopic composition obtained from destructive sampling at the measurement scale of 10 cm are well reproduced by the in situ sampling methods, indicating that in fact a localized isotopic equilibrium between the pore water phases existed and was represented in the sampling results.

The majority of samples plot closely along the global (Craig, 1961) and local meteoric water lines (GMWL and LMWL, respectively; Fig. 5), indicating coherence of obtained $\delta^{18}O_l$ and δ^2H_l values assuming a mostly unaltered meteoric origin of waters. Only few samples, obtained in close proximity to the surface, show potentially evaporative enrichment effects. Under non-equilibrium conditions of actual net evaporation, a more complex model-based interpretation of vapor data is necessary to obtain liquid water estimates. Yet, all in situ samples displayed approximately water vapor saturation within measurement precision of spatiotemporally local temperature (i.e., based on either Kelvin or Magnus Equation and taking dilution rate into account) and agree well with headspace samples from soil cores. This may indicate that no additional net evaporation was occurring at the time of in situ sampling and further, that the effects are not related to surface proximity of sampling.

A physical interpretation of the measured isotope profiles is beyond the scope of this paper. Nevertheless, it can be noted that the last major summer rainfall reaching the ground surface had a composition of ca. $-3.5\,‰$ $\delta^{18}O_l$ and $-22\,‰$ δ^2H_l, while typical winter rains are at $-9.5\,‰$ $\delta^{18}O_l$ and $-66\,‰$ δ^2H_l. Thus, the obtained results may be deemed physically plausible and reflect the common perception of an attenuated seasonal meteoric variation subject to losses by transpiration and some evaporative enrichment in the topsoil (Dansgaard, 1964; Barnes and Allison, 1988; Garvelmann et al., 2012).

4 Discussion

4.1 Monitoring system implementation and functionality

The presented system was developed to allow for unattended functional and efficient minimal invasive field monitoring of pore water stable isotope distributions with high sampling rate and limited cost. The monitoring system is set up modularly and can be easily and flexibly installed, adapted, and extended for accommodation of a multitude of probes and changing applications with relatively little extra expenses. Automated sampling is facilitated by a microprocessor controlled switching unit with solenoid valves enduring $> 10^9$ cycles. The rigid tubular design and small dimensions of the probes (diameter of 10 mm) allow for simple installation causing little disturbance and obstruction to the subsurface flow field and processes under investigation as well as for localized vapor extraction. All parts and materials were selected to be robust and weathering resistant, allowing the system to persist in the subsurface for continuous or repeated application. Additionally, all parts that come into contact with the sample air, most importantly the FEP sample lines, have low dead volume, are chemically mostly inert, and have low water sorption, porosity, and gas permeability (e.g., Mark, 2009).

Avoiding a sample line heating approach, the in situ dilution technique provides optimal control over sample mixing ratios from the subsurface sampling location to the analyzer cavity. Adequate dilution sequences allow for efficient purging of the sample lines and clear sample detection, prevent condensation and associated fractionation effects at any point of the system, and potentially reduce the effects of sorption and storage in the various compartments of the monitoring system. Along with the low pumping rate, the physicochemical and isotopic disturbance of the natural soil water phase system in the probing environment (i.e., due to imposed absolute or partial pressure gradients and soil air losses) is reduced during actual and prolonged sampling.

As stated above, all in situ samples displayed approximate water vapor saturation at extraction, independent of local soil physical states or gradients. Also, clear optima of agreement between sample and saturation vapor pressures computed based on temperature measurements could be observed when considering the spatially and temporally closest observation. Further, neither repeated nor prolonged extraction (i.e., > 30 min) moved the liquid–vapor system from chemical equilibrium or caused continuous modification of the observed isotopic composition. This is consistent with recent laboratory results of Rothfuss et al. (2013). Importantly, these observations imply that monitoring liquid pore water isotopic composition using the presented system does not necessarily demand actual measurements of soil temperature. Instead, the required liquid–vapor isotopic fractionation factors may be inferred based on observed vapor concentrations, provided that these are measured accurately (see e.g., Aemisegger et al., 2012), equilibrium conditions prevail, and effects of water potential are negligible or can be estimated. For the present application, very similar estimates of $\delta^{18}O_l$ and $\delta^{2}H_l$ could be obtained based on either measured temperatures or vapor concentrations (data not shown).

4.2 Monitoring system performance

The outlined technical and operational implementation is essential in practical field applications with continual accurate, precise, and rapid sampling for at least most of the year and for different climatic regions. The presented and evaluated excerpt of a field application employing a branching network of actually more than 30 SWIPs demonstrates these capabilities for the conditions encountered. A quick stabilization of the sample signal is attained, facilitating sampling intervals of approximately four minutes. Employing a specific identical treatment approach for calibration and drift correction, along with a vapor concentration bias correction and the well established formulations for free water liquid–vapor equilibrium fractionation by Majoube (1971), accurate and reproducible inference of normalized liquid pore water isotopic composition from vapor analysis was possible without specific prior characterization of additional sampling effects. Thereby, the two in situ methods tested (ADS and DDS) and the two modes of employment (continual and push-in) were shown to provide similar overall results in good agreement with laboratory-based direct equilibration measurements of destructively collected samples. The comparison of in situ and destructive sampling results, the physical plausibility and reproducibility of the measurements, and sample vapor concentrations indicate that, for the level of data quality given, the extracted vapor did in fact physicochemically and isotopically represent the spatiotemporally local liquid pore water.

The obtained precision of liquid water δ-value estimates is deemed sufficient for the anticipated applications. Improvements may come from further enhanced technical and operational implementation and calibration and correction approaches. Additional measures, such as a sophisticated thermal stabilization of the isotope analyzer, may also prove necessary for continuous application across seasons and years. Longer sampling and integration times and higher vapor concentrations (i.e., lower dilution) may often reduce sampling errors and uncertainties due to proceeded signal stabilization, reduced impact of background short-term noise, and instrument precision dependencies on vapor concentration (e.g., Sturm and Knohl, 2010; Aemisegger et al., 2012), while a higher pumping rate could accelerate signal stabilization. However, the former and both must be traded off for lower sample generation rate and increased actual and long-term disturbance of the natural system, respectively. Further, it is important to note in this context that slow signal stabilization would impede proper signal attainment when the temporal scale of changes in liquid source water or of other effects on the vapor isotopic composition (e.g., temperature changes or instrument drift) is exceeded, which also limits the maximum integration time.

4.3 Assumptions and potential limitations

In principle, the implementation of the monitoring system allows for continuous long-term field observations with little user interaction. Primary limits on unattended application are probably imposed by the need to refresh dry gas supply and calibration standard waters. The number of samples extractable from the calibration boxes with negligible isotopic alteration depends on the initial water reservoir and the volume of liquid water lost per sample. Ca. $0.4\,\mu$L of water are lost per sample (assuming 20 °C and 1 atm), resulting in $< 0.2\,\%$ total volume loss after 10^4 calibrations for the present setup. Secondary limits derive from the life span of individual instruments (with 10^4 h for the vacuum pump as likely minimum) or the probes themselves (expected several years). In remote field locations, power supply for the instruments can become the critical limiting factor. Application in a laboratory setting should, however, generally be straightforward.

Although not observed for the tested timescales, alteration of local soil water content and isotopic composition due to vapor extraction may become an issue during prolonged sampling that requires further investigation. A rough impression can be gained in this context by estimating that a single sample may reduce the water content in an isolated soil volume surrounding the probe to an arbitrary distance of 1 cm by $< 10^{-3}$ vol% or, at a low water content of e.g., 5 vol%, by

$< 0.02\,\%$. Thus, ca. 50 repeated local measurements could be conducted before reducing even a dry moisture state by $1\,\%$. Considering that small local water losses and isotopic enrichment effects are evened by flow and transport processes in the liquid and vapor phase over time, we would not expect continued alteration of local soil water once sampling intervals are sufficiently long and individual sampling times are sufficiently short. Thus, the question becomes more one of feasible frequency rather than total of repeated representative samples. Finally, it should be pointed out in this context that the alternative in situ method using suction cups for extraction of liquid pore water will definitely reduce the local water content significantly and hence alter the flow field, therefore, not allowing for a continuous sampling, particularly in dry soils.

A range of environmental and sampling related characteristics and processes could potentially affect the actual and observed gas-phase isotope signatures and their relation to the local target liquid-phase signatures in space and time. The employed calibration approach provides an advantage in this regard; any constant or temporally variable sampling effects independent of variable conditions diverging from the reference can be assumed to be removed or reflected in the estimated methodological precision. If necessary, systematic effects equally impacting control standard and soil probing (e.g., due to variations of vapor concentration or analyzer temperature) could also be corrected based on assimilated data. A variety of effects could, however, require additional considerations to reduce biases and avoid misinterpretation.

Such effects may be related to sampling and measurement and due to, for example, organics causing spectroscopic interference (West et al., 2010a, 2011; Schultz et al., 2011), vapor extraction causing evaporative surface cooling (Cappa et al., 2003), or variable specific isotopic fractionation associated with transport across a particular porous probing material. Although the latter was found to be negligible in initial headspace experiments comparing SWIP-based versus direct intake sampling (data not shown), such may arise under saturated conditions (Herbstritt et al., 2012) and depending upon concentration gradients across the porous walls, vapor extraction rates, and properties of the probing tube.

More fundamental are potential environmental effects in the soil. Such may include variable liquid–vapor isotopic equilibrium depending on physicochemical states such as chemical (osmotic or matric) potential of water (see e.g., Sofer and Gat, 1972; Horita, 1989; Horita and Wesolowski, 1994; but also Allison, 1987; Rothfuss et al., 2013) or temporary disequilibrium due to water and vapor movement, net phase changes, and isotopic exchange in soils (see Soderberg et al., 2012, and references therein). Note in this context that the relevant rates of liquid–vapor isotopic exchange may depend upon, for example, properties of the geologic material (Hsieh et al., 1998b; Wassenaar et al., 2008), physicochemical states (Ingraham and Criss, 1998), and saturation controlled liquid surface area to volume ratio (Ingraham and Criss, 1993; Costanza-Robinson and Brusseau, 2002), with implications for the actual resolvability of dynamic liquid pore water signatures. In addition and similar to previous sampling approaches (e.g., Landon et al., 1999; Figueroa-Johnson et al., 2007; Wassenaar et al., 2008), the process information may be blurred by unconsidered alteration of liquid soil water isotopic composition itself (e.g., due to interactions with inorganic or organic soil constituents) or unclear contribution of water stored in different compartments of the pore space (e.g., Brooks et al., 2010).

4.4 Outlook

The above enumeration of potential effects that may interact and vary depending upon the specific soil properties clarifies that further research into the underlying physical and the sampling related aspects is needed to attain complete confidence in the representativeness of results obtained using the presented or any other, particularly vapor-based pore water isotope sampling approach. While we did not observe obvious systematic impediment to multi-day measurements of control standards or deviations from methodologically similar direct equilibration measurements of soil core samples, such cannot be generally precluded based on the data set presented. Instead, relevant effects may be anticipated to emerge during long-term applications, when more extreme and variable conditions in space and time are encountered, and when the level of "identity" of natural and reference conditions present for this study cannot be practically attained. As such, the introduced monitoring techniques will certainly profit from continued laboratory experimentation, broad and prolonged field application, and further comparison with conventional methods.

Subject to its own peculiarities, the presented approach drastically decreases the expenditure of time, labor, and money for pore water stable isotope monitoring while eliminating a variety of error and uncertainty sources associated with sample collection, transport, storage, and preparation. The ability of minimal-invasive repeated sample generation at a multitude of locations and at an unprecedented rate extends the scope of stable isotope analysis for manifold applications, such as investigations of vertical and lateral subsurface flow and transport processes or plant–soil water interactions. According to the premise that an experimental design should provide the maximum information for the minimum expenditure, the choice of method will thereby depend upon the processes and anticipated conditions at the site under investigation. DDS probes are more expensive than ADS probes in terms of acquisition and maintenance (i.e., demanding additional carrier gas supply and control), but will be required when saturated conditions could be encountered. In contrast, the push-in application strongly reduces acquisition cost and will be particularly useful when spatial variations of soil water isotopic composition are of primary interest, while continuous assessment of highly dynamic processes will require permanently installed probes, and a combination of both applications with each other and/or destructive sampling can provide valuable spatiotemporal information.

5 Conclusions

This paper introduced a novel system for continual in situ monitoring of pore water stable isotope distributions. The system was technically and operationally implemented to be durable and functional, facilitating automated, minimal-invasive sampling with high frequency, yet limited in cost. Along with a specific on-site calibration approach and basic corrections for instrument bias and liquid–vapor isotopic fractionation, sufficiently accurate and precise inference of normalized liquid pore water isotopic composition was facilitated at sampling intervals of less than four minutes under the conditions encountered. Thereby, the isotopic variability along natural depth profiles was resolved at the measurement scale of ten centimeters. Comparison indicated that the presented in situ approaches may be used interchangeably with each other and with laboratory-based direct equilibration measurements of destructively collected samples. The choice of method will thus depend upon the task and anticipated conditions of sampling. While further research is clearly needed to evaluate and enhance universal reliability and applicability across variable conditions in space and time, the monitoring system has demonstrated capabilities of continual sample generation from a multitude of locations and at an unprecedented rate with minimal disturbance of natural processes. This provides for a powerful tool that can open new avenues towards a detailed quantitative understanding of subsurface water dynamics and fluxes in and out of the soil such as infiltration, recharge, evapotranspiration, and runoff.

Acknowledgements. This work is part of the project "Coupled soil–plant water dynamics – Environmental drivers and species effects" funded by the German Research Foundation (DFG). The authors gratefully thank the technicians, student assistants, and other helpers during preparation and realization of the field experiments, especially Emil Blattmann, Nicolai Dietermann, Benjamin Eisele, Barbara Herbstritt, Lara Kirn, and Lukas Neuhaus.

Edited by: H. Cloke

References

Aemisegger, F., Sturm, P., Graf, P., Sodemann, H., Pfahl, S., Knohl, A., and Wernli, H.: Measuring variations of δ^{18}O and δ^{2}H in atmospheric water vapour using two commercial laser-based spectrometers: an instrument characterisation study, Atmos. Meas. Tech., 5, 1491–1511, doi:10.5194/amt-5-1491-2012, 2012.

Aggarwal, P. K., Gat, J. R., and Froehlich, K. F. O. (Eds.): Isotopes in the water cycle: past, present and future of a developing science, Springer, Dordrecht, The Netherlands, 381 pp., 2007.

Allison, G. B.: The relationship between ^{18}O and deuterium in water in sand columns undergoing evaporation, J. Hydrol., 55, 163–169, doi:10.1016/0022-1694(82)90127-5, 1982.

Allison, G. B., Barnes, C. J., and Hughes, M. W.: The distribution of deuterium and ^{18}O in dry soils 2. Experimental, J. Hydrol., 64, 377–397, doi:10.1016/0022-1694(83)90078-1, 1983.

Allison, G. B., Colin-Kaczala, C., Filly, A., and Fontes, J. C.: Measurement of isotopic equilibrium between water, water vapour and soil CO_2 in arid zone soils, J. Hydrol., 95, 131–141, doi:10.1016/0022-1694(87)90120-x, 1987.

Barnes, C. J. and Allison, G. B.: Tracing of water movement in the unsaturated zone using stable isotopes of hydrogen and oxygen, J. Hydrol., 100, 143–176, doi:10.1016/0022-1694(88)90184-9, 1988.

Berden, G., Peeters, R., and Meijer, G.: Cavity ring-down spectroscopy: Experimental schemes and applications, Int. Rev. Phys. Chem., 19, 565–607, doi:10.1080/014423500750040627, 2000.

Bland, M. J. and Altman, D. G.: Statistical methods for assessing agreement between two methods of clinical measurement, The Lancet, 327, 307–310, doi:10.1016/S0140-6736(86)90837-8, 1986.

Brand, W. A.: Mass spectrometer hardware for analyzing stable isotope ratios, in: Handbook of stable isotope analytical techniques, edited by: de Groot, P. A., Elsevier, Amsterdam, The Netherlands, 835–856, 2004.

Brooks, J. R., Barnard, H. R., Coulombe, R., and McDonnell, J. J.: Ecohydrologic separation of water between trees and streams in a Mediterranean climate, Nat. Geosci., 3, 100–104, doi:10.1038/Ngeo722, 2010.

Buttle, J. M. and Sami, K.: Recharge processes during snowmelt: An isotopic and hydrometric investigation, Hydrol. Process., 4, 343–360, doi:10.1002/hyp.3360040405, 1990.

Cappa, C. D., Hendricks, M. B., DePaolo, D. J., and Cohen, R. C.: Isotopic fractionation of water during evaporation, J. Geophys. Res., 108, 4525, doi:10.1029/2003jd003597, 2003.

Costanza-Robinson, M. S. and Brusseau, M. L.: Air-water interfacial areas in unsaturated soils: Evaluation of interfacial domains, Water Resour. Res., 38, 1195, doi:10.1029/2001wr000738, 2002.

Craig, H.: Isotopic Variations in Meteoric Waters, Science, 133, 1702–1703, doi:10.1126/science.133.3465.1702, 1961.

Crosson, E. R.: A cavity ring-down analyzer for measuring atmospheric levels of methane, carbon dioxide, and water vapor, Appl. Phys. B: Lasers Opt., 92, 403–408, doi:10.1007/s00340-008-3135-y, 2008.

Dansgaard, W.: Stable isotopes in precipitation, Tellus, 16, 436–468, doi:10.1111/j.2153-3490.1964.tb00181.x, 1964.

Darling, W. G. and Bath, A. H.: A stable isotope study of recharge processes in the English Chalk, J. Hydrol., 101, 31–46, doi:10.1016/0022-1694(88)90026-1, 1988.

Dawson, T. E.: Hydraulic lift and water use by plants: implications for water balance, performance and plant-plant interactions, Oecologia, 95, 565–574, doi:10.1007/bf00317442 1993.

Dawson, T. E.: Determining water use by trees and forests from isotopic, energy balance and transpiration analyses: the roles of tree size and hydraulic lift, Tree Physiol., 16, 263-272, doi:10.1093/treephys/16.1-2.263, 1996.

Dawson, T. E. and Ehleringer, J. R.: Streamside trees that do not use stream water, Nature, 350, 335–337, doi:10.1038/350335a0, 1991.

Dincer, T., Al-Mugrin, A., and Zimmermann, U.: Study of the infiltration and recharge through the sand dunes in arid zones with special reference to the stable isotopes and thermonuclear tritium, J. Hydrol., 23, 79–109, doi:10.1016/0022-1694(74)90025-0, 1974.

Ehhalt, D. and Knott, K.: Kinetische Isotopentrennung bei der Verdampfung von Wasser, Tellus, 17, 389–397, doi:10.1111/j.2153-3490.1965.tb01431.x, 1965.

Ehleringer, J. R. and Dawson, T. E.: Water uptake by plants: perspectives from stable isotope composition, Plant. Cell Environ., 15, 1073–1082, doi:10.1111/j.1365-3040.1992.tb01657.x, 1992.

Figueroa-Johnson, M., Tindall, J., and Friedel, M.: A comparison of $^{18}O\delta$ composition of water extracted from suction lysimeters, centrifugation, and azeotropic distillation, Water, Air, Soil Pollut., 184, 63–75, doi:10.1007/s11270-007-9399-8, 2007.

Friedman, I., Redfield, A. C., Schoen, B., and Harris, J.: The variation of the deuterium content of natural waters in the hydrologic cycle, Rev. Geophys., 2, 177–224, doi:10.1029/RG002i001p00177, 1964.

Garvelmann, J., Külls, C., and Weiler, M.: A porewater-based stable isotope approach for the investigation of subsurface hydrological processes, Hydrol. Earth Syst. Sci., 16, 631–640, doi:10.5194/hess-16-631-2012, 2012.

Gat, J. R.: Oxygen and hydrogen isotopes in the hydrologic cycle, Annu. Rev. Earth Planet. Sci., 24, 225–262, doi:10.1146/annurev.earth.24.1.225, 1996.

Gazis, C. and Feng, X.: A stable isotope study of soil water: evidence for mixing and preferential flow paths, Geoderma, 119, 97–111, doi:10.1016/s0016-7061(03)00243-x, 2004.

Gröning, M.: Improved water δ^2H and $\delta^{18}O$ calibration and calculation of measurement uncertainty using a simple software tool, Rapid Commun. Mass Spectrom., 25, 2711–2720, doi:10.1002/rcm.5074, 2011.

Gupta, P., Noone, D., Galewsky, J., Sweeney, C., and Vaughn, B. H.: Demonstration of high-precision continuous measurements of water vapor isotopologues in laboratory and remote field deployments using wavelength-scanned cavity ring-down spectroscopy (WS-CRDS) technology, Rapid Commun. Mass Spectrom., 23, 2534–2542, doi:10.1002/rcm.4100, 2009.

Hädrich, F. and Stahr, K.: Die Böden des Breisgaus und angrenzender Gebiete, Berichte der Naturforschenden Gesellschaft Freiburg i.Br. 91, 143 pp., 2001.

Helliker, B. and Noone, D.: Novel approaches for monitoring of water vapor isotope ratios: plants, lasers and satellites, in: Isoscapes, edited by: West, J. B., Bowen, G. J., Dawson, T. E., and Tu, K. P., Springer, Dordrecht, The Netherlands, 71–88, 2010.

Herbstritt, B., Gralher, B., and Weiler, M.: Continuous in situ measurements of stable isotopes in liquid water, Water Resour. Res., 48, W03601, doi:10.1029/2011wr011369, 2012.

Horita, J.: Analytical aspects of stable isotopes in brines, Chem. Geol., 79, 107–112, doi:10.1016/0168-9622(89)90013-4, 1989.

Horita, J. and Kendall, C.: Stable isotope analysis of water and aqueous solutions by conventional dual-inlet mass spectrometry, in: Handbook of stable isotope analytical techniques, edited by: de Groot, P. A., Elsevier, Amsterdam, The Netherlands, 1–37, 2004.

Horita, J. and Wesolowski, D. J.: Liquid-vapor fractionation of oxygen and hydrogen isotopes of water from the freezing to the critical temperature, Geochim. Cosmochim. Acta, 58, 3425–3437, doi:10.1016/0016-7037(94)90096-5, 1994.

Hsieh, J. C. C., Chadwick, O. A., Kelly, E. F., and Savin, S. M.: Oxygen isotopic composition of soil water: Quantifying evaporation and transpiration, Geoderma, 82, 269–293, doi:10.1016/s0016-7061(97)00105-5, 1998a.

Hsieh, J. C. C., Savin, S. M., Kelly, E. F., and Chadwick, O. A.: Measurement of soil-water O-18 values by direct equilibration with CO_2, Geoderma, 82, 255–268, doi:10.1016/s0016-7061(97)00104-3, 1998b.

Iannone, R. Q., Romanini, D., Cattani, O., Meijer, H. A. J., and Kerstel, E. R. T.: Water isotope ratio (δ^2H and $\delta^{18}O$) measurements in atmospheric moisture using an optical feedback cavity enhanced absorption laser spectrometer, J. Geophys. Res., 115, D10111, doi:10.1029/2009jd012895, 2010.

Ingraham, N. L. and Criss, R. E.: Effects of surface area and volume on the rate of isotopic exchange between water and water vapor, J. Geophys. Res.-Atmos., 98, 20547–20553, doi:10.1029/93jd01735, 1993.

Ingraham, N. L. and Criss, R. E.: The effect of vapor pressure on the rate of isotopic exchange between water and water vapor, Chem. Geol., 150, 287–292, doi:10.1016/s0009-2541(98)00109-0, 1998.

Ingraham, N. L. and Shadel, C.: A comparison of the toluene distillation and vacuum/heat methods for extracting soil water for stable isotopic analysis, J. Hydrol., 140, 371–387, doi:10.1016/0022-1694(92)90249-u, 1992.

Johnson, L. R., Sharp, Z. D., Galewsky, J., Strong, M., Van Pelt, A. D., Dong, F., and Noone, D.: Hydrogen isotope correction for laser instrument measurement bias at low water vapor concentration using conventional isotope analyses: application to measurements from Mauna Loa Observatory, Hawaii, Rapid Commun. Mass Spectrom., 25, 608–616, doi:10.1002/Rcm.4894, 2011.

Kendall, C. and McDonnell, J. J. (Eds.): Isotope tracers in catchment hydrology, Elsevier, Amsterdam, The Netherlands, 839 pp., 1998.

Kerstel, E. and Gianfrani, L.: Advances in laser-based isotope ratio measurements: selected applications, Appl. Phys. B: Lasers Opt., 92, 439–449, doi:10.1007/s00340-008-3128-x, 2008.

Kerstel, E. R. T. and Meijer, H. A. J.: Optical isotope ratio measurements in hydrology, in: Isotopes in the Water Cycle, edited by: Aggarwal, P. K., Gat, J. R., and Froehlich, K. F. O., Springer, Dordrecht, The Netherlands, 109–123, 2005.

Koehler, G., Wassenaar, L. I., and Hendry, M. J.: An Automated Technique for Measuring δD and $\delta^{18}O$ Values of Porewater by Direct CO_2 and H_2 Equilibration, Anal. Chem., 72, 5659–5664, doi:10.1021/ac000498n, 2000.

Landon, M. K., Delin, G. N., Komor, S. C., and Regan, C. P.: Comparison of the stable-isotopic composition of soil water collected from suction lysimeters, wick samplers, and cores in a sandy unsaturated zone, J. Hydrol., 224, 45–54, doi:10.1016/S0022-1694(99)00120-1, 1999.

Laudon, H., Seibert, J., Köhler, S., and Bishop, K.: Hydrological flow paths during snowmelt: Congruence between hydrometric measurements and oxygen 18 in meltwater, soil water, and runoff, Water Resour. Res., 40, W03102, doi:10.1029/2003wr002455, 2004.

Majoube, M.: Fractionnement en oxygène-18 et en deutérium entre l'eau et sa vapeur, J. Chim. Phys., 68, 1423–1436, 1971.

Mark, J. E.: Polymer data handbook, Oxford University Press, New York, USA, 2009.

McDonnell, J. J.: A Rationale for Old Water Discharge Through Macropores in a Steep, Humid Catchment, Water Resour. Res., 26, 2821–2832, doi:10.1029/WR026i011p02821, 1990.

Munksgaard, N. C., Wurster, C. M., and Bird, M. I.: Continuous analysis of $\delta^{18}O$ and δD values of water by diffusion sampling cavity ring-down spectrometry: a novel sampling device for unattended field monitoring of precipitation, ground and surface waters, Rapid Commun. Mass Spectrom., 25, 3706–3712, doi:10.1002/rcm.5282, 2011.

Munksgaard, N. C., Wurster, C. M., Bass, A., and Bird, M. I.: Extreme short-term stable isotope variability revealed by continuous rainwater analysis, Hydrol. Process., 26, 3630–3634, doi:10.1002/hyp.9505, 2012a.

Munksgaard, N. C., Wurster, C. M., Bass, A., Zagorskis, I., and Bird, M. I.: First continuous shipboard $\delta^{18}O$ and δD measurements in sea water by diffusion sampling – cavity ring-down spectrometry, Environ. Chem. Lett., 10, 301–307, doi:10.1007/s10311-012-0371-5, 2012b.

Picarro L2120-i $\delta D/\delta^{18}O$ isotopic water analyzer: http://www.picarro.com/sites/default/files/L2120-i%20Datasheet.pdf (last access: 10 April 2013), 2012.

Rothfuss, Y., Biron, P., Braud, I., Canale, L., Durand, J. L., Gaudet, J. P., Richard, P., Vauclin, M., and Bariac, T.: Partitioning evapotranspiration fluxes into soil evaporation and plant transpiration using water stable isotopes under controlled conditions, Hydrol. Process., 24, 3177–3194, doi:10.1002/Hyp.7743, 2010.

Rothfuss, Y., Vereecken, H., and Brüggemann, N.: Monitoring water stable isotopic composition in soils using gas-permeable tubing and infrared laser absorption spectroscopy, Water Resour. Res., 49, 3747–3755, doi:10.1002/wrcr.20311, 2013.

Saxena, R.: Seasonal variations of oxygen-18 in soil moisture and estimation of recharge in esker and moraine formations, Nord. Hydrol., 15, 235–242, doi:10.2166/nh.1984.020, 1984.

Schmidt, M., Maseyk, K., Lett, C., Biron, P., Richard, P., Bariac, T., and Seibt, U.: Concentration effects on laser-based $\delta^{18}O$ and δ^2H measurements and implications for the calibration of vapour measurements with liquid standards, Rapid Commun. Mass Spectrom., 24, 3553–3561, doi:10.1002/rcm.4813, 2010.

Schultz, N. M., Griffis, T. J., Lee, X., and Baker, J. M.: Identification and correction of spectral contamination in $^2H/^1H$ and $^{18}O/^{16}O$ measured in leaf, stem, and soil water, Rapid Commun. Mass Spectrom., 25, 3360–3368, doi:10.1002/rcm.5236, 2011.

Schulze, E. D., Caldwell, M. M., Canadell, J., Mooney, H. A., Jackson, R. B., Parson, D., Scholes, R., Sala, O. E., and Trimborn, P.: Downward flux of water through roots (i.e. inverse hydraulic lift) in dry Kalahari sands, Oecologia, 115, 460–462, doi:10.1007/s004420050541, 1998.

Sklash, M. G. and Farvolden, R. N.: The role of groundwater in storm runoff, J. Hydrol., 43, 45–65, doi:10.1016/0022-1694(79)90164-1, 1979.

Soderberg, K., Good, S. P., Wang, L., and Caylor, K.: Stable isotopes of water vapor in the vadose zone: A review of measurement and modeling techniques, Vadose Zone J., 11, No. 3, doi:10.2136/vzj2011.0165, 2012.

Sofer, Z. and Gat, J.: Activities and concentrations of oxygen-18 in concentrated aqueous salt solutions: analytical and geophysical implications, Earth Planet. Sci. Lett., 15, 232–238, doi:10.1016/0012-821X(72)90168-9, 1972.

Stewart, M. K. and McDonnell, J. J.: Modeling base flow soil water residence times from deuterium concentrations, Water Resour. Res., 27, 2681–2693, doi:10.1029/91wr01569, 1991.

Sturm, P. and Knohl, A.: Water vapor δ^2H and $\delta^{18}O$ measurements using off-axis integrated cavity output spectroscopy, Atmos. Meas. Tech., 3, 67–77, doi:10.5194/amt-3-67-2010, 2010.

Thoma, G., Esser, N., Sonntag, C., Weiss, W., Rudolph, J., and Leveque, P.: New technique of in-situ soil moisture sampling for environmental isotope analysis applied at Pilat sand dune near Bordeaux, Isot. Hydrol. Proc. Symp., Neuherberg, 1979, 753–766, 1979.

Thorburn, P., Walker, G., and Hatton, T.: Are river red gums taking water from soil, groundwater or streams, Catchments of green: a national conference on vegetation & water management, conference proceedings-volume B, 63–68, 1992.

Vitvar, T., Aggarwal, P. K., and McDonnell, J. J.: A review of isotope applications in catchment hydrology, in: Isotopes in the Water Cycle, edited by: Aggarwal, P. K., Gat, J. R., and Froehlich, K. F. O., Springer, Dordrecht, The Netherlands, 151–169, 2005.

Walker, C. D. and Richardson, S. B.: The use of stable isotopes of water in characterising the source of water in vegetation, Chem. Geol., 94, 145–158, doi:10.1016/0168-9622(91)90007-j, 1991.

Walker, G. R., Hughes, M. W., Allison, G. B., and Barnes, C. J.: The movement of isotopes of water during evaporation from a bare soil surface, J. Hydrol., 97, 181–197, doi:10.1016/0022-1694(88)90114-x, 1988.

Walker, G. R., Woods, P. H., and Allison, G. B.: Interlaboratory comparison of methods to determine the stable isotope composition of soil water, Chem. Geol., 111, 297–306, doi:10.1016/0009-2541(94)90096-5, 1994.

Wang, X. F. and Yakir, D.: Using stable isotopes of water in evapotranspiration studies, Hydrol. Process., 14, 1407–1421, doi:10.1002/1099-1085(20000615)14:8<1407::aid-hyp992>3.0.co;2-k, 2000.

Wassenaar, L. I., Hendry, M. J., Chostner, V. L., and Lis, G. P.: High resolution pore water δ^2H and $\delta^{18}O$ measurements by$H_2O_{(liquid)}$-$H_2O_{(vapor)}$ equilibration laser spectroscopy, Environ. Sci. Technol., 42, 9262–9267, doi:10.1021/Es802065s, 2008.

Wels, C., Cornett, R. J., and Lazerte, B. D.: Hydrograph separation: A comparison of geochemical and isotopic tracers, J. Hydrol., 122, 253–274, doi:10.1016/0022-1694(91)90181-G, 1991.

West, A. G., Patrickson, S. J., and Ehleringer, J. R.: Water extraction times for plant and soil materials used in stable isotope analysis, Rapid Commun. Mass Spectrom., 20, 1317–1321, doi:10.1002/rcm.2456, 2006.

West, A. G., Goldsmith, G. R., Brooks, P. D., and Dawson, T. E.: Discrepancies between isotope ratio infrared spectroscopy and isotope ratio mass spectrometry for the stable isotope analysis of plant and soil waters, Rapid Commun. Mass Spectrom., 24, 1948–1954, doi:10.1002/rcm.4597, 2010a.

West, J. B., Bowen, G. J., Dawson, T. E., and Tu, K. P. (Eds.): Isoscapes: understanding movement, pattern, and process on Earth through isotope mapping, Springer, Dordrecht, The Netherlands, 2010b.

West, A. G., Goldsmith, G. R., Matimati, I., and Dawson, T. E.: Spectral analysis software improves confidence in plant and soil water stable isotope analyses performed by isotope ratio infrared spectroscopy (IRIS), Rapid Commun. Mass Spectrom., 25, 2268–2274, doi:10.1002/rcm.5126, 2011.

White, J. W. C., Cook, E. R., Lawrence, J. R., and Wallace S, B.: The D/H ratios of sap in trees: Implications for water sources and tree ring D/H ratios, Geochim. Cosmochim. Acta, 49, 237–246, doi:10.1016/0016-7037(85)90207-8, 1985.

Yakir, D. and Sternberg, L. D. S. L.: The use of stable isotopes to study ecosystem gas exchange, Oecologia, 123, 297–311, doi:10.1007/s004420051016, 2000.

Yepez, E. A., Huxman, T. E., Ignace, D. D., English, N. B., Weltzin, J. F., Castellanos, A. E., and Williams, D. G.: Dynamics of transpiration and evaporation following a moisture pulse in semiarid grassland: A chamber-based isotope method for partitioning flux components, Agric. For. Meteorol., 132, 359–376, doi:10.1016/j.agrformet.2005.09.006, 2005.

Zimmermann, U., Ehhalt, D., and Münnich, K. O.: Soil water movement and evapotranspiration: changes in the isotopic composition of the water, Proc. Symp. Isot. Hydrol., Vienna, Austria, 1967, 567–584, 1967.

2

Studying catchment storm response using event- and pre-event-water volumes as fractions of precipitation rather than discharge

Jana von Freyberg[1,2], Bjørn Studer[1], Michael Rinderer[3], and James W. Kirchner[1,2]

[1]Department of Environmental Systems Science, ETH Zurich, Zurich, Switzerland
[2]Swiss Federal Institute for Forest, Snow and Landscape Research (WSL), Birmensdorf, Switzerland
[3]Chair of Hydrology, University of Freiburg, Freiburg im Breisgau, Germany

Correspondence: Jana von Freyberg (jana.vonfreyberg@usys.ethz.ch)

Abstract. Catchment response to precipitation is often investigated using two-component isotope-based hydrograph separation, which quantifies the contribution of precipitation (i.e., event water Q_e) or water from storage (i.e., pre-event water Q_{pe}) to total discharge (Q) during storm events. In order to better understand streamflow-generating mechanisms, two-component hydrograph separation studies often seek to relate the event-water fraction Q_e/Q to storm characteristics or antecedent wetness conditions. However, these relationships may be obscured because the same factors that influence Q_e also necessarily influence total discharge Q as well. Here we propose that the fractions of event water and pre-event water relative to total precipitation (Q_e/P and Q_{pe}/P), instead of total discharge, provide useful alternative tools for studying catchment storm responses. These two quantities separate the well-known runoff coefficient (Q/P, i.e., the ratio between total discharge and precipitation volumes over the event timescale) into its contributions from event water and pre-event water. Whereas the runoff coefficient Q/P quantifies how strongly precipitation inputs affect streamflow, the fractions Q_e/P and Q_{pe}/P track the sources of this streamflow response.

We use high-frequency measurements of stable water isotopes for 24 storm events at a steep headwater catchment (Erlenbach, central Switzerland) to compare the storm-to-storm variations in Q_e/Q, Q_e/P and Q_{pe}/P. Our analysis explores how storm characteristics and antecedent wetness conditions affect the mobilization of event water and pre-event water at the catchment scale. Isotopic hydrograph separation shows that catchment outflow was typically dominated by pre-event water, although event water exceeded 50 % of discharge for several storms. No clear relationships were found linking either storm characteristics or antecedent wetness conditions with the volumes of event water or pre-event water (Q_e, Q_{pe}), or with event water as a fraction of discharge (Q_e/Q), beyond the unsurprising correlation of larger storms with greater Q_e and greater total Q. By contrast, event water as a fraction of precipitation (Q_e/P) was strongly correlated with storm volume and intensity but not with antecedent wetness, implying that the volume of event water that is transmitted to streamflow increases more than proportionally with storm size under both wet and dry conditions. Conversely, pre-event water as a fraction of precipitation (Q_{pe}/P) was strongly correlated with all measures of antecedent wetness but not with storm characteristics, implying that wet conditions primarily facilitate the mobilization of old (pre-event) water, rather than the fast transmission of new (event) water to streamflow, even at a catchment where runoff coefficients can be large.

Thus, expressing event- and pre-event-water volumes as fractions of precipitation rather than discharge was more insightful for investigating the Erlenbach catchment's hydrological behaviour. If Q_e/P and Q_{pe}/P exhibit similar relationships with storm characteristics and antecedent wetness conditions in other catchments, we suggest that these patterns may potentially be useful as diagnostic "fingerprints" of catchment storm response.

1 Introduction

Studying catchment hydrological responses to precipitation events can be useful in identifying dominant controls on streamflow generation. For decades, hydrologists have compared the volumes of precipitation (P) and discharge (Q) during storm events using the runoff coefficient Q/P (e.g., Fischer et al., 2017; Horton, 1933; Jordan, 1994; Litt et al., 2015; McGlynn et al., 2004; Munyaneza et al., 2012; Shanley and Chalmers, 1999; Sidle et al., 1995). Other studies have separated the hydrograph into base flow and quick flow (using graphical methods, low-pass digital filtering or recession curve analysis; Blume et al., 2007; Hall, 1968; Hewlett and Hibbert, 1967; Nathan and McMahon, 1990) and have compared the quick-flow fraction of Q to precipitation. Both the runoff coefficient Q/P and the ratio of quick flow to precipitation quantify how streamflow responds to precipitation inputs, but neither tracks the source of this streamflow response. In other words, both metrics characterize the celerity or the effect of an event on streamflow, not the velocity of water movement through the catchment (McDonnell and Beven, 2014). Because the runoff coefficient Q/P and the ratio of quick flow to precipitation can be calculated from hydrometric data alone, they have been estimated for many events and sites. Runoff coefficients Q/P typically increase with storm size (e.g., Fischer et al., 2017; Jordan, 1994) and antecedent wetness (e.g., Litt et al., 2015; Sidle et al., 1995) and have been found to be unaffected by catchment area (e.g., McGlynn et al., 2004). The ratio of quick flow to precipitation has often been found to increase with storm size and intensity (e.g., Blume et al., 2007; Norbiato et al., 2009), with wetter antecedent conditions (e.g., Detty and McGuire, 2010; Merz et al., 2006; Penna et al., 2011; von Freyberg et al., 2014) and with catchment area (e.g., Brown et al., 1999). However, it remains unclear whether these relationships arise because certain storm characteristics, antecedent wetness conditions, landscape properties, etc., facilitate the more efficient transmission of recent precipitation ("event water") to the stream or the more effective mobilization of pre-event water from catchment storage. This question cannot be answered with hydrometric data alone; instead it also requires using tracer data to track the flow of water through the catchment and thus to separate the runoff coefficient into its event and pre-event components.

Tracer-based, two-component hydrograph separation uses stable water isotopes (^{2}H, ^{18}O) to estimate the relative "time source components" of catchment outflow, i.e., recent precipitation (event water, Q_e/Q) and catchment storage (pre-event water, Q_{pe}/Q; Klaus and McDonnell, 2013). Particularly in humid environments, stable water isotopes are considered to be nearly conservative tracers, because isotopic fractionation effects due to evaporation are negligible, so mixing is the major process controlling the isotopic composition of stream waters. Hydrograph separation studies have related the event-water fraction of discharge (Q_e/Q) to storm characteristics (e.g., James and Roulet, 2009; Pellerin et al., 2008), antecedent catchment wetness conditions (e.g., Casper et al., 2003; James and Roulet, 2009; Penna et al., 2015; Shanley et al., 2002; von Freyberg et al., 2017), landscape properties (e.g., Buttle, 1994; Fischer et al., 2017) and catchment size (e.g., Laudon et al., 2007; Shanley et al., 2002). Antecedent wetness has frequently been identified as a major control on the relative contribution of event water to streamflow. Wetter antecedent conditions typically lead to smaller event-water fractions Q_e/Q and, by implication, larger pre-event-water fractions. This suggests that under these conditions, larger volumes of pre-event water are available in more permeable subsurface layers that can be rapidly activated by incoming precipitation (e.g., Casper et al., 2003; Klaus and McDonnell, 2013; Muñoz-Villers and McDonnell, 2012). In contrast, at sites where positive relationships between Q_e/Q and antecedent wetness have been observed, it has been hypothesized that vertical infiltration must first replenish storage deficits before event water can be rapidly transported via lateral flow pathways or surface runoff (Shanley et al., 2002; von Freyberg et al., 2017), or that the expansion of saturated areas in the catchment enhances direct runoff of rainwater (Penna et al., 2016). Still other studies have found no correlation between antecedent wetness and Q_e/Q (Ocampo et al., 2006), which has often been attributed to an insufficient number of sampled events or an insufficient range of antecedent moisture conditions (e.g., Barthold et al., 2017; Fischer et al., 2017; James and Roulet, 2009; Penna et al., 2015).

Although the runoff coefficient Q/P and the event-water fraction Q_e/Q have commonly been used in catchment studies, the ratio between Q_e and P might provide an alternative tool for studying streamflow responses to precipitation events. The ratio Q_e/P has previously been used as a surrogate for the fraction of the catchment area that generates surficial runoff (Buttle and Peters, 1997; Eshleman et al., 1993; Rodhe, 1987; Pellerin et al., 2008), but it has not been widely used to explore how catchment storm response varies with antecedent wetness and storm characteristics (but see Ocampo et al., 2006, for one example).

The conceptual differences between the ratios Q_e/Q and Q_e/P can be explained by analogy, with forward and backward transit times (e.g., Hrachowitz et al., 2016). A streamwater parcel's backward transit time is its age since its entry as precipitation at the time it exits the catchment, whereas a precipitation parcel's forward transit time is the age it will accumulate in the catchment before it eventually exits as streamflow. Analogously, Q_e/Q quantifies the fraction of total storm discharge that comes from recent precipitation ("event water") – that is, the fraction of streamflow with a short backward transit time. The ratio Q_e/P, on the other hand, quantifies the fraction of total storm precipitation that will be discharged during the same event – that is, the fraction of precipitation with a short forward transit time.

The "backward" event-water fraction Q_e/Q and the "forward" event-water fraction Q_e/P are not the same, for the simple reason that not all precipitation is discharged during the event (otherwise there would be no recharge of storage that supplies base flow between events). Thus at the event timescale, total P is typically greater than total Q, so Q_e/P will be less than Q_e/Q. To use a concrete example, a 60 mm storm event might produce 24 mm of catchment outflow, from which 6 mm is event water (Q_e) from precipitation, and the remaining 18 mm of discharge is pre-event water (Q_{pe}) mobilized from catchment storage; the 54 mm of precipitation that does not form event discharge then goes into storage. In this example, the backward event-water fraction Q_e/Q would be 6 mm / 24 mm = 0.25, and the forward event-water fraction Q_e/P would be 6 mm / 60 mm = 0.1.

In addition to the ratio Q_e/P, we can also calculate the ratio of pre-event-water volume to precipitation (Q_{pe}/P). Whereas the event- and pre-event-water fractions of discharge are mathematically linked through the relationship $Q_e/Q + Q_{pe}/Q = 1$, the ratios Q_e/P and Q_{pe}/P instead sum to the runoff coefficient Q/P. From this perspective, Q_e/P and Q_{pe}/P represent the contributions of event and pre-event water to the runoff coefficient; thus Q_e/P and Q_{pe}/P can be viewed as the "event runoff coefficient" and "pre-event runoff coefficient", respectively.

Care has to be taken with the conceptual definition of Q_{pe}/P; it quantifies how much pre-event water is mobilized by, not contained in, a unit volume of precipitation (which, by definition, cannot contain pre-event water). This highlights an important distinction between Q_{pe}/P and the other three ratios (Q_e/Q, Q_{pe}/Q and Q_{pe}/P). In the event- and pre-event-water fractions of discharge (Q_e/Q and Q_{pe}/Q) as well as in the event-water fraction of precipitation (Q_e/P), the numerator is a physical (not just a mathematical) fraction of the denominator. That is, Q_e is physically derived from P, and along with Q_{pe}, it is physically part of Q. But Q_{pe} is not physically part of P, and thus Q_{pe}/P does not represent a physical fraction of a whole.

Whereas Q_e/P has been estimated for several events at several catchments (Buttle and Peters, 1997; Eshleman et al., 1993; Rodhe, 1987; Pellerin et al., 2008), with the exception of a single figure in a single study (Fig. 7 of Ocampo et al., 2006), the behaviour of Q_{pe}/P has apparently remained entirely unexplored. From the general concepts outlined above, we hypothesize that the event and pre-event runoff coefficients Q_e/P and Q_{pe}/P may be more informative metrics for studying catchment storm responses, compared to the widely used event-water fraction of discharge Q_e/Q or the runoff coefficient Q/P. In this paper, we test this hypothesis by comparing runoff coefficients Q/P and the different ratios Q_e/Q, Q_{pe}/Q, Q_e/P and Q_{pe}/P across 24 storm events and analysing their relationships with storm characteristics and antecedent wetness. These relationships shed light on possible streamflow generation processes at our study site and highlight the potential benefits of using Q_e/P and Q_{pe}/P, rather than Q/P or Q_e/Q, to characterize catchment storm response. Our source data consist of high-frequency isotope measurements from a steep, partly forested headwater catchment in central Switzerland, collected during snow-free periods between September 2016 and October 2017.

2 Methods

2.1 Field site

The Erlenbach research catchment is located in the northern Swiss pre-Alps roughly 40 km south of Zurich. The 0.7 km^2 catchment spans an elevation range from 1100 to 1655 m a.s.l. (Fig. 1). The bedrock geology consists of alternating layers of conglomerates, clay and marl shales, limestone and calcareous sandstones, with lithological boundaries nearly perpendicular to the main valley axis (Hantke, 1967; Hsü and Briegel, 1991). Due to the layering and the diversity of the bedrock material, the landscape in the upper half of the catchment is divided into a sequence of steep hillslopes and plateaus. On the plateaus, groundwater tables are shallow, and mollic Gleysols with a carbonate-rich topsoil predominate; the hillslopes are mostly covered with macropore-rich Umbric Gleysols (Schleppi et al., 1998). Overall, the soils reach maximum depths of around 2.5 m in depressions on the plateaus, while soils are shallower (0.5–1 m) on steeper slopes (Fischer et al., 2015; Rinderer et al., 2017). Due to the high clay content of the flysch material, the bedrock and soils are generally characterized by low permeabilities and are prone to waterlogging. As a consequence, soil saturation is likely to occur on approximately 30 % of the catchment area (FOEN, 2011). The distribution of these areas agrees well with the mapped locations of numerous wet meadows that cover roughly 22 % of the catchment area (Fig. 1a). The channel network in the Erlenbach catchment is dense (around 20 km km^{-2}), partly due to open drainage ditches in meadows as well as numerous small springs emerging from the bases of hillslopes in the upper part of the catchment. The Erlenbach catchment is mainly covered with coniferous forests (53 %) on the hillslopes, while partially forested areas (22 %) and meadows (25 %) occur largely on the plateaus (Fischer et al., 2015; Fig. 1b).

Annual precipitation at Erlenbach averages roughly 1850 mm yr^{-1} (over the period 2000–2015), with monthly maxima and minima in June and October, respectively (van Meerveld et al., 2018). In years with cold winters, up to 40 % of annual precipitation falls as snow (Stähli and Gustafsson, 2006), and in summer, frequent thunderstorms generate pronounced peaks in streamflow.

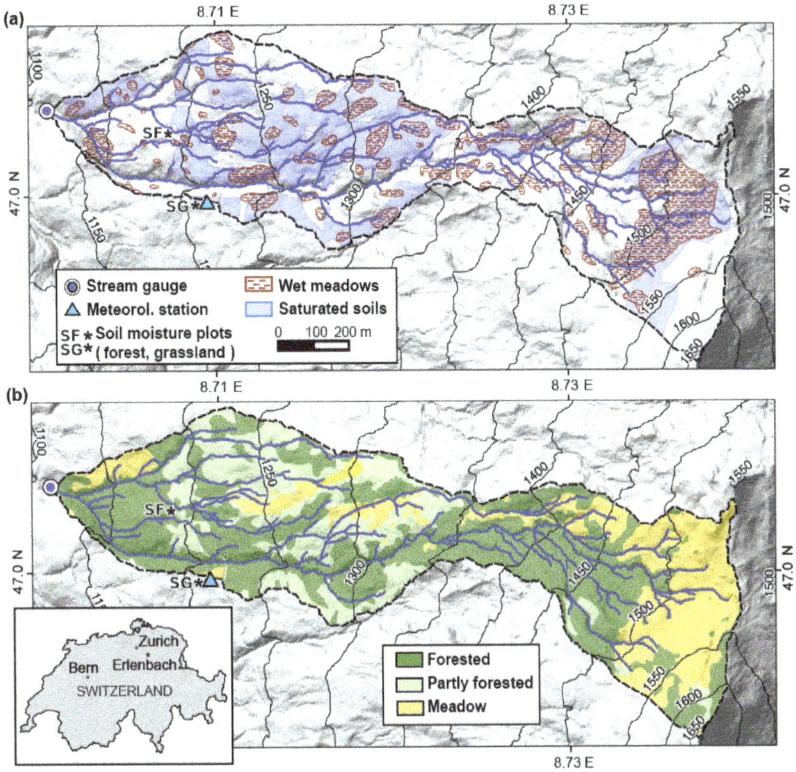

Figure 1. The Erlenbach catchment with spatial distribution of **(a)** saturated soils (FOEN, 2011), mapped wet meadows and **(b)** vegetation (Fischer et al., 2015).

2.2 Data set

A meteorological station is located at 1216 m a.s.l. on a grassland site near the south-western catchment border (Fig. 1). At this station, precipitation is measured at 1.5 m above ground with a heated tipping bucket (Joss-Tognini 15183; Lambrecht meteo GmbH, Göttingen, Germany) and groundwater table depth is measured in a fully screened pipe with a pressure transducer (BTL2-P1-1000-B-S50; Balluff GmbH, Neuhausen a.d.F., Germany). The pressure transducer readings have not been calibrated against manual measurements and thus should be considered only as a relative indication of groundwater table variations. River discharge has been monitored at the catchment outlet with a concrete flume since 1984 (Hegg et al., 2006). All measurements are recorded at 10 min intervals and were aggregated to 30 min or 1 h intervals in the following analysis.

Soil moisture content is measured every 1 min at two locations in the Erlenbach catchment, a grassland site (at the meteorological station, 1216 m a.s.l.) and a forest site (1185 m a.s.l.; Fig. 1). At each site, one ECH_2O 10HS and three ECH_2O 5TM probes (both probe types from METER Environment, Munich, Germany) were installed at roughly 50 cm depth across an area of 3 × 4 m. Soil moisture data collection at the forested plot began only in December 2016. Prior to installation, the factory-calibrated soil moisture probes were compared to each other by installing them together in a bucket filled with moist sand. Only those probes that showed similar values were installed in the field. As we were only interested in the temporal variations of soil moisture content over the course of individual events, we considered the factory calibration of the soil moisture probes sufficient for our purposes. The soil moisture measurements from the four probes at each site were averaged to smooth out anomalous variations in soil moisture at the individual sampling locations, and they were subsequently aggregated to 30 min or 1 h intervals.

Recent technological advances allow for on-site high-frequency sampling and analysis of stable water isotopes in stream water and precipitation (von Freyberg et al., 2017). With such a lab in the field, isotopic responses in streamflow can be captured over long periods spanning a wide range of hydro-climatic conditions. A field laboratory was installed just above the Erlenbach outlet to measure stable water isotopes at 30 min intervals. The set-up of the field laboratory is similar to the system described in von Freyberg et al. (2017) and will therefore be described here only briefly.

Stream water was continuously pumped from the stream to the field laboratory. Precipitation was captured in a 45 cm diameter funnel, transmitted into the field lab through a Tygon tube and collected in a 1 L teflon-lined separating funnel. Whenever the sample volume exceeded 50 mL, the field lab alternately analysed precipitation and stream water (thus yielding one precipitation and one stream-water sample every hour). During rainless periods stream water was analysed on a 30 min cycle; a drift correction check-standard was analysed every 4 h.

A new analysis cycle was started every 30 min, with an automated pumping routine that filled a 50 mL collection vessel with either stream water or precipitation. A suction tube with a PE-filter frit (20 µm pore size) led from the collection vessel to two programmable high-precision dosing pumps (800 Dosino, hereafter simply "Dosino"; Metrohm AG, Herisau, Switzerland). These alternatingly took up 30 mL from the collection vessel and injected it at a constant flow rate of $1\,\mathrm{mL\,min^{-1}}$ into a continuous water sampler module (CWS; Picarro Inc., Santa Clara, CA, USA) connected to a wavelength-scanned cavity ring-down spectrometer (CRDS; Picarro model L2130-i). During the 30 min injection cycle, the 50 mL collection vessel, the other Dosino and all tubing were flushed with Nanopure water and air and then rinsed with approximately 10 mL of the next sample, to minimize sample carryover effects.

Within the CWS, liquid samples flow through a semipermeable membrane tube that is surrounded by a flow of dry air. Thus, water vapour diffuses steadily through the membrane wall and is transported with the dry-air flow to the isotope analyser. The CRDS measures concentrations of ^{18}O and ^{2}H every 6 s, however, we averaged the last 10 min of each 30 min injection period to obtain the final isotope values. The measurement precision of the CWS coupled to the CRDS with the Dosino injection system was estimated from the standard deviations of these 10 min averages (von Freyberg et al., 2017). For δ^{18}O and δ^{2}H, the averages of these standard deviations were 0.09 ‰ and 0.21 ‰, respectively, which are used as estimates of uncertainty in the error propagation calculations (Sect. 2.4). Isotopic abundances are reported in δ^{18}O and δ^{2}H relative to Vienna Standard Mean Ocean Water (VSMOW). The isotope analyser was initially calibrated to VSMOW-SLAP standards before installation in the field laboratory; during the field deployment, instrument drift and carry-over were quantified and corrected by regularly measuring one internal isotope standard every 4 h and two internal standards approximately weekly.

2.3 Event properties: storm characteristics and initial catchment state

The following metrics were used to describe the storm characteristics: total event precipitation (P, mm), cumulative precipitation before peak flow ($P_{\mathrm{untilQpeak}}$, mm), mean precipitation intensity (P_{int}, $\mathrm{mm\,h^{-1}}$), maximum precipitation over 1 h ($P_{1\,\mathrm{h}}$, $\mathrm{mm\,h^{-1}}$) and 4 h ($P_{4\,\mathrm{h}}$, mm), maximum cumulative discharge volume over 4 h ($Q_{4\,\mathrm{h}}$, mm) and rainfall duration (T, hours). The following metrics were used to quantify antecedent catchment wetness: 3-day and 7-day antecedent precipitation (AP3 and AP7, mm) as well as the 1 h average values of discharge (Q_{ini}, $\mathrm{mm\,h^{-1}}$), groundwater table depth ($\mathrm{GW_{ini}}$, cm), and soil moisture at the grassland site ($\mathrm{SM_{ini}}$, $\mathrm{m^3\,m^{-3}}$) before the onset of the storm event.

We compared these metrics of storm characteristics and antecedent wetness conditions with the event- and pre-event-water fractions using Spearman rank correlation. Spearman rank correlation was preferred over Pearson (product-moment) correlation because it is less sensitive to extreme values and thus more suitable for smaller data sets. For our analyses, correlation p values of $p < 0.01$ were considered statistically significant.

2.4 Two-component hydrograph separation to quantify event- and pre-event-water volumes

We use the isotopic composition (δ^{2}H and δ^{18}O) of stream water and precipitation to separate the streamflow hydrograph into two components, event water and pre-event water. Event water (subscript e) is stream water that entered the catchment as precipitation during a storm event, whereas pre-event water (subscript pe) is stream water derived from pre-event catchment storage. Following Pinder and Jones (1969), tracer-based hydrograph separation is based on a mass balance for water,

$$q_{\mathrm{i}} = q_{\mathrm{e_i}} + q_{\mathrm{pe_i}}, \tag{1}$$

and tracer flux,

$$c_{\mathrm{i}} q_{\mathrm{i}} = c^*_{\mathrm{e_i}} q_{\mathrm{e_i}} + c_{\mathrm{pe_i}} q_{\mathrm{pe_i}}, \tag{2}$$

where q_{i}, q_{e_i} and $q_{\mathrm{pe_i}}$ are total, event-water and pre-event-water fluxes at time step i, and c_{i}, $c^*_{\mathrm{e_i}}$ and $c_{\mathrm{pe_i}}$ are the tracer concentrations in stream water, event water and pre-event water, respectively.

The instantaneous fraction of event water in discharge for each time step i can thus be obtained by combining Eqs. (1) and (2):

$$\frac{q_{\mathrm{e_i}}}{q_{\mathrm{i}}} = \frac{c_i - c_{\mathrm{pe_i}}}{c^*_{\mathrm{e}_i} - c_{\mathrm{pe_i}}}. \tag{3}$$

To make the $\frac{q_{\mathrm{e_i}}}{q_{\mathrm{i}}}$ time series continuous, it is linearly interpolated between measurement gaps that occurred whenever check standards or precipitation samples were analysed instead of stream-water samples.

Following general practice, we assume that the tracer concentration of pre-event water is constant over the duration of each event (Sklash and Farvolden, 1979) so that $c_{\mathrm{pe_i}}$ becomes c_{pe}. We calculate c_{pe} as the average isotopic composition of the last five stream-water samples before the beginning of each storm event (i.e., 2.5 h or 3 h, if a check standard was sampled within this period). We estimate the time series of event-water tracer concentration, $c^*_{\mathrm{e_i}}$, as an incremental weighted mean (McDonnell et al., 1990), weighted by precipitation rates over all previous time steps j since the beginning of the storm:

$$c^*_{e_i} = \frac{\sum_{j=k}^{i} P_j c_{e_j}}{\sum_{j=k}^{i} P_j}. \tag{4}$$

The total event-water volume (Q_e) is the cumulative sum of the instantaneous values q_{e_i} over the event duration, and the cumulative pre-event-water volume (Q_{pe}) was obtained by subtracting Q_e from the total discharge volume Q over the event. The beginning of a storm event was the time of first rainfall, and the end of a storm event was defined as the time that (i) event-water discharge declined to 5 % of its peak value or (ii) another precipitation event began, whichever came first; case (i) prevailed for 18 of 24 events.

The ratios of event and pre-event water relative to precipitation are

$$\frac{Q_e}{P} = \frac{Q_e}{Q} \cdot \frac{Q}{P} \tag{5}$$

and

$$\frac{Q_{pe}}{P} = \left(1 - \frac{Q_e}{Q}\right) \cdot \frac{Q}{P}. \tag{6}$$

The standard errors (SE) of $c^*_{e_i}$ and $\frac{q_{e_i}}{q_i}$ were estimated though Gaussian error propagation (Genereux, 1998):

$$\mathrm{SE}\left(c^*_{e_i}\right) = \left[\frac{\sum_{j=k}^{i} P_j \left(c_{e_j} - c^*_{e_j}\right)^2}{(j-k)\sum_{j=k}^{i} P_j}\right]^{\frac{1}{2}} \tag{7}$$

and

$$\mathrm{SE}\left(\frac{q_{e_i}}{q_i}\right) = \left\{\left[\frac{-1}{c_{pe} - c^*_{e_i}}\mathrm{SE}(c_i)\right]^2 + \left[\frac{c_i - c^*_{e_i}}{\left(c_{pe} - c^*_{e_i}\right)^2}\mathrm{SE}(c_{pe})\right]^2 + \left[\frac{c_p - c_i}{\left(c_{pe} - c^*_{e_i}\right)^2}\mathrm{SE}(c^*_{e_i})\right]^2\right\}^{\frac{1}{2}}. \tag{8}$$

The standard error of c_{pe} is estimated by pooling the uncertainty in the individual measurements and their standard deviation from one another (von Freyberg et al., 2017). Because c_i and c_{pe} are independent measurements, their errors SE_{c_i} and $\mathrm{SE}_{c_{pe}}$ are likely to be uncorrelated with each other. In contrast, errors in the calculated event-water isotope values $c^*_{e_i}$ will be highly correlated with each other over time due to the incremental volume-weighting of tracer concentrations in precipitation. Taking these correlations into account requires first-order, second-moment error propagation (Bevington and Robinson, 2003), which reduces to Gaussian error propagation in the special case of uncorrelated errors. The first-order, second-moment error propagation formula for the event-water fraction Q_e/Q, averaged over all times i in the storm event (Sect. 2.3), is

$$\mathrm{SE}\left(\frac{Q_e}{Q}\right) = \left\{\sum_i\left[\frac{q_i}{Q}\cdot\frac{\mathrm{SE}(c_i)}{c^*_{e_i} - c_{pe}}\right]^2 + \sum_i\left[\frac{q_i}{Q}\cdot\frac{c_i - c^*_{e_i}}{\left(c^*_{e_i} - c_{pe}\right)^2}\cdot\mathrm{SE}(c_{pe})\right]^2 + \left[\sum_i\left(\frac{q_i}{Q}\cdot\frac{c_{pe} - c_i}{\left(c^*_{e_i} - c_{pe}\right)^2}\cdot\mathrm{SE}\left(c^*_{e_i}\right)\right)\right]^2\right\}^{\frac{1}{2}}, \tag{9}$$

where the square brackets on the outside of the last summation, rather than the inside, reflect the conservative assumption that the errors in the event-water isotope values $c^*_{e_i}$ are perfectly correlated. The other two pairs of square brackets are inside the summations, reflecting the assumption that the errors in c_i and c_{pe} are uncorrelated. For simplicity, and because our main focus is on the event and pre-event fractions of the water fluxes rather than the fluxes themselves, we ignore any measurement errors in q_i and P.

Following Eqs. (5) and (6), the standard errors of Q_e/P and Q_{pe}/P can be estimated with the following:

$$\mathrm{SE}\left(\frac{Q_e}{P}\right) = \mathrm{SE}\left(\frac{Q_e}{Q}\right) \cdot \frac{Q}{P} \tag{10}$$

and

$$\mathrm{SE}\left(\frac{Q_{pe}}{P}\right) = \mathrm{SE}\left(\frac{Q_{pe}}{Q}\right) \cdot \frac{Q}{P}. \tag{11}$$

3 Results and Discussion

Figure 2 shows the time series of the observed variables for the roughly 8-month study period from 15 September 2016 to 1 November 2017, excluding the winter season influenced by snowfall and snow cover (6 November 2016–7 May 2017). Streamflow responds promptly to precipitation and is strongly synchronized with shallow soil moisture and groundwater table variations. The average soil moisture content at the grassland site was slightly higher and more variable than at the forested site. The values of stable water isotopes in precipitation are highly variable within and across events, ranging between −170.3 ‰ and −31.7 ‰ for δ^2H and between −22.5 ‰ and −4.2 ‰ for δ^{18}O for the storms considered here. Stream-water isotopes are strongly damped but also show distinct responses to individual storms (Fig. 2a). For the 24 events, values of δ^2H and δ^{18}O in stream water ranged from −56.1 ‰ to −84.3 ‰ and from −8.6 ‰ to −12.2 ‰, respectively. During the roughly 8-month study period, more than 9400 water samples were measured; missing values due to problems with the automatic sampling cycle or instrument malfunctioning account for roughly 8 % of the data set.

3.1 General properties of the events

Table 1 provides an overview of the storm characteristics and antecedent wetness conditions for the individual storm events. Total storm rainfall P ranged between 8.2 and 63.2 mm (25.1 ± 3.1 mm, mean ± SE), and total discharge Q ranged between 0.4 and 25.7 mm (9.8 ± 1.7). During the individual storm events, the 4 h peak discharge $Q_{4\,\mathrm{h}}$ reached values between 0.11 and 12.5 mm. Figure 3 shows that the 24 storm events used for our analysis span a wide range of flow regimes.

The various metrics of catchment antecedent wetness conditions were highly correlated with each other. Spearman rank correlation coefficients were $\rho > 0.60$ ($p < 0.002$) for all combinations of metrics except for AP3 and $\mathrm{GW_{ini}}$ ($\rho = 0.50$, $p = 0.011$). Initial soil moisture $\mathrm{SM_{ini}}$, initial groundwater levels $\mathrm{GW_{ini}}$, and 7-day antecedent precipitation (AP7) correlate strongly ($\rho > 0.83$ and $p < 0.0001$ in all cases) with initial discharge (Q_{ini}), suggesting that these measures are representative relative indicators of antecedent moisture at the catchment scale.

3.1.1 Both isotopes yield similar hydrograph separation results

Figure 4a shows that δ^{18}O and δ^{2}H yield instantaneous event-water fractions of discharge ($\frac{q_{\mathrm{e_i}}}{q_{\mathrm{i}}}$) at peak flow that do not differ significantly from one another (that is, by more than twice their pooled standard errors). For Q_{e}/Q we also observe a good agreement between both isotopes, except for the events on 25 June, 25 September, 26 and 29 October 2017, for which the differences are 178 %, 4 %, 2 % and 2 % greater than twice their pooled uncertainties, respectively (Fig. 4b, Table S1 in the Supplement). We thus assume that inferences derived from the two isotopes will be consistent with each other. Measurements of δ^{2}H were less noisy than those of δ^{18}O relative to their respective ranges of variability, so values such as $\frac{q_{\mathrm{e_i}}}{q_{\mathrm{i}}}$ and Q_{e}/Q will be more precise when derived from δ^{2}H rather than δ^{18}O. Therefore, the following analysis is performed based on δ^{2}H; δ^{18}O would yield similar results but with larger uncertainties.

3.1.2 Two-component hydrograph separation results for 24 storm events

Figure 5 and Table 2 compare the storm events' runoff coefficients Q/P and show that total storm discharge is typically less than half of total storm precipitation and, in some cases, is much less. On average, runoff coefficients are 0.34 ± 0.04 (mean ± SE), but their storm-to-storm variability is large (0.03 to 0.72), suggesting that the effectiveness with which precipitation signals are converted to streamflow responses varies considerably at Erlenbach.

The relative fractions of event water in discharge (Q_{e}/Q) are highly variable across the 24 storm events, ranging from 0.04 to 0.75, with a mean value of 0.23 ± 0.04. The relative contribution of event water to discharge exceeded 50 % for only two storms (Fig. 5), and on average, discharge at Erlenbach was comprised of roughly 77 % pre-event water. Similarly high pre-event-water fractions relative to discharge have been observed at other humid forested headwater catchments (e.g., Brown et al., 1999; Buttle, 1994; Jones et al., 2006; McGlynn and McDonnell, 2003).

For all 24 storms, the event-water fractions of precipitation Q_{e}/P are smaller than the corresponding event-water fractions of discharge, for the simple reason that P exceeds Q (Table 2). The values of Q_{e}/P range from 0.002 to 0.34 (mean ± SE 0.08 ± 0.02), while the pre-event-water volume relative to precipitation (Q_{pe}/P) ranges from 0.03 to 0.68 (mean ± SE 0.28±0.03). This suggests that, on average, each precipitation event at Erlenbach activated pre-event water equal to roughly a third of the rainfall volume, while the event-water contribution to discharge accounted for less than 10 % of the rainfall volume. Thus, precipitation had a nearly 3-fold larger effect on the activation of pre-event water than on the transmission of event water to the stream.

Relatively few stable isotope studies have analysed numerous events at high temporal resolution (e.g., Birkel et al., 2012; Fischer et al., 2017; Ocampo et al., 2006; von Freyberg et al., 2017), revealing large variations in the relative amounts of event and pre-event water from storm to storm. At Erlenbach, we find that the event-water fraction of discharge Q_{e}/Q is much more variable, relative to its mean, than the pre-event-water fraction Q_{pe}/Q (coefficients of variation CV = 0.74 and CV = 0.23, respectively). This follows as a direct consequence of Q_{e}/Q being smaller, on average, than Q_{pe}/Q and of these two quantities being complements of one another ($Q_{\mathrm{pe}}/Q = 1 - Q_{\mathrm{e}}/Q$), implying that their standard deviations must be equal. Event- and pre-event-water volumes relative to precipitation are more variable across storms, (Q_{e}/P CV = 0.96 and Q_{pe}/P CV = 0.61), suggesting that the event- and pre-event runoff coefficients (Q_{e}/P, Q_{pe}/P) might be more informative, for instance when used for correlation analyses, compared to the less variable event- and pre-event-water fractions of discharge (Q_{e}/Q, Q_{pe}/Q). More fundamentally, Q_{pe}/Q and Q_{e}/Q contain completely redundant information, because they sum to 1. By contrast, Q_{e}/P and Q_{pe}/P do not sum to a constant (instead they sum to the runoff coefficient), so they each contain distinct information.

3.1.3 Detailed description of three contrasting events

To investigate the conceptual differences of the ratios Q_{pe}/Q, Q_{e}/P and Q_{pe}/P in more detail, Fig. 6 shows the hydrograph separation results for three storm events, 2 Oc-

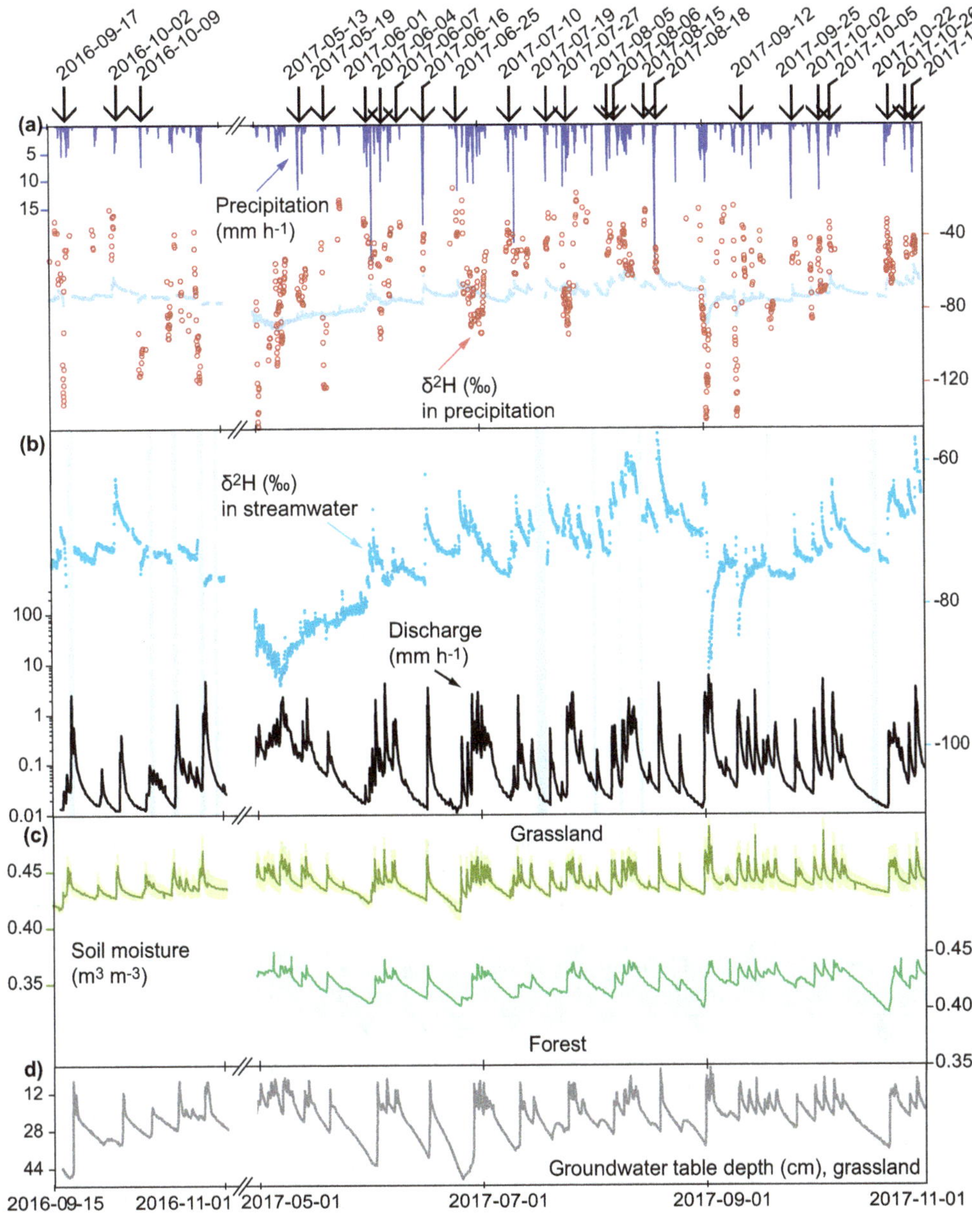

Figure 2. Time series of **(a)** precipitation and δ^2H in precipitation (δ^2H in stream water is shown for comparison), **(b)** discharge (on log axis) and δ^2H in stream water (vertical bars indicate gaps in the isotope data), **(c)** soil moisture measured at 50 cm depth at the grassland and forest sites (Fig. 1a), with shaded areas showing the standard error from averaging the measurements from the four probes at each plot, and **(d)** groundwater levels at the grassland site. The winter period with snow cover (6 November 2016–7 May 2017) was not considered in this analysis, since the individual contributions of rainfall and snowmelt to river discharge could not be distinguished sufficiently. Vertical arrows indicate the events analysed in this paper.

tober 2016, 5 October and 10 July 2017, along with the time series of precipitation, discharge, soil moisture, and δ^2H values in precipitation and stream water. During the 2 October 2016 storm, antecedent wetness conditions were dry (AP7 = 11 mm), and total precipitation (P) and discharge (Q) were 21.6 and 4.8 mm, respectively, resulting in a runoff coefficient Q/P of 0.22 (Fig. 6a). During the 5 October 2017 storm, antecedent conditions were wetter (AP7 = 69 mm), and consequently 33.5 mm of rain produced 20.5 mm of discharge, yielding a runoff coefficient of 0.61; roughly 50 % more rain generated roughly 300 % more discharge, relative to the earlier event (Fig. 6b). The response times of streamflow to incoming rainfall, measured here as the time it takes for q_i to increase by more than 30 % relative to Q_{ini}, were similar for both storm events (2 vs. 2.5 h), as were the changes in soil moisture recorded at the grassland

Table 1. Properties of the storm events: total discharge (Q), total precipitation (P), cumulative precipitation before peak flow ($P_{\text{untilQpeak}}$), mean precipitation intensity (P_{int}), maximum precipitation over 1 h ($P_{1\,\text{h}}$) and 4 h ($P_{4\,\text{h}}$), event duration (T), and 4 h peak discharge volume ($Q_{4\,\text{h}}$). The initial catchment wetness state was quantified using 3-day and 7-day antecedent precipitation (AP3 and AP7) as well as the 1 h average values of discharge (Q_{ini}), groundwater table depth (GW_{ini}) and soil moisture (SM_{ini}) before the onset of the storm event.

Event	Q (mm)	P (mm)	$P_{\text{until}Q_{\text{peak}}}$ (mm)	$P_{1\,\text{h}}\pm$SE (mm)	$P_{4\,\text{h}}\pm$SE (mm)	$T\pm1$ h (h)	P_{int} (mm h^{-1})	$Q_{4\,\text{h}}\pm$SE (mm)	AP3 (mm)	AP7 (mm)	$Q_{\text{ini}}\pm$SE (mm h^{-1})	SM_{ini} (m^3 m^{-3})	$\text{GW}_{\text{ini}}\pm$SE (cm below arbitrary level)
17 Sep 2016	0.8	9.2	8.6	6.1 ± 1.1	8.4 ± 4.0	7.5	1.23	0.25 ± 0.00	7.3	7.3	0.018 ± 0.000	0.419	47.3 ± 0.1
2 Oct 2016	4.8	21.6	21.4	5.4 ± 0.1	11.5 ± 3.0	15.5	1.39	1.23 ± 0.10	0.7	10.8	0.012 ± 0.000	0.427	33.4 ± 0.2
9 Oct 2016	0.4	12.6	12.5	9.1 ± 1.7	12.5 ± 5.6	2.5	5.04	0.11 ± 0.00	0.4	9.4	0.012 ± 0.000	0.428	30.2 ± 0.2
13 May 2017	9.5	20.8	20.2	9.0 ± 1.6	20.4 ± 6.2	5.5	3.78	6.11 ± 0.62	16.7	91.2	0.200 ± 0.000	0.444	17.3 ± 0.1
19 May 2017	5.3	17.6	17.6	4.2 ± 0.1	11.0 ± 2.8	7.0	2.51	1.48 ± 0.08	1.2	43.6	0.063 ± 0.000	0.433	28.4 ± 0.1
1 Jun 2017	7.5	28.5	26.9	23.2 ± 6.5	27.7 ± 17.4	6.5	4.38	4.98 ± 0.65	20.6	20.6	0.042 ± 0.000	0.433	43.0 ± 0.2
4 Jun 2017	17.4	33.7	30.3	10.8 ± 0.6	26.2 ± 6.5	11.5	2.93	10.57 ± 1.62	43.1	60.3	0.074 ± 0.001	0.439	20.1 ± 0.1
7 Jun 2017	8.0	11.2	11.2	3.7 ± 0.8	6.5 ± 2.5	6.5	1.72	2.53 ± 0.16	50.3	109.4	0.278 ± 0.005	0.445	16.3 ± 0.2
16 Jun 2017	17.1	46.0	39.9	18.1 ± 7.7	25.8 ± 12.4	9.5	4.83	9.70 ± 1.31	0	2.5	0.014 ± 0.000	0.424	39.4 ± 0.1
25 Jun 2017	1.7	21.2	21.2	15.2 ± 0.9	16.3 ± 9.9	6.5	3.26	1.01 ± 0.14	1.4	1.4	0.013 ± 0.000	0.414	48.4 ± 0.1
10 Jul 2017	7.0	25.4	25.4	21.4 ± 6.8	23.0 ± 17.1	6.8	3.76	5.30 ± 0.93	20.2	20.2	0.032 ± 0.000	0.437	35.3 ± 0.2
19 Jul 2017	4.0	20.2	18.1	10.3 ± 1.1	18.1 ± 6.7	10.5	1.92	1.58 ± 0.18	0.5	16.8	0.016 ± 0.000	0.427	31.0 ± 0.1
27 Jul 2017	7.7	12.9	12.7	6.8 ± 0.5	7.1 ± 4.5	11.0	1.17	1.49 ± 0.06	74.4	105.8	0.109 ± 0.001	0.438	21.6 ± 0.2
5 Aug 2017	2.7	17.1	15.8	8.5 ± 0.1	15.8 ± 10.4	9.5	1.80	1.30 ± 0.18	0	15.8	0.020 ± 0.000	0.429	29.7 ± 0.1
6 Aug 2017	4.8	11.2	10.8	5.7 ± 0.3	9.7 ± 3.2	9.5	1.18	1.86 ± 0.12	17.4	32.6	0.084 ± 0.001	0.443	17.6 ± 0.2
15 Aug 2017	2.3	8.2	8.2	3.7 ± 1.6	4.5 ± 0.3	7.5	1.09	0.37 ± 0.00	0.6	86.5	0.042 ± 0.000	0.433	25.5 ± 0.2
18 Aug 2017	25.6	56.3	34.7	31.0 ± 5.3	43.6 ± 19.6	12.0	4.68	11.71 ± 1.64	8.5	26.3	0.028 ± 0.000	0.431	28.5 ± 0.2
12 Sep 2017	9.2	19.7	18.9	5.8 ± 0.1	10.8 ± 3.3	11.5	1.71	2.76 ± 0.16	45.5	53	0.082 ± 0.000	0.440	21.3 ± 0.2
25 Sep 2017	3.5	15.2	15.2	14.9 ± 5.5	15.3 ± 12.7	1.5	10.12	1.93 ± 0.25	0.2	20.4	0.026 ± 0.000	0.434	27.6 ± 0.2
2 Oct 2017	20.9	39.1	37.9	10.9 ± 0.3	23.6 ± 5.5	19.5	2.00	12.50 ± 1.68	28	48	0.059 ± 0.000	0.440	19.8 ± 0.2
5 Oct 2017	20.5	33.5	31.9	5.3 ± 0.8	19.5 ± 1.9	9.5	3.52	7.99 ± 0.65	39.2	67.9	0.047 ± 0.000	0.438	22.0 ± 0.2
22 Oct 2017	25.5	63.2	59.4	8.1 ± 0.3	18.7 ± 4.1	51.6	1.23	2.14 ± 0.05	0	0	0.013 ± 0.000	0.429	36.5 ± 0.1
26 Oct 2017	4.3	12.5	12.3	6.1 ± 0.6	10.5 ± 3.5	6.5	1.92	1.85 ± 0.15	0	63.2	0.038 ± 0.001	0.437	22.5 ± 0.2
29 Oct 2017	25.7	44.8	27.2	8.4 ± 0.3	22.8 ± 3.0	17.0	2.64	10.19 ± 0.74	12.5	56	0.053 ± 0.000	0.441	19.3 ± 0.1

Table 2. Results of hydrograph separations based on δ^2H (results for δ^{18}O are provided in Table S1). Columns are the runoff coefficient (Q/P), event and pre-event discharge as whole-storm totals (Q_{e} and Q_{pe}), the maximum instantaneous event-water fraction $q_{\text{e,i}}/q_{\text{i}}$ and its value at peak flow, the event-water fractions (Q_{e}/Q), as well as the event and pre-event runoff coefficients (Q_{e}/P and Q_{pe}/P).

Event	Q/P (–)	$Q_{\text{e}}\pm$SE (mm)	$Q_{\text{pe}}\pm$SE (mm)	$\max(q_{\text{e,i}}/q_{\text{i}})\pm$SE (–)	$q_{\text{e,i}}/q_{\text{i}}$ at peak flow ± SE (–)	$Q_{\text{e}}/Q\pm$SE (–)	$Q_{\text{e}}/P\pm$SE (–)	$Q_{\text{pe}}/P\pm$SE (–)
17 Sep 2016	0.08	0.30 ± 0.02	0.45 ± 0.02	0.514 ± 0.047	0.453 ± 0.044	0.402 ± 0.030	0.033 ± 0.002	0.049 ± 0.002
2 Oct 2016	0.22	1.09 ± 0.04	3.68 ± 0.04	0.304 ± 0.013	0.304 ± 0.013	0.229 ± 0.009	0.051 ± 0.002	0.170 ± 0.002
9 Oct 2016	0.03	0.03 ± 0.00	0.41 ± 0.00	0.103 ± 0.007	0.062 ± 0.006	0.066 ± 0.001	0.002 ± 0.000	0.033 ± 0.000
13 May 2017	0.46	1.50 ± 0.10	7.97 ± 0.10	0.307 ± 0.055	0.208 ± 0.037	0.158 ± 0.011	0.072 ± 0.005	0.383 ± 0.005
19 May 2017	0.30	0.27 ± 0.02	5.03 ± 0.02	0.113 ± 0.034	0.085 ± 0.013	0.050 ± 0.004	0.015 ± 0.001	0.286 ± 0.001
1 Jun 2017	0.26	1.88 ± 0.07	5.66 ± 0.07	0.477 ± 0.024	0.297 ± 0.018	0.249 ± 0.009	0.066 ± 0.002	0.199 ± 0.002
4 Jun 2017	0.52	2.26 ± 0.11	15.13 ± 0.11	0.175 ± 0.018	0.135 ± 0.016	0.130 ± 0.006	0.067 ± 0.003	0.449 ± 0.003
7 Jun 2017	0.72	0.41 ± 0.01	7.63 ± 0.01	0.101 ± 0.010	0.081 ± 0.010	0.051 ± 0.002	0.037 ± 0.001	0.682 ± 0.001
16 Jun 2017	0.37	7.13 ± 0.57	9.94 ± 0.57	0.742 ± 0.127	0.458 ± 0.063	0.418 ± 0.033	0.155 ± 0.012	0.216 ± 0.012
25 Jun 2017	0.08	0.35 ± 0.00	1.37 ± 0.00	0.255 ± 0.008	0.255 ± 0.008	0.203 ± 0.003	0.016 ± 0.000	0.065 ± 0.000
10 Jul 2017	0.28	3.42 ± 0.11	3.57 ± 0.11	0.567 ± 0.020	0.563 ± 0.028	0.489 ± 0.016	0.135 ± 0.005	0.141 ± 0.005
19 Jul 2017	0.20	1.12 ± 0.02	2.91 ± 0.02	0.347 ± 0.011	0.346 ± 0.011	0.278 ± 0.004	0.055 ± 0.001	0.144 ± 0.001
27 Jul 2017	0.59	0.57 ± 0.01	7.10 ± 0.01	0.129 ± 0.007	0.100 ± 0.007	0.075 ± 0.002	0.044 ± 0.001	0.550 ± 0.001
5 Aug 2017	0.16	0.52 ± 0.02	2.15 ± 0.02	0.255 ± 0.033	0.233 ± 0.034	0.196 ± 0.007	0.031 ± 0.001	0.125 ± 0.001
6 Aug 2017	0.43	0.46 ± 0.01	4.31 ± 0.01	0.158 ± 0.011	0.145 ± 0.011	0.097 ± 0.002	0.041 ± 0.001	0.385 ± 0.001
15 Aug 2017	0.28	0.10 ± 0.00	2.22 ± 0.00	0.076 ± 0.008	0.056 ± 0.008	0.043 ± 0.002	0.012 ± 0.000	0.270 ± 0.000
18 Aug 2017	0.45	19.25 ± 1.03	6.33 ± 1.03	0.843 ± 0.021	0.841 ± 0.044	0.753 ± 0.040	0.342 ± 0.018	0.112 ± 0.018
12 Sep 2017	0.46	0.33 ± 0.03	8.82 ± 0.03	0.133 ± 0.018	0.032 ± 0.012	0.036 ± 0.003	0.017 ± 0.001	0.448 ± 0.001
25 Sep 2017	0.23	0.73 ± 0.01	2.78 ± 0.01	0.360 ± 0.011	0.264 ± 0.012	0.208 ± 0.004	0.048 ± 0.001	0.183 ± 0.001
2 Oct 2017	0.53	5.62 ± 0.69	15.25 ± 0.69	0.313 ± 0.077	0.306 ± 0.052	0.269 ± 0.033	0.144 ± 0.018	0.390 ± 0.018
5 Oct 2017	0.61	5.03 ± 0.11	15.52 ± 0.11	0.326 ± 0.015	0.298 ± 0.012	0.245 ± 0.005	0.150 ± 0.003	0.463 ± 0.003
22 Oct 2017	0.40	11.32 ± 0.32	14.17 ± 0.32	0.582 ± 0.013	0.560 ± 0.019	0.444 ± 0.013	0.179 ± 0.005	0.224 ± 0.005
26 Oct 2017	0.34	0.85 ± 0.03	3.40 ± 0.03	0.267 ± 0.019	0.265 ± 0.018	0.200 ± 0.007	0.068 ± 0.002	0.272 ± 0.002
29 Oct 2017	0.57	8.88 ± 0.17	16.85 ± 0.17	0.415 ± 0.011	0.408 ± 0.011	0.345 ± 0.007	0.198 ± 0.004	0.376 ± 0.004

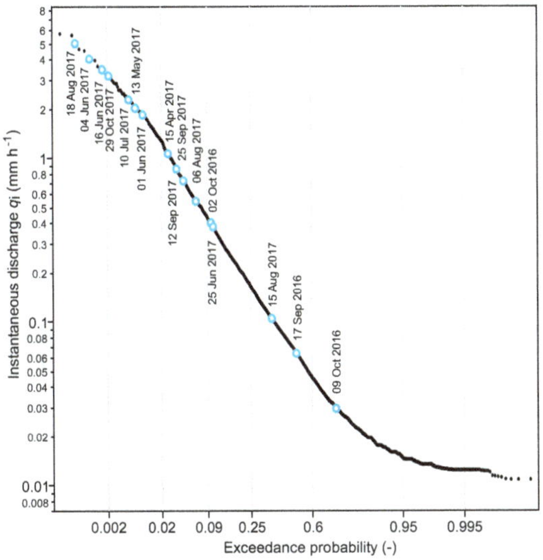

Figure 3. Flow duration curve of Erlenbach for the combined study periods 15 September to 6 November 2016 and 7 May to 1 November 2017. Blue data points indicate the hourly peak flow rates of 16 storm events analysed in this study, showing that widely varying flow conditions were captured in our data set. For the sake of readability, the remaining eight events are not included in the graph, because they share very similar hourly peak flow rates to those events depicted here.

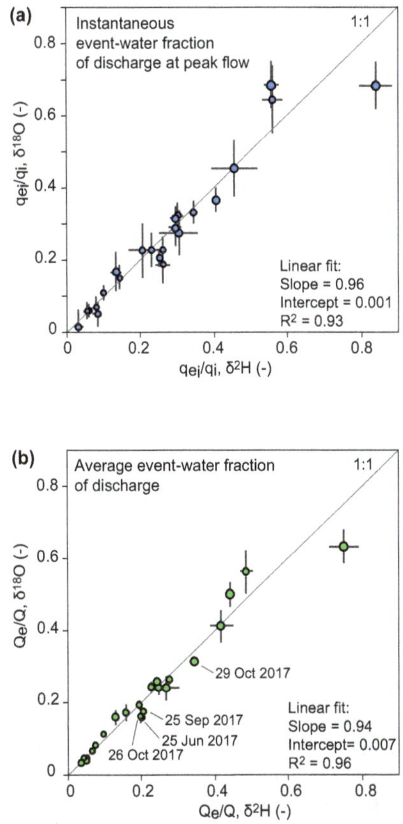

Figure 4. (a) Instantaneous event-water fractions of discharge at peak flow ($q_{e,i}/q_i$) as well as **(b)** whole-storm event-water fractions of discharge (Q_e/Q) obtained from either δ^2H or δ^{18}O. The sizes of the data points reflect the precipitation totals of the storm events, and error bars show ± 1 SE. For the 17 September 2017 storm, unrealistic results were obtained for Q_e when δ^{18}O was used as a tracer; therefore this data point is excluded from the comparison analysis. Hydrograph separation results for δ^{18}O are provided Table S1.

site. The instantaneous event-water fractions of discharge $\frac{q_{e_i}}{q_i}$ peaked at similar values in the two events (0.30 ± 0.01 and 0.33 ± 0.01, respectively), and the aggregated event-water volumes relative to discharge (Q_e/Q) were likewise similar (0.23 ± 0.01 and 0.24 ± 0.01). Thus, river discharge was predominantly pre-event water during both events, and despite the great differences in total event rainfall and antecedent wetness conditions, both storms resulted in similar event-water fractions of discharge (Table 2). In contrast, the event and pre-event runoff coefficients were roughly 3 times higher in the second storm ($Q_e/P = 0.05$ vs. 0.15, and $Q_{pe}/P = 0.17$ vs. 0.46; Table 2), suggesting that Q_e/P and Q_{pe}/P might more clearly reflect how catchments respond to variations in antecedent wetness conditions and total event rainfall.

During the 10 July 2017 storm, antecedent conditions were slightly wetter (AP7 = 20.2 mm) than on 2 October 2016, whereas the total rainfall volume and the runoff coefficient were intermediate to those of the two October events ($P = 25.4$ mm, $Q/P = 0.28$). However, because the maximum 1 h rainfall intensity during the 10 July 2017 storm was nearly 4 times larger compared to the two October events, peak flow rates during the 10 July 2017 storm were similar to the much larger 5 October 2017 storm (Fig. 6b, c). In contrast to both the 2 October 2016 and 5 October 2017 storms, event water comprised nearly 50 % of discharge during the 10 July 2017 storm. Similarly, the event and pre-event runoff coefficients ($Q_e/P = 0.135$ and $Q_{pe}/P = 0.141$, respectively) indicate that the 10 July 2017 storm event mobilized equivalent volumes of event and pre-event water. This suggests that infiltration excess during the high-intensity precipitation period enhanced the direct contribution of event water to the stream.

Across all 24 events, there is a general tendency for the instantaneous event-water fraction $\frac{q_{e_i}}{q_i}$ to peak on the rising limb, ahead of the flow peak (Figs. S1–S4 in the Supplement; for the three storms, 2 October 2016, 5 October 2017 and 10 July 2017, the peak times are indicated by grey and black vertical arrows in Fig. 6a–f). Thus the event-water fraction at the time of peak flow was typically somewhat smaller than the peak event-water fraction. This observation shows the importance of evaluating event-water fractions over the entire hydrograph rather than just at peak flow (von Freyberg et al., 2017). It also suggests that peak flows are primarily generated by mobilizing pre-event water, which dilutes the event water that is more prominent on the rising limb of the hydrograph.

Table 3. Spearman rank correlation coefficients (ρ) and p values for measures of storm characteristics, antecedent wetness and catchment storm response. Bolded, underlined values represent statistically significant correlations with $p < 0.0001$, and bolded values represent statistically significant correlations with $p < 0.01$. Measures of storm characteristics (left-hand columns in table) are total event precipitation (P), cumulative precipitation before peak flow ($P_{\mathrm{untilQpeak}}$), mean precipitation intensity (P_{int}), maximum precipitation over 1 h ($P_{1\,\mathrm{h}}$) and 4 h ($P_{4\,\mathrm{h}}$), event duration (T), and 4 h peak discharge volume ($Q_{4\,\mathrm{h}}$). Measures of initial catchment wetness state (right-hand columns in table) are 3-day and 7-day antecedent precipitation (AP3 and AP7) as well as the 1 h average values of discharge (Q_{ini}), groundwater table depth ($\mathrm{GW_{ini}}$), and soil moisture ($\mathrm{SM_{ini}}$) before the onset of the storm event. Measures of catchment storm response (rows of table) are precipitation (P), discharge (Q), event discharge (Q_{e}) and pre-event discharge (Q_{pe}), all defined as totals over the event and ratios among them. Correlations are not shown for Q_{pe}/Q, because they are just the negative of the corresponding correlations for Q_{e}/Q (since $Q_{\mathrm{pe}}/Q = 1 - Q_{\mathrm{e}}/Q$).

ρ	P	$P_{\mathrm{untilQpeak}}$	$P_{1\,\mathrm{h}}$	$P_{4\,\mathrm{h}}$	P_{int}	T	$Q_{4\,\mathrm{h}}$	AP3	AP7	Q_{ini}	$\mathrm{SM_{ini}}$	$\mathrm{GW_{ini}}$
P	**1.00**	**0.99**	**0.55**	**0.88**	0.42	**0.57**	**0.71**	0.03	−0.25	−0.21	−0.06	−0.17
Q	**0.75**	**0.74**	0.14	**0.57**	0.13	**0.60**	**0.88**	0.41	0.33	0.39	0.50	0.41
Q_{e}	**0.87**	**0.85**	0.50	**0.79**	0.32	**0.53**	**0.81**	0.04	−0.05	−0.07	0.13	0.01
Q_{pe}	**0.65**	**0.67**	−0.03	0.45	0.04	**0.57**	**0.83**	0.48	0.41	0.49	**0.58**	0.50
Q/P	0.27	0.29	−0.26	0.09	−0.14	0.39	**0.69**	**0.65**	**0.74**	**0.76**	**0.80**	**0.76**
Q_{e}/Q	**0.65**	**0.62**	**0.59**	**0.64**	0.31	0.34	0.40	−0.24	−0.51	−0.50	−0.35	−0.49
Q_{e}/P	**0.78**	**0.76**	0.41	**0.71**	0.31	0.48	**0.80**	0.04	0.03	−0.01	0.20	0.08
Q_{pe}/P	0.00	0.05	−0.46	−0.15	−0.30	0.22	0.45	**0.63**	**0.79**	**0.83**	**0.81**	**0.79**

p	P	$P_{\mathrm{untilQpeak}}$	$P_{1\,\mathrm{h}}$	$P_{4\,\mathrm{h}}$	P_{int}	T	$Q_{4\,\mathrm{h}}$	AP3	AP7	Q_{ini}	$\mathrm{SM_{ini}}$	$\mathrm{GW_{ini}}$
P	**0.0000**	**0.0000**	**0.0055**	**0.0000**	0.0407	**0.0037**	**0.0000**	0.8955	0.2336	0.3349	0.7713	0.4186
Q	**0.0000**	**0.0000**	0.5126	**0.0039**	0.5490	**0.0019**	**0.0000**	0.0487	0.1210	0.0580	0.0120	0.0479
Q_{e}	**0.0000**	**0.0000**	0.0129	**0.0000**	0.1220	**0.0073**	**0.0000**	0.8604	0.8150	0.7436	0.5408	0.9678
Q_{pe}	**0.0006**	**0.0004**	0.9053	0.0292	0.8401	**0.0038**	**0.0000**	0.0182	0.0439	0.0160	**0.0029**	0.0130
Q/P	0.2064	0.1753	0.2269	0.6654	0.5036	0.0626	**0.0002**	**0.0006**	**0.0000**	**0.0000**	**0.0000**	**0.0000**
Q_{e}/Q	**0.0006**	**0.0014**	**0.0024**	**0.0007**	0.1352	0.1029	0.0533	0.2647	0.0115	0.0128	0.0941	0.0156
Q_{e}/P	**0.0000**	**0.0000**	0.0440	**0.0001**	0.1470	0.0181	**0.0000**	0.8477	0.8877	0.9582	0.3552	0.7193
Q_{pe}/P	0.9871	0.8181	0.0221	0.4702	0.1482	0.2917	0.0272	**0.0009**	**0.0000**	**0.0000**	**0.0000**	**0.0000**

3.2 Catchment responses to storm characteristics and antecedent wetness

3.2.1 Runoff coefficients Q/P depend on antecedent wetness, not storm size

To identify the main controls on the relative contribution of event and pre-event water to catchment outflow, we analyse their correlations with storm characteristics and catchment antecedent wetness conditions (Fig. 7, Table 3). Larger, longer and more intense storms result in larger discharges Q, whereas there is no strong effect of antecedent precipitation (AP3, AP7), antecedent discharge (Q_{ini}), antecedent soil moisture ($\mathrm{SM_{ini}}$) or antecedent groundwater table depth ($\mathrm{GW_{ini}}$) on Q (Table 3). In contrast, although the runoff coefficient Q/P does not seem to be affected by storm size, it is strongly positively correlated with all metrics of catchment antecedent wetness conditions. This indicates that wetter conditions enhance the efficiency with which precipitation inputs trigger increases in streamflow. Clarifying the mechanisms behind this phenomenon requires not only hydrometric measurements but also tracers that track the water flow paths through the catchment.

3.2.2 Event-water discharge is controlled by storm characteristics, not antecedent wetness

Event-water fractions of discharge and precipitation (Q_{e}/Q and Q_{e}/P) at Erlenbach show statistically significant positive correlations with most storm characteristics, i.e., P, $P_{\mathrm{untilQpeak}}$, $P_{1\,\mathrm{h}}$ and $P_{4\,\mathrm{h}}$ (Table 3). These relationships suggest that event-water discharge increases more than proportionally with storm size. Similar results have been reported for Q_{e}/Q in forested and urban catchments (James and Roulet, 2009; Pellerin et al., 2008; Penna et al., 2015), and it has been hypothesized that more incoming rainfall eventually triggers saturation or infiltration excess, which leads to more surface runoff. Rainfall intensity has also been reported to affect Q_{e}/Q (Eshleman et al., 1993; Waddington et al., 1993), and at Erlenbach we find strong positive correlations of Q_{e}/Q and Q_{e}/P with 1 h and 4 h peak precipitation intensity ($P_{1\,\mathrm{h}}$ and $P_{4\,\mathrm{h}}$). We do not identify a strong relationship with the average rainfall intensity P_{int}, probably because its definition (total volume divided by total storm duration) makes it strongly dependent on the duration of low-intensity rainfall that contributes little to Q_{e} or Q.

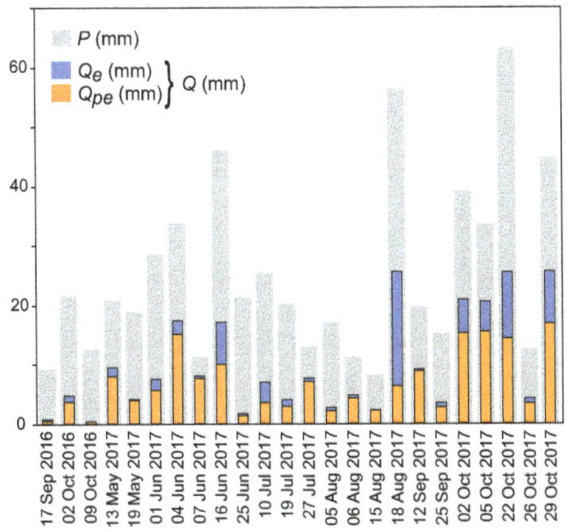

Figure 5. Volumes of precipitation (P, grey), compared to event water (Q_e, blue) and pre-event water (Q_{pe}, orange) in discharge across 24 storm events. Total discharge (Q) is the sum of event and pre-event water; relative standard errors for Q_e were between 1 % and 12 %, and they were between 0.1 % and 13 % for Q_{pe} (Table 2). For most of the storms, pre-event water comprised the major fraction of streamflow. Event water dominated streamflow only during two storms (10 July and 18 August 2017).

Perhaps surprisingly, the event-water fraction of discharge is lower, not higher, under wetter antecedent conditions; correlations between Q_e/Q and the antecedent wetness metrics range from −0.24 to −0.51, but they are not statistically significant ($p > 0.01$; Table 3). Indeed, even the volume of event water Q_e by itself (not as a fraction of total Q) does not become systematically larger under wetter conditions; the correlations between Q_e and the antecedent wetness metrics range from 0.01 to 0.13 (Table 3). Thus, wetter antecedent conditions do not lead to systematically higher event-water discharges in either absolute or relative terms. The negative correlation between antecedent wetness and Q_e/Q arises for the simple reason that wetter antecedent conditions increase total discharge (primarily by mobilizing more pre-event water), while Q_e remains largely unchanged.

3.2.3 Pre-event contributions relative to discharge correlate weakly with antecedent wetness

Because the pre-event-water fraction of discharge (Q_{pe}/Q) is defined as the complement to the event-water fraction ($Q_{pe}/Q = 1 - Q_e/Q$), its correlations with storm properties and antecedent wetness will be the opposite to those of Q_e/Q. At Erlenbach, Q_e/Q is weakly negatively correlated (and thus Q_{pe}/Q is weakly positively correlated) with our metrics of antecedent wetness. The positive correlation between antecedent wetness and Q_{pe}/Q suggests that a greater volume of pre-event water is available under wet conditions. However, the relationships between Q_{pe}/Q or Q_e/Q and antecedent wetness are highly scattered, consistent with other studies (Fischer et al., 2017; James and Roulet, 2009; Ocampo et al., 2006). Thus, these relationships are not much help in explaining why runoff coefficients at Erlenbach are strongly correlated with antecedent wetness and not with storm size and intensity.

3.2.4 Ratio of pre-event water to precipitation correlates strongly with antecedent wetness

In contrast to the event- and pre-event-water fractions of discharge (Q_e/Q, Q_{pe}/Q), the event runoff coefficient (Q_e/P) correlates strongly with metrics of storm characteristics (but not antecedent wetness), and the pre-event runoff coefficient (Q_{pe}/P) correlates strongly with the metrics of antecedent wetness conditions (but not storm characteristics; Fig. 7, Table 3). We find positive relationships between Q_e/P and most metrics of storm characteristics, such as P, $P_{\mathrm{until}Q_{\mathrm{peak}}}$ and $P_{4\,\mathrm{h}}$, similar to the relationships found for Q_e/Q. We also find that Q_{pe}/P, but not Q_e/P, is strongly (positively) correlated with all of our metrics of antecedent wetness conditions (Q_{ini}, $\mathrm{GW_{ini}}$, $\mathrm{SM_{ini}}$, AP3 and AP7). The correlations between Q_{pe}/P and storm characteristics are much weaker, suggesting that the activation of pre-event water by precipitation is primarily controlled by pre-storm wetness conditions and not by storm size.

The runoff coefficient's sensitivity to antecedent wetness, and its insensitivity to storm characteristics, can be understood through the behaviour of Q_e/P and Q_{pe}/P, which sum to the runoff coefficient itself: $Q/P = Q_e/P + Q_{pe}/P$. Because Q_{pe}/P is larger and more variable than Q_e/P (with one exception, all values of Q_e/P are less than 0.2, whereas Q_{pe}/P spans a range roughly 3 times as large), variations in the runoff coefficient Q/P will be dominated by variations in Q_{pe}/P. Thus, because Q_{pe}/P is sensitive to antecedent wetness, so is the runoff coefficient. For example, 7-day antecedent wetness AP7 is much more strongly correlated with Q_{pe}/P ($\rho = 0.79$) than with Q_e/P ($\rho = 0.03$), and because the variability of Q_{pe}/P is much greater than that of Q_e/P, it dominates the correlation between AP7 and the runoff coefficient Q/P ($\rho = 0.74$; Table 3). The same line of argument explains why the runoff coefficient is relatively insensitive to storm size and intensity. For example, the correlation between 4 h storm intensity $P_{4\,\mathrm{h}}$ and Q_e/P is 0.71, but the correlation with Q_{pe}/P is −0.15, and because the range of variation in Q_{pe}/P is roughly 3 times larger, the resulting correlation between $P_{4\,\mathrm{h}}$ and the runoff coefficient Q/P is only 0.09.

Storm characteristics and antecedent wetness generally exhibit stronger correlations with Q_e/P and Q_{pe}/P than with Q_e/Q and Q_{pe}/Q (Fig. 7, Table 3). For this reason, and because they are both components of the runoff coefficient itself, Q_e/P and Q_{pe}/P are more informative than Q_e/Q or Q_{pe}/Q in explaining how the runoff coefficient is controlled by event properties. One reason for the weaker correlations between storm characteristics and Q_e/Q or Q_{pe}/Q is that larger and more intense storms increase not only Q_e and Q_{pe}

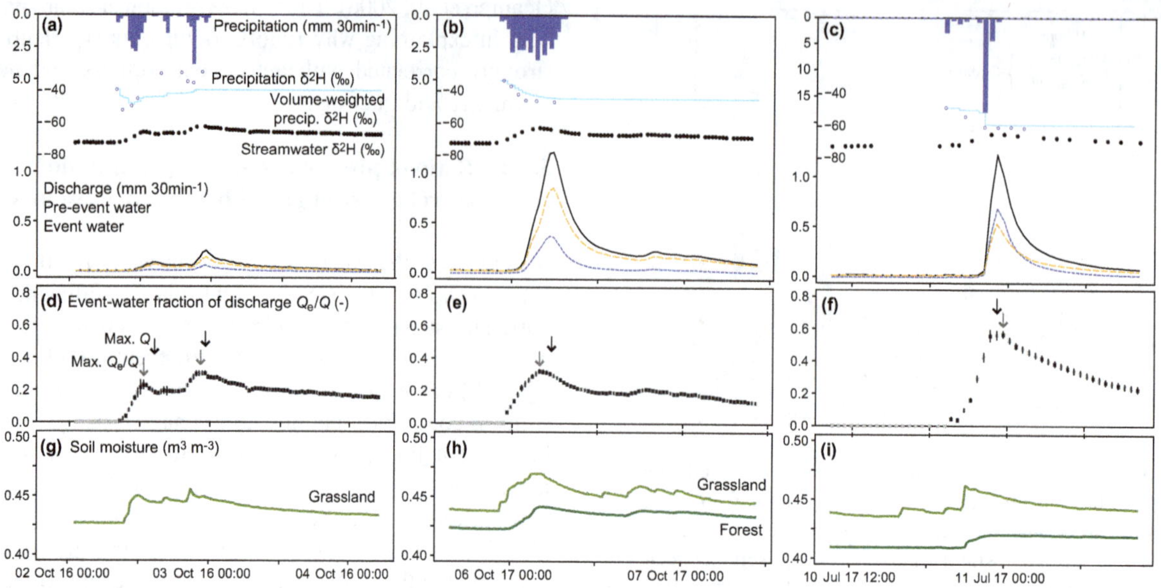

Figure 6. Time series of the storm events of 2 October 2016, 5 October and 10 July 2017. **(a–c)** precipitation hyetographs and deuterium abundance (δ^2H) in precipitation (with individual measurements in dark blue and incremental weighted means in light blue), and stream water (black) as well as discharge hydrograph separated into event and pre-event water. **(d–f)** event-water fraction of discharge; error bars indicate ±1 SE, and open bars indicate linearly interpolated event-water fractions when discharge isotope measurements are missing. **(g–i)** soil moisture at the grassland site (light green) and forest site (dark green, no data in 2016). Despite great differences in total event rainfall and antecedent wetness conditions between the two storms in the two left columns, their event-water fractions of discharge are very similar. In most cases, peaks in instantaneous event-water fractions precede peak flows (times of peak values indicated by grey and black vertical arrows in panels **(a)** to **(f)**.

but also Q; thus the ratios Q_e/Q and Q_{pe}/Q will change less than one might expect from the effects on Q_e and Q_{pe} themselves. This points to an important limitation when using Q_e/Q or Q_{pe}/Q as descriptors of catchment response during storm events and might explain why the relationships between Q_e/Q or Q_{pe}/Q and metrics of event properties often do not yield clear inferences about controlling factors for streamflow generation under different boundary conditions.

Our results provide important context for previous studies that have used the ratio between graphically estimated quick flow and precipitation as a proxy for how efficiently precipitation is translated into streamflow (e.g., Detty and McGuire, 2010; James and Roulet, 2007; Merz et al., 2006; Norbiato et al., 2009; Penna et al., 2011; Taylor and Pearce, 1982; Sidle et al., 1995). Many of these studies have shown that the ratio between quick flow and precipitation increases with antecedent wetness, suggesting that wetter conditions lead to the activation of rapid flow pathways, as groundwater levels rise to shallower, more permeable layers, or as hydrologic connectivity increases due to expansion of the river network and connection of wetlands and hillslopes to the stream (e.g., Dunne and Black, 1970; Godsey and Kirchner, 2014; McGuire and McDonnell, 2010). It should be stressed, however, that separating the hydrograph into base flow and quick flow with graphical or digital filter methods is a highly subjective process (Blume et al., 2007) that cannot resolve whether quick flow is primarily event water or pre-event water. Simply put, wetter conditions lead to more quick flow, but is this because more rainfall reaches the stream, or because more pre-event water is mobilized from catchment storage? Our results show that even at the highly dynamic Erlenbach catchment, antecedent wetness correlates with Q_{pe}/P but not Q_e/P; that is, wetter conditions lead primarily to more efficient mobilization of pre-event water, rather than to more efficient transmission of rainfall to the stream.

3.3 Controls on the event and pre-event water at Erlenbach

Previous research in the Erlenbach catchment has identified two possible sources of base flow: springs in the uppermost part of the catchment and groundwater outflow from a shallow aquifer on top of the low-permeability bedrock (van Meerveld et al., 2018). However, conceptual models of streamflow generation at Erlenbach have not considered antecedent wetness conditions as a major control on the discharge of pre-event water, possibly because tracer-based estimates of Q_{pe}/Q did not correlate with various metrics of antecedent wetness (Fischer et al., 2017 and results presented here). By contrast, when Q_{pe}/P is considered instead of Q_{pe}/Q, our data clearly show that high antecedent wetness triggers the mobilization of pre-event water. Because pre-

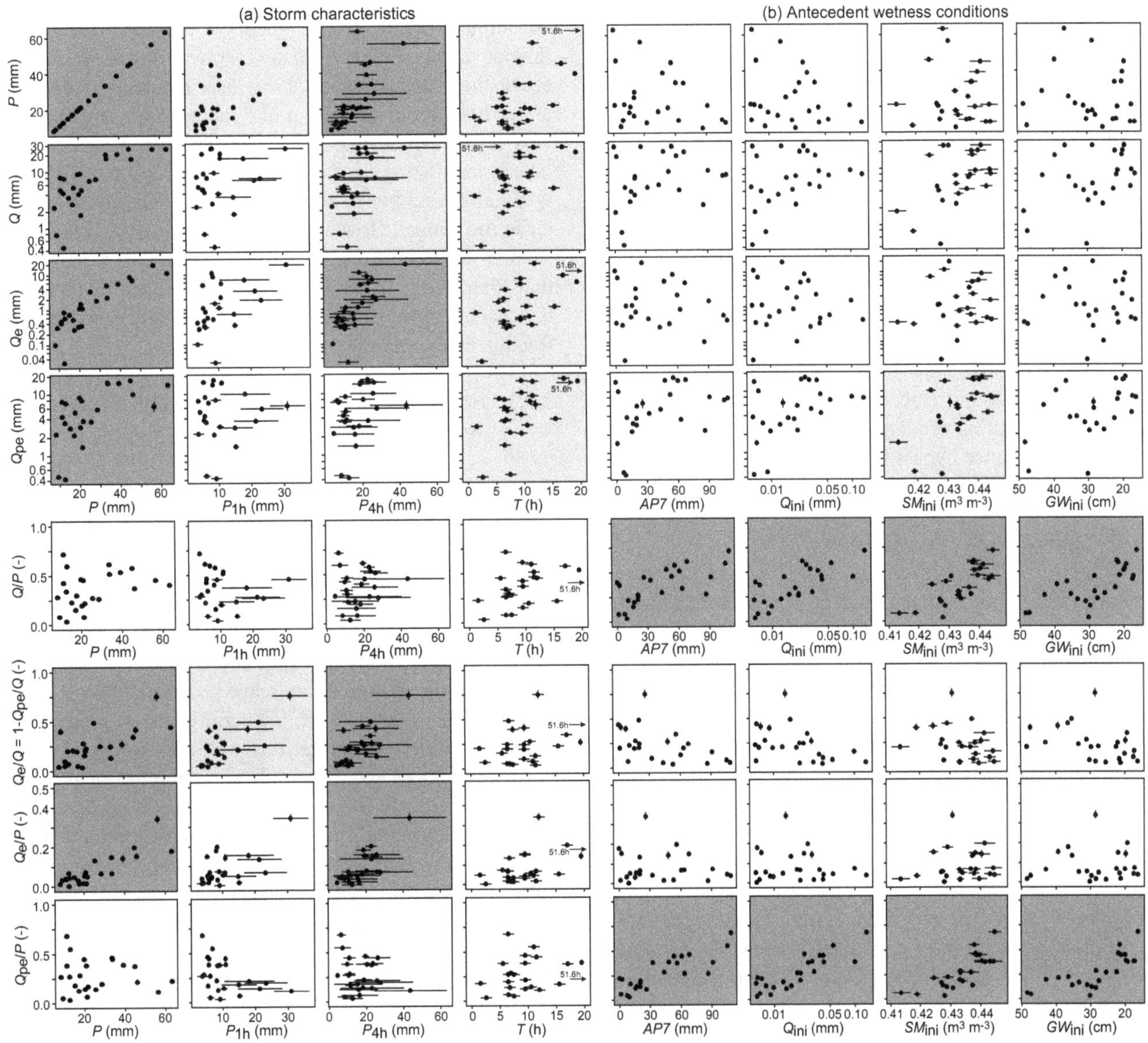

Figure 7. Total volumes of storm precipitation, discharge, event and pre-event water (Q_e, Q_{pe}), event-water fractions (Q_e/Q; please note that $Q_{pe}/Q = 1 - Q_e/Q$, and thus all relationships for Q_{pe}/Q, are the inverse of those for Q_e/Q) and event- and pre-event runoff coefficients (Q_e/P, Q_{pe}/P) of the 24 storm events, plotted against metrics of storm characteristics and catchment antecedent wetness conditions. Measures of storm characteristics (**a**) are total precipitation, peak 1 h precipitation, peak 4 h precipitation (P, $P_{1\,h}$, $P_{4\,h}$) and total event duration (T). Measures of antecedent wetness (**b**) are 7-day antecedent precipitation (AP7) and the 1 h average values of discharge (Q_{ini}), soil moisture (SM_{ini}), and groundwater table depth (GW_{ini}) before the onset of the storm event. Panels with light grey backgrounds indicate correlations that are statistically significant at $p < 0.01$; panels with dark grey backgrounds indicate correlations that are statistically significant at $p < 0.0001$. Because the rainfall duration (T) of storm 22 October 2017 was 51.6 h, it is off the scale and is thus indicated with vertical arrows.

event water comprises a large fraction of catchment outflow, even during events, antecedent wetness conditions are thus an important control on the streamflow regime at Erlenbach (along with storm size and intensity).

The several-fold increase of Q_{pe}/P with antecedent wetness implies that pre-event water is more efficiently mobilized under wetter conditions. The rapid activation of this stored pre-event water at Erlenbach (even during small storms) can be explained with generally shallow perched groundwater tables in the aquifer overlying the low-permeability bedrock. In a neighbouring catchment, groundwater tables are usually less than 0.4 m below the ground surface in low-permeability soils on plateaus and at the bottoms of the hillslopes, and they are not much deeper in the more permeable hillslope soils (Rinderer et al., 2014; Schleppi et al., 1998). Earlier studies at the Erlenbach and neighbouring catchments showed that the low-permeability soils on the plateaus saturate first during events, and as a consequence, a mixture of event and pre-event water flows off as shallow subsurface storm flow and surface runoff (Rinderer et al., 2016; van Meerveld et al., 2018).

Our isotope-based, two-component hydrograph separation results show that the relative contribution of event water to discharge plays only a minor role in the streamflow regime of the Erlenbach, despite the runoff coefficients Q/P being >0.4 for more than half of all storm events. Only for two storms with dry antecedent conditions and high-intensity rainfall did the event-water fraction of discharge Q_e/Q exceed 0.5. While wetter antecedent conditions clearly facilitated the mobilization of more pre-event water, they did not significantly enhance bypass flow of event water via surface runoff on saturated areas.

3.3.1 Source areas of event water

Several studies have used the ratio Q_e/P as a proxy for the relative catchment area that generates event water (e.g., Buttle and Peters, 1997; Ocampo et al., 2006) or have shown that Q_e/P predicts the mapped extent of saturated or impermeable areas (e.g., Eshleman et al., 1993; Rodhe, 1987; Pellerin et al., 2008). Direct runoff of event rainfall can occur on impermeable and low-permeability surfaces, on saturated areas, and through preferential flow pathways (Beven and Germann, 1982; Dunne and Black, 1970). At Erlenbach, the channel network itself and an asphalt road account for roughly 1.2 % of the total catchment area; this is a plausible lower bound for the area that can generate surface runoff. In addition, surface runoff may also occur on the saturated low-permeability soils of the plateaus as well as in locations where the water table is close to the surface (depressions, bottoms of hillslopes and river banks; Rinderer et al., 2014). Mapped wet meadows occur on approximately 22 % of the catchment area, and geostatistical analyses suggest that around 30 % of the total catchment area is prone to saturation (FOEN, 2011; Fig. 1a); 30 % is thus a plausible upper bound for the area that can generate surface runoff. The range between these upper and lower bounds is spanned across the 24 storms that we analysed, whose Q_e/P varied between 0.002 and 0.34 (mean $\pm$ SE 0.08$\pm$0.02). This suggests that the variability of Q_e/P across storms may reflect the contraction and expansion of these source areas and changes in their hydrological connectivity to the channel network.

3.3.2 Precipitation primarily mobilizes pre-event water instead of running off to the stream

For most of the analysed storms, event water is a much smaller contributor to streamflow than pre-event water. This observation makes sense if we assume that most precipitation lands directly on the more permeable hillslope soils (which constitute most of the catchment) or reaches these soils by flowing down gradient from low-permeability or saturated areas on the plateaus located above these hillslopes (Rinderer et al., 2014). Water infiltrating into the hillslope soils presumably raises the groundwater table into more permeable soil layers, facilitating the rapid downslope transport of groundwater and resulting in a mixture of event and pre-event water in the stream. However, during particularly large storms, such as those on 10 July and 18 August 2017, it is likely that event-water generating areas hydrologically connect to the channel network, and event water becomes a much larger fraction of the streamflow hydrograph.

3.3.3 Sources of pre-event water and link to antecedent wetness

Temporal and spatial variations in subsurface hydrological connectivity have been studied in a 20 ha catchment adjacent to Erlenbach, using a dense network of groundwater observation wells (Rinderer et al., 2018). That study showed that catchment areas with a subsurface hydrological connection to the channel network expand and contract during storm events. It seems likely that similar processes occur at Erlenbach, as it shares similar landscape properties with Rinderer et al.'s (2018) study site. Following this line of thought, one can speculatively infer that the infiltration of precipitation into the hillslopes and the mobilization of hillslope pre-event water significantly increase subsurface hydrologic connectivity. The amount of hillslope pre-event water that is mobilized will therefore largely depend on the pre-storm storage deficit in the hillslopes and not so much on the pre-storm storage deficit in the event-water source areas (where the storage deficit is always small). This would explain our observation that direct runoff of event water to the stream is mainly controlled by storm characteristics (and not by antecedent wetness), whereas the mobilization of pre-event water is strongly controlled by antecedent wetness (and much less by storm characteristics). However, no spatially distributed measurements of groundwater table dynamics are available to further investigate this hypothesis at Erlenbach.

3.4 Fingerprints of catchment response

Our analysis, spanning 24 storm events with contrasting characteristics, reveals at least four patterns of behaviour that could potentially be useful as "fingerprints" of catchment response if they are found to also hold in other catchments. First, the runoff coefficient Q/P is a roughly linear function of the logarithm of antecedent discharge Q_{ini} (Fig. 7). If similar linear relationships are also observed elsewhere, their slopes (which are dimensionless) could be used as indices of catchment response for catchment comparison purposes. Second, the event-water runoff coefficient Q_e/P is a roughly linear function of P itself (Fig. 7), and its slope can be considered as an index of how storm size alters the fraction of the catchment area that is connected to the stream by fast flow paths. One could even consider the hypothetical point where this linear relationship crosses the $Q_e/P = 1$ line (which would not be reached in practice) as an indicator of how much precipitation would be required to establish fast flow paths connecting the entire catchment to the stream. Third, the pre-event runoff coefficient Q_{pe}/P is a roughly linear function of antecedent precipitation AP7. Under the assumption that AP7 is a reliable surrogate for catchment antecedent wetness, the slope of the AP7–Q_{pe}/P relationship could be considered as an index of how antecedent wetness alters the fraction of the catchment in which stored, pre-event water can be efficiently mobilized during events. Fourth, the pre-event runoff coefficient Q_{pe}/P is a roughly linear function of the logarithm of antecedent discharge Q_{ini} (Fig. 7). The slope of this relationship, which is dimensionless, contains information on how antecedent discharge reflects antecedent wetness and how antecedent wetness determines the mobility of pre-event water. We emphasize that these fingerprints of catchment behaviour are necessarily speculative, unless and until they are confirmed by cross-catchment comparisons. We also note that these four ratios are logically constrained to be ≤ 1, and thus they must become asymptotic at some point.

4 Summary and outlook

Tracer-based estimates of event-water and pre-event-water fractions of discharge (Q_e/Q, Q_{pe}/Q) are often compared to catchment properties, storm characteristics and antecedent wetness conditions to identify controls on streamflow generation mechanisms. However, these relationships may be obscured, because the same factors that influence event discharge Q_e and pre-event discharge Q_{pe} also necessarily influence total discharge Q as well. We thus propose that the fractions of event water and pre-event water relative to precipitation (Q_e/P and Q_{pe}/P) provide an alternative and more insightful approach to study catchment storm responses. Here, we use 30 min stable water isotope data, collected over a period of roughly 8 months at the pre-Alpine Erlenbach catchment, to obtain robust estimates of Q_e and Q_{pe}. In total, we analysed 24 rainfall events spanning a wide range of hydro-climatic conditions. Our findings are summarized below:

1. Pre-event water dominates the streamflow hydrograph for the majority of the storms. The event-water fraction Q_e/Q correlates strongly (positively) with storm size and intensity but weakly with antecedent wetness conditions. Because Q is the sum of Q_e and Q_{pe}, Q itself is affected by storm size and antecedent wetness conditions in similar ways as Q_e and Q_{pe}, hampering the use of Q_e/Q and Q_{pe}/Q to identify major controls on streamflow generation.

2. By relating event-water and pre-event-water volumes to event precipitation P instead (i.e., Q_e/P, Q_{pe}/P), we find that the event-water fraction of precipitation (Q_e/P) correlates strongly with metrics of storm characteristics (but not antecedent wetness), and the ratio of pre-event water to precipitation (Q_{pe}/P) correlates strongly with metrics of antecedent wetness (but not storm characteristics). Thus, Q_e/P and Q_{pe}/P more clearly reflect the influence of major controls on streamflow generation compared to Q_e/Q (or Q_{pe}/Q).

3. Although numerous studies have used runoff coefficients or tracer-based event-water fractions Q_e/Q to study catchment hydrological behaviour during storm events (e.g., Klaus and McDonnell, 2013), the additional information provided by Q_e/P and Q_{pe}/P has yet not been exploited. Together, Q_e/P and Q_{pe}/P separate the runoff coefficient Q/P into its contributions from event water and pre-event water, providing a straightforward way to quantify both of these components of streamflow response.

4. At Erlenbach, Q/P, Q_e/P and Q_{pe}/P exhibit roughly linear relationships with several measures of storm size and antecedent wetness, suggesting that these relationships, particularly their slopes, may be diagnostic fingerprints that may be useful for characterizing hydrologic response across diverse catchments.

5. Looking toward the future, we anticipate that hydrograph separation studies will increasingly seek to quantify many different sources of streamflow, beyond the traditional separation of the two components Q_e and Q_{pe}. We note that if other components can be identified (e.g., streamflow originating from throughfall, soil water, deep bedrock water, etc.), studying their volumes relative to precipitation, rather than discharge, may shed important light on how they are mobilized during storm events.

Author contributions. JF and BS collected the data set. JF analysed the data set. JF prepared the paper with contributions from JK and MR.

Competing interests. The authors declare that they have no conflict of interest.

Acknowledgements. We thank Leonie Kiewiet for providing information of groundwater table variations in a catchment neighbouring our study site. The comments of Daniele Penna and two anonymous reviewers helped to improve the paper.

Edited by: Mariano Moreno de las Heras

References

Barthold, F. K., Turner, B. L., Elsenbeer, H., and Zimmermann, A.: A hydrochemical approach to quantify the role of return flow in a surface flow-dominated catchment, Hydrol. Process., 31, 1018–1033, https://doi.org/10.1002/hyp.11083, 2017.

Beven, K. and Germann, P.: Macropores and water flow in soils, Water Resour. Res., 18, 1311–1325, https://doi.org/10.1029/WR018i005p01311, 1982.

Bevington, P. R. and Robinson, D. K.: Data reduction and error analysis for the physical sciences, 3rd Edn., McGraw-Hill, Boston, 320 pp., 2003.

Birkel, C., Soulsby, C., Tetzlaff, D., Dunn, S., and Spezia, L.: High-frequency storm event isotope sampling reveals time-variant transit time distributions and influence of diurnal cycles, Hydrol. Process., 26, 308–316, https://doi.org/10.1002/hyp.8210, 2012.

Blume, T., Zehe, E., and Bronstert, A.: Rainfall-runoff response, event-based runoff coefficients and hydrograph separation, Hydrolog. Sci. J., 52, 843–862, https://doi.org/10.1623/hysj.52.5.843, 2007.

Brown, V. A., McDonnell, J. J., Burns, D. A., and Kendall, C.: The role of event water, a rapid shallow flow component, and catchment size in summer stormflow, J. Hydrol., 217, 171–190, 1999.

Buttle, J. M.: Isotope hydrograph separations and rapid delivery of pre-event water from drainage basins, Prog. Phys. Geog., 18, 16–41, https://doi.org/10.1177/030913339401800102, 1994.

Buttle, J. M. and Peters, D. L.: Inferring hydrological processes in a temperate basin using isotopic and geochemical hydrograph separation: A re-evaluation, Hydrol. Process., 11, 557–573, https://doi.org/10.1002/(SICI)1099-1085(199705)11:6<557::AID-HYP477>3.0.CO;2-Y, 1997.

Casper, M. C., Volkmann, H. N., Waldenmeyer, G., and Plate, E. J.: The separation of flow pathways in a sandstone catchment of the Northern Black Forest using DOC and a nested Approach, Phys. Chem. Earth, 28, 269–275, 2003.

Detty, J. M. and McGuire, K. J.: Threshold changes in storm runoff generation at a till-mantled headwater catchment, Water Resour. Res., 46, W07525, https://doi.org/10.1029/2009wr008102, 2010.

Dunne, T. and Black, R. D.: Partial Area Contributions to Storm Runoff in a Small New-England Watershed, Water Resour. Res., 6, 1296–1311, https://doi.org/10.1029/Wr006i005p01296, 1970.

Eshleman, K. N., Pollard, J. S., and Obrien, A. K.: Determination of Contributing Areas for Saturation Overland-Flow from Chemical Hydrograph Separations, Water Resour. Res., 29, 3577–3587, https://doi.org/10.1029/93WR01811, 1993.

Fischer, B. M. C., Rinderer, M., Schneider, P., Ewen, T., and Seibert, J.: Contributing sources to baseflow in pre-alpine headwaters using spatial snapshot sampling, Hydrol. Process., 29, 5321–5336, https://doi.org/10.1002/hyp.10529, 2015.

Fischer, B. M. C., Stähli, M., and Seibert, J.: Pre-event water contributions to runoff events of different magnitude in pre-alpine headwaters, Hydrol. Res., 48, 28–47, https://doi.org/10.2166/nh.2016.176, 2017.

FOEN: National Ecological Network (REN), Swiss Federal Office for the Environment (FOEN), Berne, Switzerland, 2011.

Genereux, D.: Quantifying uncertainty in tracer-based hydrograph separations, Water Resour. Res., 34, 915–919, 1998.

Godsey, S. E. and Kirchner, J. W.: Dynamic, discontinuous stream networks: Hydrologically driven variations in active drainage density, flowing channels and stream order, Hydrol. Process., 28, 5791–5803, https://doi.org/10.1002/hyp.10310, 2014.

Hall, F. R.: Base-Flow Recessions – a Review, Water Resour. Res., 4, 973–983, https://doi.org/10.1029/Wr004i005p00973, 1968.

Hantke, R.: Die neue geologische Karte des Kantons Zürich und seiner Nachbargebiete, in: Vierteljahrsschrift der Naturforschenden Gesellschaft in Zürich, Vol. 88, Zurich, Switzerland, 618–620, 1967.

Hegg, C., McArdell, B. W., and Badoux, A.: One hundred years of mountain hydrology in Switzerland by the WSL, Hydrol. Process., 20, 371–376, https://doi.org/10.1002/hyp.6055, 2006.

Hewlett, J. D. and Hibbert, A. R.: Factors affecting the response of small watersheds to precipitation in humid areas, in: Proceedings of the International Symposium on Forest Hydrology, edited by: Sopper, W. E. and Lull, H. W., Pergamon Press, New York, 275–290, 1967.

Horton, R. E.: The role of infiltration in the hydrologic cycle, Eos T. Am. Geophys. Un., 14, 446–460, 1933.

Hrachowitz, M., Benettin, P., van Breukelen, B. M., Fovet, O., Howden, N. J. K., Ruiz, L., van der Velde, Y., and Wade, A. J.: Transit timesthe link between hydrology and water quality at the catchment scale, Wiley Interdisciplinary Reviews-Water, 3, 629–657, https://doi.org/10.1002/wat2.1155, 2016.

Hsü , K. J. and Briegel, U.: The Flysch, in: Geology of Switzerland SE – 6, Birkhäuser, Basel, 65–82, 1991.

James, A. L. and Roulet, N. T.: Investigating hydrologic connectivity and its association with threshold change in runoff response in a temperate forested watershed, Hydrol. Process., 21, 3391–3408, https://doi.org/10.1002/hyp.6554, 2007.

James, A. L. and Roulet, N. T.: Antecedent moisture conditions and catchment morphology as controls on spatial patterns of runoff generation in small forest catchments, J. Hydrol., 377, 351–366, https://doi.org/10.1016/j.jhydrol.2009.08.039, 2009.

Jones, J. P., Sudicky, E. A., Brookfield, A. E., and Park, Y.-J.: An assessment of the tracer-based approach to quantifying groundwater contributions to streamflow, Water Resour. Res., 42, W02407, https://doi.org/10.1029/2005WR004130, 2006.

Jordan, J. P.: Spatial and temporal variability of stormflow generation processes on a Swiss catchment, J. Hydrol., 153, 357–382, https://doi.org/10.1016/0022-1694(94)90199-6, 1994.

Klaus, J. and McDonnell, J. J.: Hydrograph separation using stable isotopes: Review and evaluation, J. Hydrol., 505, 47–64, https://doi.org/10.1016/j.jhydrol.2013.09.006, 2013.

Laudon, H., Sjöblom, V., Buffam, I., Seibert, J., and Mörth, M.: The role of catchment scale and landscape characteristics for runoff generation of boreal streams, J. Hydrol., 344, 198–209, https://doi.org/10.1016/j.jhydrol.2007.07.010, 2007.

Litt, G. F., Gardner, C. B., Ogden, F. L., and Lyons, W. B.: Hydrologic tracers and thresholds: A comparison of geochemical techniques for event-based stream hydrograph separation and flowpath interpretation across multiple land covers in the Panama Canal Watershed, Appl. Geochem., 63, 507–518, https://doi.org/10.1016/j.apgeochem.2015.04.003, 2015.

McDonnell, J. J. and Beven, K.: Debates – The future of hydrological sciences: A (common) path forward? A call to action aimed at understanding velocities, celerities and residence time distributions of the headwater hydrograph, Water Resour. Res., 50, 5342–5350, https://doi.org/10.1002/2013WR015141, 2014.

McDonnell, J. J., Bonell, M., Stewart, M. K., and Pearce, A. J.: Deuterium Variations in Storm Rainfall – Implications for Stream Hydrograph Separation, Water Resour. Res., 26, 455–458, https://doi.org/10.1029/WR026i003p00455, 1990.

McGlynn, B. L. and McDonnell, J. J.: Quantifying the relative contributions of riparian and hillslope zones to catchment runoff, Water Resour. Res., 39, 1310, https://doi.org/10.1029/2003wr002091, 2003.

McGlynn, B. L., McDonnell, J. J., Seibert, J., and Kendall, C.: Scale effects on headwater catchment runoff timing, flow sources, and groundwater-streamflow relations, Water Resour. Res., 40, W07504, https://doi.org/10.1029/2003wr002494, 2004.

McGuire, K. J. and McDonnell, J. J.: Hydrological connectivity of hillslopes and streams: Characteristic time scales and nonlinearities, Water Resour. Res., 46, W10543, https://doi.org/10.1029/2010wr009341, 2010.

Merz, R., Bloschl, G., and Parajka, J.: Spatio-temporal variability of event runoff coefficients, J. Hydrol., 331, 591–604, https://doi.org/10.1016/j.jhydrol.2006.06.008, 2006.

Muñoz-Villers, L. E. and McDonnell, J. J.: Runoff generation in a steep, tropical montane cloud forest catchment on permeable volcanic substrate, Water Resour. Res., 48, W09528, https://doi.org/10.1029/2011WR011316, 2012.

Munyaneza, O., Wenninger, J., and Uhlenbrook, S.: Identification of runoff generation processes using hydrometric and tracer methods in a meso-scale catchment in Rwanda, Hydrol. Earth Syst. Sci., 16, 1991–2004, https://doi.org/10.5194/hess-16-1991-2012, 2012.

Nathan, R. J. and McMahon, T. A.: Evaluation of Automated Techniques for Base-Flow and Recession Analyses, Water Resour. Res., 26, 1465–1473, https://doi.org/10.1029/Wr026i007p01465, 1990.

Norbiato, D., Borga, M., Merz, R., Blöschl, G., and Carton, A.: Controls on event runoff coefficients in the eastern Italian Alps, J. Hydrol., 375, 312–325, https://doi.org/10.1016/j.jhydrol.2009.06.044, 2009.

Ocampo, C. J., Oldham, C. E., Sivapalan, M., and Turner, J. V.: Hydrological versus biogeochemical controls on catchment nitrate export: a test of the flushing mechanism, Hydrol. Process., 20, 4269–4286, https://doi.org/10.1002/hyp.6311, 2006.

Pellerin, B. A., Wollheim, W. M., Feng, X., and Vörörsmarty, C. J.: The application of electrical conductivity as a tracer for hydrograph separation in urban catchments, Hydrol. Process., 22, 1810–1818, https://doi.org/10.1002/hyp.6786, 2008.

Penna, D., Tromp-van Meerveld, H. J., Gobbi, A., Borga, M., and Dalla Fontana, G.: The influence of soil moisture on threshold runoff generation processes in an alpine headwater catchment, Hydrol. Earth Syst. Sci., 15, 689–702, https://doi.org/10.5194/hess-15-689-2011, 2011.

Penna, D., van Meerveld, H. J., Oliviero, O., Zuecco, G., Assendelft, R. S., Dalla Fontana, G., and Borga, M.: Seasonal changes in runoff generation in a small forested mountain catchment, Hydrol. Process., 29, 2027–2042, https://doi.org/10.1002/hyp.10347, 2015.

Penna, D., van Meerveld, H. J., Zuecco, G., Fontana, G. D., and Borga, M.: Hydrological response of an Alpine catchment to rainfall and snowmelt events, J. Hydrol., 537, 382–397, https://doi.org/10.1016/j.jhydrol.2016.03.040, 2016.

Pinder, G. F. and Jones, J. F.: Determination of the groundwater component of peak discharge from the chemistry of total runoff, Water Resour. Res., 5, 438–445, https://doi.org/10.1029/WR005i002p00438, 1969.

Rinderer, M., van Meerveld, H. J., and Seibert, J.: Topographic controls on shallow groundwater levels in a steep, prealpine catchment: When are the TWI assumptions valid?, Water Resour. Res., 50, 6067–6080, https://doi.org/10.1002/2013wr015009, 2014.

Rinderer, M., van Meerveld, I., Stähli, M., and Seibert, J.: Is groundwater response timing in a pre-alpine catchment controlled more by topography or by rainfall?, Hydrol. Process., 30, 1036–1051, https://doi.org/10.1002/hyp.10634, 2016.

Rinderer, M., McGlynn, B. L., and van Meerveld, H. J.: Groundwater similarity across a watershed derived from time-warped and flow-corrected time series, Water Resour. Res., 53, 3921–3940, https://doi.org/10.1002/2016WR019856, 2017.

Rinderer, M., Meerveld, H. J., and McGlynn, B.: From points to patterns: using groundwater time series clustering to investigate subsurface hydrological connectivity and runoff source area dynamics, Hydrol. Process., submitted, 2018.

Rodhe, A.: The origin of streamwater traced by oxygen-18, Ph.D., Dept. Phys. Geogr. Div. Hydrol., Uppsala Univ., 290 pp., 1987.

Schleppi, P., Muller, N., Feyen, H., Papritz, A., Bucher, J. B., and Fluehler, H.: Nitrogen budgets of two small experimental forested catchments at Alptal, Switzerland, Forest Ecol. Manag., 101, 177–185, https://doi.org/10.1016/S0378-1127(97)00134-5, 1998.

Shanley, J. B. and Chalmers, A.: The effect of frozen soil on snowmelt runoff at Sleepers River, Vermont, Hydrol. Process., 13, 1843–1857, https://doi.org/10.1002/(SICI)1099-1085(199909)13:12/13<1843::AID-HYP879>3.0.CO;2-G, 1999.

Shanley, J. B., Kendall, C., Smith, T. E., Wolock, D. M., and McDonnell, J. J.: Controls on old and new water contributions to stream flow at some nested catchments in Vermont, USA, Hydrol. Process., 16, 589–609, https://doi.org/10.1002/hyp.312, 2002.

Sidle, R. C., Tsuboyama, Y., Noguchi, S., Hosoda, I., Fujieda, M., and Shimizu, T.: Seasonal hydrologic response at various spatial scales in a small forested catchment, Hitachi-Ohta, Jpn. J. Hydrol., 168, 227–250, https://doi.org/10.1016/0022-1694(94)02639-s, 1995.

Sklash, M. G. and Farvolden, R. N.: The role of groundwater in storm runoff, J. Hydrol., 43, 45–65, https://doi.org/10.1016/0022-1694(79)90164-1, 1979.

Stähli, M. and Gustafsson, D.: Long-term investigations of the snow cover in a subalpine semi-forested catchment, Hydrol. Process., 20, 411–428, https://doi.org/10.1002/hyp.6058, 2006.

Taylor, C. H. and Pearce, A. J.: Storm runoff processes and subcatchment characteristics in a new zealand hill country catchment, Earth Surf. Proc. Land, 7, 439–447, https://doi.org/10.1002/esp.3290070505, 1982.

van Meerveld, H. J. I., Fischer, B. M. C., Rinderer, M., Stähli, M., and Seibert, J.: Runoff generation in a pre-Alpine catchment: A discussion between a tracer and a shallow groundwater hydrologist, Cuadernos de Investigación Geográfica, 44, 429–452, https://doi.org/10.18172/cig.3349, 2018.

von Freyberg, J., Radny, D., Gall, H. E., and Schirmer, M.: Implications of hydrologic connectivity between hillslopes and riparian zones on streamflow composition, J. Contam. Hydrol., 169, 62–74, https://doi.org/10.1016/j.jconhyd.2014.07.005, 2014.

von Freyberg, J., Studer, B., and Kirchner, J. W.: A lab in the field: high-frequency analysis of water quality and stable isotopes in stream water and precipitation, Hydrol. Earth Syst. Sci., 21, 1721–1739, https://doi.org/10.5194/hess-21-1721-2017, 2017.

Waddington, J. M., Roulet, N. T., and Hill, A. R.: Runoff Mechanisms in a Forested Groundwater Discharge Wetland, J. Hydrol., 147, 37–60, https://doi.org/10.1016/0022-1694(93)90074-J, 1993.

3

What controls the stable isotope composition of precipitation in the Mekong Delta? A model-based statistical approach

Nguyen Le Duy[1,2], **Ingo Heidbüchel**[1], **Hanno Meyer**[3], **Bruno Merz**[1,4], **and Heiko Apel**[1]

[1]GFZ German Research Centre for Geosciences, Section 5.4 – Hydrology, Potsdam, Germany
[2]SIWRR Southern Institute of Water Resources Research, Ho Chi Minh City, Vietnam
[3]Alfred Wegener Institute for Polar and Marine Research, Research Unit Potsdam, Potsdam, Germany
[4]University Potsdam, Institute of Earth and Environmental Science, Potsdam, Germany

Correspondence: Nguyen Le Duy (duy@gfz-potsdam.de)

Abstract. This study analyzes the influence of local and regional climatic factors on the stable isotopic composition of rainfall in the Vietnamese Mekong Delta (VMD) as part of the Asian monsoon region. It is based on 1.5 years of weekly rainfall samples. In the first step, the isotopic composition of the samples is analyzed by local meteoric water lines (LMWLs) and single-factor linear correlations. Additionally, the contribution of several regional and local factors is quantified by multiple linear regression (MLR) of all possible factor combinations and by relative importance analysis. This approach is novel for the interpretation of isotopic records and enables an objective quantification of the explained variance in isotopic records for individual factors. In this study, the local factors are extracted from local climate records, while the regional factors are derived from atmospheric backward trajectories of water particles. The regional factors, i.e., precipitation, temperature, relative humidity and the length of backward trajectories, are combined with equivalent local climatic parameters to explain the response variables $\delta^{18}O$, $\delta^{2}H$, and d-excess of precipitation at the station of measurement.

The results indicate that (i) MLR can better explain the isotopic variation in precipitation ($R^2 = 0.8$) compared to single-factor linear regression ($R^2 = 0.3$); (ii) the isotopic variation in precipitation is controlled dominantly by regional moisture regimes ($\sim 70\,\%$) compared to local climatic conditions ($\sim 30\,\%$); (iii) the most important climatic parameter during the rainy season is the precipitation amount along the trajectories of air mass movement; (iv) the influence of local precipitation amount and temperature is not significant during the rainy season, unlike the regional precipitation amount effect; (v) secondary fractionation processes (e.g., sub-cloud evaporation) can be identified through the d-excess and take place mainly in the dry season, either locally for $\delta^{18}O$ and $\delta^{2}H$, or along the air mass trajectories for d-excess. The analysis shows that regional and local factors vary in importance over the seasons and that the source regions and transport pathways, and particularly the climatic conditions along the pathways, have a large influence on the isotopic composition of rainfall. Although the general results have been reported qualitatively in previous studies (proving the validity of the approach), the proposed method provides quantitative estimates of the controlling factors, both for the whole data set and for distinct seasons. Therefore, it is argued that the approach constitutes an advancement in the statistical analysis of isotopic records in rainfall that can supplement or precede more complex studies utilizing atmospheric models. Due to its relative simplicity, the method can be easily transferred to other regions, or extended with other factors.

The results illustrate that the interpretation of the isotopic composition of precipitation as a recorder of local climatic conditions, as for example performed for paleorecords of water isotopes, may not be adequate in the southern part of the Indochinese Peninsula, and likely neither in other regions affected by monsoon processes. However, the presented approach could open a pathway towards better and seasonally differentiated reconstruction of paleoclimates based on isotopic records.

1 Introduction

Precipitation is typically composed of regional contributions, where atmospheric moisture has been transported over large distances, and local contributions, where the moisture has been provided by evapotranspiration within the close vicinity. Understanding the sources of precipitation and their relative contribution is critical for basin-wide water balance studies (Ingraham, 1998). Stable isotopes ($\delta^{18}O$ and $\delta^{2}H$) offer the possibility to identify the sources of precipitation and to quantify the contribution of regional and local sources (Gat, 1996). Furthermore, they can be used to investigate hydrological processes such as mechanisms responsible for streamflow generation (e.g., Kendall and Caldwell, 1998), in groundwater studies (e.g., Gonfiantini et al., 1998), and rainfall–runoff studies (e.g., Genereux and Hooper, 1998).

Isotopic variation in precipitation has been correlated with climatic parameters such as precipitation amount, air temperature, and air mass history (Dansgaard, 1964; Rozanski et al., 1992; Gat, 1996), termed amount effect, temperature effect (Dansgaard, 1964), and circulation effect (Tan, 2009, 2014), respectively. The circulation effect describes the changes in isotopic composition of precipitation because of air moisture originating from different areas of the ocean.

Delineating the present-day relationship between climatic factors and stable isotope variation in precipitation can also help to understand past climatic conditions on regional and global scales. However, the factors controlling isotopic variation in precipitation are numerous and complex; hence a better understanding of the climatic influences on isotopic values would improve the use of precipitation isotopes as a proxy to reconstruct paleoclimates (Yang et al., 2016).

In the Asian monsoon region, the isotopic signature of precipitation has been found to correlate with large-scale climatic parameters such as sea surface temperature and relative humidity of the air masses (Dansgaard, 1964; Merlivat and Jouzel, 1979; Clark and Fritz, 1997; Lachniet, 2009), ENSO (Ichiyanagi and Yamanaka, 2005; Tan, 2014; Yang et al., 2016) and the vertical wind shear index (Vuille et al., 2005). Other processes were also identified as relevant for monsoon regions: distillation during vapor transport (Araguás-Araguás et al., 1998; Yoshimura et al., 2003; Vuille et al., 2005; Dayem et al., 2010; Lee et al., 2012; Liu et al., 2014), re-evaporation, and rain–vapor interactions (Risi et al., 2008b; Chakraborty et al., 2016).

Relations between climate and water isotopes have been analyzed by univariate statistical regression methods (e.g., Araguás-Araguás et al., 1998; Bowen, 2008), isotope-enabled global climate models (GCMs; Yoshimura et al., 2008, 2014; Risi et al., 2010b; Okazaki and Yoshimura, 2017), isotope-incorporated Lagrangian models (Pfahl and Wernli, 2008; Sodemann et al., 2008), or the combination of GCMs (or Lagrangian models) with statistical analysis (Vuille et al., 2003, 2005; LeGrande and Schmidt, 2009; Tindall et al., 2009; Ishizaki et al., 2012; Conroy et al., 2013). Statistical models are not able to explain the actual processes causing a phenomenon (e.g., the physical controls of isotope variations in precipitation), in contrast to the physical model representation of the processes like GCMs or Lagrangian models. They can, however, detect the results of a process, and thus help to identify the responsible processes. Both approaches, statistical and physical models, have their advantages and disadvantages and hence coexist supplementing each other. We argue that taking into consideration the limitations and benefits of both statistical and physical models can enhance their power to interpret the relations between climate and water isotopes.

As illustrated in previous studies (e.g., Noone and Simmonds, 2002) and discussed in Sturm et al. (2010), the inherent limitations of empirical (or statistical) climate reconstructions from precipitation isotopes can lead to incorrect paleoclimate reconstructions. A significant limitation is the assumption that a single climatic factor controls the isotopic signal and that the stationary relationship (e.g., between temperature and $\delta^{18}O$) remains valid over the entire proxy record. This mono-factorial relationship does not consider the interplay of different climatic factors and is possibly biased. Another limitation is the assumption of a constant precipitation source or similar isotopic signatures of different moisture sources throughout the study period when using only local parameters (e.g., local precipitation) to interpret precipitation isotopes. In real cases, these assumptions are rarely fulfilled and often unrealistic because of the changes in seasonality and atmospheric circulation patterns. This is particularly true in those parts of the Asian monsoon region located in the transition zone between the Indian and western North-Pacific monsoons, where precipitation originates from both the Indian and Pacific Oceans (Delgado et al., 2012). The isotopic signatures of air masses arising from the Indian Ocean differ considerably from those of the Pacific Ocean, where the average $\delta^{18}O$ of the latter is about 2.5 ‰ more negative (Araguás-Araguás et al., 1998). Seasonally varying sources of precipitation have also been observed in China (Tan, 2014, and references therein), India (e.g., Breitenbach et al., 2010; Chakraborty et al., 2016), South Korea (Lee et al., 2003), Thailand (Ishizaki et al., 2012), and elsewhere (Araguás-Araguás et al., 1998).

Additionally, many studies have presented evidence that large-scale monsoon circulation is the primary driver of variations in precipitation isotopes instead of local controls (e.g., local precipitation amount or temperature) in some parts of the Asian monsoon region. This evidence has been found on different temporal scales including daily isotopic variability (Yoshimura et al., 2003, 2008), seasonal isotopic variability (Araguás-Araguás et al., 1998; Kurita et al., 2009; Dayem et al., 2010; Peng et al., 2010; Baker et al., 2015), and/or interannual isotopic variability (Vuille et al., 2005; LeGrande and Schmidt, 2009; Ishizaki et al., 2012; Tan, 2014; Kurita et al., 2015). This means that the isotopic composition of rainfall cannot be described adequately by mono-factorial statistical

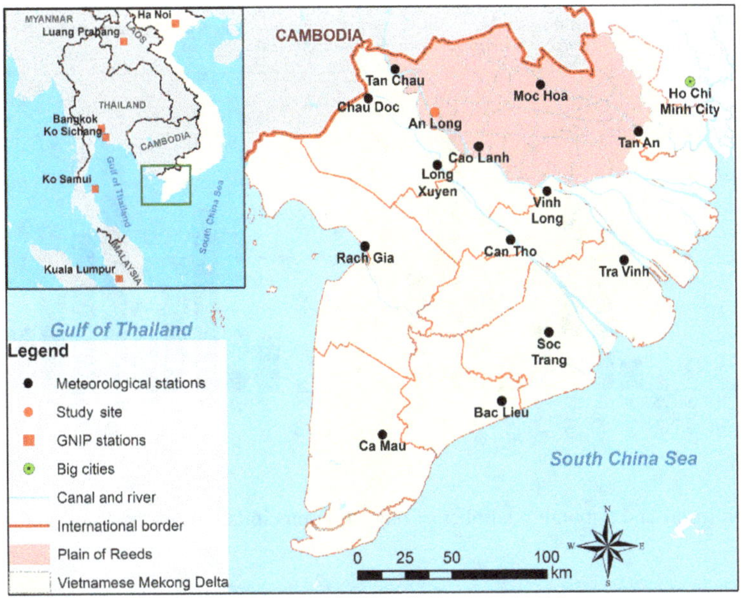

Figure 1. Sampling and monitoring sites in the study area.

approaches. Quantitative multi-factorial statistical studies of the isotopic composition are, however, rare. Only Ishizaki et al. (2012) present a multi-factorial analysis, which is limited to two factors (local precipitation amount and distillation of the moisture along its transport trajectories). To our best knowledge, there is no study considering the interplay of several local and regional factors available quantitatively, whereas the need for a simultaneous consideration of local and regional factors for the explanation of isotopic signatures has already been formulated in other studies (e.g., Johnson and Ingram, 2004). Therefore a quantitative multi-factorial statistical approach is developed in this study, utilizing multiple linear regression (MLR) and relative factor importance analysis to determine the most important drivers of isotopic composition in rainfall and their individual contributions.

We hypothesize that using multiple factors in a single linear model can explain a more significant share of the observed variance in isotopic composition, compared to monofactorial analyses. Furthermore, we also hypothesize that through importance analysis of the factors considered in the MLR models, the most important factors can be identified and their contribution quantified, thus enabling a better process understanding. Such a multi-factorial statistical method could also be applied in paleoclimate studies, separating and quantifying the impacts of local and regional factors on the isotopic composition of local precipitation (Sturm et al., 2010), thus overcoming the shortcomings of single factor analysis. The advantages of the statistical models are their simplicity and ease of interpretation, in contrast to complex circulation models (e.g., GCMs), which can also distinguish between local and regional drivers, but which require more extensive expertise and are more difficult to interpret. Circulation models can simulate the fractionation processes along the transport pathways of water in the atmosphere, but they cannot be used in a straightforward way to extract the impact of dominant factors and weigh their relative importance for the variability in the observed isotopic signal. For this, statistical techniques are required, in order to quantify the correlation between observed isotopic signal variability and regional climate change patterns (Sturm et al., 2010). Thus, using statistical models straight away might be an attractive alternative to obtain first indications about the dominant processes controlling the isotopic composition of rainfall, both for contemporary and paleoclimate studies.

This study uses the Vietnamese Mekong Delta (VMD) as a test case, for which isotopic data of precipitation has been collected for the first time. The rainfall samples ($\delta^{18}O$ and $\delta^{2}H$) were collected comparatively frequently (bi-weekly to weekly) over a period of 18 months. This data set enables a better analysis of the temporal dynamics of the isotopic composition as compared to the typical monthly Global Network of Isotopes in Precipitation (GNIP) data (IAEA/WMO, 2016). The sampled data were used to test the proposed approach for the identification and quantification of the controls on the isotopic variation in precipitation.

2 Study area

The study area, the Plain of Reeds (Fig. 1), is located in the northern part of the VMD, between latitudes 10°42′7″ to 10°48′9″ N and longitudes 105°22′45″ to 105°33′54″ E. With an area of 697 000 ha, it accounts for 17.7 % of the total area of the VMD. About 95 % of the Plain of Reeds is primarily used for rice paddies, but also for vegetable culti-

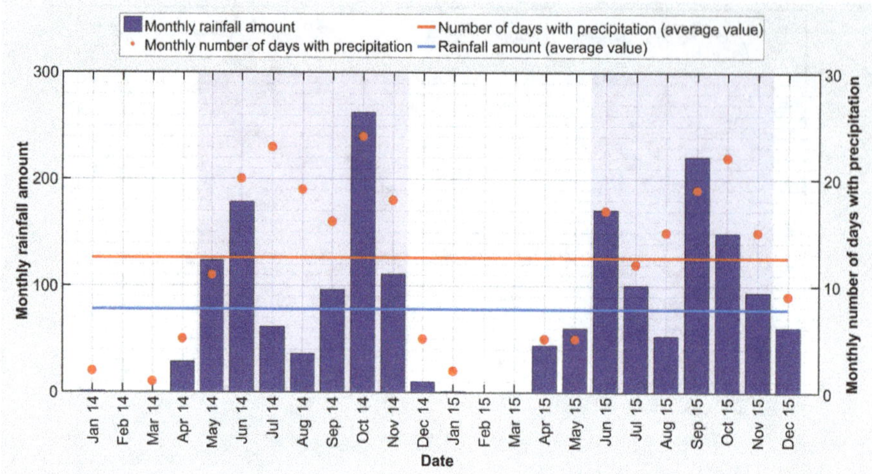

Figure 2. Monthly precipitation (mm) and a monthly number of days with precipitation for the Cao Lanh station. Light blue background indicates rainy season.

vating and shrimp and fish farming (Hung et al., 2014). The average elevation ranges from 1–4 m a.s.l. (above sea level).

Located in a tropical monsoon region, the climate of the VMD has a distinct seasonality with two seasons: the rainy season (May to November) resulting from the flow of moisture from the Indian Ocean and western North-Pacific Ocean accounting for approximately 80–90 % of the annual rainfall (Tri, 2012) and the dry season (December to April) controlled by high-pressure systems over the Asian continent (Wang et al., 2001). Precipitation from the Indian monsoon is forced by the convective heat sources over the Bay of Bengal (Wang et al., 2001) and arrives earlier than precipitation from the western North-Pacific monsoon (Delgado et al., 2012), forced by a convective heat source over the South China Sea–Philippine Sea. The average annual rainfall is 1400–2200 mm, characterized by an uneven distribution, both spatially and temporally (Renaud and Kuenzer, 2012; GSO, 2014).

The local climate of An Long is described by data from Cao Lanh station. Cao Lanh is the closest national meteorological station to An Long with continuous climate records, located approximately 37 km south-east of An Long. It is assumed that the climatic conditions of An Long and Cao Lanh are similar. The proximity of the two locations without any orographic obstacles in between, the same elevation, and the similar geographical setting (flat topography, located at the eastern bank of the Mekong River) justify this assumption.

The two distinct seasons were defined by the monthly precipitation amounts and the monthly number of days with precipitation recorded at Cao Lanh (Fig. 2). The dry season is set as the months with rainfall amount smaller than the overall average (blue line in Fig. 2), and a monthly number of days with precipitation smaller than the overall average (red line in Fig. 2). All other months are included in the rainy season. The definition used here is mainly developed for the local climatic conditions, the problem to be solved, and the data available. Other definitions could cause some data points to be assigned to the other season. However, those data points will most likely be from the transition period from one season to the other, i.e., other definitions would affect samples that have the least explanatory value for the actual dry and wet seasons.

The data indicate that the rainy season in 2014 lasted from May to November, and in 2015 from June to November. The dry season is thus defined from December 2014 to May 2015 and starts again in December 2015. The study period was somewhat dry with an annual rainfall of 985 mm compared to the long-term average of 1550 mm at the Cao Lanh station. This anomaly needs to be considered in the interpretation of the results.

The annual average temperature is 27 °C with the small interannual variability of about 1 °C. Variation in temperature throughout the year is small with monthly averages in the range of 25 to 29 °C (Fig. 3). The average annual relative humidity ranges from 82 to 85 %, with a seasonal variation of 80 to 88 % during the rainy season and 77 to 83 % during the dry season (Fig. 3). The mean annual evaporation is 984 mm with a significant difference between the rainy season and the dry season. The monthly evaporation rate ranges from 67 to 80 mm and from 76 to 109 mm in the rainy and dry season, respectively. Daily sunshine duration is about 8.7 to 9.6 h in the dry season and 5.5 to 5.9 h in the rainy season (Renaud and Kuenzer, 2012; GSO, 2014).

3 Methodology

An overview of the methodology is given in Fig. 4. For the derivation of local factors relevant for the isotopic composition of precipitation, climate data from the nearby mete-

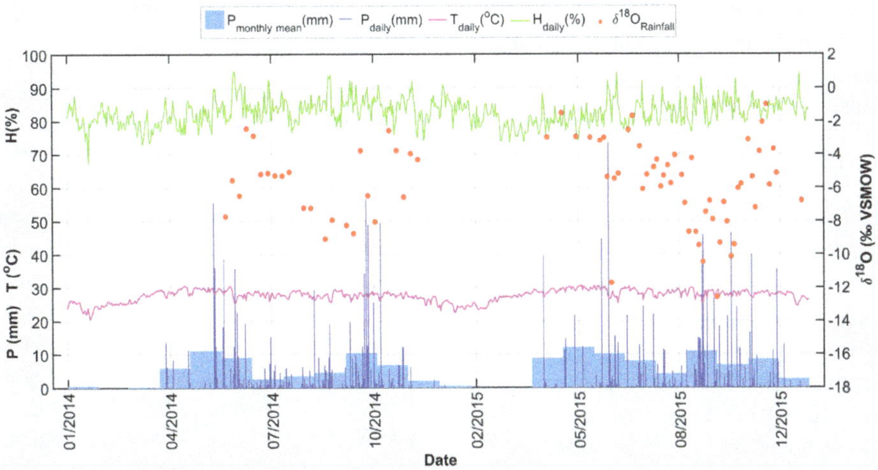

Figure 3. Climate data from the Cao Lanh meteorological station for the study period. Daily temperature (T) is given together with monthly and daily precipitation (P) and daily relative humidity (H). Weekly and bi-weekly $\delta^{18}O$ (‰ VSMOW) values of rainwater are presented as red circles.

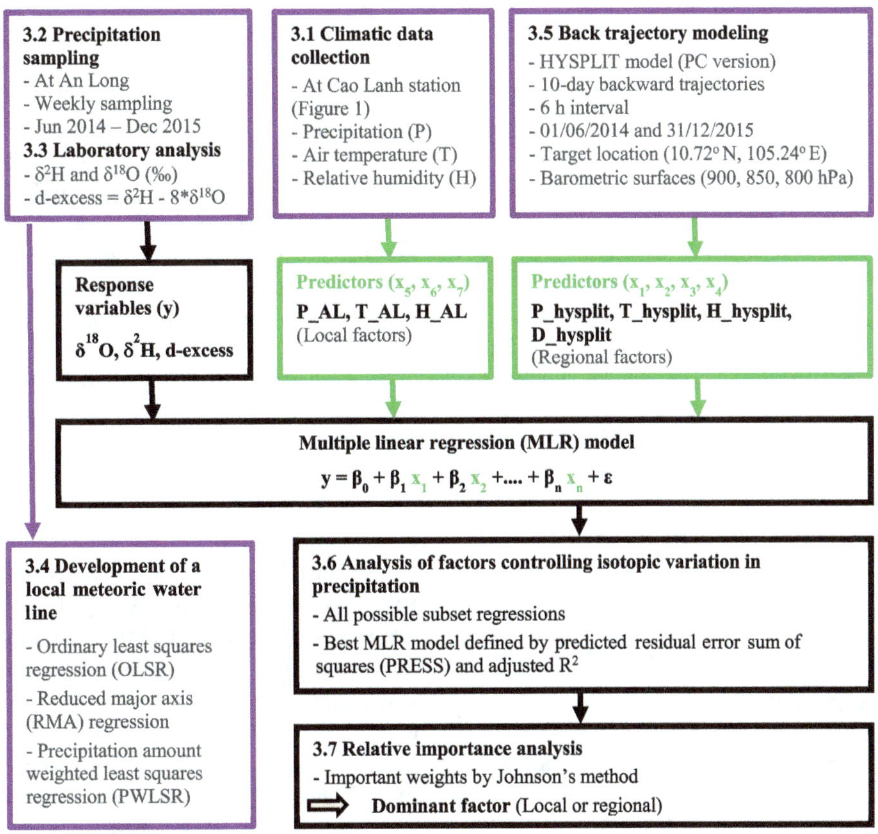

Figure 4. Methodology used in the study. Local precipitation (P_AL), air temperature (T_AL), and relative humidity (H_AL) at An Long. Precipitation amount (P_hysplit), mean temperature (T_hysplit) and relative humidity (H_hysplit) along the transport pathways, and the length of backward trajectories (D_hysplit).

orological station of Cao Lanh were collected (Sect. 3.1). At the test location, precipitation samples were analyzed for their isotopic composition (Sects. 3.2 and 3.3). Local meteoric water lines (LMWLs) were derived from the isotopic data (Sect. 3.4). The regional factors were derived from atmospheric back trajectory modeling (Sect. 3.5). All possible combinations of local and regional predictors were included in multiple linear regressions, and their ability to explain the observed variance in the isotopic composition of precipitation was determined with performance statistics (MLR; Sect. 3.6). Finally, the influence of the different factors on the explained variance in isotopic composition was determined by relative importance analysis (Sect. 3.7).

3.1 Climatic and isotopic data collection

Daily precipitation, air temperature, and relative humidity were obtained from the National Centre for Hydro-Meteorological Forecasting (NCHMF) of Vietnam at Cao Lanh for the period 2012–2015. Long-term (more than 30 years) climatic data at this station were provided by SIWRP (2014). Precipitation isotopic data from six selected GNIP stations (IAEA/WMO, 2016) located in the Indochinese Peninsula (Fig. 1) were collected for comparison with the isotopic data sampled in this study in order to investigate regional similarities or differences in isotope composition in rainfall in Southeast Asia. This was tested by the Levene test (Levene, 1960) for equality of variances in the isotopic data of the different stations, and the comparison of LMWLs.

3.2 Precipitation sampling at An Long

Precipitation at An Long in the Plain of Reeds (Fig. 1) was sampled on a weekly basis between June 2014 and May 2015 and twice a week between June 2015 and December 2015. The rain collector was a dip-in sampler type as described in the guidelines of the IAEA technical procedure for precipitation sampling (IAEA, 2014). It consists of a 5 L accumulation glass bottle fitted with a vertical 14 cm diameter plastic funnel that reaches almost to the bottom to prevent evaporative losses, and a pressure equilibration plastic tube (2 mm in diameter and 15 m in length) to minimize evaporation. All collected samples were stored in 30 mL plastic sample bottles with tight screw caps to avoid evaporation effects. Between collection and laboratory analysis, the samples were stored in the dark.

3.3 Isotopic laboratory analysis

All stable isotope samples were analyzed at the laboratory of the Alfred-Wegener-Institute (AWI) in Potsdam, Germany. The measurements were performed with a Finnigan MAT Delta-S mass spectrometer using equilibration techniques to determine the ratio of stable oxygen ($^{18}O:^{16}O$) and hydrogen ($^{2}H:^{1}H$) isotopes. Analytical results were reported as δ^2H and $\delta^{18}O$ (‰, relative to Vienna Standard Mean Ocean Water – VSMOW) with internal 1σ errors of better than 0.8 and 0.1 ‰ for δ^2H and $\delta^{18}O$, respectively. The measuring procedure is described in detail in Meyer et al. (2000). The deuterium excess (d-excess) was calculated following Eq. (1) (Dansgaard, 1964):

$$\text{d}-\text{excess} = \delta^2\text{H} - 8 \cdot \delta^{18}\text{O}. \tag{1}$$

3.4 Development of local meteoric water lines

LMWLs were derived by least-squares regression of $\delta^{18}O$ against δ^2H values of the samples. The quality-of-fit of the LMWLs was evaluated based on the coefficient of determination R^2 (also referred to as explained variance), the standard error SE, and the statistical significance value (p value). The regression model indicates a good fit to the data when R^2 is close to 1.0, the standard error is small in relation to the magnitude of the data, and the p value is smaller than 0.0001 (Helsel and Hirsch, 2002). The slope and intercept of the LMWL of An Long were compared to those of GNIP stations to put the data from An Long in a regional context.

3.5 Back trajectory modeling

The potential locations of atmospheric moisture sources and the direction of the air mass causing precipitation before reaching An Long station were investigated via back-trajectory analysis. This investigation was performed using the PC Windows-based HYSPLIT (Hybrid Single Particle Lagrangian Integrated Trajectory) model developed by NOAA (National Oceanic and Atmospheric Administration) at the Air Resources Laboratory (ARL) (http://www.arl.noaa.gov/HYSPLITinfo.php). The model builds on the Lagrangian approach, using a moving frame of reference for the advection and diffusion calculation as the air parcels move from their initial location (Draxler and Rolph, 2003; Stein et al., 2015).

The $1° \times 1°$ climatic data set generated by the global data assimilation system (GDAS) was used as input to the HYSPLIT model. The 10-day backward trajectory analysis was performed every 6 h between 1 June 2014 and 31 December 2015 at the sampling site (10.72° N, 105.24° E) for three levels at 1000, 1500, and 2000 m above ground (corresponding to barometric surfaces of approximately 900, 850, and 800 hPa, respectively). These barometric surfaces were chosen because the 850 hPa vorticity is highly indicative of the strength of the boundary layer moisture convergence and rainfall in regions away from the equator (Wang et al., 2001), hence rainfall is expected to mostly originate from these altitudes. Consequently, the combination of 800 and 850 hPa barometric surfaces accounts for the fact that rainfall is expected to mostly originate between 1500 and 2000 m a.g.l. (above ground level). Correspondingly, the combination of the barometric surfaces of 800, 850, and 900 hPa means that rainfall is expected to mostly originate between 1000 and 2000 m a.g.l. In total, 6948 backward trajectories were computed. The HYSPLIT outputs, i.e., precipitation, temperature, relative humidity along the backward trajectories, and the length of trajectories (the distance of moisture sources traveled), were used as regional factors potentially controlling the variation in the isotopic composition of precipitation at An Long. Accumulated precipitation, mean values of temperature, and humidity of the hourly HYSPLIT output were calculated along the trajectory, as well as the length of the trajectory. All these factors were used as predictors in the MLR.

A cluster analysis of a large number of trajectories arriving at the study site was applied to group similar trajectories and thus source regions, in order to better interpret the trajectories and to confirm the accuracy of the trajectory analysis. More information about the HYSPLIT cluster analysis can be found at https://ready.arl.noaa.gov/documents/Tutorial/html/.

3.6 Analysis of factors controlling isotopic variation in precipitation

MLR was used to assess how the isotopic variation in precipitation is related to regional and local controlling factors. As indicators of regional factors, the output of the HYSPLIT model was used consisting of the accumulated precipitation amount along the transport pathways (hereafter P_hysplit), mean temperature (T_hysplit), and mean relative humidity (H_hysplit) along the trajectory, and the distance of moisture sources traveled within the time frame of 10 days (D_hysplit). The local climatic factors are weekly precipitation amount (P_AL) at An Long station, and weekly mean air temperature (T_AL) and weekly mean relative humidity (H_AL) taken from the nearby Cao Lanh station during the sampling period. These seven predictors were related to isotopic values (δ^{18}O, δ^{2}H, and d-excess) defined as response variables in the MLR. Pearson linear correlation coefficients were computed to show intercorrelations between response and predictor variables and then used to determine the importance of predictors in the MLR.

All possible subset regression models consisting of all possible combinations of predictors ($2^7 - 1 = 127$ models) were applied separately for δ^{18}O, δ^{2}H, and d-excess. The coefficient of determination R^2 for the MLR was calculated for each subset regression. The goodness of each MLR model was evaluated based on the prediction residual error sum of squares (PRESS; Eq. 2) and adjusted R^2 (R_a^2; Eq. 3) (Helsel and Hirsch, 2002).

$$\mathrm{PRESS} = \sum_{i=1}^{n} e_{(i)}^2, \tag{2}$$

$$R_a^2 = R^2 - \left(1 - R^2\right)\frac{p}{(n-p-1)}. \tag{3}$$

In Eq. (2), the PRESS residuals are defined as $e_{(i)} = y_i - \hat{y}_{(i)}$, where $\hat{y}_{(i)}$ is the regression estimate of y_i based on a regression equation computed leaving out the ith observation. The process is repeated for all n observations. The selection of best models with PRESS is equivalent to a leave-one-out cross-validation, which tests the regression models for robustness and reduces the chances of model over-fitting, i.e., the chances of finding spurious regression models that provide good results for the given combination of factors and selected period only.

In Eq. (3), p is the total number of predictors in the MLR model and n is the number of observations. The statistical significance of all linear regression was evaluated based on the p value for the F test as part of a one-way ANOVA analysis. A good MLR model is thus characterized by the following:

i. PRESS approaching zero,

ii. adjusted R^2 (R_a^2) approaching 1.0,

iii. a p value smaller than 0.0001.

For each response variable, six pressure layers (800, 850, 900 hPa, and mean values of their combinations) and 10 durations of backward trajectories (from 1 to 10 days backward) were used. The different pressure levels and combinations were chosen to tackle the inherent uncertainty regarding the pressure levels from which the rainfall actually stems. Similarly, various durations of the trajectories were chosen in order to avoid fixing the a priori unknown travel time of precipitation reaching An Long. Overall, this resulted in 7620 MLR models for each response variable δ^{18}O, δ^{2}H, and d-excess (6 pressure levels times 10 trajectory durations times 127 predictor sets). The best MLR model was then identified by the smallest PRESS value (Eq. 2). Furthermore, the goodness-of-fit of the MLR models was characterized based on the adjusted R^2 values.

3.7 Relative importance analysis

Relative importance analysis determines the proportion of the variance explained by the individual predictors in the regression. However, this is difficult when predictors are correlated, since multicollinearity can lead to a high sensitivity of regression coefficients caused by small changes in the model. This means that the importance can strongly shift from one predictor to another well correlated one if the data set is changed even only slightly. The leave-one-out cross-validation may be particularly vulnerable to this effect.

While several methods such as zero-order bivariate correlation (e.g., Pearson correlation), standardized regression weights (β weights), structure coefficient (the correlation between an observed predictor and the predicted criterion scores), or all possible subsets regression can be applied to interpret and understand the contributions of predictors playing in forming regression effects, these traditional estimates of importance fail to appropriately partition variance to the various predictors when they are correlated (Tonidandel and LeBreton, 2011). In response, two useful techniques for interpreting a regression equation and for quantifying predictor importance are available: dominance analysis (Budescu, 1993) and relative weight analysis (Johnson, 2000), which has been developed for more accurately quantifying the power of predictors when they are correlated. Both of these methods are advantageous over traditional measures of importance (LeBreton et al., 2007). Although both of these methods yield qualitatively similar results, Johnson's relative weight analysis is much more computationally efficient, especially with large numbers (> 10) of predictors (Johnson, 2000). For a review of approaches to estimate predictor importance, readers are referred to Tonidandel and LeBreton (2011) and Kraha et al. (2012).

Because of its computational efficiency, the relative weight analysis (Johnson, 2000) was applied in this study to quantify the proportion of the variance explained by the individual regression factors in the MLR, and thus identifies the dominant controls on the isotopic composition of rainfall. Specifically, relative weight analysis approximates the relative importance of a set of predictors by creating a set of variables that are highly related to the original set of variables but are uncorrelated with each other. The response variable is then regressed on the uncorrelated set of predictors to approximate the relative weight of the original set of predictors, defined as the relative contribution of each predictor to R^2. Relative weights of correlated predictors are therefore analogous to squared standardized regression coefficients when predictors are uncorrelated. Details are given in Johnson (2000) and Tonidandel et al. (2009). In this study, relative weights were also reported as percentages of the regression R^2.

4 Results and discussion

4.1 Variability in moisture sources

Single backward trajectory computations by the HYSPLIT model are not definite, but can contain some uncertainty. The horizontal uncertainty of the trajectory calculations by HYSPLIT has been estimated to be 10–20 % of the travel distance (Draxler and Hess, 1998). Errors in trajectory calculation computed from analyzed wind fields seem to be typically of the order of 20 % of the distance traveled (Stohl, 1998). However, the statistical analysis of a large number of trajectories arriving at a study site would increase the accuracy of the trajectory analysis (Cabello et al., 2008). In this study, several quality control measures were applied, as recommended in Stohl (1998), to increase confidence in the HYSPLIT-generated back trajectories and to improve the validity of the air mass history. Firstly, trajectories were computed for three pressure levels (900, 850, and 800 hPa). Similar origins of atmospheric moisture for these pressure levels suggest that resolution errors and atmospheric shearing instabilities are negligible, which increases the confidence in the results. Secondly, we use the shortest possible integration time step (i.e., 1 h) and a small value for the parameter TRATIO (0.25), which is the fraction of a grid cell that a trajectory is permitted to transit in one advection time step. Smaller values of TRATIO help to minimize the trajectory computation error using the HYSPLIT model. Thirdly, the statistical analysis of a large number of trajectories (e.g., trajectory cluster analysis) arriving at the study site was applied to confirm the accuracy of the trajectory analysis. The trajectory cluster analysis is conducted by the HYSPLIT model to group trajectories with similar pathways. The cluster analysis merges these trajectories that are near each other and represents those clusters by their mean trajectory. Differences between trajectories within a cluster are minimized while differences between clusters are maximized. Computationally, trajectories are combined to decrease the number of clusters until the total spatial variance (TSV) starts to increase significantly. This occurs when disparate clusters are combined. This number of clusters is then selected as the optimal cluster number for sorting and combining similar trajectories.

Figure 5 shows back-calculated trajectories of atmospheric moisture prior to rainy days at An Long for the sampling period from June 2014 to December 2015. Left and right columns show the results of 850 hPa trajectories for 2014 and 2015, respectively, and the upper, middle, and lower panels show the results for the early (June–September) and late (October–November) rainy season and the dry season (December–May), respectively. Figure 6 shows the spatial distribution of vapor trajectories (cluster means) for precipitation days at An Long for three barometric surfaces (800, 850, 900 hPa) between June 2014 and December 2015, and the change in TVS for different cluster numbers. The TSV was used to identify the optimum number of clusters. The similarity of back-calculated trajectories (Fig. 5) and trajectory cluster analysis (Fig. 6) at three barometric surfaces (900, 850, and 800 hPa) illustrates that the trajectories and thus the source regions do not differ between different atmospheric layers. This indicates a barotropic atmosphere, with the consequence that it is unlikely that the selection of the pressure layer for the HYSPLIT trajectories modifies the results of the MLR significantly.

Figures 5 and 6 demonstrate that the dry-season precipitation (from December to May) in the Plain of Reeds mainly originates from the moisture sources of the Asian continental air masses and the oceanic air masses carried by the equatorial easterlies. Whereas during the rainy season (from June to November) air masses travel a longer distance over the tropical Indian Ocean (from June to September) and the South Pacific Ocean (October to November).

These findings for An Long agree with the general characterization of monsoonal circulation and precipitation over the Southeast Asia region, with moisture from the Indian Ocean dominating during the initial stage of monsoon evolution, and the Pacific Ocean dominating in the later stages (Lim et al., 2002; Aggarwal et al., 2004; Delgado et al., 2012). This indicates that the HYSPLIT model provides valid trajectories to be used in the MLR.

The mean $\delta^{18}O$ values for the five clusters are noted in Fig. 6 (in blue). The mean cluster values are similar for the three pressure levels. Also, the mean values of the two clus-

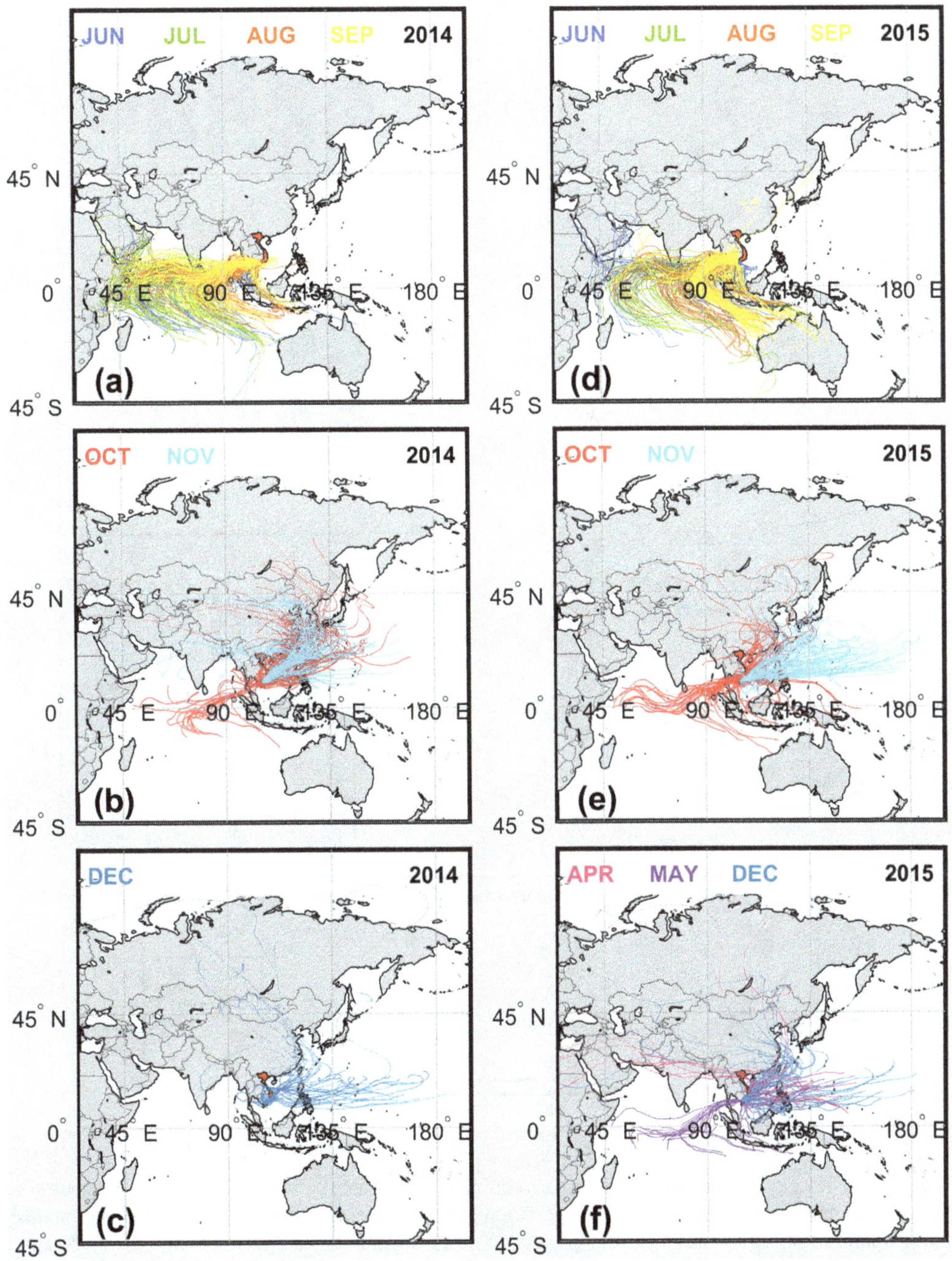

Figure 5. Back-trajectories indicating potential moisture sources of precipitation (plotted only for days with precipitation) at An Long station for the barometric surfaces at 850 hPa between June 2014 and December 2015. **(a)–(c)** show the results for 2014, **(d)–(f)** for 2015; top row **(a, d)** early rainy season (June–September), middle row **(b, e)** late rainy season (October–November), bottom row **(c, f)** dry season (December–May). In January, February, and March 2015 no rainfall was recorded.

ters from the Indian Ocean, as well as the two clusters from the Pacific, are similar. For a fingerprinting, one also has to consider the variation in the values within the clusters, which partly overlap. This means that the δ^{18}O values of precipitation in the Mekong Delta cannot be used to uniquely identify the origin of the trajectory. However, they provide a coarse indication of their origin.

4.2 Isotopic composition of precipitation

4.2.1 Meteoric water lines

The ordinary least squares linear regression (OLSR) of 74 pairs of δ^{18}O and δ^{2}H values at An Long yield the following LMWL for the Plain of Reeds:

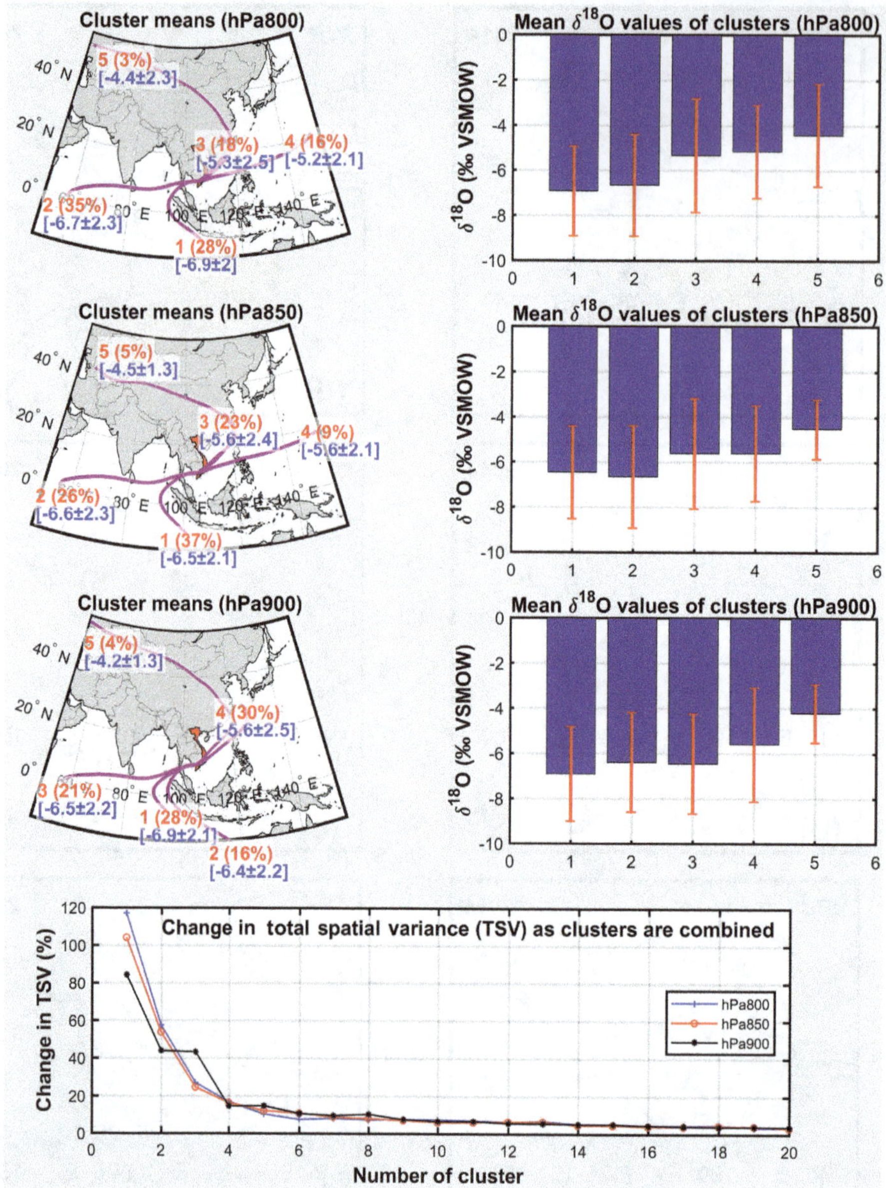

Figure 6. Spatial distribution of vapor trajectories (cluster means) for precipitation days at An Long for three barometric surfaces (800, 850, 900 hPa) between June 2014 and December 2015, and change in total spatial variance (TVS) for different cluster numbers. The TSV was used to identify the optimum number of clusters (here 5 clusters). Red numbers indicate the cluster number (1–5) and the percent of all trajectories assigned to each of the five clusters. Blue numbers and bar charts indicate the mean δ^{18}O values for each cluster plus/minus the standard deviation of each cluster.

$$\delta^2\mathrm{H} = (7.56 \pm 0.11) \cdot \delta^{18}\mathrm{O} + (7.26 \pm 0.67)$$
$$(\mathrm{SE} = 2.26;\ r^2 = 0.99;\ p < 0.0001;\ n = 74).$$

The numbers in brackets indicate the estimates of slope and intercept plus/minus the standard deviation, indicating the parameter uncertainty. The regression is stable against different regression techniques, as shown in the Supplement S1.

The close fit of all considered regressions indicates a very good linear relationship between δ^{18}O and δ^2H in the study. On large temporal scales, a good linear relationship between δ^{18}O and δ^2H is usually observed at sites where secondary fractionation processes, e.g., sub-cloud evaporation, are insignificant (Crawford et al., 2014). The LMWL for An Long is slightly different from the global meteoric water line (GMWL; defined by $\delta^2\mathrm{H} = 8 \cdot \delta^{18}\mathrm{O} + 10$; Craig, 1961) (Fig. 7) and the LMWLs derived for six selected GNIP stations (IAEA/WMO, 2016) located in the Indochinese Peninsula (Table 1). The small difference in slope between the LMWL compared to that of GMWL, and the distribution

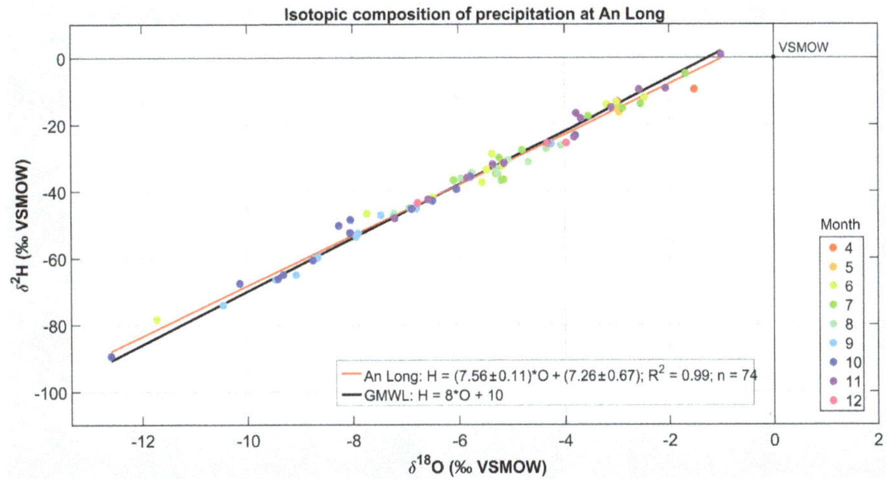

Figure 7. The LMWL of An Long in comparison to the GMWL.

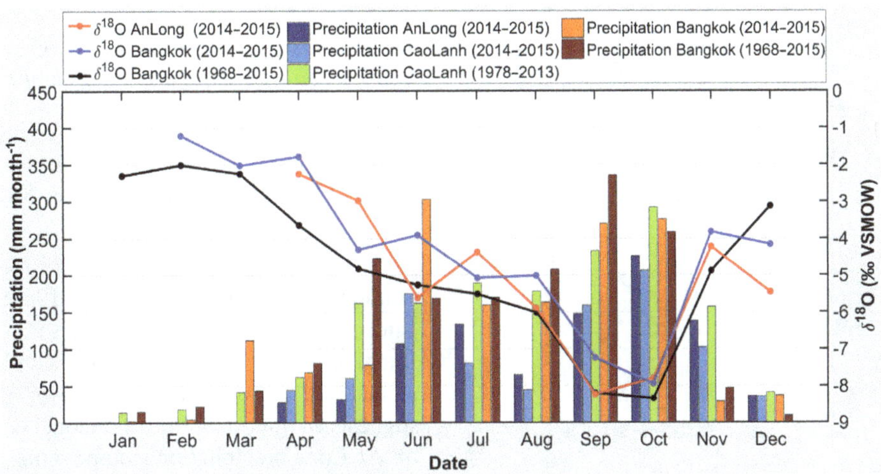

Figure 8. Seasonal variation in the average monthly precipitation for An Long and Cao Lanh and $\delta^{18}O$ values of precipitation for An Long (for the period of observation – red) and Bangkok (both for the period of observation – blue – and the long-term mean – black).

of isotope values along the GMWL indicate that evaporative isotopic enrichment during rainfall is not significant. However, the less positive intercepts of the LMWL (< 10 ‰) (Fig. 7) may reflect smaller kinetic effects during evaporation (Ingraham, 1998) over the Mekong Delta compared to the worldwide average.

4.2.2 Seasonal variation and spatial homogeneity of the isotope composition

The 74 precipitation samples at An Long showed that $\delta^{18}O$ ranges between −12.6 and −1.0 ‰, with an arithmetic mean value and standard deviation of -5.8 ± 2.5 ‰, and δ^2H ranges between −89.3 and 0.9 ‰, with an arithmetic mean value and standard deviation of -36.2 ± 18.7 ‰. Generally, less negative isotopic values are observed in the dry-season precipitation samples. The most negative values occur in the second half of the rainy season (September and October), whereas the least negative values are observed in the late dry season in April and May (Figs. 7 and 8). This shows that the isotopic composition of precipitation at An Long station exhibits marked seasonal variations, which in turn indicates different dominant moisture sources and/or processes in the different seasons. A comparison of the seasonal variation in $\delta^{18}O$ with the short-term (2014–2015) and long-term (1968–2015) monthly averages of Bangkok (Fig. 8) reveals very similar seasonality, both in terms of timing and magnitude. The small differences between $\delta^{18}O$ for An Long and Bangkok are likely caused by the exceptional low rainfall in the study period compared to the long-term monthly values, particularly during May and July. But considering the similarity of the isotopic signatures and general factors controlling stable isotopic composition of precipitation between the two stations, i.e. annual rainfall amount, air temperature,

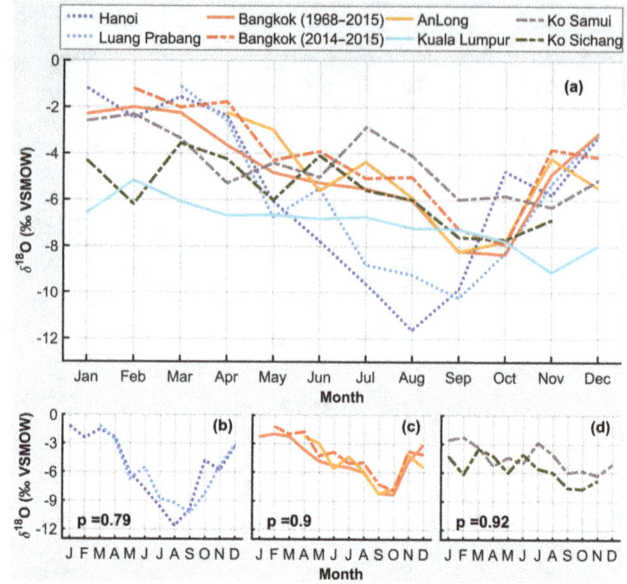

Figure 9. Seasonal monthly mean $\delta^{18}O$ values for An Long and GNIP data from the Indochinese Peninsula. The data are grouped according to similar variability tested with the Levene test. The *p* values given in **(b)–(d)** are the test statistics. High values indicate similar variance. The time series of Bangkok is plotted for short-term (2014–2015) and long-term (1968–2015) periods.

Table 1. Isotopic composition of precipitation at An Long and six selected GNIP stations throughout the Indochinese Peninsula.

Station name	Location	Country	AL	*P* (mm)	*T* (°C)	Period	$\delta^{18}O$ WM	$\delta^{18}O$ *M*	δ^2H WM	δ^2H *M*	d-excess WM	LMWL (OLSR) *S*	LMWL (OLSR) Int.	LMWL (RMA[a]) *S*	LMWL (RMA[a]) Int.	LMWL (PWLSR[b]) *S*	LMWL (PWLSR[b]) Int.
An Long	105.24° E–10.72° N	Vietnam	2	985* (1550)	27.4* (27.2)	Jun 2014–Dec 2015	−6.4 ± 1.5	−5.8 ± 2.5	−40.9 ± 11.5	−36.2 ± 18.7	10.4 ± 1.8	7.56 ± 0.11	7.26 ± 0.67	7.61 ± 0.11	7.58 ± 0.67	7.61 ± 0.11	7.87 ± 0.73
Hanoi**	105.84° E –21.02° N	Vietnam	10	1659 ± 257	24.8 ± 0.5	2004–2007	−8.8 ± 0.7	−5.9 ± 0.5	−56.9 ± 4.2	−33.8 ± 3.6	13.5 ± 1.5	7.91 ± 0.10	12.45 ± 1.25	7.99 ± 0.18	12.90 ± 1.22	7.77 ± 0.21	10.92 ± 1.91
Bangkok**	100.50° E –13.73° N	Thailand	2	1558 ± 314	28.5 ± 0.6	1968–2015	−6.5 ± 1.0	−5.2 ± 1.0	−42.6 ± 7.6	−33.2 ± 7.2	9.4 ± 1.6	7.35 ± 0.04	5.36 ± 0.47	7.53 ± 0.08	6.29 ± 0.47	7.68 ± 0.07	7.25 ± 0.49
Ko Samui**	100.03° E –09.28° N	Thailand	7	1265 ± 611	27.9 ± 0.2	1979–1983	−5.8 ± 1.4	−4.8 ± 0.9	−28.8 ± 7.1	−24.1 ± 4.4	10.8 ± 0.0	7.18 ± 0.10	6.89 ± 1.20	7.30 ± 0.25	7.41 ± 1.16	7.45 ± 0.25	7.89 ± 1.26
Ko Sichang**	100.80° E –13.17° N	Thailand	26	877 ± 320	27.9 ± 0.6	1983–1995	−6.2 ± 0.6	−6.2 ± 1.1	−39.3 ± 5.1	−39.7 ± 8.8	10.2 ± 0.6	7.62 ± 0.06	7.61 ± 1.15	7.72 ± 0.18	8.16 ± 1.12	7.77 ± 0.23	8.65 ± 1.44
Luang Prabang**	102.13° E –19.88° N	Lao PDR	305	1228 ± 178	25.7 ± 0.7	1961–1967	−7.8 ± 1.2	−6.7 ± 0.3	−54.2 ± 7.6	−45.9 ± 0.9	8.4 ± 1.9	7.90 ± 0.13	7.97 ± 2.00	8.01 ± 0.27	8.70 ± 1.93	7.80 ± 0.28	7.52 ± 2.29
Kuala Lumpur**	101.68° E –03.13° N	Malaysia	26	1801 ± 787	–	1993–2012	−7.3 ± 0.8	−7.0 ± 0.7	−46.6 ± 7.7	−45.1 ± 6.8	11.8 ± 4.1	7.63 ± 0.07	8.10 ± 1.93	8.26 ± 0.26	12.53 ± 1.92	7.73 ± 0.29	8.95 ± 2.24

Note: * measured at An Long in 2015; numbers in parentheses show mean values of long-term measurements at Cao Lanh. ** Data are from https://nucleus.iaea.org/wiser/gnip.php (IAEA/WMO, 2016). *P*: annual precipitation (mm yr^{-1}); *T*: average temperature (°C); AL: altitude (m a.s.l. – meter above sea level); WM: weighted mean value; *M*: mean value; *S*: slope; Int.: intercept; [a] RMA: reduced major axis regression; [b] PWLSR: precipitation amount weighted least squares regression.

altitude, and latitude (Dansgaard, 1964; Ingraham, 1998), it can be concluded that the isotopic variations in An Long and Bangkok follow the same dynamics and controls, both on an annual and seasonal scale. Moreover, the overall similarity suggests an important role of regional or larger-scale controls on An Long precipitation isotope ratios.

In order to test the representativeness of the An Long data for a wider area, the variability in the monthly mean $\delta^{18}O$ data of An Long was compared to the available GNIP data of the Indochinese Peninsula (Table 1). As shown in Fig. 9, the test results in four distinct groups of data series with similar variances: the northern part of the Indochinese Peninsula (Hanoi and Luang Prabang; Fig. 9b), the southern part of the Indochinese Peninsula (Bangkok and An Long; Fig. 9c), the islands in the Gulf of Thailand (Ko Samui and Ko Sichang; Fig. 9d), and finally Kuala Lumpur showing only little seasonal variability. The northern and southern parts of the Indochinese Peninsula generally show a similar seasonal behavior with a distinct higher depletion during the rainy season, but in the northern part the highest depletion is 1 month earlier (August) than in the southern part, and the magnitude of the depletion is larger. The seasonal $\delta^{18}O$ variability in precipitation on the islands is much lower than at the stations located on the continent. This is likely due to the maritime setting and could indicate a continental effect in the data of the mainland stations. In addition, the short-term time series of Bangkok and An Long (i.e., 2014–2015) show similar variances, resulting in a highly significant Levene test

Table 2. Pairwise correlation coefficients between regional factors (P_hysplit, T_hysplit, H_hysplit, D_hysplit) and local factors (P_AL, T_AL, H_AL) and stable isotopic values (δ^{18}O, δ^{2}H, and d-excess). Bold and italic numbers denote significance at the 0.01 and 0.05 level (2-tailed), respectively. The meteorological data are aggregated to weekly values corresponding to the precipitation sampling at An Long.

(a)	P_hysplit	H_hysplit	T_hysplit	D_hysplit	P_AL	H_AL	T_AL	Isotopic values
δ^{18}O	**−0.74**	**−0.45**	**−0.38**	*0.24*	**−0.34**	**−0.53**	**0.45**	vs. regional
δ^{2}H	**−0.76**	**−0.47**	**−0.39**	0.20	−0.32	**−0.53**	**0.45**	and local
d-excess	0.18	0.04	0.07	**−0.36**	*0.27*	0.20	−0.15	factors
(b)	P_hysplit	H_hysplit	T_hysplit	D_hysplit				
P_AL	0.13	0.23	0.04	0.03				
H_AL	**0.38**	0.17	0.21	0.10	regional factors vs. local factors			
T_AL	−0.21	0.05	0.17	**−0.33**				
(c)	P_hysplit	H_hysplit	T_hysplit	D_hysplit				
P_hysplit	1							
H_hysplit	**0.77**	1			regional factors vs. regional factors			
T_hysplit	**0.59**	**0.67**	1					
D_hysplit	−0.10	−0.17	**−0.49**	1				
(d)	P_AL	H_AL	T_AL					
P_AL	1							
H_AL	0.20	1			local factors vs. local factors			
T_AL	−0.14	**−0.78**	1					

statistic of 0.98. The variation in the short-term time series of Bangkok and An Long is also very similar to the long-term time series, again shown by a highly significant Levene test statistic of 0.90 (Fig. 9c). This indicates that the variability in the isotopic signature of the An Long time series is almost identical to the one from Bangkok. In summary, the analyzed GNIP data suggest that the data and results from this study are likely to be representative of the southern continental part of the Indochinese Peninsula.

4.3 Factors controlling isotopic composition of precipitation

Prior to the MLR, the correlation of the predictors was analyzed (Table 2). The absolute values of the correlation coefficients between local (P_AL, T_AL, H_AL) and regional (P_hysplit, T_hysplit, H_hysplit, D_hysplit) climatic parameters are relatively small and mostly not significant ($|r| < 0.4$, Table 2b). However, the correlation coefficients between regional predictors are in most cases high and significant (Table 2c). Highest correlations are found between temperature and humidity for local factors, and between humidity and precipitation for regional factors. Interestingly, the correlation between P_AL and H_AL is quite low. This indicates that the local precipitation is influenced by other factors than local humidity (which is rather uniform throughout the year), e.g. large-scale circulation. The correlation between the predictors underlines the necessity to consider multicollinearity when investigating how the predictors control the response variables δ^{18}O and δ^{2}H.

4.3.1 Local factors and isotopic composition of precipitation

Typically, in tropical regions subject to a monsoon climate, the correlation between δ^{18}O and δ^{2}H values of precipitation and air temperature is virtually nonexistent, whereas a strong relation between δ^{18}O and amount of precipitation has been observed (Rozanski et al., 1992; Araguás-Araguás et al., 1998). Our data show that the correlation of local precipitation amount (P_AL) and local temperature (T_AL) with isotopic values (δ^{18}O and δ^{2}H) are both low ($|r| < 0.45$, Table 2a). This suggests that δ^{18}O and δ^{2}H variation is neither dominated by local precipitation amount nor by local temperature during the sampling period. This lack of a significant correlation ($|r| < 0.5$) between δ^{18}O and local rainfall amount was also observed in other regions affected by the Asian monsoon climate such as Bangkok, Hong Kong, New Delhi (Ishizaki et al., 2012), and Cherrapunji, India (Breitenbach et al., 2010). This again supports the statement that δ^{18}O may not be an adequate proxy for local climatic conditions (e.g., temperature or rainfall amount) in the Asian monsoon region (Aggarwal et al., 2004; Vuille et al., 2005).

Secondary fractionation processes such as sub-cloud evaporation or secondary evaporation from open water bodies tend to decrease d-excess in the residual rainwater (Stewart, 1975) and enrich it in the heavy isotopes (Guan et al., 2013). The negative correlation of humidity (H_AL) with δ^{18}O and δ^{2}H ($r = -0.53$, Table 2a) combined with a positive correlation with d-excess ($r = 0.2$, Table 2a) indicates that some sec-

Table 3. Results of the linear regression analysis between local relative humidity (H_AL) and isotopic values at An Long. Regressions that are statistically significant at the 0.05 level are marked in bold.

	Linear regression line	r	R^2	p value	n	Period
δ^2H-δ^{18}O	$\delta^2H = 7.56 \cdot \delta^{18}O + 7.26$	**0.99**	**0.99**	**0.000**	74	full year
	$\delta^2H = 7.62 \cdot \delta^{18}O + 7.74$	**0.99**	**0.99**	**0.000**	67	rainy season (Jun–Nov)
	$\delta^2H = 7.58 \cdot \delta^{18}O + 7.21$	**0.99**	**0.98**	**0.000**	42	early monsoon (Jun–Sep)
	$\delta^2H = 7.68 \cdot \delta^{18}O + 8.6$	**0.99**	**0.99**	**0.000**	25	late monsoon (Oct–Nov)
	$\delta^2H = 6.9 \cdot \delta^{18}O + 3.98$	**0.98**	**0.96**	**0.000**	7	dry season (Dec–May)
δ^{18}O – Humidity	$\delta^{18}O = -0.51 \cdot H\_AL + 36.05$	**−0.53**	**0.28**	**0.000**	74	full year
	$\delta^{18}O = -0.46 \cdot H\_AL + 32.09$	**−0.47**	**0.22**	**0.000**	67	rainy season (Jun–Nov)
	$\delta^{18}O = -0.33 \cdot H\_AL + 21.84$	**−0.42**	**0.17**	**0.006**	42	early monsoon (Jun–Sep)
	$\delta^{18}O = -0.83 \cdot H\_AL + 63.12$	**−0.61**	**0.37**	**0.001**	25	late monsoon (Oct–Nov)
	$\delta^{18}O = -0.56 \cdot H\_AL + 41.34$	**−0.88**	**0.77**	**0.010**	7	dry season (Dec–May)
d-excess – Humidity	d-excess $= 0.2 \cdot H\_AL - 6.36$	0.20	0.04	0.090	74	full year
	d-excess $= 0.13 \cdot H\_AL - 0.46$	0.13	0.02	0.301	67	rainy season (Jun–Nov)
	d-excess $= 0.18 \cdot H\_AL - 5.35$	0.21	0.04	0.211	42	early monsoon (Jun–Sep)
	d-excess $= -0.08 \cdot H\_AL + 17.44$	−0.07	0.01	0.734	25	late monsoon (Oct–Nov)
	d-excess $= 0.34 \cdot H\_AL - 19.42$	0.31	0.10	0.455	7	dry season (Dec–May)

ondary fractionation processes (Risi et al., 2008b; Crawford et al., 2017) may occur during some months at An Long. To examine in which month secondary fractionation processes are likely significant, amount-weighted mean and arithmetic mean, for both δ^{18}O and d-excess, are compared. The rationale is that if secondary fractionation processes are important (with the assumption that the moisture sources of different events within the month are the same), the arithmetic mean should have a δ^{18}O value that is more enriched in heavy isotopes, and a much smaller d-excess than the weighted mean (Guan et al., 2013). Figure 10 shows that secondary fractionation processes may take place mainly during the dry season, in December 2014, and in April, May, and November 2015, because in these months less negative δ^{18}O values and lower d-excess values compared to the overall arithmetic mean are observed, while at the same time the monthly arithmetic means are higher for δ^{18}O and lower for d-excess compared to the monthly weighted means.

To further corroborate this finding, linear regression was performed for different seasons to derive seasonal LMWLs and relations between local humidity, δ^{18}O, and d-excess. Table 3 suggests that secondary fractionation processes are likely to take place in the dry season between December 2014 and May 2015. This is depicted by a slope of lower than 8 (slope $= 6.9$) for the dry season, the slightly negative correlation between δ^{18}O and local relative humidity, and the markedly positive correlation between humidity and d-excess. This is a distinctly different behavior compared to the rainy season as a whole, but also for the first (early monsoon) and second (late monsoon) parts of the rainy season. In summary, these findings indicate that secondary fractionation processes influence the isotopic composition of precipitation primarily in the dry season, which is characterized by lower humidity and higher temperature in the Plain of Reeds. While this conclusion is plausible due to the climatic conditions and low rainfall amounts, one has to consider the low number of rainfall samples during the dry season, which associates some uncertainty to this analysis.

4.3.2 Regional factors and isotopic composition of precipitation

In comparison to other regional and local parameters, the precipitation amount along the transport pathways of moisture sources (P_hysplit) shows the strongest correlation with δ^{18}O and δ^2H as depicted by a correlation coefficient of -0.76 (Table 2a). Other predictors show weaker correlations with $|r| < 0.55$. This, however, does not exclude that these predictors do have some predictive power for the isotopic composition of precipitation in An Long when used in combination with other predictors. Although δ^{18}O and δ^2H are rather well correlated with some climatic parameters, d-excess (which is a function of both) is not well correlated. This is because of the relative difference of the variation in δ^{18}O and d-excess, which is expressed by a low correlation coefficient between these two variables ($r = -0.44$). The weak correlation between d-excess and all climatic parameters ($|r| < 0.36$) indicates that the selected predictors (i.e., selected climatic parameters) are not sufficient to explain the processes responsible for the variability in the d-excess. On a global scale, drivers controlling d-excess variation are likely sea surface temperature or near-surface relative humidity of moisture sources (Pfahl and Wernli, 2008; Uemura et al., 2008; Pfahl and Sodemann, 2014), which are not considered in this study. In tropical areas, a major contribution to the

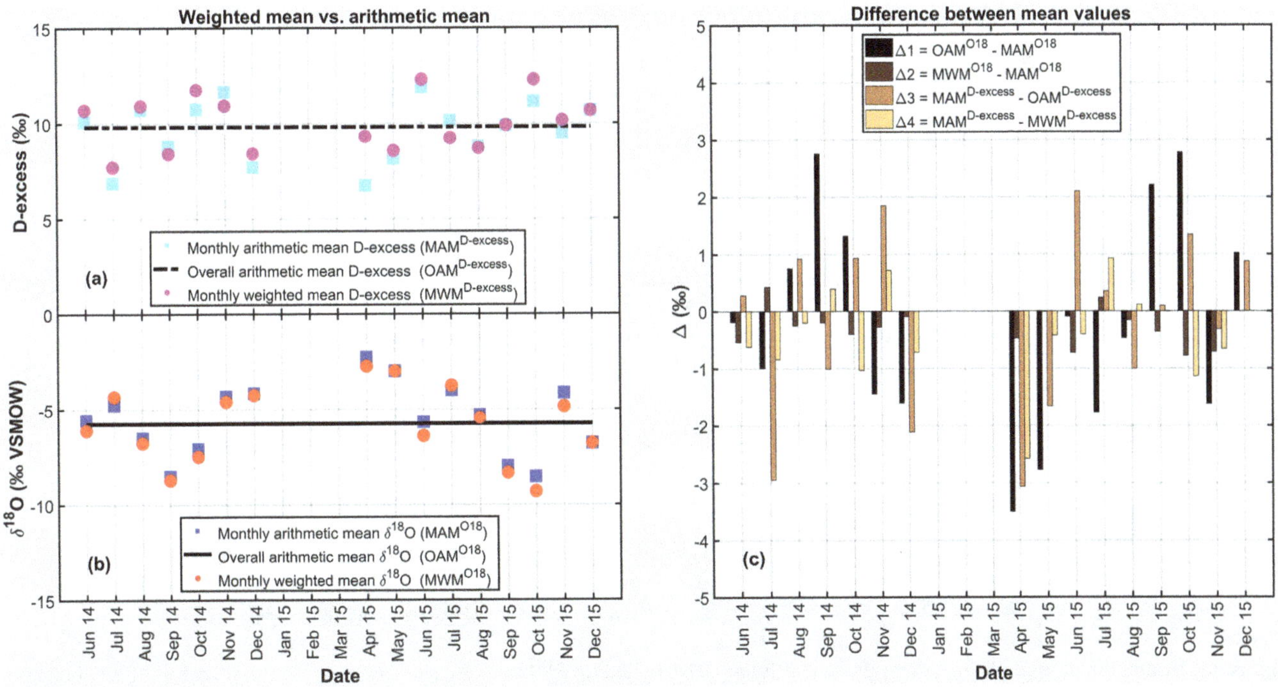

Figure 10. Monthly arithmetic mean (MAM) vs. overall arithmetic mean (OAM), and MAM vs. monthly amount-weighted mean (MWM) for d-excess **(a)** and δ^{18}O **(b)** at An Long during the sampling period June 2014 to December 2015. The difference between these mean values **(c)** including $\Delta^1 = \mathrm{OAM}^{18O} - \mathrm{MAM}^{18O}$, $\Delta^2 = \mathrm{MWM}^{18O} - \mathrm{MAM}^{18O}$ for δ^{18}O, and $\Delta^3 = \mathrm{MAM}^{\mathrm{D-excess}} - \mathrm{OAM}^{\mathrm{D-excess}}$, $\Delta^4 = \mathrm{MAM}^{\mathrm{D-excess}} - \mathrm{MWM}^{\mathrm{D-excess}}$ for d-excess are shown to examine in which month secondary fractionation processes are likely significant (defined by all values of Δ in that month smaller than zero).

seasonal variation in d-excess can be convective processes, e.g. re-evaporation and rain–vapor interactions (Risi et al., 2008a, 2010a), or the influence of large-scale processes, e.g. conditions at the vapor source, convection, and recycling of moisture along trajectories (Landais et al., 2010). A complete investigation of factors controlling d-excess in precipitation is thus not possible by the presented study design and selected predictors. However, some conclusions about the factors controlling the d-excess can be obtained with the presented method (see below).

4.4 MLR and relative importance analysis

The results of the MLR with all predictor combinations indicate that the δ^{18}O signal in precipitation at An Long is best explained by moisture sources of 5-day backward trajectories (Fig. 11). The MLRs using the predictors of these trajectories produce the lowest PRESS and highest R^2 values, indicating that about 80 % of the variability in precipitation (δ^{18}O, Fig. 11 and δ^2H, Fig. S1 in the Supplement) at An Long can be explained by the best MLR model. However, the explained variance differs only slightly between the different pressure levels used. The best performance regarding the lowest PRESS value was obtained by the mean backward trajectories of the 800 and 850 hPa levels.

Contrary to δ^{18}O and δ^2H, the MLR fails to explain the variation in d-excess over the whole study period to a large extent, with a maximal R^2 of 0.3 (Fig. 11). This indicates that the climatic parameters used in our MLR models only have little impact on the annual d-excess variation, which corroborates the findings of the linear correlation analysis in Sect. 4.3.2.

In the next step, the importance of the MLR predictors was analyzed. Figure 12 shows the results applying Johnson's relative weight analysis for the best performing MLR models, i.e. using the mean of the 800 and 850 hPa 5-day backward trajectories. In general, the predictive power of the MLR models increases with increasing number of predictors.

The results indicate that regional factors are always more important than local factors if the R^2 value is above 0.5. The local factors dominate only in MLR models with low performance, or when no regional factors are used as predictors. This is also highlighted by the sum ratio line (black line in Fig. 12), defined as the fraction of R^2 explained by regional factors normalized to the overall R^2. In the best MLR model (124th model) with the lowest PRESS value and an R^2 of 0.80, which is equivalent to an explained variance of 80 %, the regional factors explain 56 % of the absolute δ^{18}O variance (which is equivalent to 70 % relative to $R^2 = 0.80$), while local factors explain only 24 % (30 % relative to $R^2 = 0.80$). This result agrees with the two-factor

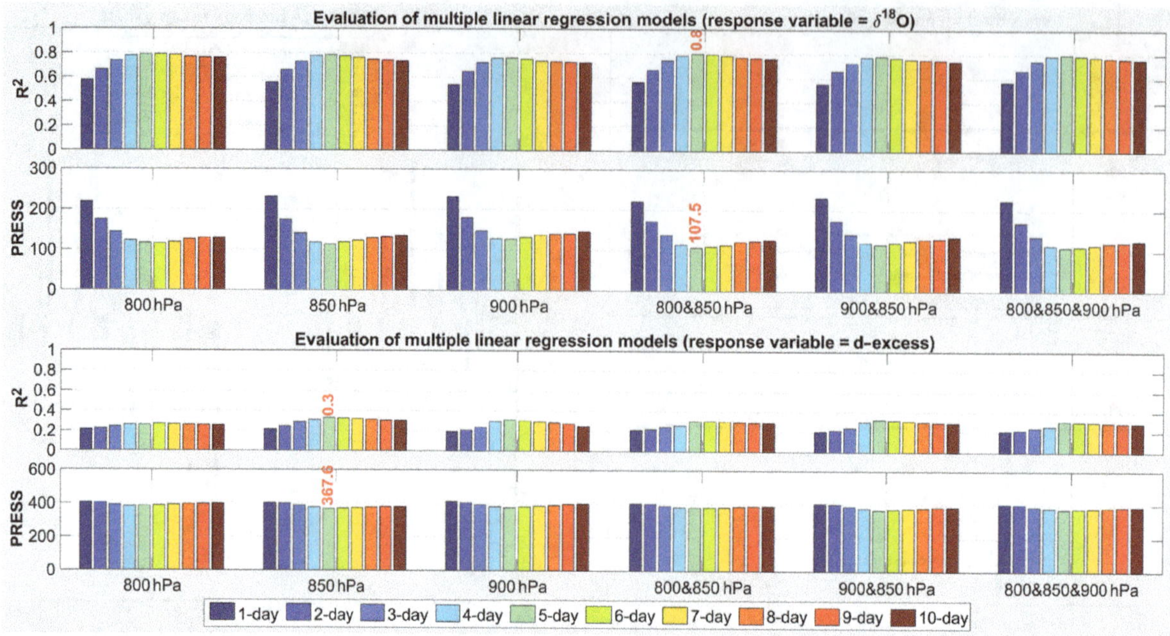

Figure 11. Evaluation of multiple linear regression (MLR) models applied for δ^{18}O and d-excess as response variables for different pressure levels used for three HYSPLIT backward trajectories and their combinations (mean values of the different levels). The best MLR model is marked with red text.

analysis of Ishizaki et al. (2012) who stated that distillation during transport from source regions is the dominant contributor to interannual variability in δ^{18}O precipitation in Bangkok, Bombay, and Hong Kong, accounting for 70, 60, and 70 %, respectively, relative to the overall explained variance, while the amount of local precipitation contributed the remaining 27, 33, and 25 % of the explained variance.

In all models where precipitation amount along transport pathways from moisture source regions (P_hysplit) is included, this factor explains the highest proportion of R^2, which is always at least double and up to triple of the explained variance in other factors (Fig. 12). In turn, the absence of P_hysplit as a predictor in the MLR model considerably decreases the R^2, indicating that P_hysplit is the most dominant factor. In the best MLR model (124th model) the most important predictor is P_hysplit, explaining 47 % of the total δ^{18}O variance (partial $R^2 = 0.47$; Fig. 12). The second dominant factor is T_AL, accounting for 21 % of the explained total variance. The remaining factors account for less than 13 % of the δ^{18}O variance. This result indicates that the regional amount effect is a dominant process in controlling isotopic variation, whereas the local amount effect is not important in the VMD. Similar findings are reported for other regions in Asia (e.g. Rozanski et al., 1992; Araguás-Araguás et al., 1998). The local temperature T_AL, however, can be regarded as a modulating factor for the isotopic composition on top of P_hysplit.

In a next step, the predictor importance analysis is performed for different seasons, in order to analyze if seasonal differences in the dominating factors for the isotopic composition exist, as the correlation analysis of local factors and isotopic composition suggests (Sect. 4.3.1). The samples were split into dry season and rainy season subsets, for which the MLR was applied individually. The definition of the seasons follows the analysis in Sect. 2, i.e., the dry season lasts from December to May. However, due to the low number of samples during this period, the dry season samples were taken from mid-November to mid-June in order to increase the sample number, thus enabling a more robust MLR fitting. This selection can be justified because the delineation of the dry and wet season above is based on monthly data, the "sharp" distinction between the rainy and dry season is forced by the temporal resolution of the data used. In reality, the transition between rainy and dry season is somewhat gradual. Thus, the delineation between the rainy and dry season should rather be regarded as fuzzy. Using data from the last 2 weeks of November and the first 2 weeks of June can be seen as one way to consider this.

Furthermore, the rainy season was subdivided according to the different moisture source regions shown in Sect. 4.1: the Indian Ocean, dominating during the initial and high stage of the Indian monsoon from June to September/mid-October, and the South China Sea–Philippine Sea and the north-west Pacific Ocean from October to May during the late rainy and dry seasons, with some contribution from continental Asia (Fig. 5). In order to test if the factors have different importance caused by different source regions during the rainy season, the MLR models and relative importance analysis were

Table 4. Explained variance (partial R^2) in regional and local factors of the best MLR model according to the PRESS value. The first value indicates the absolute partial R^2, the second value the relative contribution to the overall explained variance.

	Whole period	Early rainy season	Late rainy season	Dry season
Regional factors	0.56/70 %	0.51/68 %	0.76/79 %	0.14/22 %
Local factors	0.24/30 %	0.24/32 %	0.20/21 %	0.51/78 %
Total	0.80/100 %	0.75/100 %	0.96/100 %	0.65/100 %

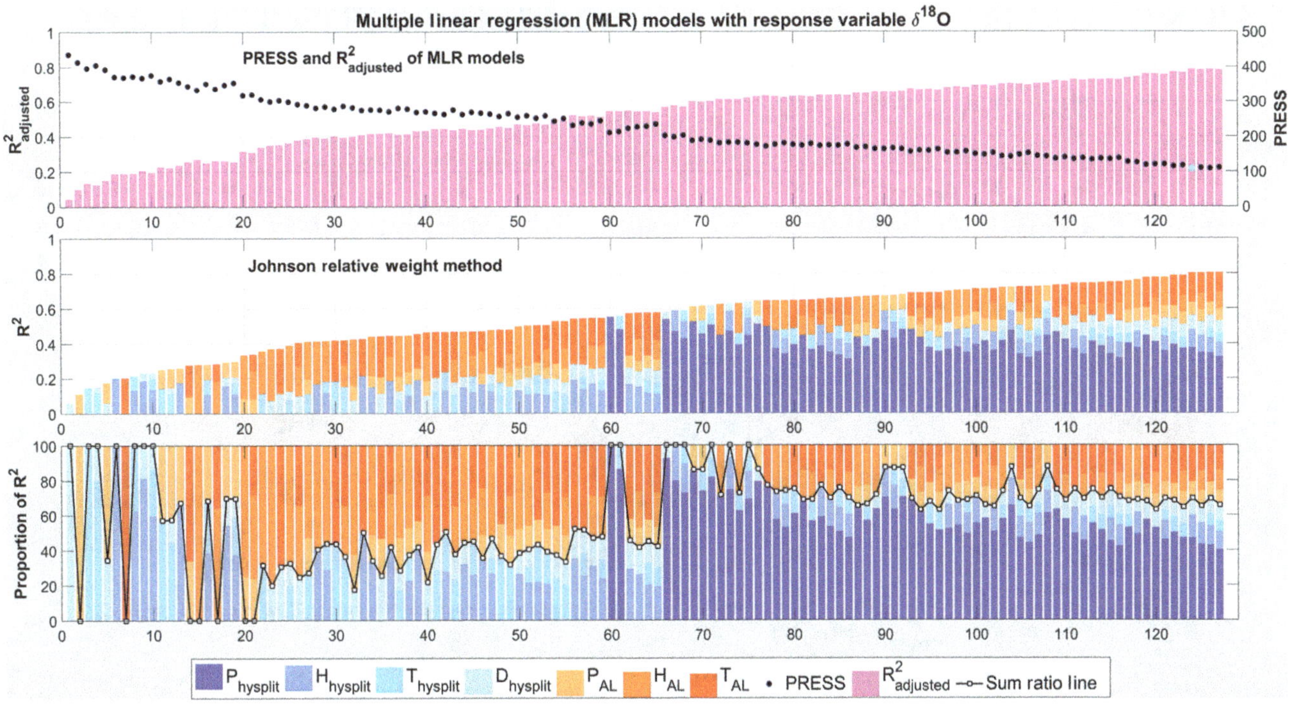

Figure 12. MLR with response variable δ^{18}O and relative importance analysis applied for all possible subsets. The 127 MLR models are sorted according to their R^2 values in ascendant order. Colors represent the relative contribution (in %) of the predictors. The sum ratio line separates the contribution of local (in red and orange) and regional (in blue) factors. PRESS and adjusted R^2 values indicate the quality of the MLR model. The best MLR model depicted by the lowest PRESS (model 124, highlighted by the cyan dot) explains 80 % of the δ^{18}O variation ($R^2 = 0.8$).

applied for these two time periods in addition to the dry season. The number of samples for the different subsets was 42, 18, and 14 for the early rainy season, late rainy season, and dry season, respectively. This reduced number of data points needs to be considered when interpreting the results of the seasonal MLR analysis, because the degree of freedom of the regression (i.e. the difference between the number of data points and the number of predictors) is getting rather small, particularly for the dry season.

Figure 13 shows the results of the MLR and importance analysis for the three seasonal subsets for δ^{18}O. The sorting of the models is the same as in Fig. 12. At a first glance, the results for the rainy season subsets (Fig. 13a and b) are quite similar to each other and to the overall data set. The best performing model in terms of the lowest PRESS value is in all cases model 124. However, in terms of R^2, the performance of the early rainy season is somewhat lower compared to the overall data set, while for the late rainy season it is significantly better, with $R^2 = 0.96$. This increase in explained variance is caused by an increased contribution of the regional factors. In the late rainy season, the regional factors alone contribute 76 % to the overall R^2 of 0.96 of the best PRESS model, which equals 79 % of the explained variance (Table 4). The final best models (both for annual and seasonal analyses) are presented in Table 5.

This is a much larger contribution compared to the partial R^2 values of 56 and 51 % for the whole data set and the early rainy season, respectively. The increase stems from a more significant importance of the other regional factors H_hysplit and/or T_hysplit. While their contribution to the whole data

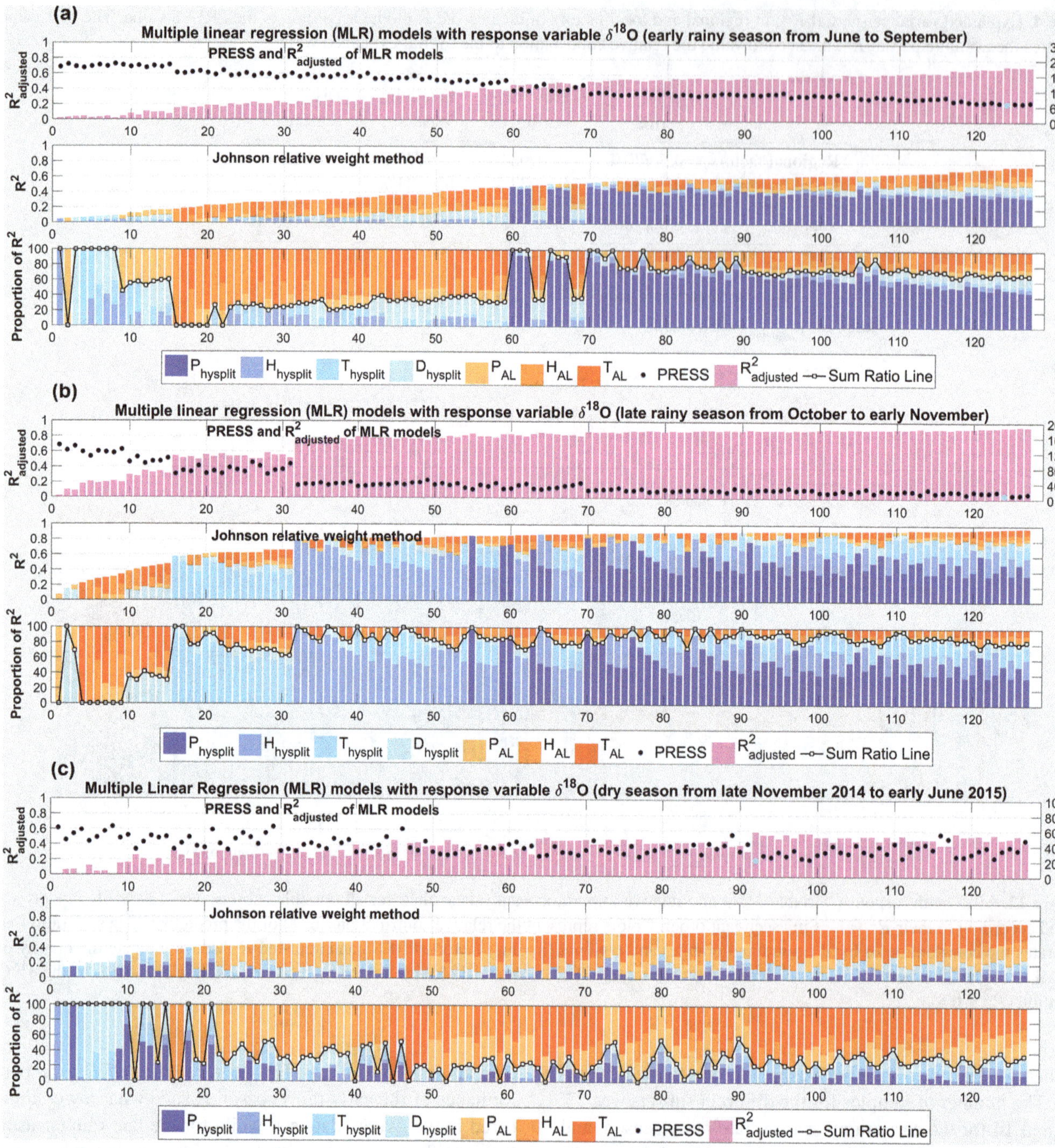

Figure 13. MLR with response variable $\delta^{18}O$ and relative importance analysis applied for all possible subsets (127 MLR models) for different seasons: **(a)** early monsoon from June to September, **(b)** late monsoon from October to mid-November, and **(c)** the dry season from mid-November to mid-June.

set and the early rainy period is rather low and P_hysplit dominates the contribution of the regional factors, it is raised to about 30 % in the late rainy season, either individually or in combination. For the best PRESS model marked with the cyan dot in Fig. 13b, T_hysplit contributes 27 % to the overall R^2 of 0.96. It indicates that temperature and humidity play a more significant role in the isotopic fractionation along the trajectories of water stemming from the north-west Pacific or South China Sea and continental Asia compared to water originating from the Indian Ocean during the boreal summer months. The broad regional and thus climatic heterogeneity of water sources during the late rainy season offers a plausible explanation for this result. The source regions during this period are located in oceans and continental areas of higher

Table 5. The final best models (both for annual and seasonal analyses) for $\delta^{18}O$, $\delta^{2}H$, and d-excess as the response variable in MLR.

Response variable	Multiple linear regression (MLR) models	R^2	R^2 adjusted	p value	n	Period
$\delta^{18}O$	$\delta^{18}O = (-0.116)\cdot P\_hysplit + (-0.033)\cdot P\_AL + (0.118)\cdot H\_hysplit + (1.043)\cdot T\_AL + (0.078)\cdot D\_hysplit - 40.64$	0.80	0.79	1.9×10^{-22}	74	full year
	$\delta^{18}O = (-0.101)\cdot P\_hysplit + (-0.041)\cdot P\_AL + (0.311)\cdot T\_hysplit + (1.128)\cdot T\_AL + (0.104)\cdot D\_hysplit - 41.42$	0.75	0.72	4.3×10^{-10}	42	early monsoon (Jun–Sep)
	$\delta^{18}O = (-0.113)\cdot P\_hysplit + (-0.018)\cdot P\_AL + (0.461)\cdot H\_AL + (-0.404)\cdot T\_hysplit + (1.603)\cdot T\_AL - 77.59$	0.96	0.94	7.1×10^{-8}	18	late monsoon (Oct–Nov)
	$\delta^{18}O = (-0.369)\cdot T\_hysplit + (2.493)\cdot T\_AL - 68.43$	0.65	0.58	3.3×10^{-3}	14	dry season (Dec–May)
$\delta^{2}H$	$\delta^{2}H = (-0.894)\cdot P\_hysplit + (-0.224)\cdot P\_AL + (0.826)\cdot H\_hysplit + (7.744)\cdot T\_AL + (0.513)\cdot D\_hysplit - 289.23$	0.79	0.78	5.9×10^{-22}	74	full year
	$\delta^{2}H = (-0.792)\cdot P\_hysplit + (-0.313)\cdot P\_AL + (2.220)\cdot T\_hysplit + (8.390)\cdot T\_AL + (0.679)\cdot D\_hysplit - 295.19$	0.74	0.71	9.4×10^{-10}	42	early monsoon (Jun–Sep)
	$\delta^{2}H = (-0.914)\cdot P\_hysplit + (-0.094)\cdot P\_AL + (3.045)\cdot H\_AL + (-2.615)\cdot T\_hysplit + (12.044)\cdot T\_AL - 547.97$	0.97	0.95	1.7×10^{-8}	18	late monsoon (Oct–Nov)
	$\delta^{2}H = (-3.446)\cdot T\_hysplit + (19.923)\cdot T\_AL - 529.90$	0.67	0.61	2.2×10^{-3}	14	dry season (Dec–May)
d-excess	$\text{d-excess} = (0.034)\cdot P\_hysplit + (0.029)\cdot P\_AL + (0.206)\cdot H\_AL + (-0.421)\cdot T\_hysplit + (-0.148)\cdot D\_hysplit + 3.05$	0.33	0.28	4.2×10^{-5}	74	full year
	$\text{d-excess} = (0.346)\cdot H\_AL + (-0.311)\cdot T\_hysplit + (-0.162)\cdot D\_hysplit - 8.00$	0.38	0.33	3.7×10^{-4}	42	early monsoon (Jun–Sep)
	$\text{d-excess} = (0.049)\cdot P\_AL + (-0.532)\cdot H\_AL + (-1.989)\cdot T\_AL + (-0.228)\cdot D\_hysplit + 113.98$	0.65	0.54	6.0×10^{-3}	18	late monsoon (Oct–Nov)
	$\text{d-excess} = (0.105)\cdot P\_AL + (-0.300)\cdot H\_hysplit + (1.013)\cdot T\_AL + (0.202)\cdot D\_hysplit - 6.41$	0.66	0.51	3.1×10^{-2}	14	dry season (Dec–May)

latitudes outside the tropics, where vast climatic differences may occur during the transport along the trajectories. Therefore, fractionation processes caused not only by the rainfall amount but also evaporation and condensation are likely to have a significant effect on the final isotopic composition of rainfall reaching An Long during this period. This climatic variability along the transport pathways is much more substantial compared to the low climatic variability of the tropical Indian Ocean source region, where the rainfall during the early rainy seasons originates.

An entirely different picture reveals the MLR fitting and importance analysis for the dry seasons shown in Fig. 13c. While the overall performance in terms of R^2 is comparable to the early rainy season, the importance of the local and regional factors is very different from the other seasons. For the dry season, the local factors dominate. In the best performing MLR model with the lowest PRESS value (cyan dot in Fig. 13c), T_AL contributes 78 % of the explained variance. Similar results are obtained for almost all of the MLR models. For the models with $R^2 > 0.5$, T_AL is the most important factor, followed by P_AL and H_AL with similar importance. The regional factors do not contribute more than 22 % of the explained variance if $R^2 > 0.6$. This finding corroborates the assumed higher importance of secondary fractionation processes during the dry season in the VMD, as already hypothesized in Sect. 4.3.1. However, in combination with other predictors, T_AL seems to be a better predictor of the secondary fractionation processes compared to H_AL, which was used in Sect. 4.3.1. As T_AL and H_AL are closely correlated (Table 2), the findings of Sect. 4.3.1 and the MLR of the dry season presented in this section agree well. With regard to the low degrees of freedom for the dry season MLR models, it can be stated that this consistent high importance of the local factors in various MLR models reduces the uncertainty of this analysis stemming from the low number of data points to some extent.

The MLR modeling of δ^2H shows very similar results to δ^{18}O leading to the same conclusions (Figs. S1 and S2). The MLR modeling of seasonal d-excess also shows an improved fit for the late rainy and dry seasons (Figs. S3 and S4), but for the early rainy season the results are not as satisfying as for the whole data set. In contrast to δ^{18}O and δ^2H, regional factors explain the bulk of the d-excess variance also for the dry season. Among the regional factors, P_hysplit has the lowest importance for d-excess, while the other factors T_hysplit, H_hysplit, and D_hysplit explain about 65 % of the best R^2 of 0.66. This is also a distinctively different result compared to δ^{18}O and δ^2H, where P_hysplit always dominated the regional factor contribution. The remaining explained variance stems mainly from the local precipitation P_AL, with some contribution of T_AL. This finding is in line with the rationale outlined in Sect. 4.3.1, that evaporation along the transport pathway decreases the d-excess (Stewart, 1975). This effect is much more variable during the late rainy and dry season due to the transport pathways from higher latitudes, as compared to the rather uniform climatic conditions along the transport pathways during the rainy season, as already argued in the previous paragraph for the late rainy season results of δ^{18}O. Particularly the water stemming from continental Asia is very likely prone to evaporation and thus changes in d-excess along its transport pathways as it crosses some semi-arid areas (e.g. Tibet, Mongolia; cf. Fig. 5). This means, in summary, that the MLR and relative importance analysis of d-excess for the late rainy and dry season corroborate the hypothesis that secondary fractionation processes caused by evaporation are relevant during the dry season, respectively for rainfall stemming from the Pacific region and continental Asia. However, for δ^{18}O and δ^{18}H, local factors describing evaporation are more critical, while for d-excess regional factors and thus evaporation processes along the transport pathways dominate. But again, the low number of data points for the dry season needs to be taken into account for the interpretation of the results. Although the results are similar over a wide range of different MLR models corroborating the findings and although the mechanisms explaining the results are plausible, the results for the dry seasons still contain some uncertainty stemming from the low data volume.

Nevertheless, it can be stated that applying all possible subset regressions, MLR models can better explain the isotopic variation in rainfall compared to approaches considering only one predictor, i.e., a simple correlation analysis. Moreover, the associated relative importance analysis enables the identification of the dominant factors, thus offering interpretation aids for the identification of the processes responsible for the isotopic signature of local rainfall. The presented analysis illustrates that the investigation of dominant factors controlling isotopic composition of precipitation with simple correlation analyses may lead to wrong conclusions, particularly when predictors are correlated. Additionally, MLR is able to consider the combination of different local and regional factors, thus enabling a better identification and interpretation of the manifold processes controlling the isotopic composition of rainfall.

5 Conclusions

This study analyzes the influence of local and regional meteorological factors on the isotopic composition of precipitation, expressed as δ^{18}O, δ^2H, and d-excess in the VMD. For this purpose, rainfall samples were taken on a weekly to bi-weekly basis for 1.5 years at An Long in the north-eastern part of the VMD and analyzed for stable water isotopes. The regional factors potentially influencing isotopic composition were derived by back-tracing of water particles up to 10 days from the target location using the HYSPLIT model, while the local factors were derived from local climate records. The influence of the different factors on the isotopic condition was quantified by MLR of all factor combinations combined with relative importance analysis. This method is novel for the interpretation of isotopic records of rainfall and opens a pathway for a better understanding of the controlling processes with a rather simple statistical method.

The MLR showed that up to 80 % of the total variation in δ^{18}O could be explained by linear combinations of the selected factors. Similar results are obtained for δ^2H. Contrary to this, only about 30 % of the total variation in d-excess can be explained by the selected factors if the whole data series is used. General considerations regarding the controls of d-excess in tropical areas suggest that additional factors, like sea surface temperatures of the source region, need to be taken into account for an improved modeling of d-excess variation by MLR.

The study showed that local climatic factors, specifically rainfall amount and temperature, play a minor role in controlling the isotopic composition of the rainfall at An Long. However, there is evidence that sub-cloud evaporation has a small but notable effect during the dry season. Regional factors, on the contrary, dominate the isotopic composition of rainfall at An Long. Some 70 % of the explained variance, i.e., a partial R^2 of up to 0.56, can be attributed to regional factors, among which precipitation amount along the transport pathway can explain most of the variance. The remaining 30 % of the explained variance is attributed to local factors, among which the temperature plays the most important role. These findings indicate that local secondary fractionation processes like sub-cloud evaporation play a small additional role for the isotopic composition, which is otherwise dominated by the rainout along the transport pathway of the precipitation.

Furthermore, the analysis of transport durations implies that the moisture-producing precipitation reaching An Long travels about 4–6 days from its source, as the best regression results are obtained for these travel durations. For longer travel durations, the explained variability in the regression decreases, suggesting that the moisture is recycled, i.e., precipitated and evaporated again, when the travel time exceeds 6 days.

If the data set is divided into seasonal subsets defined by precipitation amount and water source regions, the MLR and importance analysis enables a better identification of factors and thus processes controlling the isotopic composition in the different seasons. For the late rainy and dry seasons (i.e., October to May), the importance of regional (late rainy season) and local (dry season) factors increases, respectively, compared to the overall dominating factor P_hysplit. Because of this, the explained variance is raised, particularly for the late rainy season. The source regions (Pacific Ocean and continental Asia) and the associated transport pathways, as well as local processes, are more critical for the late rainy and dry season, indicating that secondary fractionation processes by evaporation, either along the pathway (for d-excess) or locally (for $\delta^{18}O$ and $\delta^{2}H$), are more important than the amount effect, which is dominant during the Indian monsoon (early rainy season) period. This is reasonable because moisture transported to the Mekong Delta from the Pacific region and continental Asia passes through different climatic regimes, compared to the more uniform climatic conditions along the pathway from the Indian Ocean during the Indian summer monsoon.

In summary, it can be concluded that the proposed approach, consisting of simultaneous testing of all possible factors by MLR combined with relative importance analysis, is able to detect the relevant factors controlling the isotopic composition of rainfall as well as their individual contributions. If applied to seasonal data subsets, the predictions can be improved, and the seasonal differences in controlling factors and processes can be identified. The validity of the approach is confirmed by similar, but mainly qualitative results obtained in other studies. The comparable results provide a strong indication that the method is able to identify the dominant factors responsible for the isotopic composition of rainfall records. In contrast to previous studies, however, the presented approach and results provide a quantitative assessment of the impact of different factors, and thus information about the dominant processes of isotopic fractionation. It can support the interpretation of processes responsible for observed patterns of isotopic composition. The rather simple approach can, of course, not provide detailed information about atmospheric dynamics but it offers a relatively simple and easy to apply approach supplementing or preceding more complex studies of isotopic composition with circulation models. Due to the simplicity, the method can be easily applied to investigate factors controlling isotopic composition of precipitation at any given study area around the world, without the requirement of setting up or in-depth knowledge about running complex numerical atmospheric circulation models. Furthermore, the approach is easily reproducible and contains a rigorous quantitative analysis of the interplay of different driving factors. The analysis can easily be extended to other factors and processes of importance, e.g. in order to capture the d-excess better by including sea surface temperatures at the source regions. However, a sufficient number of data points (multiple times the number of predictors) is required to obtain robust regression results.

The similarity of isotopic signatures and their variability over Southeast Asia, as well as similar general climatic conditions, allows the conclusion that the findings are representative for a larger area. Particularly the similarity of the variability in the monthly isotopic composition of rainfall, and climatic conditions of the VMD and Bangkok, suggests that the results are representative for the whole Mekong Delta, and possibly for large areas of the southern tip of the continental Indochinese Peninsula.

The results have direct implications for the interpretation of paleorecords of stable water isotopes in terms of past climate conditions for Southeast Asia. This study shows that the factors controlling the isotopic signature of precipitation are changing between and even within seasons, and that regional factors have substantial impacts on the local isotopic composition of rainfall. This needs to be considered in the reconstruction of past climates based on isotopic records. For the presented study area, $\delta^{18}O$ and $\delta^{2}H$ values are likely to be representative of the local climatic conditions during the dry season. However, regional factors dominate during most of the rainy season receiving the bulk of the total annual rainfall. In this case, reconstructions of past climates based on paleo-isotopic records would have to be carefully interpreted. The proposed approach might open a pathway for an improved reconstruction of paleoclimates based on isotopic records. It may, for example, be used for identifying suitable variables to improve the performance of proxy data assimilation in paleoclimate reconstruction by circulation models. Moreover, assuming that the general circulation is stable over the period considered in paleoclimate reconstruction, which is reasonable for the Holocene for example, the presented findings can be used to infer moisture source regions and the strength of the two monsoonal regimes influencing Southeast Asia from paleo-isotopic records.

Competing interests. The authors declare that they have no conflict of interest.

Acknowledgements. We acknowledge the German Academic Exchange Service (DAAD) for the scholarship that enabled the first author to pursue a PhD degree at the University of Potsdam, Germany. We acknowledge the International Atomic Energy Agency (IAEA) for their on-going maintenance of the Global Network of Isotopes in Precipitation. The authors gratefully acknowledge the NOAA Air Resources Laboratory (ARL) for provision of the HYSPLIT transport and dispersion model and/or READY website (http://www.ready.noaa.gov) used in this publication. We would also like to thank the staff at An Long station for isotope sample collection of precipitation. Our thanks extend to the staff who analyzed part of the isotope samples at the laboratory of the Alfred-Wegener-Institute (AWI) in Potsdam, Germany. We thank the anonymous reviewers for their constructive comments to improve the manuscript.

Edited by: Lixin Wang

References

Aggarwal, P. K., Fröhlich, K., Kulkarni, K. M., and Gourcy, L. L.: Stable isotope evidence for moisture sources in the Asian summer monsoon under present and past climate regimes, Geophys. Res. Lett., 31, L08203, https://doi.org/10.1029/2004GL019911, 2004.

Araguás-Araguás, L., Froehlich, K., and Rozanski, K.: Stable isotope composition of precipitation over southeast Asia, J. Geophys. Res.-Atmos., 103, 28721–28742, 1998.

Baker, A. J., Sodemann, H., Baldini, J. U., Breitenbach, S. F., Johnson, K. R., Hunen, J., and Zhang, P.: Seasonality of westerly moisture transport in the East Asian summer monsoon and its implications for interpreting precipitation $\delta^{18}O$, J. Geophys. Res.-Atmos., 120, 5850–5862, 2015.

Bowen, G. J.: Spatial analysis of the intra-annual variation of precipitation isotope ratios and its climatological corollaries, J. Geophys. Res.-Atmos., 113, D05113, https://doi.org/10.1029/2007JD009295, 2008.

Breitenbach, S. F., Adkins, J. F., Meyer, H., Marwan, N., Kumar, K. K., and Haug, G. H.: Strong influence of water vapor source dynamics on stable isotopes in precipitation observed in Southern Meghalaya, NE India, Earth Planet. Sc. Lett., 292, 212–220, 2010.

Budescu, D. V.: Dominance analysis: A new approach to the problem of relative importance of predictors in multiple regression, Psycholog. Bull., 114, 542–551, 1993.

Cabello, M., Orza, J., Galiano, V., and Ruiz, G.: Influence of meteorological input data on backtrajectory cluster analysis? a seven-year study for southeastern Spain, Adv. Sci. Res., 2, 65–70, 2008.

Chakraborty, S., Sinha, N., Chattopadhyay, R., Sengupta, S., Mohan, P., and Datye, A.: Atmospheric controls on the precipitation isotopes over the Andaman Islands, Bay of Bengal, Scient. Rep., 6, 19555, https://doi.org/10.1038/srep19555, 2016.

Clark, I. D. and Fritz, P.: Environmental isotopes in hydrogeology, CRC Press, New York, 1997.

Conroy, J. L., Cobb, K. M., and Noone, D.: Comparison of precipitation isotope variability across the tropical Pacific in observations and SWING2 model simulations, J. Geophys. Res.-Atmos., 118, 5867–5892, 2013.

Craig, H.: Isotopic variations in meteoric waters, Science, 133, 1702–1703, 1961.

Crawford, J., Hughes, C. E., and Lykoudis, S.: Alternative least squares methods for determining the meteoric water line, demonstrated using GNIP data, J. Hydrol., 519, 2331–2340, 2014.

Crawford, J., Hollins, S. E., Meredith, K. T., and Hughes, C. E.: Precipitation stable isotope variability and subcloud evaporation processes in a semi-arid region, Hydrol. Process., 31, 20–34, https://doi.org/10.1002/hyp.10885, 2017.

Dansgaard, W.: Stable isotopes in precipitation, Tellus, 16, 436–468, 1964.

Dayem, K. E., Molnar, P., Battisti, D. S., and Roe, G. H.: Lessons learned from oxygen isotopes in modern precipitation applied to interpretation of speleothem records of paleoclimate from eastern Asia, Earth Planet. Sc. Lett., 295, 219–230, 2010.

Delgado, J. M., Merz, B., and Apel, H.: Monsoon variability and the Mekong flood regime, in: The Mekong Delta System, Springer, Dordrecht, 233–244, 2012.

Draxler, R. R. and Hess, G.: An overview of the HYSPLIT_4 modelling system for trajectories, Aust. Meteorol. Mag., 47, 295–308, 1998.

Draxler, R. R. and Rolph, G.: HYSPLIT (HYbrid Single-Particle Lagrangian Integrated Trajectory) model access via NOAA ARL READY website (http://ready.arl.noaa.gov/HYSPLIT.php), NOAA Air Resources Laboratory, Silver Spring, MD, 2003.

Gat, J. R.: Oxygen and hydrogen isotopes in the hydrologic cycle, Annu. Rev. Earth Planet. Sci., 24, 225–262, 1996.

Genereux, D. P. and Hooper, R. P.: Oxygen and hydrogen isotopes in rainfall–runoff studies, in: Isotope tracers in catchment hydrology, Elsevier Science, the Netherlands, 319–346, 1998.

Gonfiantini, R., Fröhlich, K., Araguas-Araguas, L., and Rozanski, K.: Isotopes in groundwater hydrology, in: Isotope tracers in catchment hydrology, Elsevier Science, the Netherlands, 203–246, 1998.

GSO: Statistical Yearbook of Vietnam 2013, General Statistics Office (GSO) Vietnam, Hanoi, 2014.

Guan, H., Zhang, X., Skrzypek, G., Sun, Z., and Xu, X.: Deuterium excess variations of rainfall events in a coastal area of South Australia and its relationship with synoptic weather systems and atmospheric moisture sources, J. Geophys. Res.-Atmos., 118, 1123–1138, 2013.

Helsel, D. R. and Hirsch, R. M.: Statistical methods in water resources, US Geological survey, Reston, VA, 2002.

Hung, N. N., Delgado, J. M., Güntner, A., Merz, B., Bárdossy, A., and Apel, H.: Sedimentation in the floodplains of the Mekong Delta, Vietnam. Part I: suspended sediment dynamics, Hydrol. Process., 28, 3132–3144, 2014.

IAEA: IAEA/GNIP precipitation sampling guide, International Atomic Energy Agency, Vienna, Austria, 2014.

IAEA/WMO: Global Network of Isotopes in Precipitation, The GNIP Database, available at: http://www.iaea.org/water (last access: 5 August 2017), 2016.

Ichiyanagi, K. and Yamanaka, M. D.: Interannual variation of stable isotopes in precipitation at Bangkok in response to El Niño Southern Oscillation, Hydrol. Process., 19, 3413–3423, 2005.

Ingraham, N. L.: Isotopic variations in precipitation, in: Isotope Tracers in Catchment Hydrology, Elsevier Science, the Netherlands, 87–118, 1998.

Ishizaki, Y., Yoshimura, K., Kanae, S., Kimoto, M., Kurita, N., and Oki, T.: Interannual variability of $H_2^{18}O$ in precipitation over the Asian monsoon region, J. Geophys. Res.-Atmos., 117, D16308, https://doi.org/10.1029/2011JD015890, 2012.

Johnson, J. W.: A heuristic method for estimating the relative weight of predictor variables in multiple regression, Multivar. Behav. Res., 35, 1–19, 2000.

Johnson, K. R., and Ingram, B. L.: Spatial and temporal variability in the stable isotope systematics of modern precipitation in China: implications for paleoclimate reconstructions, Earth Planet. Sc. Lett., 220, 365–377, 2004.

Kendall, C. and Caldwell, E. A.: Fundamentals of isotope geochemistry, in: Isotope tracers in catchment hydrology, Elsevier Science, the Netherlands, 51–86, 1998.

Kraha, A., Turner, H., Nimon, K., Zientek, L., and Henson, R.: Tools to Support Interpreting Multiple Regression in the Face of Multicollinearity, Front. Psychol., 3, 44, https://doi.org/10.3389/fpsyg.2012.00044, 2012.

Kurita, N., Ichiyanagi, K., Matsumoto, J., Yamanaka, M. D., and Ohata, T.: The relationship between the isotopic content of precipitation and the precipitation amount in tropical regions, J. Geochem. Explor., 102, 113–122, 2009.

Kurita, N., Fujiyoshi, Y., Nakayama, T., Matsumi, Y., and Kitagawa, H.: East Asian Monsoon controls on the inter-annual variability in precipitation isotope ratio in Japan, Clim. Past, 11, 339–353, https://doi.org/10.5194/cp-11-339-2015, 2015.

Lachniet, M. S.: Sea surface temperature control on the stable isotopic composition of rainfall in Panama, Geophys. Res. Lett., 36, L03701, https://doi.org/10.1029/2008GL036625, 2009.

Landais, A., Risi, C., Bony, S., Vimeux, F., Descroix, L., Falourd, S., and Bouygues, A.: Combined measurements of 17 O excess and d-excess in African monsoon precipitation: Implications for evaluating convective parameterizations, Earth Planet. Sc. Lett., 298, 104–112, 2010.

LeBreton, J. M., Hargis, M. B., Griepentrog, B., Oswald, F. L., and Ployhart, R. E.: A multidimensional approach for evaluating variables in organizational research and practice, Person. Psychol., 60, 475–498, 2007.

Lee, J. E., Risi, C., Fung, I., Worden, J., Scheepmaker, R. A., Lintner, B., and Frankenberg, C.: Asian monsoon hydrometeorology from TES and SCIAMACHY water vapor isotope measurements and LMDZ simulations: Implications for speleothem climate record interpretation, J. Geophys. Res.- Atmos., 117, D15112, https://doi.org/10.1029/2011JD017133, 2012.

Lee, K.-S., Grundstein, A. J., Wenner, D. B., Choi, M.-S., Woo, N.-C., and Lee, D.-H.: Climatic controls on the stable isotopic composition of precipitation in Northeast Asia, Clim. Res., 23, 137–148, 2003.

LeGrande, A. N. and Schmidt, G. A.: Sources of Holocene variability of oxygen isotopes in paleoclimate archives, Clim. Past, 5, 441–455, https://doi.org/10.5194/cp-5-441-2009, 2009.

Levene, H.: Robust tests for equality of variances, Contrib. Probabil. Stat., 1, 278–292, 1960.

Lim, Y.-K., Kim, K.-Y., and Lee, H.-S.: Temporal and spatial evolution of the Asian summer monsoon in the seasonal cycle of synoptic fields, J. Climate, 15, 3630–3644, 2002.

Liu, Z., Wen, X., Brady, E., Otto-Bliesner, B., Yu, G., Lu, H., Cheng, H., Wang, Y., Zheng, W., and Ding, Y.: Chinese cave records and the East Asia summer monsoon, Quaternary Sci. Rev., 83, 115–128, 2014.

Merlivat, L. and Jouzel, J.: Global climatic interpretation of the deuterium–oxygen 18 relationship for precipitation, J. Geophys. Res.-Oceans, 84, 5029–5033, 1979.

Meyer, H., Schönicke, L., Wand, U., Hubberten, H.-W., and Friedrichsen, H.: Isotope studies of hydrogen and oxygen in ground ice-experiences with the equilibration technique, Isotop. Environ. Health Stud., 36, 133–149, 2000.

Noone, D. and Simmonds, I.: Associations between $\delta^{18}O$ of Water and Climate Parameters in a Simulation of Atmospheric Circulation for 1979–95, J. Climate, 15, 3150–3169, 2002.

Okazaki, A. and Yoshimura, K.: Development and evaluation of a system of proxy data assimilation for paleoclimate reconstruction, Clim. Past, 13, 379–393, https://doi.org/10.5194/cp-13-379-2017, 2017.

Peng, T.-R., Wang, C.-H., Huang, C.-C., Fei, L.-Y., Chen, C.-T. A., and Hwong, J.-L.: Stable isotopic characteristic of Taiwan's precipitation: A case study of western Pacific monsoon region, Earth Planet. Sc. Lett., 289, 357–366, 2010.

Pfahl, S. and Sodemann, H.: What controls deuterium excess in global precipitation?, Clim. Past, 10, 771–781, https://doi.org/10.5194/cp-10-771-2014, 2014.

Pfahl, S. and Wernli, H.: Air parcel trajectory analysis of stable isotopes in water vapor in the eastern Mediterranean, J. Geophys. Res.-Atmos., 113, D20104, https://doi.org/10.1029/2008JD009839s, 2008.

Renaud, F. G. and Kuenzer, C.: The Mekong Delta system: Interdisciplinary analyses of a river delta, Springer Science & Business Media, Dordrecht, 2012.

Risi, C., Bony, S., and Vimeux, F.: Influence of convective processes on the isotopic composition ($\delta^{18}O$ and δD) of precipitation and water vapor in the tropics: 2. Physical interpretation of the amount effect, J. Geophys. Res.-Atmos., 113, D19306, https://doi.org/10.1029/2008JD009943, 2008a.

Risi, C., Bony, S., Vimeux, F., Descroix, L., Ibrahim, B., Lebreton, E., Mamadou, I., and Sultan, B.: What controls the isotopic composition of the African monsoon precipitation? Insights from event-based precipitation collected during the 2006 AMMA field campaign, Geophys. Res. Lett., 35, L24808, https://doi.org/10.1029/2008GL035920, 2008b.

Risi, C., Bony, S., Vimeux, F., Chong, M., and Descroix, L.: Evolution of the stable water isotopic composition of the rain sampled along Sahelian squall lines, Q. J. Roy. Meteorol. Soc., 136, 227–242, 2010a.

Risi, C., Bony, S., Vimeux, F., and Jouzel, J.: Water-stable isotopes in the LMDZ4 general circulation model: Model evaluation for present-day and past climates and applications to climatic interpretations of tropical isotopic records, J. Geophys. Res.-Atmos., 115, 2010b.

Rozanski, K., Araguas-Araguas, L., and Gonfiantini, R.: Relation between long-term trends of oxygen-18 isotope composition of precipitation and climate, Science, 258, 981–985, 1992.

SIWRP: Flood Planning in Vietnamese Mekong Delta until 2020, SIWRP, Ho Chi Minh, Vietnam, 2014.

Sodemann, H., Schwierz, C., and Wernli, H.: Interannual variability of Greenland winter precipitation sources: Lagrangian moisture diagnostic and North Atlantic Oscillation influence, J. Geophys. Res.-Atmos., 113, D03107, https://doi.org/10.1029/2007JD008503, 2008.

Stein, A., Draxler, R., Rolph, G., Stunder, B., Cohen, M., and Ngan, F.: NOAA's HYSPLIT atmospheric transport and dispersion modeling system, B. Am. Meteorol. Soc., 96, 2059–2077, 2015.

Stewart, M. K.: Stable isotope fractionation due to evaporation and isotopic exchange of falling waterdrops: Applications to atmospheric processes and evaporation of lakes, J. Geophys. Res., 80, 1133–1146, 1975.

Stohl, A.: Computation, accuracy and applications of trajectories – a review and bibliography, Atmos. Environ., 32, 947–966, 1998.

Sturm, C., Zhang, Q., and Noone, D.: An introduction to stable water isotopes in climate models: benefits of forward proxy modelling for paleoclimatology, Climate of the Past, 6, 115-129, 2010.

Tan, M.: Circulation effect: climatic significance of the short term variability of the oxygen isotopes in stalagmites from monsoonal China – dialogue between paleoclimate records and modern climate research, Quaternary Sci., 29, 851–862, 2009.

Tan, M.: Circulation effect: response of precipitation $\delta^{18}O$ to the ENSO cycle in monsoon regions of China, Clim. Dynam., 42, 1067–1077, 2014.

Tindall, J., Valdes, P., and Sime, L. C.: Stable water isotopes in HadCM3: Isotopic signature of El Niño–Southern Oscillation and the tropical amount effect, J. Geophys. Res.-Atmos., 114, D04111, https://doi.org/10.1029/2008JD010825, 2009.

Tonidandel, S. and LeBreton, J. M.: Relative importance analysis: A useful supplement to regression analysis, J. Business Psychol., 26, 1–9, 2011.

Tonidandel, S., LeBreton, J. M., and Johnson, J. W.: Determining the statistical significance of relative weights, Psycholog. Meth., 14, 387–399, 2009.

Tri, V. K.: Hydrology and hydraulic infrastructure systems in the Mekong Delta, Vietnam, in: The Mekong Delta System, Springer, Dordrecht, 49–81, 2012.

Uemura, R., Matsui, Y., Yoshimura, K., Motoyama, H., and Yoshida, N.: Evidence of deuterium excess in water vapor as an indicator of ocean surface conditions, J. Geophys. Res.-Atmos., 113, D19114, https://doi.org/10.1029/2008JD010209, 2008.

Vuille, M., Bradley, R., Werner, M., Healy, R., and Keimig, F.: Modeling $\delta^{18}O$ in precipitation over the tropical Americas: 1. Interannual variability and climatic controls, J. Geophys. Res.-Atmos., 108, 4174, https://doi.org/10.1029/2001JD002038, 2003.

Vuille, M., Werner, M., Bradley, R., and Keimig, F.: Stable isotopes in precipitation in the Asian monsoon region, J. Geophys. Res.-Atmos., 110, D23108, https://doi.org/10.1029/2005JD006022, 2005.

Wang, B., Wu, R., and Lau, K.: Interannual variability of the Asian summer monsoon: contrasts between the Indian and the Western North Pacific-East Asian Monsoons, J. Climate, 14, 4073–4090, 2001.

Yang, H., Johnson, K., Griffiths, M., and Yoshimura, K.: Interannual controls on oxygen isotope variability in Asian monsoon precipitation and implications for paleoclimate reconstructions, J. Geophys. Res.-Atmos., 121, 8410–8428, 2016.

Yoshimura, K., Oki, T., Ohte, N., and Kanae, S.: A quantitative analysis of short-term ^{18}O variability with a Rayleigh-type isotope circulation model, J. Geophys. Res.-Atmos., 108, 4647, https://doi.org/10.1029/2003JD003477, 2003.

Yoshimura, K., Kanamitsu, M., Noone, D., and Oki, T.: Historical isotope simulation using reanalysis atmospheric data, J. Geophys. Res.-Atmos., 113, D19108, https://doi.org/10.1029/2008JD010074, 2008.

Yoshimura, K., Miyoshi, T., and Kanamitsu, M.: Observation system simulation experiments using water vapor isotope information, J. Geophys. Res.-Atmos., 119, 7842–7862, 2014.

4

Inter-laboratory comparison of cryogenic water extraction systems for stable isotope analysis of soil water

Natalie Orlowski[1,2,12], Lutz Breuer[2,3], Nicolas Angeli[4], Pascal Boeckx[5], Christophe Brumbt[6], Craig S. Cook[7], Maren Dubbert[8,22], Jens Dyckmans[9], Barbora Gallagher[10], Benjamin Gralher[11], Barbara Herbstritt[12], Pedro Hervé-Fernández[5,6,13], Christophe Hissler[14], Paul Koeniger[15], Arnaud Legout[16], Chandelle Joan Macdonald[7], Carlos Oyarzún[6], Regine Redelstein[17], Christof Seidler[18], Rolf Siegwolf[19], Christine Stumpp[11,20], Simon Thomsen[21], Markus Weiler[12], Christiane Werner[8,22], and Jeffrey J. McDonnell[1]

[1]Global Institute for Water Security, School of Environment and Sustainability, University of Saskatchewan, Saskatoon, Canada
[2]Institute for Landscape Ecology and Resources Management (ILR), Research Centre for BioSystems, Land Use and Nutrition (IFZ), Justus Liebig University Giessen, Giessen, Germany
[3]Centre for International Development and Environmental Research, Justus Liebig University Giessen, Giessen, Germany
[4]INRA-UHP Ecologie et Ecophysiologie Forestières, INRA Centre de Nancy, Champenoux, France
[5]Isotope Bioscience Laboratory (ISOFYS), Faculty of Bioscience Engineering, Ghent University, Ghent, Belgium
[6]Instituto de Ciencias de la Tierra, Universidad Austral de Chile, Valdivia, Chile
[7]Department of Ecosystem Science and Management, Stable Isotope Facility, University of Wyoming, Laramie, Wyoming, USA
[8]BayCEER, University of Bayreuth, Bayreuth, Germany
[9]Institute of Soil Science and Forest Nutrition, Centre for Stable Isotope Research and Analysis (KOSI), University of Göttingen, Göttingen, Germany
[10]Institute for Environmental Research, Australia Nuclear Science and Technology Organization, Sydney, Australia
[11]Institute of Groundwater Ecology, German Research Center for Environmental Health, Helmholtz Zentrum München, Neuherberg, Germany
[12]Hydrology, Faculty of Environment and Natural Resources, University of Freiburg, Freiburg, Germany
[13]Laboratory of Hydrology and Water Management, Faculty of Bioscience Engineering, Ghent University, Ghent, Belgium
[14]Luxembourg Institute of Science and Technology (LIST), Department of Environmental Research and Innovation (ERIN), Esch-sur-Alzette, Luxembourg, Luxembourg
[15]German Federal Institute for Geosciences and Natural Resources (BGR), Hanover, Germany
[16]INRA UR1138 Biogéochimie des Ecosystèmes Forestiers, INRA Centre de Nancy, Champenoux, France
[17]Plant Ecology and Ecosystems Research, University of Göttingen, Göttingen, Germany
[18]Ecophysiology of Plants, Technical University of Munich, Munich, Germany
[19]Stable Isotope Research Facility, Paul Scherrer Institute (PSI), Villigen, Switzerland
[20]Institute of Hydraulics and Rural Water Management (IHLW), University of Natural and Life Sciences (BOKU), Vienna, Austria
[21]Institute of Soil Science, University of Hamburg, Hamburg, Germany
[22]Ecosystem Physiology, University of Freiburg, Freiburg, Germany

Correspondence: Natalie Orlowski (natalie.orlowski@hydrology.uni-freiburg.de)

Abstract. For more than two decades, research groups in hydrology, ecology, soil science, and biogeochemistry have performed cryogenic water extractions (CWEs) for the analysis of δ^2H and $\delta^{18}O$ of soil water. Recent studies have shown that extraction conditions (time, temperature, and vacuum) along with physicochemical soil properties may affect extracted soil water isotope composition. Here we present results from the first worldwide round robin laboratory intercomparison. We test the null hypothesis that, with identical soils, standards, extraction protocols, and isotope analyses, cryogenic extractions across all laboratories are identical. Two standard soils with different physicochemical characteristics along with deionized (DI) reference water of known isotopic composition were shipped to 16 participating laboratories. Participants oven-dried and rewetted the soils to 8 and 20 % gravimetric water content (WC), using the deionized reference water. One batch of soil samples was extracted via predefined extraction conditions (time, temperature, and vacuum) identical to all laboratories; the second batch was extracted via conditions considered routine in the respective laboratory. All extracted water samples were analyzed for $\delta^{18}O$ and δ^2H by the lead laboratory (Global Institute for Water Security, GIWS, Saskatoon, Canada) using both a laser and an isotope ratio mass spectrometer (OA-ICOS and IRMS, respectively). We rejected the null hypothesis. Our results showed large differences in retrieved isotopic signatures among participating laboratories linked to soil type and soil water content with mean differences compared to the reference water ranging from +18.1 to −108.4 ‰ for δ^2H and +11.8 to −14.9 ‰ for $\delta^{18}O$ across all laboratories. In addition, differences were observed between OA-ICOS and IRMS isotope data. These were related to spectral interferences during OA-ICOS analysis that are especially problematic for the clayey loam soils used. While the types of cryogenic extraction lab construction varied from manifold systems to single chambers, no clear trends between system construction, applied extraction conditions, and extraction results were found. Rather, observed differences in the isotope data were influenced by interactions between multiple factors (soil type and properties, soil water content, system setup, extraction efficiency, extraction system leaks, and each lab's internal accuracy). Our results question the usefulness of cryogenic extraction as a standard for water extraction since results are not comparable across laboratories. This suggests that defining any sort of standard extraction procedure applicable across laboratories is challenging. Laboratories might have to establish calibration functions for their specific extraction system for each natural soil type, individually.

1 Introduction

The interpretation of the stable isotope signatures of water (δ^2H and $\delta^{18}O$) from soils in many research disciplines relies on accurate, high-precision measurements (Wassenaar et al., 2012). To extract water from soils for isotopic analysis, cryogenic water extraction (CWE) is the most widely used laboratory-based removal technique (Araguás-Araguás et al., 1995; Orlowski et al., 2016a). The ability to obtain measurable amounts of water from small sample sizes (i.e., < 10 g) makes this method attractive. However, CWE is also accompanied by high capital and operating costs. Despite its widespread use, recent work has identified several extraction artifacts leading to uncertain isotopic signature identification (Gaj et al., 2017a; Orlowski et al., 2016b). Studies have shown that extraction conditions (i.e., extraction time, temperature, and vacuum) need to be adapted specifically to the soil used (Araguás-Araguás et al., 1995; Gaj et al., 2017a; Meißner et al., 2014; Orlowski et al., 2016a). Notwithstanding, isotope effects triggered by physicochemical soil properties (e.g., clay minerals; soil organic carbon content; and water content, WC) can occur (Araguás-Araguás et al., 1995; Gaj et al., 2017a; Meißner et al., 2014; Oerter et al., 2014; Orlowski et al., 2013). However, the ecohydrology and soil science communities currently lack clear recommendations for standardized water extraction conditions from soils. Although there seems to be an agreement on the need to control the extraction yield of cryogenic extraction facilities (recovery rate in percentage of previously added water), there exists a large variability in the applied extraction conditions between laboratories. Moreover, extraction systems vary in terms of heating elements, size of extraction containers, or throughput, in addition to the aforementioned extraction conditions (Goebel and Lascano, 2012; Koeniger et al., 2011; Orlowski et al., 2013). Thus, no standard system setup or methodology exists.

Despite the work to date and the extensive application of stable water isotope analysis, no formal interlaboratory comparison between different cryogenic systems has been published. Here we present the first worldwide interlaboratory comparison between 16 different cryogenic extraction facilities. CWE procedures were conducted with two standard soils with different physicochemical characteristics (silty sand and clayey loam), spiked with a known isotopic label at different gravimetric water contents (WC of 8 and 20 %). The null hypothesis guiding this work was that all laboratories would yield the same results independent of soil type and water content. In addition, we addressed the following research questions:

1. How does the cryogenic system configuration affect resulting soil water isotopic composition?

2. How do soil type and soil water content affect the isotope data?

Table 1. Description of the respective extraction systems that participated in the cryogenic inter-laboratory comparison, the applied extraction parameters for extraction approach I, and the amount of sample material used for both extraction methods (lab procedure: I; predefined: II). Note that not every lab provided the same detailed information.

Lab no.	Country	Description of CWE facility	Number of extraction slots	Extraction parameters for approach I	Amount of sample material used (g)
1	Germany	Similar to lab no. 8; pair of Valco Exetainer® vials connected with a 1.56 mm stainless steel capillary as extraction–collection unit; a hot plate, LN_2 cold trap	9	Temperature: 100 °C, vacuum: 1–6 Pa, time: 60 min (silty sand) and 120 min (clayey loam)	10–12
2	Canada	Mainly composed of different types of Swagelok® fittings (Swagelok Company, Solon, OH, US), flanges, and flexible hoses (Rettberg®, Rettberg Inc., Göttingen, Germany); vacuum applied or shut off via diaphragm valves and monitored via DCP 3000 and VSK 3000 (Vacuubrand Inc., Wertheim, Germany), glass tubes as extraction and collection units, LN_2 cold trap, water bath/sand bath	24	Temperature: on average 96 °C, vacuum: 3.3–7.3 Pa, time: 90 min (silty sand) and 240 min (clayey loam)	20
3	Germany	Heating lamps; LN_2 cold trap	5	Temperature: ∼ 115 °C, vacuum: 1 Pa, time: 90 min	20
4	Germany	A septum-sealed 70 mL vial (extraction) and a Valco Exetainer® vial (collection) connected with a stainless steel capillary as extraction–collection unit; heating block (aluminum), LN_2 cold trap	6	Temperature: 125 °C; vacuum: 50 Pa; time: 33 min (silty sand at 8 % WC), 56 min (silty sand at 20 % WC), 67 min (clayey loam at 8 % WC), and 83 min (clayey loam at 20 % WC)	20
5	France	Cold trap: mixture of LN_2 and EtOH	4	Temperature: 65 °C, cold trap: −50 to −70 °C, vacuum: 0.1–1 Pa (static vacuum), time: 60–90 min	10
6	Australia	Heating tape, glassware for extraction–collection unit; LN_2 cold trap	4	Temperature: 95–100 °C, starting with sealed vacuum of 0.3 Pa, time: 150–180 min	20
7	Chile	Heating element: reactor HI 839800 (Hanna Instruments); size of extraction container: 22 mL; precision measured with VD81 Thyracont model	9	Temperature: 105 °C, vacuum: 12–23 Pa, time: 240 min	20
8	Germany	Pair of Valco Exetainer® vials connected with a 1.56 mm stainless steel capillary as extraction–collection unit; an aluminum block on a hot plate, LN_2 cold trap	12	Temperature: 200 °C, vacuum: 50 Pa, time: 15 min	10
9	Germany	Stainless steel manifold (five vials each), glass tubes as extraction–collection unit: 18 mm w, 150 mm l, LN_2 cold trap, water bath	20	Temperature: 95 °C, vacuum: 0.8 Pa, time: 90 min	On average 43
10	Switzerland	Glass tubes (Vacutainer), LN_2 cold trap, water bath	20	Temperature: 80 °C	Not specified
11	USA	Pyrex culture tubes (25 mm × 150 mm), volume: 75 mL; heaters: electric coil (only allow to heat two-thirds of the tube)	10	Temperature: 102 °C, vacuum: < 0.1–2.7 Pa, time: on average 81 min (silty sand) and 134 min (clayey loam)	10 for 20 % WC, 20 for 8 % WC
12	Germany	Glass tubes, LN_2 cold trap, water bath	8	Temperature: 80 °C, vacuum: 600 Pa, time: 60 min	23

Table 1. Continued.

Lab no.	Country	Description of CWE facility	Number of extraction slots	Extraction parameters for approach I	Amount of sample material used (g)
13	Germany	Glass tubes (Schott GL 18), LN_2 cold trap, sensor-regulated tube-shaped heating element	10	Temperature: 100 °C, vacuum: 6.7–13.3 Pa, time: 15–266 min	10 for 20 % WC, 20 for 8 % WC
14	Germany	Glass tubes as extraction units, vacuum is generated by a rotary vane pump (RZ 2.5, Vacuubrand, Wertheim) and monitored via DCP 3000 with VSP 3000 (Vacuubrand), LN_2 cold trap, water bath	20	Temperature: 80 °C, vacuum: 2–46 Pa, time: 30 min (silty sand) and 40 min (clayey loam)	10
15	Germany	The septa of Labco exetainers are pierced with a cannula (1.2 mm diameter) and connected to the vacuum system, vacuum is generated by a rotary vane pump (RZ 2.5, Vacuubrand, Wertheim, Germany) and monitored via DCP 3000 with VSP 3000 (Vacuubrand), LN_2 cold trap, water bath	20	Temperature: 80 °C, vacuum: 10–350 Pa, time: 30 min (silty sand) and 40 min (clayey loam)	10
16	Germany	Mainly composed of different types of Swagelok® fittings (Swagelok Company, Solon, OH, US), flanges, and flexible hoses (Rettberg®, Rettberg Inc., Göttingen, Germany); vacuum applied or shut off via diaphragm valves and monitored via DCP 3000 and VSK 3000 (Vacuubrand Inc., Wertheim, Germany), glass tubes as extraction and collection units, LN_2 cold trap, water bath/sand bath, high-purity nitrogen purging system	18	Temperature: 100 °C, vacuum: 3.1–0.9 Pa, time: 45 min (silty sand) and 240 min (clayey loam)	20

3. How do results differ when extracted soil water stable isotopic compositions are measured via off-axis integrated cavity output spectroscopy (OA-ICOS) vs. isotope ratio mass spectrometry (IRMS)?

4. What do we learn from this exercise for standardization of cryogenic extraction facilities?

2 Methods

2.1 Experimental design

Table 1 provides a description of the respective extraction systems that participated in the intercomparison. In total, 16 independent laboratories from seven countries took part in the trial.

Before the commencement of the round robin test, participants were asked to fill out a questionnaire (see Appendix A) to characterize their cryogenic extraction system in terms of number of extraction slots or amount of sample material usually introduced into the system (size of extraction unit). Two standard soils with different physicochemical properties (clayey loam and silty sand) from the German State Research Institute for Agriculture (LUFA Speyer, 2015) (Table 2) were used for the interlaboratory comparison.

We chose a silty sand from which we expected water extractions to be relatively easy for each laboratory without cation ion exchange problems and a clayey loam soil, which is known to be challenging for CWE extraction systems. Clayey soils can be difficult due to interactions with the clay fraction and different types of clay minerals – the so-called adsorbed cation effect (Oerter et al., 2014). Clay soils also present challenges with regard to the tightness of water bound to mineral surfaces, which causes an additional isotope effect (Ingraham and Shadel, 1992; Oerter et al., 2014; Walker et al., 1994).

Soil samples were sieved to a grain size < 2 mm. Soils were pre-dried at 105 °C for 48 h, homogenized, and shipped in tightly sealed glass bottles to the 16 independent laboratories along with deionized (DI) reference water of known isotopic composition, measured on both an IWA-45EP analyzer (OA-ICOS, Los Gatos Research Inc., Mountain View, US) (δ^2H: -59.8 ± 0.2 ‰ and δ^{18}O: -8.5 ± 0.1 ‰, $n = 6$) and via a Delta V™ Advantage mass spectrometer (Thermo Fisher Scientific, Waltham, MA, US) (δ^2H: -60.5 ± 0.2 ‰ and δ^{18}O: -8.7 ± 0.1 ‰, $n = 6$). All bottles containing either soils or DI water were filled, capped tightly, and wrapped

Table 2. Soil characteristics of clayey loam and silty sand (means ± SD). The clay mineral composition of soil samples was determined via X-ray powder diffraction (XRD, Philips X'Pert PW 1830 equipped with a PW2273/20 tube and a theta/theta-goniometer) following Poppe et al. (2016). Values were not corrected for reference intensity ratios (RIR). Alternating strata can occur for illite, smectite, or vermiculite. X-ray fluorescence (XRF) characterization of the chemical composition (in % w/w) was performed using an Axios spectrometer (PANalytical, EA Almelo, the Netherlands). Loss of ignition was 12.8 for the clayey loam and 1.3 for the silty sand.

Parameter	Clayey loam	Silty sand
pH value	7.2 ± 0.2	5.0 ± 0.3
Water holding capacity (g 100 g^{-1})	43.4 ± 0.8	32.1 ± 1.4
Organic carbon (%)	2.0 ± 0.2	0.7 ± 0.1
Cation exchange capacity (cmol(+) kg^{-1})	30.6 ± 5.1	4.1 ± 0.6
Particle size (mm) distribution according to German DIN (%)		
< 0.002 (clay)	26	2.6
0.002–0.063 (silt)	46.4	12.7
0.063–2 (sand)	27.6	84.7
XRD analysis (relative %)		
Kaolinite	18.8	18.8
Illite	18	27.7
Chlorite	1.2	19.8
Vermiculite	43.4	2.9
Smectite	0.5	19.8
Mixed layered clays/alternating strata (illite, smectite, or vermiculite)	18.1	11.1
XRF analysis (%)		
SiO_2	65.1	92.3
TiO_2	0.4	0.1
Al_2O_3	8.8	3.3
Fe_2O_3	3.1	0.5
MnO	0.1	0.0
MgO	1.5	0.1
CaO	5.3	0.2
Na_2O	0.9	0.3
K_2O	1.7	1.7
P_2O_5	0.2	0.1
SO_3	0.1	< 0.01
Cl	< 0.002	< 0.002
F	< 0.05	< 0.05

with Parafilm® to prevent water loss. We decided not to ship ready-to-use rehydrated soils to avoid evaporation fractionation effects and to give participants the opportunity to adjust, for example, samples sizes to the specific requirements of their extraction system. Water loss and evaporative enrichment from the shipped DI water was checked by isotopic comparison of shipped and non-shipped DI water (shipment test 1: Giessen to Freiburg (Germany) to Saskatoon (Canada) vs. non-shipped DI water samples; shipment test 2: Giessen (Germany) to Saskatoon (Canada) vs. non-shipped DI water samples). After this simple experiment, isotope fractionation effects due to shipment were excluded.

As a reliability test, each participant in the intercomparison performed water cryogenic extractions (defined here as simply extracting pure water, i.e., without any soil material present) using their extraction facility. This was done in order to determine the capability of the respective extraction apparatus to recapture water of known isotopic composition. After showing the operational reliability, CWEs with the rehydrated soil samples were performed following a predefined protocol.

2.2 Sample preparation protocol

Before starting the rewetting of the pre-dried soil samples with the DI water, participants oven-dried (at 105 °C for 48 h) the provided soils again to remove any potential water that could be present (e.g., remoistening of the soil samples during shipment). Afterwards, soils were placed in a desiccator for cooling and to prevent remoistening of the dried soil samples with ambient water vapor (Orlowski et al., 2016b; Van

De Velde and Bowen, 2013). For rehydration, two different amounts of reference DI water were added to the respective soil types (to create 8 and 20 % gravimetric WC). Exposure of the dried soil samples to ambient conditions was kept as brief as possible. Participants adjusted the amount of respective soil material and water for rewetting the samples according to the specific requirements of their extraction system, e.g., size of extraction containers. Sample preparation was performed separately for OA-ICOS and IRMS analysis but in the identical way as specified below.

1. Soil and DI water were added alternately. A quarter of soil material (clayey loam or silty sand) and a quarter of DI water were alternatively added to the pre-weighed extraction tube to facilitate soil–water homogenization.
2. This rewetting procedure was completed by adding a quarter of soil material to the extraction tube to avoid supernatant water and to obtain the best possible mixing.
3. Samples were weighed again.
4. Finally, an inert cover (Fackelmann Inc, Hersbruck, Germany) was placed on top of the soil sample to avoid the spread of sample material throughout the respective cryogenic extraction line. The inert material was proven to not cause isotope effects during soil water extraction (Orlowski et al., 2013).
5. Extraction tubes were plugged and sealed with Parafilm® to ensure an airtight system.
6. Rehydrated soils in their respective extraction containers were placed in vertical position in a refrigerator (5 °C for 72 h), which further allowed the liquid and solid phase to equilibrate.

2.3 Cryogenic extraction approaches

Since different extraction times and temperatures were applied in past studies, we decided that participating laboratories should follow two different extraction approaches. (I) For the first subset of rehydrated soil samples, participants applied the CWE procedure considered routine in their laboratory for the specific soil type and soil water content. (II) With the second subset, CWE under predefined conditions for all labs was performed: for silty sand, a 45 min extraction time was used while 240 min was applied to clayey loam samples, both at an extraction temperature of 100 °C and a vacuum of 0.3 Pa. These predefined extraction parameters were identical for all participating laboratories. For comparison, in past studies extraction times from 2.5 min (Koeniger et al., 2011), over 30 min (West et al., 2006) to 7 h (Araguás-Araguás et al., 1995) for sandy soils and from 30 min (Goebel and Lascano, 2012), over 40 min (West et al., 2006) to 8 h (Araguás-Araguás et al., 1995) for clayey soils were reported.

Three replicates per soil type and soil water content resulting in 24 samples per extraction procedure (predefined and laboratory specific) and isotope analysis method (OA-ICOS and IRMS) were processed ($n = 48$ in total). Pre- and post-oven-dried (105 °C for 24 h) soil sample weights were used to determine water recovery rates. All extracted water samples were transferred to 2 mL amber glass vials capped with solid lids (Th. Geyer Inc., Renningen, Germany), tightly sealed with Parafilm®, labeled, and shipped to the GIWS for isotope analysis. If the amount of extracted water was not sufficient to entirely fill the 2 mL vial, inserts (0.2 mL) were used (Th. Geyer Inc., Renningen, Germany) to minimize sample vial headspace, following standard procedures as outlined by the IAEA (2014).

2.4 Isotope analyses

For cross-checking isotope data and ruling out potential lab analytical differences, the isotopic composition of the extracted water samples was analyzed via both OA-ICOS and IRMS. OA-ICOS samples were analyzed on an IWA-45EP analyzer (Los Gatos Research Inc., Mountain View, US). The accuracy of OA-ICOS analyses was ± 0.5 ‰ for δ^2H and ± 0.1 ‰ for δ^{18}O. IRMS samples were analyzed on a Delta V™ Advantage mass spectrometer (Thermo Fisher Scientific, Waltham, MA, US) and an H/Device peripheral using a Cr-reduction method for δ^2H analysis (Morrison et al., 2001). For δ^{18}O analysis, a GasBench II peripheral was utilized. Using mass spectrometry, a conversion from the water into a light gas suitable for mass spectrometry (H_2, CO_2, CO, O_2) is necessary. This conversion step often turns out to limit the achievable precision of IRMS (Brand et al., 2009). In our case, IRMS results are accurate to ± 1 ‰ for δ^2H and to ± 0.2 ‰ for δ^{18}O, respectively. All isotope ratios are reported in per mil (‰) relative to Vienna Standard Mean Ocean Water (VSMOW) ($\delta^2\text{H}$ or $\delta^{18}\text{O} = (R_{\text{sample}}/R_{\text{standard}} - 1) \times 1000$ ‰), where R is the isotope ratio of the sample and the known reference (i.e., VSMOW) (Craig, 1961). In-house standards, calibrated against VSMOW2 and SLAP2, were run as samples to allow the results to be reported against VSMOW (Nelson, 2000).

OA-ICOS isotope data of soil water extracts were checked but not corrected for spectral interferences (caused by potentially co-extracted organics such as methanol or ethanol) using the Spectral Contamination Identifier post-processing software (LWIA-SCI, Los Gatos Research Inc.) when measured via OA-ICOS. This software compares recorded spectra from unknown samples with those from known non-contaminated samples (standards) to produce a metric of contamination from either narrowband (e.g., methanol, MeOH) or broadband (e.g., ethanol, EtOH) absorbers, which indicates the likelihood or degree of spectral interference (Schultz et al., 2011). IRMS results are generally not affected by organic contaminants.

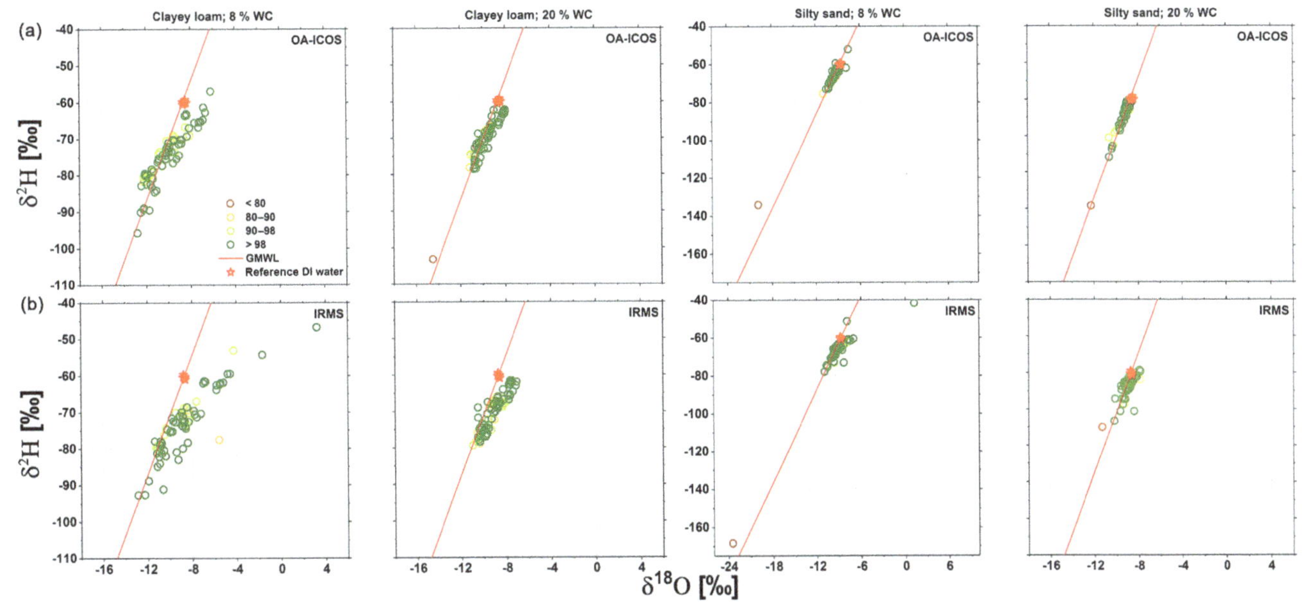

Figure 1. Water recovery rates (grouped from < 80 to > 98 %) for both soil types (clayey loam and silty sand), WCs (8 and 20 %), and OA-ICOS and IRMS (panels **a, b**, respectively) isotope data in comparison to the spiked reference DI water (red asterisks) shown in dual isotope space. For reference, plots include the global meteoric water line (GMWL, solid red line). Water recovery rates are shown for those labs that provided the complete set of soil weight data (in % of previously added water).

2.5 Statistical evaluation

We used R for statistical analyses (R version 3.3.2; R Core Team, 2014). All data were tested for normality using the Shapiro–Wilk test for quantifying laboratory variances, differences between predefined and laboratory-specific extraction procedures, effects of soil type and WC, and differences between OA-ICOS and IRMS. Homoscedasticity was tested using either the Levene's test for normally distributed data or the Fligner-Killeen test for non-normally distributed data. Cook's distance was determined in order to identify outliers ($D > 1$). Depending on the type of data (normally distributed and homoscedastic), either Kruskal–Wallis rank sum tests or analyses of variance (ANOVAs) were applied and post hoc tests (e.g., Nemenyi tests) were run to determine which groups were significantly different ($p \leq 0.05$). P-value adjustments via the FDR method (false discovery rate) were applied to reduce the family-wise error rate (Ziefler et al., 2012).

For graphical comparisons, a target standard deviation (TSD) for acceptable performance was set to ± 2 ‰ for δ^2H and ± 0.2 ‰ for δ^{18}O, similar to Orlowski et al. (2016b), which is considered reasonable for hydrologic studies (Wassenaar et al., 2012). The TSD does not account for errors associated with the extraction method itself, weighing errors, and volumetric water additions to the sample, or any standard deviations (1 SDs) related to the isotope analysis. Statistically significant ($p \leq 0.05$) linear regressions were added to dual isotope plots as references as well as the global meteoric water line (GMWL: $\delta^2\text{H} = 8.2 \times \delta^{18}\text{O} + 11.3$ ‰, as defined by Rozanski et al., 1993).

3 Results

3.1 Cryogenic extraction systems and water extraction efficiencies

Cryogenic extraction systems varied greatly from lab to lab: from manifold, high-throughput devices (as described by Orlowski et al., 2013) to small, single chamber systems (as in Koeniger et al., 2011, and West et al., 2006) (for details see Table 1). The systems showed differences in terms of the extraction containers (form, size, volume, and material), the heating module and its application temperature (heating tapes or lamps, water baths or hot plates), the type of fittings and connections (glass, stainless steel), and in the vacuum-producing units (Table 1). In relation to the amount of used sample material, most labs either introduced 10 or 20 g to their system no matter the extraction approach (I or II), soil type, or WC. Only labs 11 and 13 chose different weights with respect to the WC, e.g., 10 g for the higher WC (20 %) and 20 g for 8 % WC for extraction approach I.

To determine the degree of extraction efficiency for each lab's samples, water recovery rates were calculated for those labs that provided the complete set of soil weight data (in % of previously added water). When comparing water recovery rates against δ^2H and δ^{18}O values, the clayey soil showed no clear trend (Fig. 1). Even if water recovery rates were higher

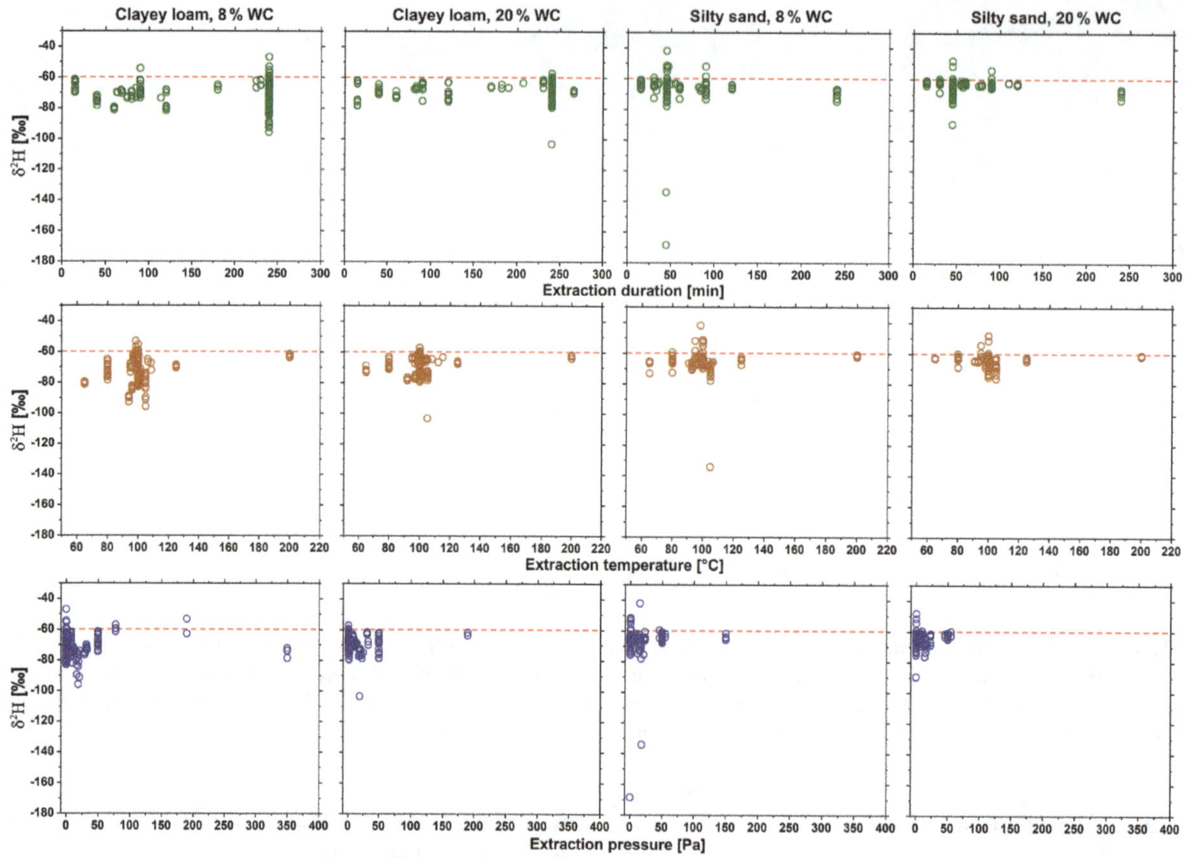

Figure 2. Effect of cryogenic extraction parameters (duration, temperature, and pressure) on δ^2H results of both soil types (clayey loam and silty sand) and WCs (8 and 20 %) shown for all labs. The mean reference DI water δ^2H value is shown as a red dotted line.

than 98 % (following the definition of Araguás-Araguás et al., 1995), extracted isotope values differed from the reference DI water (Fig. 1). For example, at 8 % soil water content (WC), recovery rates of above 98 % were achieved, but isotope values were depleted in comparison to the reference DI water (Fig. 1, left panels).

For the silty sand, recovery rates were generally higher in comparison to the clayey soil. Only a few samples showed extraction efficiencies lower than 98 % (Fig. 1, right panels). Surprisingly, we observed some recovery rates higher than 100 %. This was especially an issue for soils at 8 % WC (Fig. 1).

Correlation analysis was performed in order to relate extraction parameters (i.e., time, temperature, or vacuum) to OA-ICOS and IRMS isotope data.

We found no significant correlations between the extraction parameters and the respective isotope data, e.g., shown for δ^2H results (Fig. 2) (e.g., $R^2 = 0.0$ for δ^2H vs. duration or temperature).

3.2 Laboratory performance with respect to water content and soil type

Figures 3 and 4 show the mean differences between the extracted samples via the lab procedure's extraction approach I and the predefined extraction approach II compared to the reference DI water δ^2H and δ^{18}O values, respectively.

For the 8 % WC tests, mean differences for the clayey loam ranged from $+13.1$ to -32.8 ‰ for δ^2H. For the individual lab procedure's extraction approach I at 8 % WC for the clayey loam, two laboratories (lab 3 and 8) were able to get back to the reference δ^2H value based on no statistically significant differences ($p > 0.05$) (Fig. 3, upper left plot). For the predefined extraction approach II at 8 % WC, two other labs recovered the δ^2H value from the clayey loam (lab 9 and 15).

For soil samples with 20 % WC, variation among laboratories was smaller but only one laboratory (lab 9) recovered the reference DI water δ^2H value applying the predefined extraction approach for the clayey loam. Mean differences between the clayey loam extracts and the reference DI water ranged from $+2.8$ to -19.5 ‰ (Fig. 3, upper right plot).

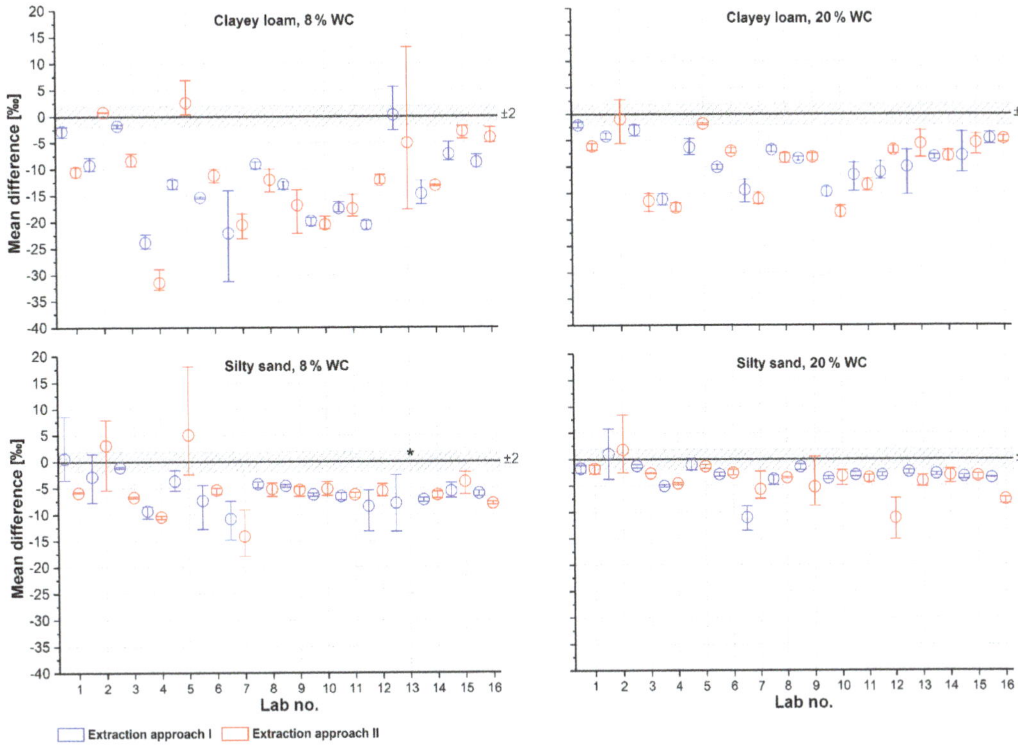

Figure 3. Mean differences from reference DI water for δ^2H OA-ICOS results of water extracts from both extraction methods (lab procedure: I; predefined: II), soil types, and water contents (8 and 20 % WC) including TSD of ±2 for δ^2H (Asterisk: −108.4 for δ^2H). Symbols represent the mean of the three replicates and y error bars stand for the isotopic variation of the replicates. There were no significant differences between the two extraction approaches over all labs.

Mean differences between the silty sand water extraction and the reference δ^2H signature were in a smaller range of ±18 ‰ than clayey loam extracts from the same treatment (8 % WC).

For the individual lab procedure's extraction approach I at 8 % WC, five laboratories recovered the added label from the silty sand (Fig. 3, lower left plot) with no statistical differences between the reference DI water ($p > 0.05$) (labs 6, 8, 9, 13, and 15), whereas for the predefined extraction approach II at 8 % WC, three labs got back to the added δ^2H value (labs 9, 12, and 15).

For silty sand at 20 % WC, most laboratories' results even fell close to the range of the TSD of ±2 ‰. Mean differences compared to the reference DI water δ^2H signature ranged from +8.5 to −15.1 ‰ (Fig. 3, lower right plot). However, extraction approach I was statistically not successful in recovering the added label ($p < 0.05$), but five laboratories (6, 9, 10, 14, and 15) showed no significant differences compared to the reference DI water when applying extraction approach II to the silty sand at 20 % WC.

Laboratories performed better for δ^{18}O signature recovery, especially with extraction approach I. For both clayey loam WC treatments, labs 13 and 15 were the most successful. Again, mean differences compared to the reference DI water were larger for the 8 % WC than for the 20 % WC (Fig. 4, upper plots). However, for the clayey loam at 20 % WC with the predefined approach II only lab 13 and 14 did not show statistically significant differences compared to the added δ^{18}O signature (Fig. 4, upper right plot) ($p > 0.05$).

For the silty sand, most laboratories were able to get back the known value with no statistically significant differences compared to the reference δ^{18}O value (Fig. 4, lower plots). For both WC treatments of the silty sand, extraction approach II seemed to work better in recovering the added label.

Across both soil types, WC treatments, and extraction approaches, lab 13 was the most successful in recovering the reference δ^{18}O value, whereas for δ^2H recovery lab 9 gained back the added label in most of the cases.

In general, δ^2H and δ^{18}O values were neither comparable between laboratories nor between one laboratory at different soil types or WCs, meaning that a specific laboratory, for example, successfully recovered the added DI water value for silty sand but was not able to gain back the known label for clayey loam. Moreover, recovery results differed between both isotopes. For example, lab 13 was the most successful for δ^{18}O but not for δ^2H signature recovery. In terms of lab internal reproducibility, some labs showed small standard deviations for the replicates of the same soil type at a given WC (Figs. 3 and 4); even so, resulting isotope values differed in a

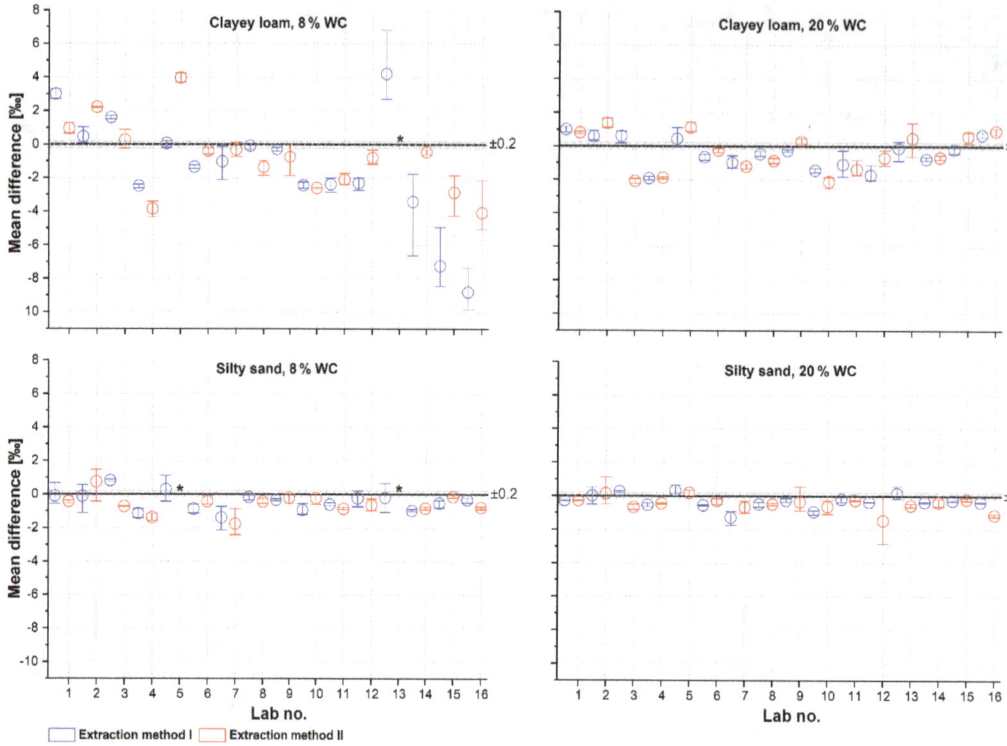

Figure 4. Mean differences from reference DI water for $\delta^{18}O$ OA-ICOS results of water extracts from both extraction methods (lab procedure: I; predefined: II), soil types, and water contents (8 and 20 % WC) including TSD of ±0.2 for $\delta^{18}O$. Asterisks represent outliers. Symbols represent the mean of the three replicates and *y* error bars stand for the isotopic variation of the replicates. There were no significant differences between the two extraction approaches over all labs.

statistically significant manner from the introduced reference DI water.

3.3 Differences between OA-ICOS- and IRMS-based measurements

Figures 5 (clayey loam) and 6 (silty sand) illustrate data variability for each laboratory and WC with respect to the labeled reference DI water added to each soil type in dual isotope space. Significant differences were observed between OA-ICOS and IRMS isotope data sets ($p \leq 0.05$). The clay soil isotope data at 8 % WC showed the greatest differences between OA-ICOS and IRMS measurements (mean differences of 1.3 and 1.2 for δ^2H and $\delta^{18}O$, respectively). The smallest differences between isotope analyzers were observed between both WC treatments of the silty sand (Fig. 6). The data sets with the lowest SD for both isotopes across labs and extraction approaches were the silty sand samples at 20 % WC measured via OA-ICOS and IRMS (SD of ±3.1 for δ^2H measured via OA-ICOS and ±4.2 for IRMS, respectively). However, those data sets still did not reach the TSD of ±2 ‰ for δ^2H and ±0.2 ‰ for $\delta^{18}O$.

For comparison, apart from soil water regression lines, the GMWL is also given in each subplot. Interestingly, isotope data across laboratories plot on slopes lower than the GMWL. For both soil types, regression lines of the IRMS measurements showed better correlations (for the silty sand $R^2 = 0.8$ and 0.9 for 8 and 20 % WC, respectively) than those of OA-ICOS measurements ($R^2 = 0.7$ for 8 and 20 % WC) (Fig. 6). Silty sand's soil water regression lines showed greater slopes (5.4–7.2 across both WCs and isotope analysis) than clayey loam's soil water regression lines (2.8–5.2 across both WCs and isotope analysis) (Figs. 5 and 6). The clayey loam regression lines for the higher WC also showed greater slopes than those of the lower WC (Fig. 5). Isotopic fractionation due to evaporation leads to a stronger kinetic effect for ^{18}O compared to ^2H, resulting in evaporative enrichment of the water along an evaporation water line with a lower slope relative to the original water (Gonfiantini, 1986). Benettin et al. (2018) recently revised the widely used concept of evaporation lines. The authors question that the trend line passing through fractionated soil water samples correctly identifies their source water and emphasize that trend lines through evaporated samples can differ widely from true evaporation lines.

For the clay soil type, the IRMS data sets (8 and 20 % WC) plot closer to the GMWL and the analyzed values showed a smaller SD in comparison to the OA-ICOS assays (SD of ±8.4 for the OA-ICOS δ^2H data vs. ±7.5 for the OA-ICOS data at 8 % WC) (Fig. 6).

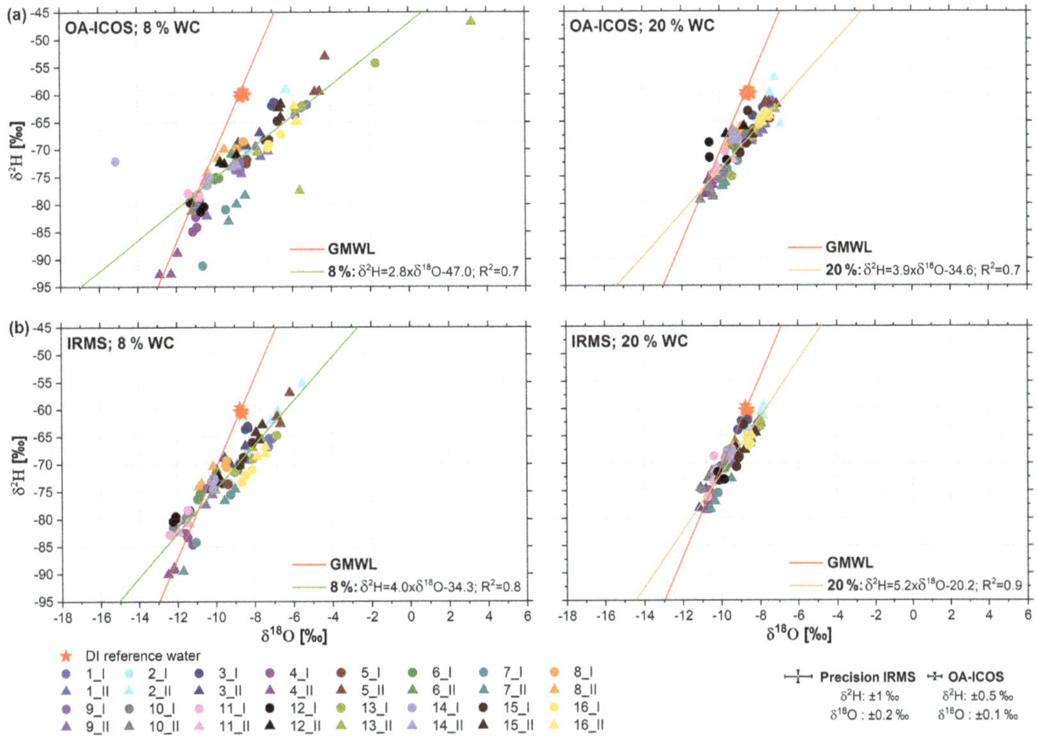

Figure 5. Dual isotope plots of clayey loam extracts for 8 and 20 % WC in comparison to reference DI water (red asterisks) for OA-ICOS and IRMS data (panels **a, b**, respectively) from the 16 participating labs (different colors represent different labs) and both extraction methods (lab procedure: I; predefined: II). For reference, plots include the global meteoric water line (GMWL, solid red line) and soil water regression lines for 8 and 20 % WC (solid green and orange lines, respectively).

In general, the spread of the isotope data decreased from 8 to 20 % WC and from OA-ICOS to IRMS measurement results (Figs. 5 and 6). The OA-ICOS isotope analyses showed more outliers than those of IRMS. Moreover, fewer outliers were found among the silty sand data when compared to that of the clayey loam soil. Overall, IRMS results for all soil types and WCs were slightly more depleted than those of OA-ICOS. However, differences were not significant ($p > 0.05$). In general, most of the water extracts were depleted in comparison to the reference DI water, which is especially true for δ^2H.

Examination of the differences between OA-ICOS and IRMS data prompted the testing of the OA-ICOS data for spectral interferences. Figure 7 shows that for the clayey loam soil, differences between OA-ICOS and IRMS data might be due to co-extracted alcoholic compounds, which caused erroneous OA-ICOS data.

Few samples among the 8 % WC versions of clay water extracts showed issues with both broadband and narrowband absorbers. This contamination by both methanol and ethanol explained the outliers found at 8 % WC in the clayey loam data (Fig. 7, upper left plot). Among these data, only a small number of samples showed no contamination, which were interestingly more depleted in comparison to data flagged as affected by narrowband absorbers. For the silty sand soil, only a few samples were contaminated and flagged as affected by narrowband absorbers. Interestingly, outliers in the silty sand soil data set at 8 % WC could not be explained by narrow- or broadband absorbers.

4 Discussion

4.1 Why are the cryogenic extraction results different across the participating laboratories?

We rejected our null hypothesis that all laboratories would yield the same results independent of soil type and water content. We showed that cryogenic extraction results were not comparable among laboratories. We also observed differences in the ability of individual labs to recover both isotope values (δ^2H and δ^{18}O) of the added reference DI water. Some laboratories were able to get back to the reference δ^2H value but were not successful for δ^{18}O.

Each extraction system's setups were different. Therefore, it was difficult to give any recommendation with regard to a high-performance and accurate extraction system that would lead to overall successful extractions. As a quality control, we checked water recovery rates, which were in some cases even higher than 100 % (Fig. 1). This could be attributed either to leaky vacuum systems (which might allow atmo-

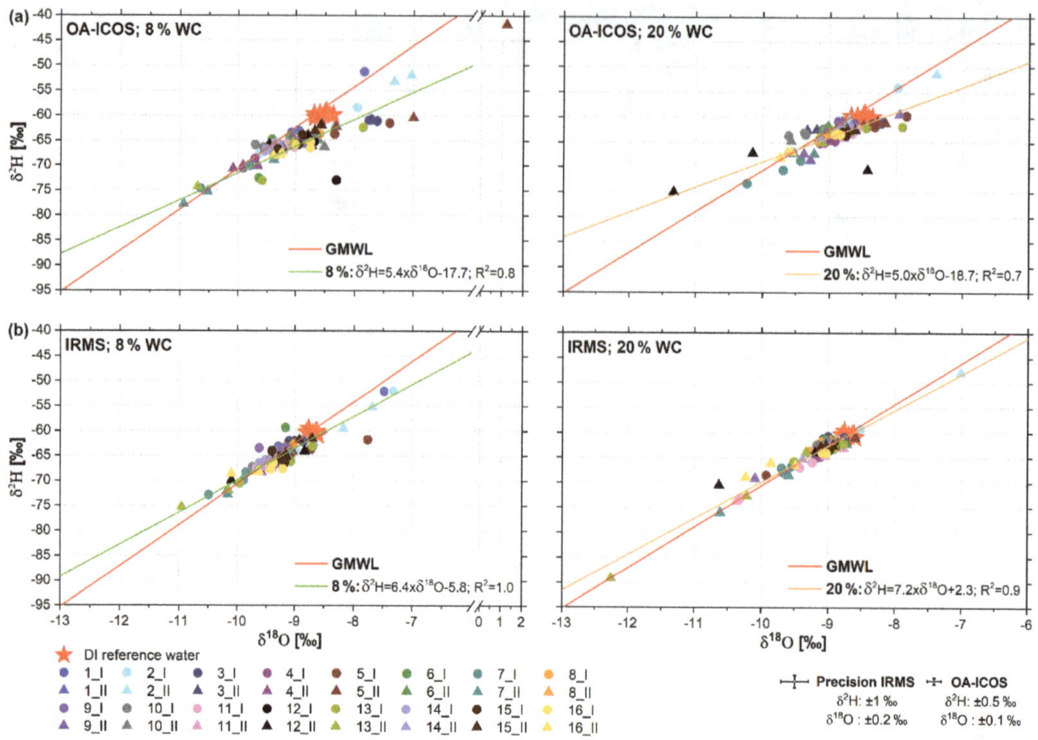

Figure 6. Dual isotope plots of silty sand extracts for 8 and 20 % WC in comparison to reference DI water (red asterisks) for OA-ICOS and IRMS data (panels **a, b**, respectively) from the 16 participating labs (different colors represent different labs) and both extraction methods (lab procedure: I; predefined: II). For reference, plots include the global meteoric water line (GMWL, solid red line) and soil water regression lines for 8 and 20 % WC (solid green and orange lines, respectively).

spheric water vapor to enter the system) or to a remoistening of the oven-dried soil samples before water extraction. Remoistening of oven-dried soil samples might be a general problem of such spiking experiments. In our case, sample preparation was not performed under an inert gas flow and, unfortunately, data on temperature and relative humidity conditions under which sample preparation took place are unavailable from the respective labs. Ambient water vapor isotopic composition measurements would have also been a relevant additional information. Contamination could also occur when an extraction system is not dried or cleaned after each extraction run, leaving moisture and/or soil material behind which would affect the next sample's results. Other measurement uncertainties during the extraction protocol could arise from weighing errors (scale calibration and precision), the accuracy of the volume of water additions to the soil samples, transfer of the samples, loss of water vapor during evacuation of the extraction system, unsteady heating temperatures, condensation of water vapor in the extraction system, and a lack of precision of analytical and laboratory equipment.

It is also possible that participating labs did not follow the predefined extraction procedure (approach II) in the exactly same ways. Even extraction results from some individual labs for the same soil type and WC showed high SDs (Figs. 3 and 4), which questions the overall repeatability of individual water extraction results. For the first, in-house extraction approach I, not all laboratories indicated the precise extraction conditions (extraction temperature, time, and vacuum) that they used for the specific soil types and WCs.

As an additional performance test, laboratories were asked to perform simple water extractions to show their ability to recover water of known isotopic composition prior to soil-based tests. For example, some laboratories, like lab 2, showed a high accuracy for these water extractions of ± 0.4 ‰ for δ^2H and ± 0.1 ‰ for δ^{18}O ($n = 119$), as well as lab 16. They performed extraction tests with tap water, which resulted in no significant differences between the initial, untreated (-56.7 ± 0.4 ‰ for δ^2H and -9.3 ± 0.1 ‰ for δ^{18}O) and extracted tap water (-57.5 ± 0.6 ‰ for δ^2H and -9.4 ± 0.1 ‰ for δ^{18}O). These examples show that these labs among others were able to reach the TSD with simple water extractions, but with soils they were unsuccessful. This indicates that differences between the reference DI water and water spiked and extracted from soils are likely caused by interactions with soil particles.

Given our findings, we now question the standard quality controls (e.g., water recovery rate calculations and water extractions without soil material). Quality controls with spiked soil samples may be a more effective way to demon-

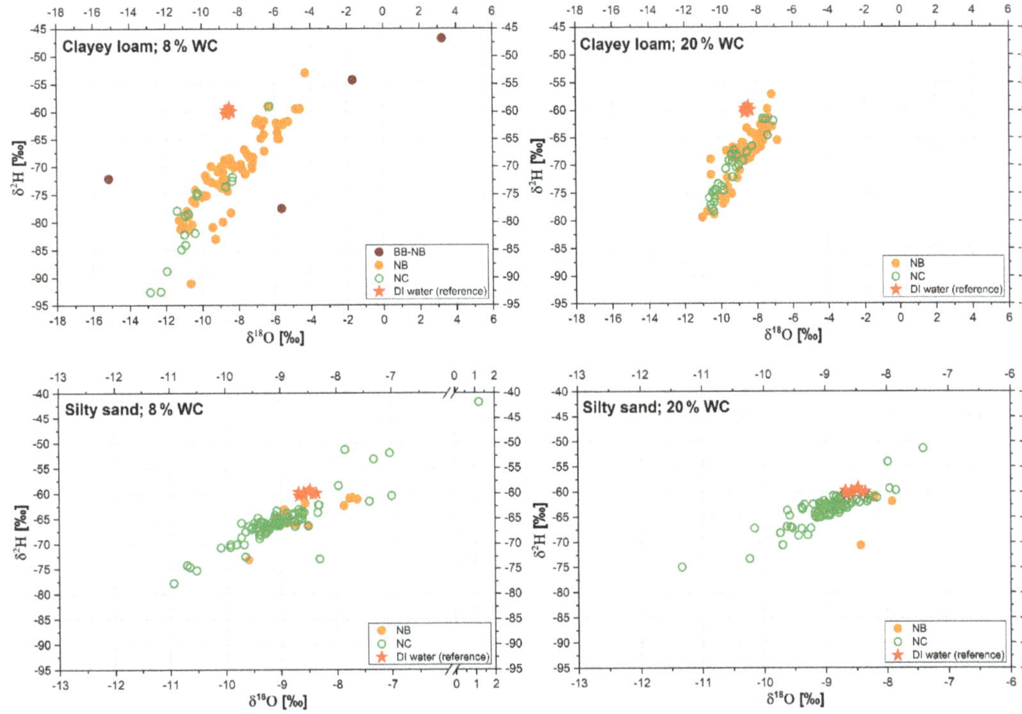

Figure 7. Dual isotope plots of clayey loam and silty sand extracts for 8 and 20 % WC in comparison to reference DI water (red asterisks) for OA-ICOS analyses flagged by spectral contamination using the Spectral Contamination Identifier (LWIA-SCI) post-processing software (Los Gatos Research Inc.). BB-NB: Broad- and narrowband absorbers (ethanol and methanol); NB: narrowband absorber (methanol); NC: no contamination detected.

strate lab's internal accuracy. However, such spiking experiments as performed in our study come along with other issues as recently outlined by Gaj et al. (2017b) and Sprenger et al. (2015). Gaj et al. (2017a) applied the Rayleigh equation (using stable isotope signatures) to calculate how much water was cryogenically extracted from pure clay minerals. They found that for samples from which water has been extracted to 100 % (determined gravimetrically), the Rayleigh equation showed that only 72 % of water was extracted at a temperature of 105 °C. When using an extraction temperature of 205 °C, the Rayleigh-estimated amount of water extracted was close to 90 %, but still not 100 %. This result clearly shows that despite the gravimetric quality control suggesting that all water has been extracted, isotopic differences may still exist.

Overall, laboratories 9 (for δ^2H) and 13 (for δ^{18}O) were the most successful in getting back to the DI reference water over all soil types and WCs. For the lab's in-house procedure, laboratory 9 extracted both soils for 90 min at 95 °C and 0.8 Pa. Their reported water extraction efficiency was 99–100 %. Glass tubes were used as extraction containers and a water bath as heating element. Laboratory 13 used different extraction parameters, which also varied slightly from sample to sample: for the clayey loam at 8 % WC, extractions were conducted for 75–114 min at 150–100 °C and 8–13.3 Pa. For the 20 % WC, they used 266 min at 100 °C and 6.7–13.3 Pa as in-house extraction parameters. For the silty sand at 8% WC, their extraction time was 15 min at 100 °C and 7.3–13.3 Pa. For the 20 % WC, they extracted for 30 min at 100 °C and 6.7–10.7 Pa. Lab 13 further specified that their extraction times were dictated by a decline in the pressure level indicating that no more water was evaporating from the respective sample. Extraction efficiencies for lab 13 varied between 93 and 127 %. Glass tubes were used as extraction containers along with a sensor-regulated tube-shaped heating element. This example shows that even for the relatively successful laboratories, extraction parameters did not seem to play a major role in achieving the reference DI water isotopic signature.

4.2 How do soil type and water content affect the results?

The adsorbed and interlayer water occurring in clayey soils can complicate the interpretation of obtained isotope data. Clay water-sorption capacity is well known (Schuttlefield et al., 2007; White and Pichler, 1959). White and Pichler (1959) found early on that montmorillonite adsorbs more water than kaolinite, illite, and chlorite, while chlorites and illites have similar water-sorption properties. The amount of water absorbed/adsorbed by clay minerals ranges from 800 to 500 % for Na montmorillonite (Kaufhold and Dohrmann, 2008;

White and Pichler, 1959) to as low as 60 % of the initial dry weight for biotite (White and Pichler, 1959). The clayey loam in our study was a vermiculite-rich (43 relative %) 2 : 1 clay type, while the silty sand had a negligible clay fraction (2.6 %) where illite (2 : 1 clay type) occurred with 28 relative % (Table 2).

Since Grim and Bradley (1940), we know that the absorbed/adsorbed water is difficult to remove. Savin and Epstein (1970) as well as Van De Velde and Bowen (2013) have demonstrated that the removal of interlayer and adsorbed water on clay soils can occur when they are heated at 100 to 300 °C under vacuum conditions. After clay minerals lose all their water, their structure changes. Hence, care should be taken in order to remove clay minerals' water, but keep their structure. Otherwise, rewetting experiments as presented here in our intercomparison might not be valid.

Savin and Epstein (1970) also observed that atmospheric vapor exchanged isotopically with interlayer water (almost completely) and Aggarwal et al. (2004) showed that this can occur within hours. This demonstrated that the isotopic composition of clay interlayer and adsorbed water can reflect the isotopic composition of atmospheric water vapor at the storage location. However, once the soil has been heated under vacuum and the interlayer water removed, the remaining water showed no evidence of isotopic exchange. Again, it should be stressed here that for our intercomparison, soil samples were oven-dried twice (before and after shipment) prior to any rewetting and labs were advised to store the dried samples in a desiccation chamber until use. However, oven-drying was performed at an intermediate temperature (105 °C for 48 h) and not under vacuum as per Savin and Epstein (1970), and different indoor laboratory "climatic conditions" at the participating laboratories were observed. Thus, it might be possible that not all of the clay interlayer and adsorbed water was removed or made isotopically non-exchangeable, and that non-equilibrium isotopic fractionation occurring at different temperatures during heating might be responsible for some of the differences we observed. Thus, sample preparation might have played its role, when it comes to discrepancies in the labs' results, especially those at low water contents. At these low water contents, the available water fraction is small, and exchange with interlayer and adsorbed water would be proportionally higher. In hindsight, repeating this work with soils dried under vacuum and at higher temperatures (i.e., 300 °C following Savin and Epstein, 1970) may help to clarify and to isolate the effect of remaining water in clay minerals. However, so far, regular oven-drying of soils is standard practice (Koeniger et al., 2011) for such rewetting experiments in the literature.

We also observed water content effects on the recovered isotope data as per Meißner et al. (2014). Cryogenically extracted isotope data across labs were closer to the added reference water isotopic composition at higher WCs. However, this isotope effect cannot be considered independent from other soil property effects such as clay mineral water interactions or effects caused by cation exchange capacity (CEC). Oerter et al. (2014) demonstrated that isotope effects due to soil type are more common in soils with high cation exchange capacity at low WCs. This can be further exacerbated by the cations present in the soil. Those soils with high ionic potential (e.g., Ca^{2+} and Mg^{2+}) can create large amounts of structured water surrounding them (hydrated radii) compared to the bulk water in the system. From an oxygen isotope perspective, O'Neil and Truesdell (1991) showed that those cations are capable of causing fractionation between bound and bulk soil water. Moreover, soils higher in potassium ions may have a greater effect on hydrogen isotopes, while sodium soils demonstrate non-fractionating effects (Oerter et al., 2014). These cation fractionation effects for montmorillic soils, in particular, can result in a depletion of up to 1.55 ‰ in dry soils and 0.49 ‰ for $\delta^{18}O$ for wet soils. In our study, chemical and salinity effects – which occur due to the fractionation of water molecules into hydration spheres around fully solvated cations compared to the pure water used to make the solutions – can be ignored for the silty sand due to a low CEC of 4.1 cmol(+) kg^{-1}. The high CEC (30.6 cmol(+) kg^{-1}) of the clayey loam soil may have caused some of the detrimental effects seen across laboratories. This is especially the case for low WCs due to ion hydration effects among the cations present (Table 2).

Gaj et al. (2017a) found out that the higher the abundance of Al_2O_3 or Fe_2O_3, commonly found in clay-rich soils, the lower the ability to isotopically recover added water during spiking experiments. Our clayey loam contained 65 % of SiO_2 but still 9 % of Al_2O_3, which might have affected the obtained isotope composition in general but cannot be an explanation for the high variability across labs.

4.3 Differences between OA-ICOS- and IRMS-based measurements

Our OA-ICOS vs. IRMS comparison showed that isotope data were significantly different between the two isotope measurement methods.

Others have found differences in isotope data obtained from laser-based OA-ICOS and CRDS systems (cavity ring-down spectroscopy) in comparison to IRMS isotope data (Martín-Gómez et al., 2015; Wassenaar et al., 2012). In a recently performed test, 235 international laboratories conducting water isotope analyses by OA-ICOS, CRDS, and IRMS were evaluated. Wassenaar et al. (2018) could show that inaccuracy or imprecise performance stemmed mainly from skill- and knowledge-based errors including calculation mistakes, inappropriate or compromised laboratory calibration standards, poorly performing instrumentation, lack of vigilance to contamination, or inattention to unreasonable isotopic outcomes. For the analysis of $\delta^{18}O$ and $\delta^{2}H$ via OA-ICOS, Penna et al. (2012) showed that between-sample memory effects can be an additional problem. Memory effects ranged from 14 and 9 % for $\delta^{18}O$ and $\delta^{2}H$ measurements, respectively, but declined to 0.1 and 0.3 % when the first 10 injections of each sample were discarded.

An additional source of error in our study might be that sample preparation for water extraction was performed separately for OA-ICOS and IRMS analysis, but labs were instructed to follow the exact same procedure. Nevertheless, extractions were performed on independent samples, which might have led to differences in the extracts' isotope composition.

Leen et al. (2012) and West et al. (2010) have observed effects of co-extracted organic compounds leading to sample contamination. This can have a knock-on effect on isotope measurements via OA-ICOS. In our study, we found effects caused by organic contamination producing spectral interferences during OA-ICOS measurements (Fig. 7). This was mainly a problem for the clay soil water extracts, where we found narrow- and broadband absorbers to be responsible for some of the outliers in the data sets. It did not seem to be a major issue for the silty sand soil water extracts. However, some labs applied longer extraction times to the clayey loam samples (see Fig. 2) which might have favored the co-extraction of organics. Martín-Gómez et al. (2015) introduced an online oxidation method for organic compounds for samples measured via isotope ratio infrared spectroscopy. The authors showed that this method was able to effectively remove methanol interference, but was not efficient for high concentrations of ethanol.

During an intercomparison water recovery experiment, Walker et al. (1994) faced difficulties in retrieving the added reference water from dry and wet clays, sand, and gypseous sand. They assumed that decomposition of organic matter or extraction of clay structural water could have caused isotope effects. Recently, Orlowski et al. (2016a) observed that δ^2H values correlated significantly and became progressively lighter with increasing organic carbon content when using CWE. In environmental organic matter, the different existing exchangeable (i.e., labile) hydrogen fractions (O-, N-, and S-bonded or aromatic hydrogen) can easily interact with ambient water or water vapor (Ruppenthal et al., 2010) and thus are assumed to be the cause of the isotope effects.

Nevertheless, the less expensive, rapid option of the OA-ICOS is still a viable alternative for routine isotope analyses if no organic contamination issues are found, six or more injections are performed, and the first two or more are discarded (Penna et al., 2012). If organics are present, proper correction schemes as per Martín-Gómez et al. (2015) need to be applied, especially when OA-ICOS data are used in ecohydrological studies. However, so far, correction procedures only account for contamination caused by methanol or ethanol but plant and soil water extracts can contain a variety of different contaminants. Our work showed that the silty sand soil water extracts were mainly free from organic contamination (Fig. 7). Still, data post-processing is highly recommended to detect issues occurring from co-extracted alcoholic compounds.

4.4 Take-home messages about cryogenic water extraction

Our lab intercomparison did not find significant correlations between (i) extraction condition parameters such as temperature, time, and applied vacuum and (ii) the obtained isotope data (Fig. 2). Others have shown that extraction time and temperature have significant effects on the CWE isotope data (Goebel and Lascano, 2012; Koeniger et al., 2011; Orlowski et al., 2013, 2016a; West et al., 2006). Gaj et al. (2017b) showed clear relationships between temperature and the release of water from interlayer cations and organics during CWE, which affected isotope values. They suggested using temperatures between 200 and 300 °C for clay water extractions. However, higher temperatures could cause a release of water by oxidation of organics and dihydroxylation of hydroxide-containing minerals, and the co-extraction of organics could become more important at harsher extraction conditions leading to spectral interferences when OA-ICOS is used. Orlowski et al. (2018) recently explored the effect of CWE for tracing plant source water. The authors tested the ability to match plant water to its putative soil water source(s) by using different CWE conditions (30–240 min, 80–200 °C, 0.1 Pa) for a clayey loam (same as in this study) and a pure sand type. They showed that with higher extraction temperatures and longer extraction times, gradually more enriched soil water was extracted, which surprisingly reflected the plants' source water.

Our interlaboratory comparison was not able to provide any recommendations with regard to higher temperatures or longer extraction times leading to possibly better extraction results. Little is known about how the applied extraction pressure affects the CWE isotope data. But one thing is clear: that CWE is a "brute force technique" (Orlowski et al., 2016a) in the sense that it is not able to distinguish between waters held at different soil tensions being of different importance for the ecohydrological water cycle. New instrumentation to sample discretely along the moisture release curve is desperately needed (McDonnell, 2014). For most past studies, possible fractionation effects associated with CWE remain unknown and the applied extraction parameters or cryogenic system specifications are often not indicated. Orlowski et al. (2018) recently stated that observed isotopic fractionation effects potentially lead to errors when CWE isotope data are used for plant water source calculation. This miscalculation in plants' water source could be quite large and could lead to misinterpretations of the role different plant species play in hydrologic processes at the ecosystem or larger scales. Millar et al. (2018) used the most common water extraction methods (centrifugation, microwave extraction, direct vapor equilibration, high-pressure mechanical squeezing, and two different CWE systems) for their intercomparison study on spring wheat (*Triticum aestivum* L.). The authors showed that all methods yielded markedly different isotopic signatures. The various methods also produced differing concentrations of co-extracted organic compounds. Again, CWE was outperformed by other extraction methods.

We found significant differences between extraction approach I (lab in-house procedure) and II (predefined extraction parameters). Both approaches showed significant differences compared to the added reference water for the OA-ICOS results, but in different ways. For example, for δ^2H signature recovery from silty sand, extraction approach II worked better. The same was true for $\delta^{18}O$ signature recovery for both WC treatments. However, for other settings, it was difficult to identify the ideal extraction approach that got closer to the reference DI water isotopic composition. We found no clear tendency for which approach should be applied, thus at present, and much to our dismay, we cannot define any standard protocol for CWE. In the light of our experience with other soil water extraction techniques (Orlowski et al., 2016b), we argue that the success of any of these methods may depend more on the specific understanding and operation leading to internal reproducibility of each individual technique's results than an inherent superiority of one technique over another.

We could show with our interlaboratory comparison that a number of factors affect CWE results among which soil properties such as clay mineral composition and concomitant release of interlayer water seemed to be important. It is therefore essential to obtain detailed soil property information to be able to apply post-corrections as per Gaj et al. (2017a). Further research is urgently needed to analyze the full extent of soil organic matter effects (i.e., exchangeable bonded hydrogen; Meißner et al., 2014) in organic-rich soils on the cryogenically extracted isotopic composition.

Future studies should test clay mineral fractionation effects on $\delta^{18}O$ and δ^2H during CWEs individually. We further recommend running individual CWE spiking tests on each natural soil material originating from field studies, also considering spatial variability of soil physicochemical properties over depth. Comparing the isotopic deviation of results from such spiking experiments with results from standardized soils will help to establish system-specific transfer functions. This will require considerable effort. However, it seems to be the only way to have some sort of calibration function for each extraction system and different soil types with their clay mineral composition.

5 Conclusions

This work presents results from a worldwide round robin laboratory intercomparison test of cryogenic extraction systems for soil water isotopic analysis. We tested the null hypothesis that, with identical soils, standards, and isotope analyses, cryogenic extraction across laboratories should yield identical isotopic composition. The 16 participating laboratories used the same two standard soils along with reference water of known isotopic composition for CWEs. With our interlaboratory comparison, we showed that multiple factors influence extracted isotopic signatures. Soil type, water content, and the applied type of isotope analysis (OA-ICOS vs. IRMS) showed major impacts, whereas applied extraction parameters (time, temperature, and vacuum) interestingly did not affect CWE isotope data across laboratories. Laboratory internal quality and water recovery rates showed additional effects.

Although the applied extraction system setups were different (e.g., size of extraction container, heating unit), we could not show a major impact of the system's design on the obtained isotope data, as laboratories were successful for the one soil type and water content but failed for the other. However, internal reproducibility for the replicates of the same soil type at a given WC was given for most of the labs. Nevertheless, different results were obtained for $\delta^{18}O$ and δ^2H.

Our intercomparison work showed that defining any sort of standard extraction procedure for CWEs across laboratories is challenging. Our results question the usefulness of this method as a standard for water extraction since results are not intercomparable across laboratories. A possible option might be that CWE labs establish system-specific calibration functions for each natural soil type, individually, to correct for the given offset to a set of reference soils.

New method intercomparison work on plant material showed that direct vapor equilibration is probably the most suitable extraction technique to be used when investigating plant water sourcing, at least for wheat. However, an inter-laboratory comparison is still lacking and should be addressed for plants in the future to account for possible effects. New continuous, in situ measurements of soil and plant water isotopic composition might overcome isotope fractionation issues we observed with CWE.

Author contributions. NO, LB, and JJ McD designed the experiment. All co-authors conducted the laboratory work. NO analyzed the data statistically with the support of Nathalie Steiner. All authors were involved in the data interpretation. NO prepared the manuscript and all co-authors were asked to review the manuscript.

Competing interests. The authors declare that they have no conflict of interest.

Acknowledgements. This interlaboratory test would not have been possible without the generous cooperation of the researchers and technical staff in our 16 stable water isotope laboratories. We especially thank Kim Janzen, Cody Millar, and Anna Winkler for their laboratory support and Nathalie Steiner for statistical support. The Gibson laboratory from Alberta Innovates Technology Futures is thanked for IRMS analyses. This research was supported by an NSERC Discovery Grant and Accelerator Award to Jeffrey J. McDonnell.

Edited by: Thom Bogaard

References

Aggarwal, P. K., Dillon, M. A., and Tanweer, A.: Isotope fractionation at the soil-atmosphere interface and the ^{18}O budget of atmospheric oxygen, Geophys. Res. Lett., 31, 1–4, https://doi.org/10.1029/2004GL019945, 2004.

Araguás-Araguás, L., Rozanski, K., Gonfiantini, R., and Louvat, D.: Isotope effects accompanying vacuum extraction of soil water for stable isotope analyses, J. Hydrol., 168, 159–171, https://doi.org/10.1016/0022-1694(94)02636-P, 1995.

Benettin, P., Volkmann, T. H. M., von Freyberg, J., Frentress, J., Penna, D., Dawson, T. E., and Kirchner, J. W.: Effects of climatic seasonality on the isotopic composition of evaporating soil waters, Hydrol. Earth Syst. Sci., 22, 2881–2890, https://doi.org/10.5194/hess-22-2881-2018, 2018.

Brand, W. A., Geilmann, H., Crosson, E. R., and Rella, C. W.: Cavity ring-down spectroscopy versus high-temperature conversion isotope ratio mass spectrometry; a case study on δ^2H and $\delta^{18}O$ of pure water samples and alcohol/water mixtures, Rapid Commun. Mass Sp., 23, 1879–1884, https://doi.org/10.1002/rcm.4083, 2009.

Craig, H.: Standard for Reporting Concentrations of Deuterium and Oxygen-18 in Natural Waters, Science, 133, 1833–1834, https://doi.org/10.1126/science.133.3467.1833, 1961.

Gaj, M., Kaufhold, S., Koeniger, P., Beyer, M., Weiler, M., and Himmelsbach, T.: Mineral mediated isotope fractionation of soil water, Rapid Commun. Mass Sp., 31, 269–280, https://doi.org/10.1002/rcm.7787, 2017a.

Gaj, M., Kaufhold, S., and McDonnell, J. J.: Potential limitation of cryogenic vacuum extractions and spiked experiments, Rapid Commun. Mass Sp., https://doi.org/10.1002/rcm.7850, 2017b.

Goebel, T. S. and Lascano, R. J.: System for high throughput water extraction from soil material for stable isotope analysis of water, J. Anal. Sci. Methods Instrum., 2, 203–207, https://doi.org/10.4236/jasmi.2012.24031, 2012.

Gonfiantini, R.: Environmental isotopes in lake studies, in: Handbook of environmental isotope geochemistry: The terrestrial environment, B, edited by: Fritz, P. and Fontes, J. C., 113–168, Elsevier, New York, USA, 1986.

Grim, R. and Bradley, W.: Investigation of the Effect of Heat on the Clay Minerals Illite and Montmorillonite, J. Am. Ceram. Soc., 23, 242–248, https://doi.org/10.1111/j.1151-2916.1940.tb14263.x, 1940.

IAEA (International Atomic Energy Agency): IAEA/GNIP precipitation sampling guide, available at: http://www-naweb.iaea.org/napc/ih/documents/other/gnip_manual_v2.02_en_hq.pdf (last access: 15 June 2014), 2014.

Ingraham, N. L. and Shadel, C.: A comparison of the toluene distillation and vacuum/heat methods for extracting soil water for stable isotopic analysis, J. Hydrol., 140, 371–387, https://doi.org/10.1016/0022-1694(92)90249-U, 1992.

Kaufhold, S. and Dohrmann, R.: Comparison of the traditional Enslin-Neff method and the modified dieng method for measuring water-uptake capacity, Clay. Clay Miner., 56, 68600692, https://doi.org/10.1346/CCMN.2008.0560609, 2008.

Koeniger, P., Marshall, J. D., Link, T., and Mulch, A.: An inexpensive, fast, and reliable method for vacuum extraction of soil and plant water for stable isotope analyses by mass spectrometry, Rapid Commun. Mass Sp., 25, 3041–3048, https://doi.org/10.1002/rcm.5198, 2011.

Leen, J. B., Berman, E. S. F., Liebson, L., and Gupta, M.: Spectral contaminant identifier for off-axis integrated cavity output spectroscopy measurements of liquid water isotopes, Rev. Sci. Instrum., 83, 044305, https://doi.org/10.1063/1.4704843, 2012.

LUFA Speyer: German State Research Institute for Agriculture, Speyer, DE, available at: http://www.lufa-speyer.de/index.php/dienstleistungen/standardboeden/8-dienstleistungen/artikel/57-standard-soils (last access: 15 April 2015), 2015.

Martín-Gómez, P., Barbeta, A., Voltas, J., Peñuelas, J., Dennis, K., Palacio, S., Dawson, T. E., and Ferrio, J. P.: Isotope-ratio infrared spectroscopy: a reliable tool for the investigation of plant-water sources?, New Phytol., 207, 914–927, https://doi.org/10.1111/nph.13376, 2015.

McDonnell, J. J.: The two water worlds hypothesis: ecohydrological separation of water between streams and trees?, Wiley Interdiscip. Rev. Water, 1, 323–329, https://doi.org/10.1002/wat2.1027, 2014.

Meißner, M., Köhler, M., Schwendenmann, L., Hölscher, D., and Dyckmans, J.: Soil water uptake by trees using water stable isotopes (δ^2H and $\delta^{18}O$) – a method test regarding soil moisture, texture and carbonate, Plant Soil, 376, 327–335, https://doi.org/10.1007/s11104-013-1970-z, 2014.

Millar, C., Pratt, D., Schneider, D. J., and McDonnell, J. J.: A comparison of extraction systems for plant water stable isotope analysis, Rapid Commun. Mass Sp., 32, 1031–1044, https://doi.org/10.1002/rcm.8136, 2018.

Morrison, J., Brockwell, T., Merren, T., Fourel, F., and Phillips, A. M.: On-line high-precision stable hydrogen isotopic analyses on nanoliter water samples, Anal. Chem., 73, 3570–3575, 2001.

Nelson, S. T.: A simple, practical methodology for routine VSMOW/SLAP normalization of water samples analyzed by continuous flow methods, Rapid Commun. Mass Sp., 14, 1044–1046, https://doi.org/10.1002/1097-0231(20000630)14:12<1044::AID-RCM987>3.0.CO;2-3, 2000.

Oerter, E., Finstad, K., Schaefer, J., Goldsmith, G. R., Dawson, T., and Amundson, R.: Oxygen isotope fractionation effects in soil water via interaction with cations (Mg, Ca, K, Na) adsorbed to phyllosilicate clay minerals, J. Hydrol., 515, 1–9, https://doi.org/10.1016/j.jhydrol.2014.04.029, 2014.

O'Neil, J. R. and Truesdell, A. H.: Oxygen isotope fractionation studies of solute-water interactions, in: Stable Isotope Geochemistry: A Tribute to Samuel Epstein, Vol. 3, edited by: Taylor Jr., H., O'Neil, J. R., and Kaplan, I. R., 17–25, Geochem. Soc, Washington, D.C., USA, 1991.

Orlowski, N., Frede, H.-G., Brüggemann, N., and Breuer, L.: Validation and application of a cryogenic vacuum extraction system for soil and plant water extraction for isotope analysis, J. Sens. Sens. Syst., 2, 179–193, https://doi.org/10.5194/jsss-2-179-2013, 2013.

Orlowski, N., Breuer, L., and McDonnell, J. J.: Critical issues with cryogenic extraction of soil water for stable isotope analysis, Ecohydrol., 9, 1–5, https://doi.org/10.1002/eco.1722, 2016a.

Orlowski, N., Pratt, D. L., and McDonnell, J. J.: Intercomparison of soil pore water extraction methods for stable isotope analysis, Hydrol. Process., 30, 3434–3449, https://doi.org/10.1002/hyp.10870, 2016b.

Orlowski, N., Winkler, A., McDonnell, J. J., and Breuer, L.: A simple greenhouse experiment to explore the effect of cryogenic water extraction for tracing plant source water, Ecohydrology, e1967, https://doi.org/10.1002/eco.1967, 2018.

Penna, D., Stenni, B., Šanda, M., Wrede, S., Bogaard, T. A., Michelini, M., Fischer, B. M. C., Gobbi, A., Mantese, N., Zuecco, G., Borga, M., Bonazza, M., Sobotková, M., Cejková, B., and Wassenaar, L. I.: Technical Note: Evaluation of between-sample memory effects in the analysis of δ^2H and δ^{18}O of water samples measured by laser spectroscopes, Hydrol. Earth Syst. Sci., 16, 3925–3933, https://doi.org/10.5194/hess-16-3925-2012, 2012.

Poppe, L. J., Paskevich, V. F., Hathaway, J. C., and Blackwood, D. S.: USGS, U.S. Geological Survey, Coastal and Marine Geology Program, Open-File Report 01-041: A Laboratory Manual for X-Ray Powder Diffraction, available at: http://pubs.usgs.gov/of/2001/of01-041/htmldocs/intro.htm (last access: 12 October 2016), 2016.

R Core Team: R: A language and environment for statistical computing, R Foundation for Statistical Computing, Vienna, Austria, 2014.

Rozanski, K., Araguás-Araguás, L., and Gonfiantini, R.: Isotopic Patterns in Modern Global Precipitation, in: Climate Change in Continental Isotopic Records, edited by: Swart, P. K., Lohmann, K. C., Mckenzie, J., and Savin, S., 1–36, American Geophysical Union, Washington, D.C., USA, 1993.

Ruppenthal, M., Oelmann, Y., and Wilcke, W.: Isotope ratios of nonexchangeable hydrogen in soils from different climate zones, Geoderma, 155, 3–4, https://doi.org/10.1016/j.geoderma.2009.12.005, 2015.

Savin, S. M. and Epstein, S.: The oxygen and hydrogen isotope geochemistry of clay minerals, Geochim. Cosmochim. Ac., 34, 25–42, https://doi.org/10.1016/0016-7037(70)90149-3, 1970.

Schultz, N. M., Griffis, T. J., Lee, X., and Baker, J. M.: Identification and correction of spectral contamination in ^{2}H/^{1}H and ^{18}O/^{16}O measured in leaf, stem, and soil water, Rapid Commun. Mass Sp., 25, 3360–3368, https://doi.org/10.1002/rcm.5236, 2011.

Schuttlefield, J., Cox, D., and Grassian, V.: An investigation of water uptake on clays minerals using ATR-FTIR spectroscopy coupled with quartz crystal microbalance measurements, J. Geophys. Res.-Atmos., 112, 1–14, https://doi.org/10.1029/2007JD008973, 2007.

Sprenger, M., Herbstritt, B., and Weiler, M.: Established methods and new opportunities for pore water stable isotope analysis, Hydrol. Process., 29, 5174–5192, https://doi.org/10.1002/hyp.10643, 2015.

Van De Velde, J. H. and Bowen, G. J.: Effects of chemical pretreatments on the hydrogen isotope composition of 2 : 1 clay minerals, Rapid Commun. Mass Sp., 27, 1143–1148, https://doi.org/10.1002/rcm.6554, 2013.

Walker, G. R., Woods, P. H., and Allison, G. B.: Interlaboratory comparison of methods to determine the stable isotope composition of soil water, Chem. Geol., 111, 297–306, https://doi.org/10.1016/0009-2541(94)90096-5, 1994.

Wassenaar, L., Ahmad, M., Aggarwal, P., van Duren, M., Pöltenstein, L., Araguas, L., and Kurttas, T.: Worldwide proficiency test for routine analysis of δ^2H and δ^{18}O in water by isotope-ratio mass spectrometry and laser absorption spectroscopy, Rapid Commun. Mass Sp., 26, 1641–1648, https://doi.org/10.1002/rcm.6270, 2012.

Wassenaar, L. I., Terzer-Wassmuth, S., Douence, C., Araguas-Araguas, L., Aggarwal, P. K., and Coplen, T. B.: Seeking excellence: An evaluation of 235 international laboratories conducting water isotope analyses by isotope-ratio and laser-absorption spectrometry, Rapid Commun. Mass Sp., 32, 393–406, https://doi.org/10.1002/rcm.8052, 2018.

West, A. G., Patrickson, S. J., and Ehleringer, J. R.: Water extraction times for plant and soil materials used in stable isotope analysis, Rapid Commun. Mass Sp., 20, 1317–1321, https://doi.org/10.1002/rcm.2456, 2006.

West, A. G., Goldsmith, G. R., Brooks, P. D., and Dawson, T. E.: Discrepancies between isotope ratio infrared spectroscopy and isotope ratio mass spectrometry for the stable isotope analysis of plant and soil waters, Rapid Commun. Mass Sp., 24, 1948–1954, https://doi.org/10.1002/rcm.4597, 2010.

White, W. A. and Pichler, E.: Water Sorption Properties of Clay Minerals (No. 208), available at: https://www.ideals.illinois.edu/bitstream/handle/2142/44988/watersorptioncha266whit.pdf?sequence=2 (last access: 10 October 2015), 1959.

Zieffler, A. S., Harring, J. R., and Long, J. D.: 12. Unplanned Contrasts, in Comparing Groups: Randomization and Bootstrap Methods Using R, 255–286, John Wiley & Sons, New York, NY, USA, 2012.

5

Evaluation of lacustrine groundwater discharge, hydrologic partitioning, and nutrient budgets in a proglacial lake in the Qinghai–Tibet Plateau: using ^{222}Rn and stable isotopes

Xin Luo[1,2,3], Xingxing Kuang[4], Jiu Jimmy Jiao[1,2,3], Sihai Liang[5], Rong Mao[1,2,4], Xiaolang Zhang[1,2,4], and Hailong Li[4]

[1]Department of Earth Sciences, The University of Hong Kong, Hong Kong, P. R. China
[2]The University of Hong Kong, Shenzhen Research Institute (SRI), Shenzhen, P. R. China
[3]The University of Hong Kong-Zhejiang Institute of Research and Innovation (HKU-ZIRI), Hangzhou, PR China
[4]School of Environmental Science and Engineering, Southern University of Science and Technology, 1088 Xueyuan Rd., Shenzhen, China
[5]School of Water Resources & Environment, China University of Geosciences, 29 Xueyuan Road, Beijing, China

Correspondence: Jiu Jimmy Jiao (jjiao@hku.hk)

Abstract. Proglacial lakes are good natural laboratories to investigate groundwater and glacier dynamics under current climate conditions and to explore biogeochemical cycling under pristine lake status. This study conducted a series of investigations of ^{222}Rn, stable isotopes, nutrients, and other hydrogeochemical parameters in Ximen Co Lake, a remote proglacial lake in the east of the Qinghai–Tibet Plateau (QTP). A radon mass balance model was used to quantify the lacustrine groundwater discharge (LGD) of the lake, leading to an LGD estimate of $10.3 \pm 8.2\,\mathrm{mm\,d^{-1}}$. Based on the three-endmember models of stable ^{18}O and Cl^-, the hydrologic partitioning of the lake is obtained, which shows that groundwater discharge only accounts for 7.0 % of the total water input. The groundwater-derived DIN and DIP loadings constitute 42.9 % and 5.5 % of the total nutrient loading to the lakes, indicating the significance of LGD in delivering disproportionate DIN into the lake. This study presents the first attempt to evaluate the LGD and hydrologic partitioning in the glacial lake by coupling radioactive and stable isotopic approaches and the findings advance the understanding of nutrient budgets in the proglacial lakes of the QTP. The study is also instructional in revealing the hydrogeochemical processes in proglacial lakes elsewhere.

1 Introduction

High-altitude and high-latitude areas are intensively influenced by the melting of glaciers due to climatic warming. Of particular importance are the proglacial areas, such as proglacial lakes and moraines, because they are particularly affected by climatic-change-induced glacier retreating and thawing of permafrost (Heckmann et al., 2016; Barry, 2006; Slaymaker, 2011). The proglacial lakes are usually located close to the ice front of a glacier, ice cap, or ice sheet, with the vicinity to the ice front sometimes defined as the areas with subrecent moraines and formed by the last significant glacier advances at the end of the Little Ice Age (Heckmann et al., 2016; Barry, 2006; Slaymaker, 2011; Harris et al., 2009). These lakes are located in the transition zones from glacial to non-glacial conditions, and can serve as natural laboratories to explore hydrological processes, biogeochemical cycles, and geomorphic dynamics under current climatic conditions (Dimova et al., 2015; Heckmann et al., 2016).

The Qinghai–Tibet Plateau (QTP), the third pole of the world, serves as the water tower of most of the major rivers in Asia (Qiu, 2008). Unique landscapes such as endorheic lakes, permafrost, glaciers, and headwater fluvial networks are developed due to the intensive interaction between the atmosphere, hydrosphere, biosphere, and cryosphere (Lei et al., 2017; Zhang et al., 2017a, b; Yao et al., 2012, 2013). Distributed mountainous glaciers and lakes are the most representative landscapes and are highly sensitive to the climate changes. In the past decade, the lakes in the interior of the QTP show overall expansion with respect to an overall increase in precipitation, accelerated glacier melting, and permafrost degradation (Zhang et al., 2013, 2017a, b; Heckmann et al., 2016; Yang et al., 2014). Some recent studies have made efforts to depict the hydrologic partitioning of the majority of the lakes in the QTP based on long-term observation of climatological parameters and remote sensing approaches. However, so far a quantitative evaluation of the water balance and hydrologic partitioning, especially the groundwater component of the lakes in the QTP, has been limited due to the scarcity of observational data. Therefore, there is a great need to conduct refined and systematical field observation to provide a groundtruth dataset and tenable models to depict the water balance and hydrologic partitioning of the lakes, especially proglacial lakes in the QTP (Yang et al., 2014; Zhang et al., 2017b).

Mountainous proglacial lakes, formed by glacial erosion and filled by melting glaciers, are widely distributed in the Qinghai–Tibet Plateau (QTP), especially along the substantial glacier retreating areas of the Himalaya Mountains (MT.), Qilian MT., Tienshan MT., etc. Characterized by higher elevations, small surface areas but relatively large depths, mountainous proglacial lakes in QTP lack systematic field-based hydrological studies due to their remote locations and difficulty in conducting fieldwork (Yao et al., 2012; Farinotti et al., 2015; Bolch et al., 2012).

There has been extensive recognition of the importance of groundwater discharge to various aquatic systems for decades (Dimova and Burnett, 2011; Valiela et al., 1978; Johannes, 1980). Very recently, the topic of lacustrine groundwater discharge (LGD), which is comprehensively defined as groundwater exfiltration from lakeshore aquifers to lakes (Lewandowski et al., 2013, 2015; Rosenberry et al., 2015; Blume et al., 2013), has been introduced. LGD is analogous to submarine groundwater discharge (SGD) in coastal environments. LGD also plays a vital role in lake hydrologic partitioning, which is defined as the separation of groundwater discharge/exfiltration, riverine inflow, riverine outflow infiltration, surface evaporation, and precipitation for the hydrological cycle of the lake (Luo et al., 2017; Good et al., 2015). LGD also serves as an important component in delivering solutes to lakes since groundwater is usually concentrated in nutrients, CH_4, dissolved inorganic/organic carbon (DIC/DOC), and other geochemical components (Paytan et al., 2015; Lecher et al., 2015; Belanger et al., 1985; Dimova et al., 2015). Nutrients and carbon loading from groundwater greatly influence ratios of dissolved inorganic nitrogen (DIN) to dissolved inorganic phosphate (DIP) (referred to as $N:P$ ratios thereafter), ecosystem structure, and the primary productivity of the lake aquatic system (Nakayama and Watanabe, 2008; Belanger et al., 1985; Hagerthey and Kerfoot, 1998).

The approaches to investigate LGD include (1) direct seepage meters (Shaw and Prepas, 1990; Lee, 1977), (2) geotracers such as radionuclides and stable ^2H and ^{18}O isotopes (Gat, 1995; Kluge et al., 2007; Kraemer, 2005; Lazar et al., 2008), (3) heat and temperature signatures (Liu et al., 2015; Sebok et al., 2013), (4) numerical modelling (Winter, 1999; Smerdon et al., 2007; Zlotnik et al., 2009, 2010), and (5) remote sensing (Lewandowski et al., 2013; Wilson and Rocha, 2016; Anderson et al., 2013). Recently, some researchers started to investigate groundwater dynamics in periglacial and proglacial areas, mostly based on the approaches of numerical modelling (Lemieux et al., 2008a, b, c; Andermann et al., 2012; Scheidegger and Bense, 2014). However, the quantification of groundwater and surface water exchange in proglacial lakes is still challenging due to limited hydrogeological data and extreme seasonal variability of aquifer permeability (Dimova et al., 2015; Callegary et al., 2013; Xin et al., 2013).

^{222}Rn, a naturally occurring inert gas nuclide highly concentrated in groundwater, can be more applicable in fresh aquatic systems and has been widely used as a tracer to quantify groundwater discharge in freshwater lakes (Luo et al., 2016; Corbett et al., 1997; Dimova et al., 2013, 2015; Dimova and Burnett, 2011; Kluge et al., 2007, 2012; Schmidt et al., 2010) and terrestrial rivers and streams (Burnett et al., 2010; Cook et al., 2003, 2006; Batlle-Aguilar et al., 2014). Of particular interest are investigations of temporal ^{222}Rn distribution in lakes, since it can be used to quantify groundwater discharge and reflect the locally climatological dynamics (Dimova and Burnett, 2011; Luo et al., 2016). Temporal radon variations give high-resolution estimates of groundwater discharge to lakes over diel cycles, allowing evaluation of LGD and the associated chemical loadings. However, there has been no study of radon-based groundwater discharge in mountainous proglacial lakes, especially for those lakes in the QTP.

This study aims to investigate the groundwater–surface water interactions for the proglacial lake of Ximen Co, by estimating the LGD and evaluating the hydrologic partitioning of the lake. LGD is estimated with the ^{222}Rn mass balance model, and the hydrologic partitioning of the lake is obtained with the three-endmember model coupling the mass balance of water, stable isotopes, and Cl^-. Moreover, LGD-derived nutrients are estimated and the nutrient budgets of the lake are depicted. This study, to our knowledge, makes the first attempt to quantify the LGD, hydrologic partition, and groundwater-borne nutrients of the proglacial lake in the QTP and elsewhere via the approach of integrating multiple tracers. This study provides insights into hydrologic partitioning in a typical mountainous proglacial lake under current climate conditions and reveals groundwater-borne chemical loadings in this proglacial lake in the QTP and elsewhere.

2 Methodology

2.1 Site descriptions

The Nianbaoyeze MT., located at the eastern margin of the QTP and being the easternmost part of NW–SW trending Bayan Har Shan, is situated at the main water divide of the upper reaches of the Yellow River and Yangtze River (Fig. 1). With a peak elevation of 5369 m, the mountain rises about 500–800 m above the surrounding peneplain and displays typical Pleistocene glacial landscapes such as moraines, U-shaped valleys, and cirques (Lehmkuhl, 1998; Schlutz and Lehmkuhl, 2009; Wischnewski et al., 2014). The present snow line is estimated to be at an elevation of 5100 m (Schlutz and Lehmkuhl, 2009). Controlled by the South Asia and East Asia monsoons, the mountain has an annual precipitation of 975 mm in the southern part and 582 mm in the northwestern part, with 80 % occurring during May and October (Yuan et al., 2014; Zhang and Mischke, 2009). The average temperature gradient is about 0.55 °C per 100 m, and the closest weather station, located in Jiuzhi (N: 33.424614°, E: 101.485998°) at the lower plains of the mountain, recorded a mean annual temperature of 0.1 °C. Snowfalls occur in nearly 10 months of the entire year and there is no free frost all year around (Böhner, 1996, 2006; Schlutz and Lehmkuhl, 2009). The precipitation, daily bin-averaged wind speed, and temperature in August 2015 were recorded to be 90 mm, 0.7 $m\,s^{-1}$, and 9.5 °C from Jiuzhi weather station (Fig. 2). The water surface evaporation was recorded to be 1429.8 mm in 2015 from Jiuzhi weather station.

Among the numerous proglacial lakes developed in the U-shaped valleys of the Nianbaoyeze MT., Ximen Co Lake is located at the northern margin of the mountain with an elevation of 4030 m a.s.l., and is well studied and easily accessible (Lehmkuhl, 1998; Zhang and Mischke, 2009; Schlutz and Lehmkuhl, 2009; Yuan et al., 2014). The lake was formed in a deep, glacially eroded basin with a catchment area of 50 km^2, and has a mean and a maximum depth of 40 and 63.2 m, and a surface area of 3.6 km^2. The vegetation around the lake is dominated by pine meadows with dwarf shrubs, rosette plants, and alpine cushion (Schlutz and Lehmkuhl, 2009; Zhang and Mischke, 2009; Yuan et al., 2014). Mostly recharged by the glacial and snowpack melting water and regional precipitation, the lake is stratified with an epilimnion depth of about 4.4 m in the summertime. The lake is usually covered by ice in the wintertime (Zhang and Mischke, 2009). The superficial layer within the U-shaped valley is characterized by peat, clay, and fluvial gravels with a depth of about 1–3.5 m. Discontinuous and isolated permafrost is present at the slope of the valley above an elevation of about 4150 m. The maximum frozen depth is about 1.5 m for the seasonal frozen ground around the lake. The seasonal frozen ground serves as an unconfined aquifer during the unfrozen months from July to October, and groundwater discharges into the epilimnion of the lake (Wang, 1997; Schlutz and Lehmkuhl, 2009; Zhang and Mischke, 2009).

2.2 Sampling and field analysis

The field campaign to Ximen Co Lake was conducted in August 2015, when it is warm enough to take the water samples of different origins as the studied site is seasonally frozen. A ^{222}Rn continuous monitoring station was set up in the southeast part of the lake, where is fairly flat for setting up our tent and monitoring system. Surface water samples were collected around the lake and rivers at the upstream and downstream. Porewater samples were collected at one side of the lake as the other side is steep and rocky. The basic water quality parameters of conductivity (EC), dissolved oxygen (DO), TDS, ORP, and pH in the water were recorded with the multi-parameter meter (HANNA, Co.). Relative humidity was recorded with a portable thermo-hydrometer (KTH-2, Co.). Lake water samples were taken with a peristaltic pump into 2.5 L glass bottles for ^{222}Rn measurement with the Big Bottle system (Durridge, Co.). Surface water samples were filtered with 0.45 µm filters (Advantec, Co.) in situ and taken into 5, 15, 15, and 50 ml Nalgene centrifugation tubes for stable isotope, major anion, cation, and nutrient analysis. Porewater samples were taken from the lakeshore aquifers with a push point sampler (M.H.E, Co.) connected to a peristaltic pump (Solinst, Co.) (Luo et al., 2014, 2016); 100 ml raw surface water or porewater was titrated with a 0.1 µM H_2SO_4 cartridge (Hach, Co.) in situ to measure total alkalinity (Hasler et al., 2016; White et al., 2016; Warner et al., 2013). Porewater was filtered with 0.45 µm syringe filters in situ and taken into 5, 15, 15, and 50 ml Nalgene centrifugation tubes for stable isotope, major anion, cation, and nutrient analysis; 250 ml porewater was taken for ^{222}Rn measurement with RAD7 H_2O (Durridge, Co.). Samples for major cation analysis were acidified with distilled HNO_3 immediately after the sampling.

A ^{222}Rn continuous monitoring station was set up in the northwest of the lake, close to the downstream of the lake (Fig. 1b). Lake water (about 0.5 m in depth) was pumped with a DC pump (12 V) driven by lithium batteries (100 Ah) and sprinkled into the chamber of RAD7 AQUA with a flow rate $>2\,L\,min^{-1}$, where ^{222}Rn in water vapour was equilibrated with the air ^{222}Rn. The vapour in the chamber was delivered into two large dry units (Drierite, Co) to remove the moisture and circulated into a RAD7 monitor, where ^{222}Rn activities were recorded every 5 min. A temperature probe (HOBO$^{@}$) was inserted into the chamber to record the temperature of the water vapour. The monitoring was performed from 11:31, 22 August to 06:30, 24 August 2015. During the period of 13:50–16:30 on 22 August, a sudden blizzard occurred, leading to an hourly precipitation of about 0.6 mm onto the lake area. Daily and hourly climatological data such as wind speed, air temperature, and precipitation were retrieved from the nearest weather station in Jiuzhi (N:

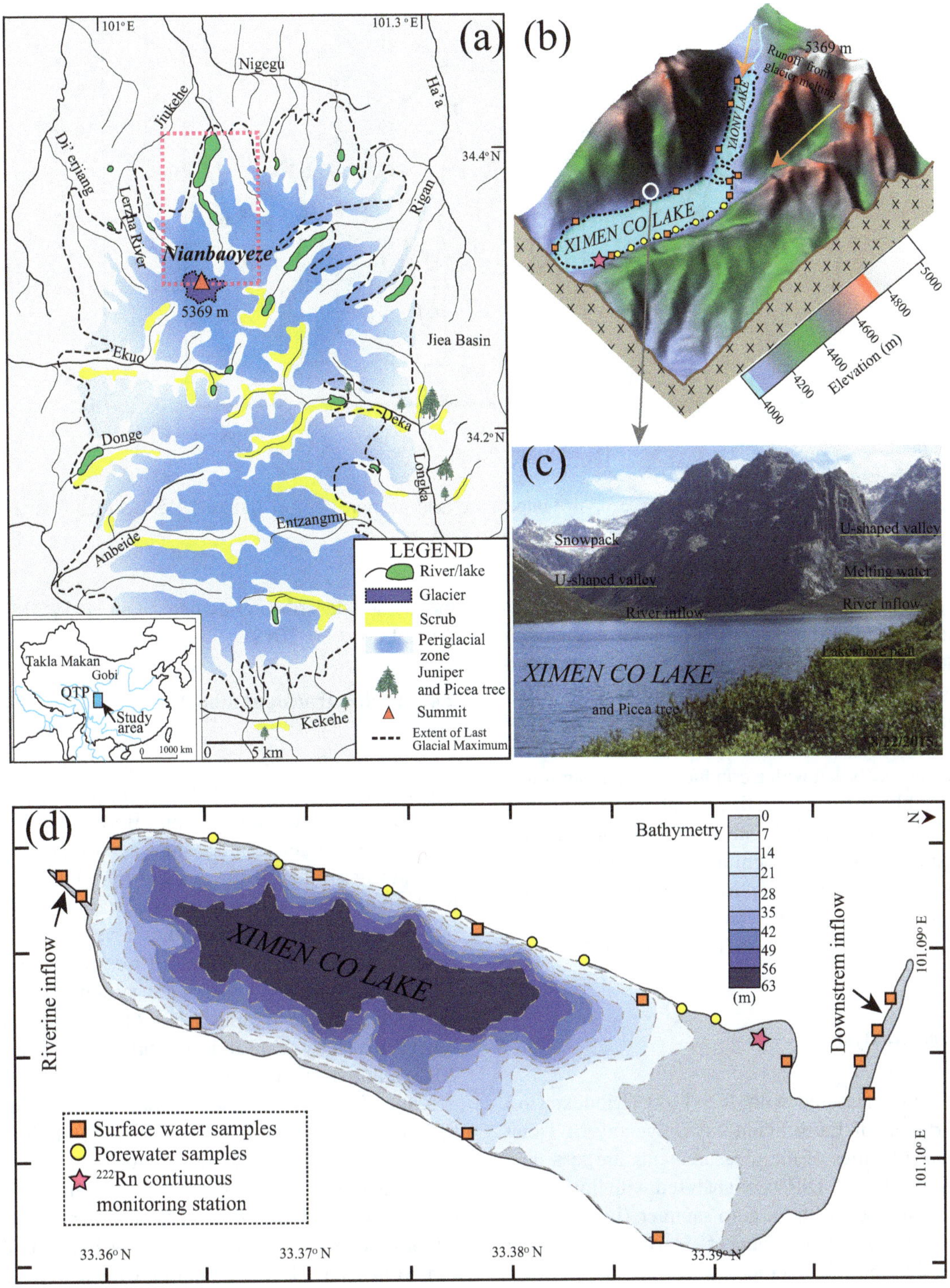

Figure 1. The geological and topographic map of the Yellow River source region, Nianbaoyeze glacial mountains **(a)**, and sampling settings of Ximen Co Lake **(b)**, with the bathymetry map of the lake **(d)**. **(c)** Photograph of Ximen Co Lake and the surrounding geomorphic settings looking in a northeasterly direction on 22 August 2015, showing the late-laying snowpack in the U-shaped valleys of the northern part of the Nianbaoyeze mountains.

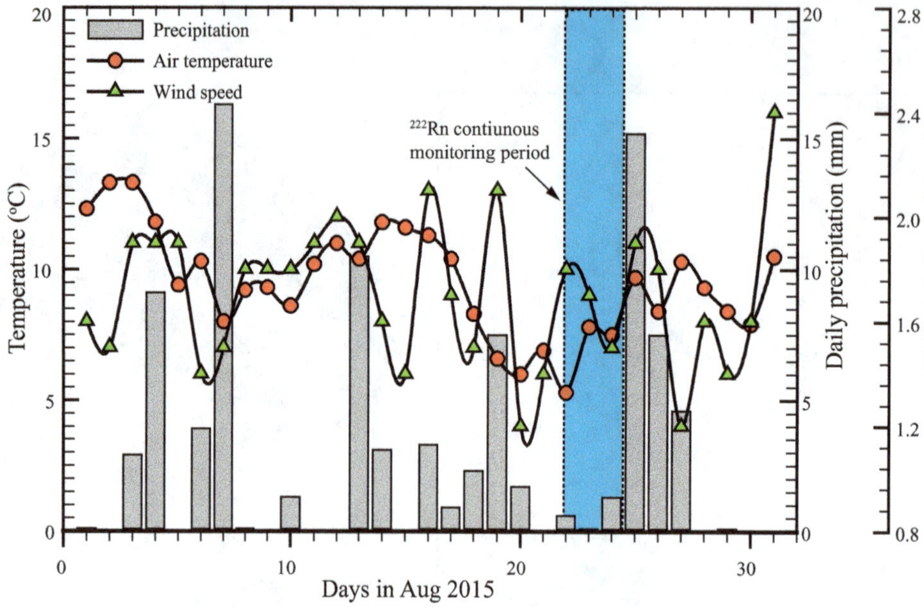

Figure 2. The climatological parameters (wind speed, air temperature, and precipitation) in August 2015 recorded from Jiuzhi weather station.

33.424614°, E: 101.485998°). Moreover, another RAD7 was placed at the lake shore to measure ^{222}Rn in the ambient air around the lake. Due to extremely low activities, the monitoring period was conducted only for 4 h, and the mean activity was adopted as the background radon-222 activity to be used in the mass balance model. Water level and temperature fluctuations were recorded with a conductivity–temperature–depth diver (Schlumberger, Co.) fixed at about 20 cm below the lake surface and calibrated with local atmospheric pressure recorded by a baro-diver (Schlumberger, Co.) above the lake. To correct for dissolved ^{226}Ra supported ^{222}Rn, one radium sample was extracted from 100 L lake water with MnO_2 fibre as described elsewhere (Luo et al., 2014; Moore, 1976).

2.3 Chemical analysis

Major ions were measured with ICS-1100 (Dionex. Co.) in the Department of Earth Sciences, University of Hong Kong. The uncertainties of the measurements are less than 5 %. Nutrients, DIN, and DIP were analysed with flow injection analysis equipped with an auto-sampler (Lachat. Co.) in the School of Biological Sciences, University of Hong Kong. Stable ^{18}O and ^{2}H isotopes were measured with a MOA-ICOS laser absorption spectrometer (Los Gatos Research (LGR) Triple Isotope Water Analyzer (TIWA-45EP)) at the State Key Laboratory of Marine Geology, Tongji University, Shanghai. The stable isotopic standards and the recovery test have been fully described elsewhere (Luo et al., 2017). The measurement uncertainty is better than 0.1 % for ^{18}O and 0.5 % for ^{2}H. ^{226}Ra was detected with RAD7 with the method described elsewhere (Kim et al., 2001; Lee et al., 2012; Luo et al., 2018).

2.4 Radon transient model

Previous studies employed a steady-state radon-222 mass balance model to quantify LGD to lentic systems such as lakes and wetlands (Dimova and Burnett, 2011; Luo et al., 2016). This model assumes that radon inputs derived from groundwater inflow, diffusion, and river inflow are balanced by the radon losses of atmospheric evasion, decay, and river outflow. However, recent studies revealed that the steady state is mainly reached after 2–15 days of constant metrological conditions, and most lentic systems cannot be treated as steady state due to rapid radon-222 degassing to the atmosphere driven by wind-induced turbulence (Gilfedder et al., 2015; Dimova and Burnett, 2011).

Ximen Co Lake has been demonstrated to be highly stratified, with an epilimnion of 4.4 m (Zhang and Mischke, 2009). The lake was formed by glacier erosion and the lake bed is characterized by granite bedrock with a thin sedimentary clay layer. Previous studies have indicated that sediment consisting of clay, soils, and gravels has developed on the bedrock and forms the lakeshore aquifer with a thickness of 0.7–3.3 m (Schlutz and Lehmkuhl 2009). Porewater sampled in the aquifer immediately behind the lake shore can well represent groundwater discharging into the lake, as suggested previously (Lewandowski et al., 2015; Rosenberry et al., 2015; Schafran and Driscoll, 1993). LGD has been widely considered to occur within the first few metres of the lake shore (Schafran and Driscoll, 1993; Rosenberry et al., 2015; Lee et al., 1980) and groundwater is considered to predominately discharge into the epilimnion since deep groundwater flow is highly limited by the Precambrian bedrock (Einarsdottir et al., 2017). Due to a negligible hydrological connection between the epilimnion and hypolimnion, the ^{222}Rn mass balance model is established to quantify LGD to the epilimnion from the lake shore.

The governing equation of the radon-222 transient mass balance model within $1 \times 1 \times z$ cm (where z is the depth in centimetres) can be expressed as (Gilfedder et al., 2015)

$$z\frac{\partial I_{\mathrm{w}}}{\partial t} = F_{\mathrm{gw}} + (I_{^{226}\mathrm{Ra}} - I_{\mathrm{w}}) \times z \times \lambda_{222} + F_{\mathrm{diff}} - F_{\mathrm{atm}}, \quad (1)$$

where F_{gw}, F_{diff}, and F_{atm} (Bq m^{-2} d^{-1}) are ^{222}Rn loadings from LGD, water–sediment diffusion, and water–air evasion, respectively; z (m) is the lake water level depth recorded by the diver. λ_{222} is the decay constant of ^{222}Rn with a value of 0.186 d^{-1}. $\lambda_{222} \times I_{^{226}\mathrm{Ra}}$ and $\lambda_{222} \times I_{\mathrm{w}}$ account for the production and decay of ^{222}Rn (Bq m^{-2} d^{-1}) in the water column, respectively. I_{w} and $I_{^{226}\mathrm{Ra}}$ (Bq m^{-2}) represent ^{222}Rn and ^{226}Ra inventories in the epilimnion, and are expressed as $I_{\mathrm{w}} = H \times C_{\mathrm{w}}$ and $I_{^{226}\mathrm{Ra}} = H \times C_{^{226}\mathrm{Ra}}$, respectively, where H (m) is the depth of the epilimnion; C_{w} and $C_{^{226}\mathrm{Ra}}$ are the ^{222}Rn and ^{226}Ra activity (Bq m^{-3}), respectively.

The model is valid under the following assumptions. (1) The epilimnion is well mixed, which is the actual condition for most natural boreal and high-altitude glacial lakes (Zhang and Mischke, 2009; Åberg et al., 2010). (2) ^{222}Rn input from riverine water inflow, and loss from the lake water outflow and infiltration into the lakeshore aquifer, are negligible compared to the groundwater-borne ^{222}Rn, because ^{222}Rn concentration of groundwater is 2–3 orders of magnitude larger than that of lake water (Dimova and Burnett, 2011; Dimova et al., 2013). Generally, ^{222}Rn in the epilimnion is sourced from LGD and decay input from the parent isotope of ^{226}Ra under secular equilibrium, and is mainly lost via atmospheric evasion and radioactive decay.

F_{atm} is the key sinking component of the transient model and is finally a function of wind speed and water temperature, both of which are temporal variant variables (information in the Supplement). Lake water level z is also a temporal variant variable which represents the fluctuations of water volume of the epilimnion. This equation is discretized by the forward finite difference method, and the groundwater flux at each time step can be solved as follows:

$$[^{222}\mathrm{Rn}_{t+\Delta t}] = \frac{[z \times {}^{222}\mathrm{Rn}_t + [F_{\mathrm{diff}} + F_{\mathrm{gw}} - F_{\mathrm{atm}} - {}^{222}\mathrm{Rn}_t \times \lambda \times z] \times \Delta t}{z}, \quad (2)$$

where $^{222}\mathrm{Rn}_{t+\Delta t}$ and $^{222}\mathrm{Rn}_{t+\Delta t}$ (Bq m^{-3}) are the ^{222}Rn activity at the current time step and at the previous time steps, respectively, and Δt (min) is the time step which is set to be 5 min, consistent with the ^{222}Rn record interval. With the inverse calculation based on Eq. (2), the groundwater inflow at each time step can be obtained. However, large errors of the final LGD calculation will be induced by even a small amount of noise in the measured ^{222}Rn data due to the $^{222}\mathrm{Rn}_{t+\Delta t} - {}^{222}\mathrm{Rn}_t$ term being with the measured uncertainty. To reduce the random errors of the measured ^{222}Rn concentrations, the time window with a width of 1 h is proposed to smooth the curve (information in the Supplement).

3 Results

3.1 Time series data

Figure 2 shows the basic climatological parameters of the lake catchment during the campaign month. There are discrete rainfall events occurring throughout the month with an average rainfall of 3.1 mm d^{-1}. The temperature during the month ranges from 5.0 to 12.5 °C, with an average of 9.3 °C. The daily averaged wind speed ranges from 0.7 to 2.5 m s^{-1}, with an average of 1.7 m s^{-1}. ^{222}Rn temporal distribution and other time series data are shown in Fig. 3a and listed in Table S1 in the Supplement. Generally, ^{222}Rn concentration varies from 32.2 to 273 Bq m^{-3}, with an average of 144.2 ± 27.7 Bq m^{-3}. ^{222}Rn over the monitoring period shows a typical diel cycle, much higher at nighttime and lower in the daytime. Figure 3b–d show the time series data of temperature (5 min interval), nearshore lake water level (1 min interval), and wind speed (1 h interval). Temperature and lake water level also show typical diel cycles, but with antiphase fluctuations with each other. Temperature is higher during the daytime and lower at nighttime. However, a sudden decrease in temperature was recorded due to the sudden blizzard (Fig. 3b). Water level is higher at nighttime and lower during the daytime, with a strong fluctuation due to the turbulence caused by the blizzard (Fig. 3c). The variability might reflect the dynamics of groundwater input and surface water inflow. The air temperature of the lake area is in phase with the water temperature. Wind speed is normally higher during the daytime and lower at nighttime (Fig. 3d).

The variation of ^{222}Rn is nearly in antiphase with the fluctuations of lake water temperature and air temperature, indicating that the dominated controlling factors of ^{222}Rn fluctuations are water temperature and wind speed (Fig. 3a). This phenomenon is reasonable as lake water ^{222}Rn is predominately lost via atmospheric evasion, which is the function of wind speed and water temperature (Dimova et al., 2013, 2015; Dimova and Burnett, 2011). High water temperature and wind speed lead to elevated atmospheric evasion and cause the decline of ^{222}Rn concentration in the lake water. However, there is a sudden reduction of radon activity from 14:00 to 16:00 (GMT + 8) on 22 July 2015, when the snow event led to a sudden decrease in water temperature, increase

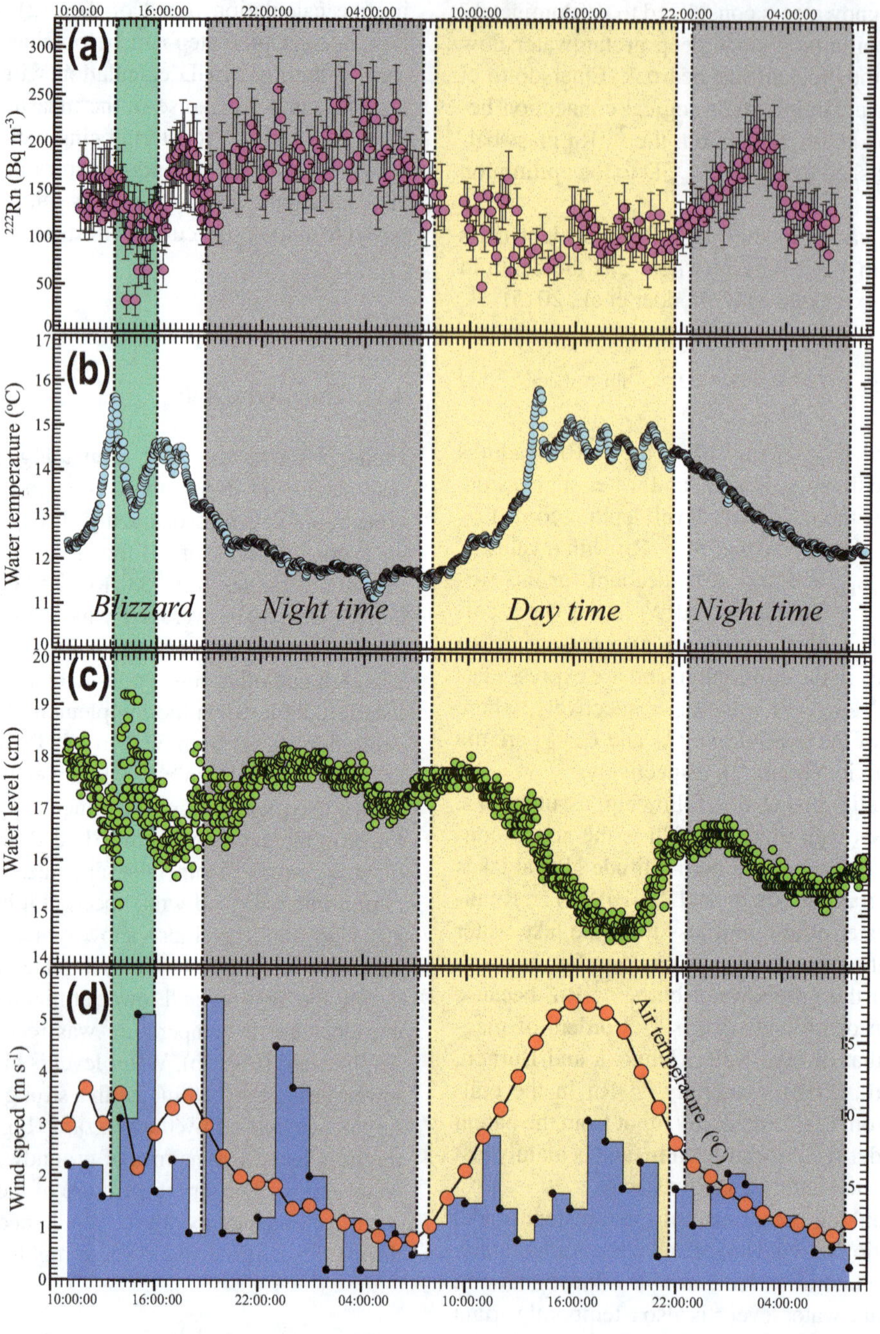

Figure 3. The temporal distributions of ^{222}Rn **(a)**, water temperature **(b)**, water level fluctuation recorded by the divers **(c)**, and hourly wind speed and air temperature recorded in Jiuzhi weather station **(d)**.

in wind speed, and large surface water turbulence as indicated by water level fluctuations (Fig. 3a–d). ^{222}Rn in the porewater is 2–3 orders of magnitude larger than ^{222}Rn in the lake water, suggesting that ^{222}Rn is an ideal tracer to estimate the LGD (Table S1). ^{222}Rn concentrations in surface water range from 22.2 to 209 Bq m^{-3}, with an average of 92.5 Bq m^{-3} ($n = 12$), which is in the range of ^{222}Rn continuous monitoring results, suggesting reliable ^{222}Rn measurements (Table S2).

3.2 Geochemical results

The results of major ions, nutrients, and stable isotopes in different water endmembers are shown in Figs. 4 and 5. Cl^- ranges from 0.6 to 2.1 mg L^{-1} in the surface water (including riverine inflow water, lake water, and downstream water) and 0.4 to 2.7 mg L^{-1} in porewater, and has a much higher concentration of 5.9 mg L^{-1} in rainfall water. Na^+ ranges from 1.6 to 3.4 mg L^{-1} in the surface water and 1.2 to 4.4 mg L^{-1}

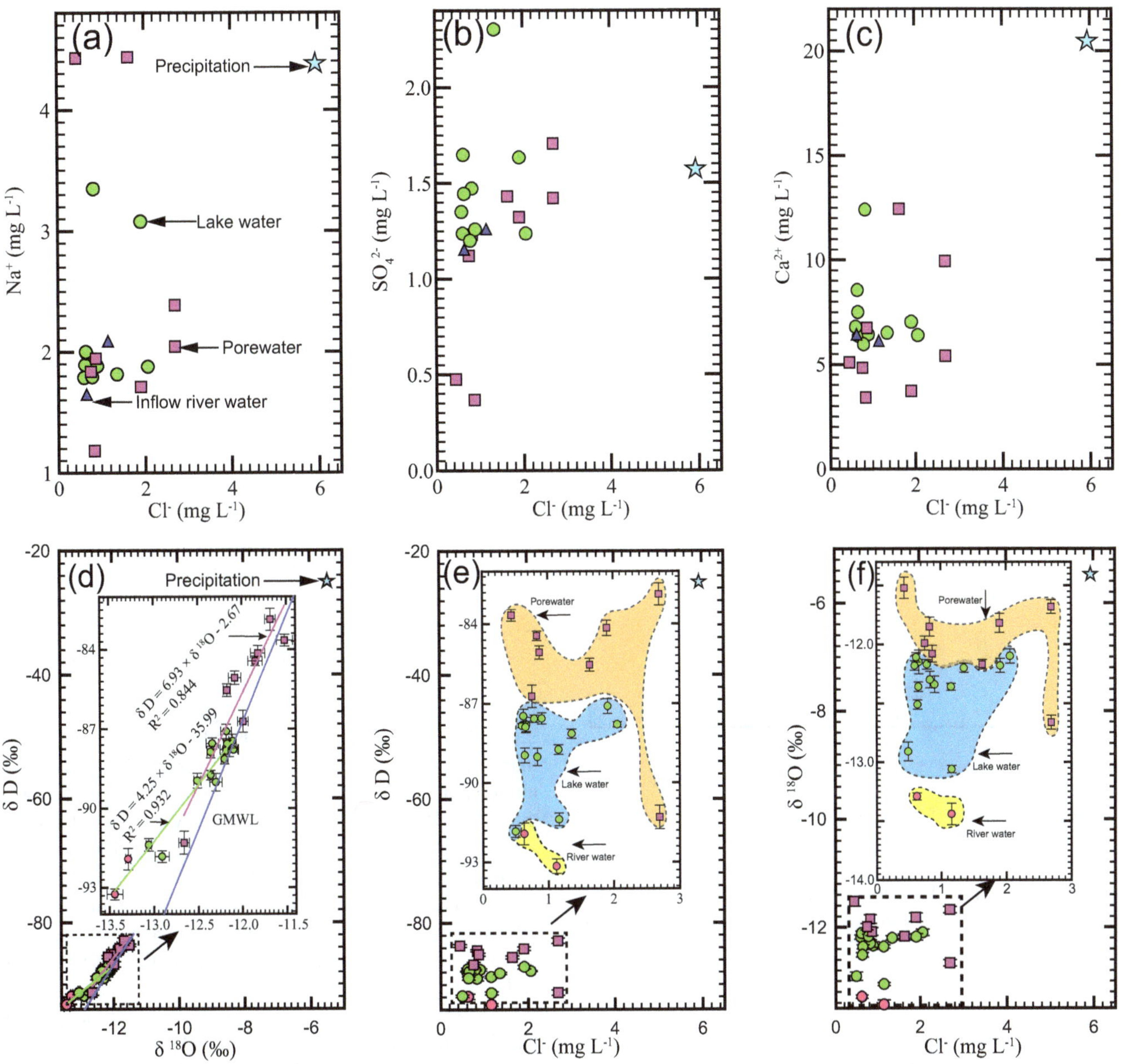

Figure 4. The cross plots of Cl^- versus Na^+ **(a)**, SO_4^{2-} versus Cl^- **(b)**, and Ca^{2+} versus Cl^- **(c)**. The relations of ^2H versus ^{18}O **(d)**, Cl^- versus ^2H **(e)**, and Cl^- versus ^{18}O **(f)**.

in porewater and has a concentration of 4.4 mg L^{-1} in rainfall water. SO_4^{2-} ranges from 1.2 to 2.3 mg L^{-1} in the surface water and 0.4 to 1.7 mg L^{-1} in porewater, and has a significant low concentration of 0.01 mg L^{-1} in rainfall water. Ca^{2+} ranges from 3.0 to 12.4 mg L^{-1} in lake water and 3.4 to 12.5 mg L^{-1} in porewater, and has a significantly high concentration of 20.5 mg L^{-1} in rainfall water. Other concentrations of major ions are listed in Table S2. As shown in Fig. 4d and Table S2, $\delta^{18}O$ in the lake water ranges from -13.06 ‰ to -12.11 ‰, with an average of -12.41 ‰ ($n = 7$), and δ^2H ranges from -91.83 ‰ to -87.47 ‰, with an average of -89.0 ‰ ($n = 7$). $\delta^{18}O$ in the riverine inflow water ranges from -13.44 ‰ to -13.29 ‰, with an average of -13.37 ‰ ($n = 2$), and δ^2H ranges from -93.25 ‰ to -91.92 ‰, with an average of -92.59 ‰ ($n = 2$). $\delta^{18}O$ in the downstream water ranges from -12.51 ‰ to -12.18 ‰, with an average of -12.35 ‰ ($n = 3$), and δ^2H ranges from -88.96 ‰ to -87.1 ‰, with an average of -87.98 ‰ ($n = 3$). $\delta^{18}O$ in the porewater ranges from -12.66 ‰ to -11.52 ‰, with an average of -11.97 ‰ ($n = 8$), and δ^2H ranges from -91.3 ‰ to -82.87 ‰, with an average of -85.5 ‰ ($n = 8$). DIN in the surface water (including riverine inflow water, lake water, and downstream water) ranges from 6.6 to 16.9 µM, with an average of 10.3 µM, and DIP from 0.36 to 0.41 µM, with an

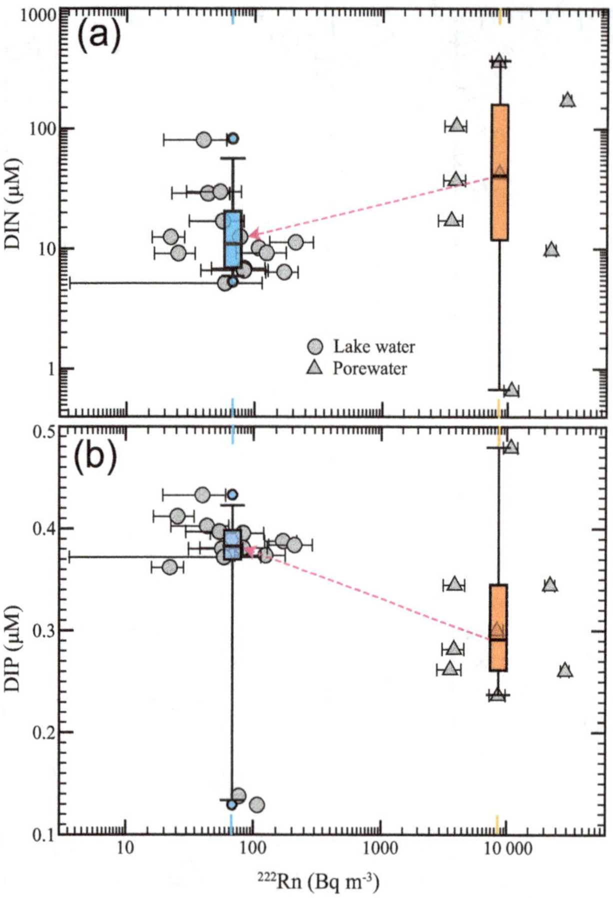

Figure 5. Cross plots of ^{222}Rn versus DIN **(a)** and DIP **(b)**.

average of 0.38 μM. DIN for the porewater ranges from 0.7 to 358.8 μM, with an average of 92.8 μM, and DIP from 0.18 to 0.44 μM, with an average of 0.31 μM (Fig. 5).

4 Discussion

4.1 Proglacial hydrologic processes and geochemical implications

Generally, major ion concentrations in the lake water and porewater of Ximen Co Lake are significantly lower than those in major rivers, streams, and other tectonic lakes in the QTP (Yao et al., 2015; Wang et al., 2010, 2016b), and are similar to those of snow and glaciers (Liu et al., 2011), suggesting that the lake water mainly originated from glacier and snow melting. Ion concentrations in the lake and porewater of Ximen Co Lake are much lower than those of rainfall collected in Jiuzhi. This suggests that lake water is less influenced by precipitation (Fig. 4a–c). The concentrations of major ions in the porewater are high compared to the lake water, indicating weathering effects from the aquifer grains. The ratios of Ca^{2+} / Na^{+} in the porewater and groundwater are >1, also suggesting influences of weathering digenesis of major ions from the seasonal frozen ground at the lakeshore aquifer (Weynell et al., 2016; Yao et al., 2015; Wang et al., 2010).

The isotopic compositions of the lake water and porewater are significantly isotopically depleted, with values close to the compositions of glaciers and surface snow in the QTP, suggesting the lake is dominantly recharged from snow and glacier melting (Cui et al., 2014; Wang et al., 2016a; Zongxing et al., 2015). The relation of $\delta^{18}O$ versus $\delta^{2}H$ for the lake water is $\delta^{2}H = 4.25 \times \delta^{18}O - 35.99$, with a slope much lower than that of the global meteoric water line (GMWL) (Fig. 4d), suggesting the effects of lake surface evaporation. The relation of $\delta^{18}O$ versus $\delta^{2}H$ for the porewater is $\delta^{2}H = 6.93 \times \delta^{18}O - 2.67$, overall on the GWML (Fig. 4d). Deuterium excess is defined as $\Delta D = \delta D - 8 \times \delta^{18}O$ (Dansgaard, 1964). The value of ΔD is dependent on airmass origins, altitude effect, and the kinetic effects during evaporation (Hren et al., 2009). Global meteoric water has a ΔD of $+10$ ‰. In the QTP, glacier/snowpack melting water usually has a large positive ΔD, while the precipitations derived from the warm and humid summer monsoon have a lower ΔD (Ren et al., 2013, 2017). In this study, ΔD of surface water, lake, and porewater ranges from $+8.5$ to $+11.8$ ‰, close to the glacier melting water but much smaller than that of the local precipitation of $+18.8$ ‰. This indicates the stream and lake waters mainly originated from glacial/snowpack melting rather than precipitation (Gat, 1996; Wang et al., 2016a; Lerman et al., 1995). The slopes of $\delta^{2}H$ versus $\delta^{18}O$ in lake water and porewater are 4.25 and 6.93, both of which are lower than that of the GMWL due to surface evaporation. Lake water is more intensively influenced by evaporation compared to porewater. The plots of $\delta^{18}O$ versus Cl^{-}, and $\delta^{2}H$ versus Cl^{-}, are well clustered for porewater endmember (orange area), lake water endmember (blue area), riverine inflow water endmember (yellow area), and precipitation water (Fig. 4e and f), suggesting stable $\delta^{18}O$ and $\delta^{2}H$ isotopes and Cl^{-} can serve as tracers to quantify the hydrologic partitioning of the lake by setting three-endmember models.

The concentrations of DIN and DIP are all within the ranges of other glacial melting water and proglacial lake water (Hawkings et al., 2016; Hodson, 2005, 2007; Hudson et al., 2000; Tockner et al., 2002). Briefly, rainfall and upstream lake water such as YN-4 have the highest DIN concentrations, indicating the glacier melting and precipitation could be important DIN sources in proglacial areas (Dubnick et al., 2017; Anderson et al., 2017). DIN in porewater is overall higher compared to the lake water, suggesting the porewater to be an effective DIN source, and DIP concentration is higher in the lake water compared to porewater, suggesting the porewater is a DIP sink (Fig. 5). The $N : P$ ratios in the lake water and porewater are averaged to be 27.1 and 320.5, respectively, both much larger than the Redfield ratio ($N : P = 16 : 1$) in water and organisms in most aquatic systems, and within the range of other proglacial lakes (Anderson et al., 2017). This also suggests that the lake water and porewater are under phosphate-limited conditions. The $N:P$ ratio in the rainfall water is 30.4, similar to the lake water. The average $N:P$ ratio of porewater is much higher than that of lake water, indicating DIN enrichment in the lakeshore aquifers (Fig. 5). In pristine groundwater, NO_3^- is the predominant form of dissolved nitrogen and is highly mobile within the oxic aquifers, leading to much higher DIN concentrations in the porewater; DIP has a high affinity with the aquifer grains, resulting in much lower DIP concentrations in the porewater (Lewandowski et al., 2015; Rosenberry et al., 2015; Slomp and Van Cappellen, 2004). Thus, analogously to surface runoff from glacier/snowpack melting, LGD can also be regarded as an important DIN source for the proglacial lakes. Because of very high DIN and $N:P$ ratios in the porewater, a relatively small portion of LGD delivers considerable nutrients into the glacial lake, shifting the aquatic $N:P$ ratios and affecting the proglacial aquatic ecosystem (Anderson et al., 2017).

4.2 Estimation of LGD

Figure 6a shows all the sinks and sources of radon with the epilimnion of the lake. Within the ^{222}Rn transient mass balance model, the dominant ^{222}Rn loss is atmospheric degassing/evasion. Generally, the ^{222}Rn degassing rate is the function of the radon-222 concentration gradient at the water–air interface and the parameter of gas piston velocity k, which is finally the function of wind speed and water temperature (Dimova and Burnett, 2011; Gilfedder et al., 2015). To evaluate the ^{222}Rn evasion rate, this study employs the widely used method proposed by MacIntyre et al. (1995) which is also detailed in the Supplement. Based on the field data of ^{222}Rn concentration in the lake water, wind speed, and temperature log, the radon degassing rate is calculated in a range of 0.8 to 265.2 $Bq\,m^2\,d^{-1}$, with an average of 42.0 $Bq\,m^2\,d^{-1}$.

In addition to the atmospheric loss and sedimentary diffusion inputs, ^{222}Rn is also sinked via radioactive decay, and sourced from decay of the parent isotope of ^{226}Ra. The decay loss of ^{222}Rn fluctuates in phase with the distribution of ^{222}Rn concentration monitored by RAD 7 AQUA. The equations to estimate benthic fluxes are shown in the Supplement. The decay loss is calculated to be 26.4 to 223.4 $Bq\,m^{-2}\,d^{-1}$, with an average of 118.0 ± 22.7 $Bq\,m^{-2}\,d^{-1}$. ^{226}Ra concentration is 0.01 $Bq\,m^{-3}$ for the lake water. Under secular equilibrium, the ^{226}Ra decay input can be calculated by multiplying ^{226}Ra concentration in the lake water by λ_{222} (Corbett et al., 1997; Kluge et al., 2007; Luo et al., 2016). ^{226}Ra decay input is calculated to be 0.83 $Bq\,m^{-2}\,d^{-1}$, which is significantly low compared to other ^{222}Rn sources to the epilimnion.

With the obtained sinks and sources of ^{222}Rn in the lake, and the constants given in Table 1, LGD rate can be obtained by dividing the groundwater-derived ^{222}Rn by its concentration in the groundwater endmember. The obtained LGD rate ranges from -23.7 to 90.0 $mm\,d^{-1}$, with an average of 10.3 ± 8.2 $mm\,d^{-1}$ (Fig. 7). The LGD rate range is relatively smaller than the daily lake water level variations (≈ 50 mm), indicating that the lake water level variation could be a combined effect of surface runoff and LGD (Hood et al., 2006). The negative values of LGD rate reflect the return groundwater flow due to infiltration into the porewater. Normally, the dominant values are positive, indicating LGD rate is significant compared to water infiltration into the lakeshore aquifer. The temporal variation of LGD rate could be attributed to the fluctuations of the hydraulic gradient in the proglacial areas (Hood et al., 2006; Levy et al., 2015). As indicated by ΔD (mostly >10) of surface water, the lake and the upstream water are considered to be mainly recharged from glacial/snowpack melting rather than other precipitations.

To assess the magnitude of uncertainty of the ^{222}Rn transient model, the sensitivity of estimated LGD to changes in other variables is examined. A sensitivity coefficient f is proposed to evaluate this uncertainty according to Langston et al. (2013):

$$f = (\Delta F_{LGD}/F_{LGD})/(\Delta y_i/y_i), \quad (3)$$

where ΔF_{LGD} is the amount of change in F_{LGD} from the original value. Δy_i is the amount of change in the other variable of y_i from the original value. Thus, higher f indicates a large uncertainty of the final LGD estimate. The uncertainty mainly stems from ^{222}Rn measurements in different water endmembers, the atmospheric loss, and the water level record. The uncertainties of ^{222}Rn measurement are about 10 % and 15 %–20 % in groundwater and lake water endmembers, respectively. The uncertainty of atmospheric loss is derived from uncertainty of ^{222}Rn in lake water (with an uncertainty of 15 %–20 %), temperature (with an uncertainty of ≈ 5 %), and wind speed (with an uncertainty of ≈ 5 %). Thus, the final LGD estimate has an integrated uncertainty of 35 %–40 %.

4.3 Hydrologic partitioning

Compared to the groundwater labelled radionuclide of ^{222}Rn, stable ^{18}O / ^{2}H isotopes are advantageous in the investigation of evaporation processes due to their fractionations from water to vapour and have been widely used to investigate the hydrologic cycle of lakes in various environments (Stets et al., 2010; Gat, 1995; Gonfiantini, 1986; Gibson et al., 1993). With the field data of stable isotopic composition and Cl^- concentrations in different water endmembers, groundwater input, surface water input, lake water outflow and infiltration, and evaporation can be partitioned by coupling the stable isotopic mass balance model with the Cl^- mass balance model (Fig. 6b).

The model, consisting of the budgets of stable isotopes and Cl^-, and water masses for the epilimnion, is used to quantify riverine inflow, lake water outflow and infiltration, and evaporation (LaBaugh et al., 1995, 1997; Gibson et al., 2016).

Table 1. Parameters used to establish the mass balance model of ^{222}Rn in Ximen Co Lake.

Parameters	Units or values	Estimated uncertainty (%)	Evaluation
Wind speed (ω_{10})	0.2-5.4 $m\,s^{-1}$	n.a.	From Jiuzhi weather stations
Water–air temperature	11.2–15.6 °C	n.a.	Recorded with probe in the chamber; sensitive to temperature results
Molecular diffusion of ^{222}Rn in water (D_m)	9.2×10^{-6}–1.0×10^{-5} $cm^2\,s^{-1}$	n.a.	1.16×10^{-6} at 20 °C; adjustable for temperature
Molecular diffusion of ^{222}Rn in sediments (D_s)	2.2×10^{-6}–2.5×10^{-5} $cm^2\,s^{-1}$	n.a.	Adjusted for temperature, sediment porosity
Dynamic viscosity (μ)	1.1×10^{-3}–1.3×10^{-3} $cm^2\,s^{-1}$	n.a.	Calculated based on water temperature, density, and salinity
Schmidt number (S_c)	1078.6–1371.6 (–)	n.a.	Calculated as the ratio of υ to D_m
Water depth (H)	4.4 m	n.a.	Epilimnion depth of Ximen Co Lake
Decay constant ^{222}Rn (λ_{222})	0.186 d^{-1}	n.a.	Constant
Groundwater endmember ^{222}Rn (C_{gw})	11 200 ± 1200 $Bq\,m^{-3}$	8 site dependent for Ximen Co Lake	Measured; final result for water flux inversely proportional to ^{222}Rn groundwater concentration
Lake water endmember ^{222}Rn (C_l)	21.6–418.8 $Bq\,m^{-3}$	15 %–25 %	Measured with RAD 7 AQUA
Ambient air ^{222}Rn (C_{air})	1.51 ± 0.97 $Bq\,m^{-3}$	15 %–25 %	Measured with RAD 7 under open-loop conditions
Atmospheric ^{222}Rn (C_a)	1.5 ± 1.0 $Bq\,m^{-3}$	20 %–25 %	Measured or assumed value, model not sensitive to radon in air variation
$K_{air/water}$	0.29–0.33 (–)	n.a.	Calculated based on temperature in the chamber and salinity in lake water
Porosity n	0.31	n.a.	Assumed based on the literature
Tortuosity θ	2.05	n.a.	Calculated based on porosity
Piston velocity (κ)	0.004–1.11 $m\,d^{-1}$	20 %–25 %	Calculated from Eq. (S3)
^{226}Ra concentration in lake waters ($C_{^{226}Ra}$)	0.01 $Bq\,m^{-3}$	≈ 10 %	Measured with RAD7
Diffusive flux of ^{222}Rn (F_{diff})	0.68–213.5 $Bq\,m^{-2}\,d^{-1}$	n.a.	Calculated from Eq. (S9)
Atmospheric flux of ^{222}Rn (F_{atm})	0.7–213.5 $Bq\,m^{-2}\,d^{-1}$	n.a.	Calculated from Eq. (S1)
Groundwater flux of ^{222}Rn (F_{gw})	14.7–349.8 $Bq\,m^{-2}\,d^{-1}$	n.a.	Calculated from Eq. (1)
Inventory of ^{222}Rn (I)	$Bq\,m^{-2}$	n.a.	Measured with RAD7 AQUA
Groundwater discharge (Q_{gw})	10.3 ± 8.2 (3.5–38.6) $mm\,d^{-1}$	n.a.	Calculated from Eq. (1)

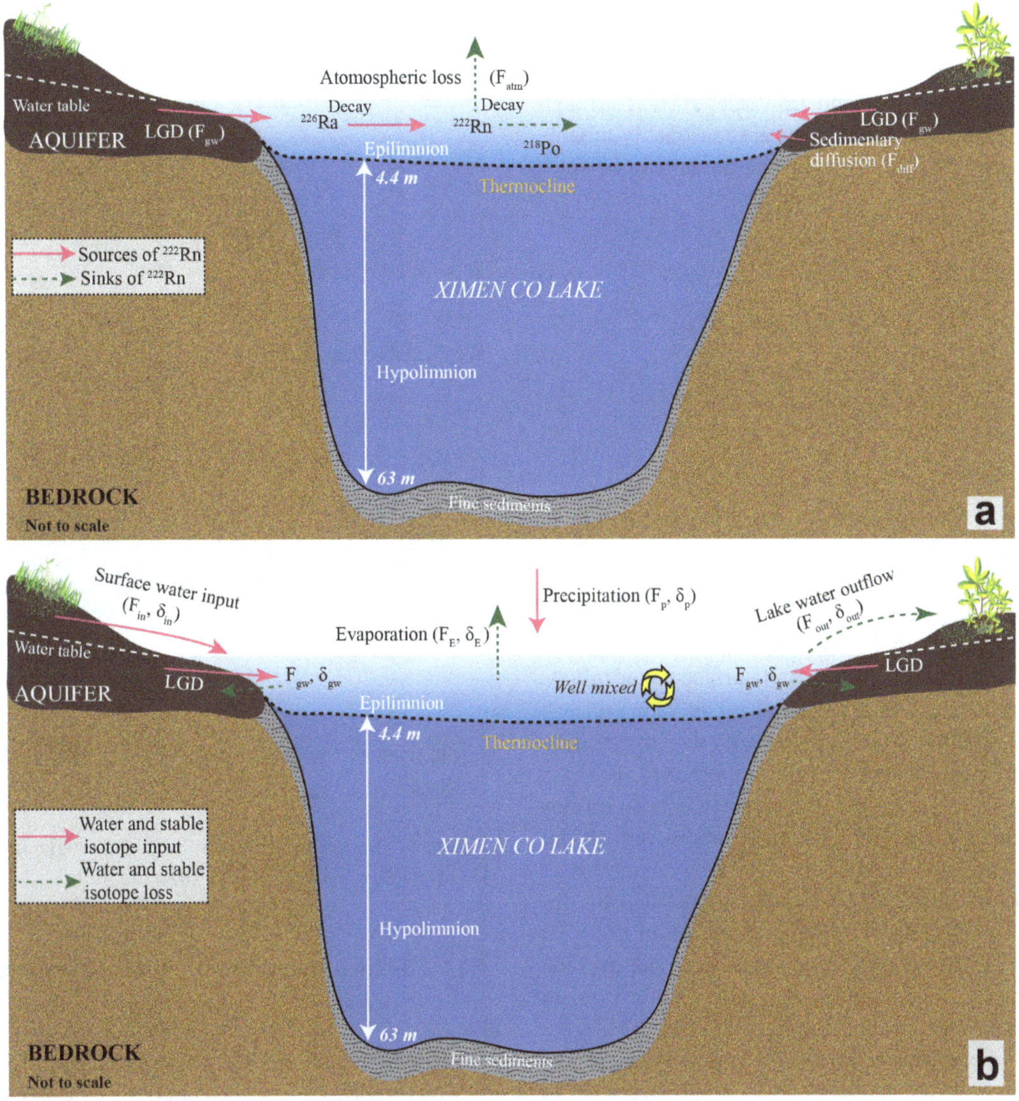

Figure 6. The conceptual model of the ^{222}Rn transient model **(a)** and the three-endmember model **(b)**.

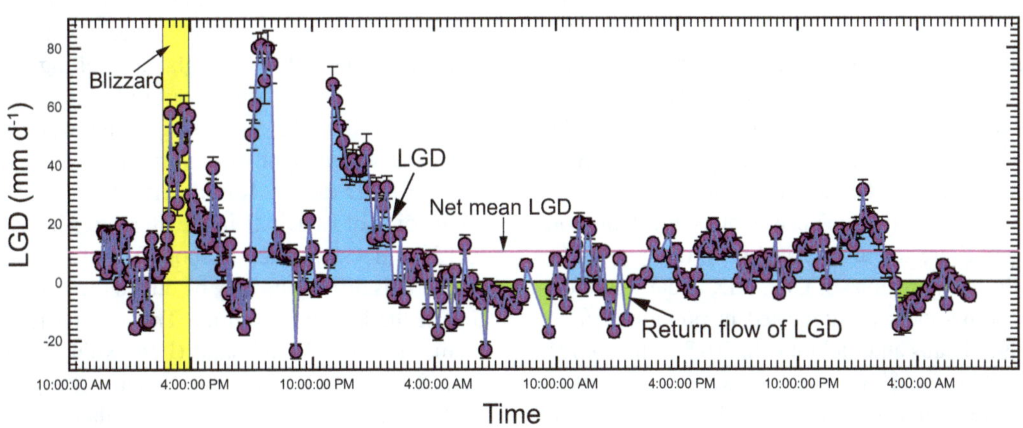

Figure 7. The results of the final LGD derived from the ^{222}Rn transient model.

The model is valid under the following assumptions: (1) constant density of water; (2) no long-term storage change in the reservoir; (3) well mixed for the epilimnion (Gibson, 2002; Gibson et al., 2016; Gibson and Edwards, 2002; LaBaugh et al., 1997). The above assumptions are reasonably tenable during the short monitoring period. The model can be fully expressed as

$$F_{\text{in}} + F_{\text{LGD}} + F_{\text{p}} = F_{\text{E}} + F_{\text{out}}, \tag{4}$$

$$F_{\text{in}} \times \delta_{\text{in}} + F_{\text{LGD}} \times \delta_{\text{gw}} + F_{\text{p}} \times \delta_{\text{p}} = F_{\text{E}} \times \delta_{\text{E}} + F_{\text{out}} \times \delta_{\text{L}}, \tag{5}$$

$$F_{\text{in}} \times [\text{Cl}^-]_{\text{in}} + F_{\text{LGD}} \times [\text{Cl}^-]_{\text{gw}} + F_{\text{p}} \times [\text{Cl}^-]_{\text{p}} = F_{\text{out}} \times [\text{Cl}^-]_{\text{L}}, \tag{6}$$

where F_{in} (mm d^{-1}) is the surface water inflow to the lake; F_{gw} (mm d^{-1}) is the LGD rate. F_{p} (mm d^{-1}) is the mean daily rainfall rate during the sampling period. F_{E} (mm d^{-1}) is the lake evaporation. F_{out} (mm d^{-1}) is the lake water outflow via runoff and infiltration into the lakeshore aquifer. δ_{in}, δ_{gw}, δ_{E}, and δ_{p} are the isotopic compositions of surface water inflow, LGD, and evaporative flux, respectively. The values of δ_{in}, δ_{gw}, and δ_{p} are obtained from field data and the composition of δ_{E} are calculated as shown in the Supplement. $[\text{Cl}^-]_{\text{in}}$, $[\text{Cl}^-]_{\text{gw}}$, $[\text{Cl}^-]_{\text{L}}$, and $[\text{Cl}^-]_{\text{p}}$ are the chloride concentrations in the inflow water, porewater, lake water, and precipitation, respectively.

The components of the mass balance model can be obtained from the field data of isotopic composition and Cl^- concentrations in different water endmembers. The average ^{18}O composition -13.37 ‰ of riverine inflow water is taken as the value of the input parameter δ_{in}. $\delta^{18}\text{O}$ and $\delta^2\text{H}$ in the groundwater endmember and lake water endmember are calculated to be -12.41 ‰ and -87.18 ‰, respectively. $\delta^{18}\text{O}$ and $\delta^2\text{H}$ in the rainfall are measured to be -5.47 ‰ and -24.98 ‰, respectively. With the measured values of δ_{L}, h, and δ_{in}, and the estimated ε and δ_{a}, the isotopic composition of δ_{E} is calculated to be -35.11 ‰, which is in line with the results of alpine and arctic lakes elsewhere (Gibson, 2002; Gibson et al., 2016; Gibson and Edwards, 2002). The values of $[\text{Cl}^-]_{\text{in}}$, $[\text{Cl}^-]_{\text{gw}}$, and $[\text{Cl}^-]_{\text{L}}$ are calculated to be $0.91\,\text{mg L}^{-1}$, $1.48\,\text{mg L}^{-1}$, and $1.02\,\text{mg L}^{-1}$, respectively. All the parameters used in the model are shown in Table 2.

According to Eqs. (4)–(6), the uncertainties of calculations of F_{in}, F_{out}, and E are mainly derived from the uncertainty of F_{LGD} and the compositions of Cl^-, ΔD, and $\delta^{18}\text{O}$ in different water endmembers as suggested in previous studies (Genereux, 1998; Klaus and McDonnell, 2013). The compositions of Cl^-, ΔD, and $\delta^{18}\text{O}$ in surface water–groundwater endmembers have an uncertainty of 5 %. The uncertainty of δ_{E} is reasonably assumed to be ≈ 20 %. Thus, considering the uncertainty propagation of all the above parameters, the uncertainties of F_{in}, F_{out}, and E would be scaled up to 70 %–80 % of the final estimates.

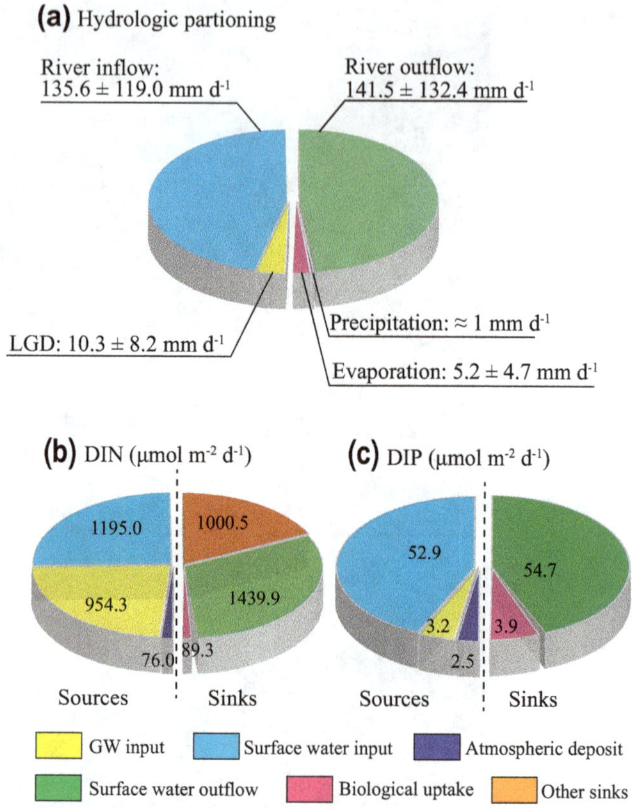

Figure 8. The hydrologic partition of the proglacial lake of Ximen Co **(a)** and the budgets of DIN **(b)** and DIP **(c)**.

4.4 The hydrologic partitioning of the glacial lake

Based on the three-endmember model of ^{18}O and Cl^-, the riverine inflow rate was calculated to be $135.6 \pm 119.0\,\text{mm d}^{-1}$, and the lake outflow rate is estimated to be $141.5 \pm 132.4\,\text{mm d}^{-1}$; the evaporation rate is calculated to be $5.2 \pm 4.7\,\text{mm d}^{-1}$. The summary of the hydrologic partitioning of the lake is shown in Fig. 8a. Generally, the proglacial lake is mostly recharged by the riverine inflow from the snowpack or the glacier melting. The groundwater discharge contributes about only 7.0 % of the total water input to the lake, indicating groundwater input does not dominate water input to the proglacial lake. The recent review of LGD rate by Rosenberry et al. (2015) suggests that the median of LGD rate in the literatures is $7.4\,\text{mm d}^{-1}$ (0.05 to $133\,\text{mm d}^{-1}$), which is about 2/3 of the LGD rate in this study. This difference may be due to the hydrogeological setting of the lakeshore aquifer. This aquifer is formed by grey loam, clayey soil, and sand (Lehmkuhl, 1998; Schlutz and Lehmkuhl, 2009), which has a relatively high permeability.

Previous studies have indicated that groundwater forms a key component of proglacial hydrology (Levy et al., 2015). However, there have been limited quantitative studies of groundwater contribution to the hydrologic budget of proglacial areas. This study further summarizes the ground-

Table 2. Input parameters for the three-endmember model of Ximen Co Lake.

Input parameter	Description	Values (using ^{18}O as a tracer)	Parametric sources
h	Relatively humidity	0.63	Measured by the humidity meter
T (°C)	Water temperature	15.66	Monitored with divers
$\delta_{surface}$ (^{18}O) ‰	Surface water isotopic compositions	−12.45	Average value of surface inflow samples
δ_{gw} (^{18}O) ‰	Groundwater isotopic compositions	−11.97	Average value of porewater samples
δ_L (^{18}O) ‰	Lake water isotopic compositions	−12.54	Average value of Ximen Co Lake water samples
F_{gw} (mm d^{-1})	LGD rates	14.18	Calculated based on the ^{222}Rn mass balance model
ε^* (^{18}O) ‰	Effective equilibrium isotopic enrichment factor	10.12	Eqs. (S13)–(S14)
C_k (^{18}O) ‰	Kinetic constant for ^{18}O	14.2	Constants based on the evaporating experiment
ε_k (^{18}O) ‰	Kinetic enrichment factor	5.2	From Eq. (S15)
ε (^{18}O) ‰	Total isotopic enrichment factor	15.33	The sum of ε^* and C_k
α^* (^{18}O) ‰	Effective isotopic equilibrium factor	1.01	$\alpha^* = 1 + \varepsilon^*$
δ_a (^{18}O) ‰	Isotopic composition of ambient air	−23.12	Estimated with δ_{in} and δ_a
δ_{in} (^{18}O) ‰	Isotopic composition of surface inflow water	−13.41	Average value of surface inflow water
δ_E (^{18}O) ‰	Isotopic compositions of evaporating vapour	−35.1	From Eq. (S12)
$[Cl^-]_{in}$ (mgL^{-1})	Chloride concentrations in surface inflow water	0.91	Filed data
$[Cl^-]_L$ (mgL^{-1})	Chloride concentrations in lake water	1.02	Filed data
$[Cl^-]_{gw}$ (mgL^{-1})	Chloride concentrations in groundwater	1.48	Filed data

water discharge studies over the glacial forefield areas. Based on long-term hydrological and climatological parameter monitoring on Nam Co Lake in the QTP, Zhou et al. (2013) estimated the LGD to be 5–8 mm d^{-1}, which is comparable to the surface runoff input and LGD of this study. Brown et al. (2006) investigated the headwater streams in the proglacial areas of Taillon Glacier in France and found that groundwater contributes 6 %–10 % of the stream water immediately downwards of the glacier. Using the water mass balance model, Hood et al. (2006) show that groundwater inflow is substantial in the hydrologic partitioning of the proglacial Lake O'Hara in front of Opabin Glacier in Canada and comprised 30 %–74 % of the total inflow. Roy and Hayashi (2008) studied the proglacial lakes of Hungabee Lake and Opabin Lake at the glacier forefield of Opabin Glacier and found that the groundwater component is a predominant water source of the lakes and consisted of 35 %–39 % of the total water input of the lakes. Langston et al. (2013) further investigated a tarn immediately in front of Opabin Glacier and indicated the tarn is predominantly controlled by groundwater inflow/outflow, which consisted of 50 %–100 % of the total tarn volume. Magnusson et al. (2014) studied the streams in the glacier forefield of the Damma Glacier, Switzerland, and revealed that groundwater contributed only 1 %–8 % of the total surface runoff. Groundwater contribution in this study is similar to those obtained from the mountainous proglacial areas in Europe, but much lower than those obtained in the proglacial areas of polar regions. It is concluded that proglacial lakes/streams in front of mountainous glaciers are mainly recharged by surface runoff from glacier/snowpack melting. This might be due to well-developed stream networks and limited deep groundwater flow (Einarsdottir et al., 2017; Brown et al., 2006; Magnusson et al., 2014). However, proglacial tarns and lakes in the polar areas are predominantly controlled by groundwater discharge, due to less connectivity of surface runoff and high shallow and deep groundwater connectivity (Langston et al., 2013; Hood et al., 2006; Roy and Hayashi, 2008).

The evaporation constitutes a relatively small ratio ($\approx$ 3.5 %) of total water losses. The annual evaporation rate was recorded to be 1429.8 mm (equivalent to 3.92 mm d^{-1}) in 2015 by the Jiuzhi weather station, lower than the obtained evaporation in this study. This may be due to much higher evaporation in August during the monitoring period. The estimation of evaporation in this study generally represents the upper limit of the lake, as the sampling campaign was conducted during the summer time when the highest evaporation might occur. The lake surface evaporation derived from the pan evaporation in the QTP ranges from $\sim$ 700 mm yr^{-1} in the eastern QTP to over 1400 mm yr^{-1} in the interior lakes of the QTP (Zhang et al., 2007; Ma et al., 2015; Yang et al., 2014). The evaporation of this study is rather in line with the previous evaporation observation in the eastern QTP, stressing the tenability of evaporation in this study. The runoff input is the predominant recharge component (>90 %) compared to other components, with an area normalized value comparable to previous studies of runoff input in other glacial melting dominant lakes in the QTP (Zhou et al., 2013; Zhang et al., 2011; Biskop et al., 2016). The runoff input and the lake evaporation of the study area, however, are subject to high daily, seasonal, and interannual variability as indicated by previous studies in the QTP

(Zhou et al., 2013; Lei et al., 2017; Ma et al., 2015; Lazhu et al., 2016). Therefore, further investigations of long-term and high-resolution climatological and isotopic data are required to provide precise constraints on hydrologic partitioning of the lakes in the QTP.

4.5 LGD-derived nutrient loadings, nutrient budget, and ecological implications

Compared to extensive studies of SGD-derived nutrient loadings in the past decade (Luo and Jiao, 2016; Slomp and Van Cappellen, 2004), studies of LGD-derived nutrient loadings have received limited attention, even given the fact that groundwater in lakeshore aquifers is usually concentrated in nutrients (Lewandowski et al., 2015; Rosenberry et al., 2015). Even fewer studies focus on chemical budgets in the proglacial lakes which are often difficult to access for sampling. Groundwater-borne DIN and DIP across the sediment–water interface in this study are determined with an equation coupling the advective, or LGD-derived, and diffusive solute transport (Lerman et al., 1995; Hagerthey and Kerfoot, 1998):

$$F_j = -nD_j^{\mathrm{m}}\frac{\mathrm{d}C_j}{\mathrm{d}x} + v_{\mathrm{gw}}C_j, \tag{7}$$

where $-nD_j^{\mathrm{m}}\frac{\mathrm{d}C_j}{\mathrm{d}x}$ is the diffusion input and $v_{\mathrm{gw}}C_j$ is the LGD-derived fluxes; F_j ($\mu\mathrm{M\,m^{-2}\,d^{-1}}$) is the mol flux of nutrient species j (representing DIN or DIP). n is the sediment porosity. D_j^{m} is the molecular diffusion coefficient of nutrient species j, which is given to be $4.8\times10^{-5}\,\mathrm{m^2\,d^{-1}}$ for DIP (Quigley and Robbins, 1986) and $8.8\times10^{-5}\,\mathrm{m^2\,d^{-1}}$ for DIN (Li and Gregory, 1974), respectively. C_j (μM) is the concentration of nutrient species j. x (m) is the sampling depth. v_{gw} is LGD rate estimated by the ^{222}Rn mass balance model and has a value of $10.3\pm8.2\,\mathrm{mm\,d^{-1}}$. $\frac{\mathrm{d}C_j}{\mathrm{d}x}$ is the concentration gradient of nutrient species j across the water–sediment interface.

Substituting the constants and the field data of DIN and DIP into Eq. (6), LGD-derived nutrient loadings are calculated to be 954.3 and $3.2\,\mu\mathrm{mol\,m^{-2}\,d^{-1}}$ for DIN and DIP, respectively. Riverine inflow brings $1195.0\,\mu\mathrm{mol\,m^{-2}\,d^{-1}}$ DIN and $52.9\,\mu\mathrm{mol\,m^{-2}\,d^{-1}}$ DIP into the lake. Lake water outflow-derived nutrient loss is estimated to be 1439.9 and $54.7\,\mu\mathrm{mol\,m^{-2}\,d^{-1}}$ for DIN and DIP, respectively. Nutrients in the lake can also be sourced from atmospheric deposit (mostly in the form of precipitation). With the nutrient concentrations in the rainwater during the monitoring period, the wet deposit is calculated to be 76 and $2.5\,\mu\mathrm{mol\,m^{-2}\,d^{-1}}$ for DIN and DIP, respectively. The loadings of DIN to the lakes are mainly from surface runoff and LGD, which comprised 42.9 % and 53.7 % of the total DIN loadings. Groundwater-derived DIP input, however, constitutes only 6.3 % of the total DIP inputs to the lake, indicating groundwater-borne DIP is less contributive to the nutrient budget of the lake compared to DIN. Very recent studies on polar regions have indicated that the glacier/snowpack water is the main N source to the proglacial lakes (Anderson et al., 2013; Dubnick et al., 2017). However, they do not consider the contribution of groundwater-borne N, in spite of the high groundwater connectivity in the proglacial areas (Roy and Hayashi, 2008). This study stresses that groundwater-borne DIN could be comparable to the surface runoff-derived DIN.

Based on nutrient results, the lake is considered to be an oligotrophic lake, similar to other glacier melting dominant lakes in the QTP (Mitamura et al., 2003; Liu et al., 2011). Phytoplankton are good dissolved organic phosphate (DOP) recyclers and will overcome inorganic P limitation though DOP cycling in most template lakes (Hudson et al., 2000). However, this may be not applicable for the glacial melting water and the periglacial/proglacial lake water. Previous studies show that phosphate nutrients are dominated by DIP and particulate phosphate, and the DOP contributes less than 10 % of the dissolved phosphate (Cole et al., 1998; Hawkings et al., 2016; Hodson, 2007). Thus, DOP recycling is not likely to lower the $N:P$ ratio under these conditions. Thus, the primary production (PP) is considered to be controlled by the DIP loadings. The sum of DIN and DIP inputs minus the calculated DIN and DIP outputs leads to surpluses of 785.4 and $3.9\,\mu\mathrm{mol\,m^{-2}\,d^{-1}}$ for DIN and DIP, respectively. The surpluses are expected to be consumed by the phytoplankton and converted into the PP under phosphate-limited conditions. As primary producers in the fresh lacustrine system consume the nutrient under variant $N:P$ ratios (7.1 to 44.2, mean: 22.9) (Downing and McCauley, 1992), the biological uptake of DIN is roughly estimated to be $89.3\,\mu\mathrm{M\,m^2\,d^{-1}}$. Therefore, the nutrient budgets for DIN and DIP can be finally conceptualized in Fig. 8b and c.

4.6 Implications, prospects, and limitations

Mountainous proglacial lakes are readily developed in glacier forefields of the QTP and other high mountainous glacial areas such as the European Alps and Pamir in central Asia (Heckmann et al., 2016). The proglacial lakes are always trapping systems of sediment and sinks for water and chemicals originating from glacier/snowpack melting and groundwater. Analogously to cosmogenic isotopes such as ^{10}Be serving as a tool to quantify the sediment sources, approaches integrating ^{222}Rn and stable isotopes provide both qualitative and quantitative evaluations of groundwater contributions and hydrologic partitioning in these remote and untapped lacustrine systems. Thus, it is expected that the multiple aqueous isotopes will be considered to be effective tools to investigate the LGD and hydrologic partitioning in other proglacial lakes. This study is mainly limited by the relatively short sampling and monitoring period. As a special hydrologic regime, the lakeshore aquifers of the proglacial lakes are experiencing frozen–unfrozen transition seasonally, and the dominant recharge of glacial melting could fluctuate significantly due to air temperature variation. Therefore, future groundwater and hydrological studies can be extended to longtime sampling and monitoring of stable isotopes and ^{222}Rn in different water endmembers to reveal the seasonal hydrological and hydrogeological dynamics and their impacts on local biogeochemical cycles and ecological systems. Special concerns would be placed on how surface–groundwater interactions and the associated biogeochemical processes respond to the seasonal frozen ground variations and glacier/snowpack melting intensity.

5 Conclusions

A ^{222}Rn continuous monitoring is conducted at Ximen Co Lake, a proglacial lake located in the eastern QTP. A dynamic ^{222}Rn mass balance model constrained by radium mass balance and water level fluctuation is used to quantify temporal distribution of the LGD of the lake. The obtained LGD over the monitoring time ranges from -23.7 to $80.9\,\mathrm{mm\,d^{-1}}$, with an average of $10.3 \pm 8.2\,\mathrm{mm\,d^{-1}}$. Thereafter, a three-endmember model consisting of the budgets of water, stable isotopes, and Cl^- is used to depict the hydrologic partitioning of the lake. Riverine inflow, lake water outflow via surface runoff, and surface evaporation are estimated to be 135.6, 141.5, and $5.2\,\mathrm{mm\,d^{-1}}$, respectively. LGD-derived nutrient loading is estimated to be 785.4 and $3.2\,\mathrm{\mu mol\,m^{-2}\,d^{-1}}$ for DIN and DIP, respectively. This study also implies that LGD constitutes a relatively small portion of the proglacial hydrologic partitioning; however, it delivers nearly half of the nutrient loadings to the proglacial lake.

This study presents the first attempt to quantify LGD and the associated nutrient loadings to the proglacial lake of the QTP. To our knowledge, there is almost no study on the groundwater–lake water interaction in the high-altitude proglacial lakes in the QTP. This study demonstrates that the ^{222}Rn-based approach can be used to investigate the groundwater dynamics in the high-altitude proglacial lakes. The method is instructional to similar studies in other proglacial lakes in the QTP and elsewhere. For a comprehensive understanding of the hydrological and biogeochemical dynamics in the QTP, interdisciplinary and multi-approach integrated studies are greatly needed. Of particular importance are the lake hydrology and groundwater–surface water interaction studies based on multiple approaches such as remote sensing products, long-term and high-resolution observation of climatological parameters, and isotopic data.

Author contributions. JJJ and XL conceived the study. XL, XXK, XLZ and SHL conducted the fieldwork; all the authors contributed to the data interpretation and manuscript construction.

Competing interests. The authors declare that they have no conflict of interest.

Acknowledgements. This study was supported by grants from the National Natural Science Foundation of China (NSFC, no. 41572208, and NSFC, 91747204), the Research Grants Council of Hong Kong Special Administrative Region, China (HKU17304815), and the seed fund programme granted by HKU-ZIRI. The authors thank Buming Jiang for his kind help in the fieldwork during the campaign and Ergang Lian for his help in stable isotope analysis. The authors thank Jessie Lai for her help in FIA analysis in the School of Biological Sciences, HKU.

Edited by: Bill X. Hu

References

Åberg, J., Jansson, M., and Jonsson, A.: Importance of water temperature and thermal stratification dynamics for temporal variation of surface water CO_2 in a boreal lake, J. Geophys. Res.-Biogeo., 115, 2005–2012, https://doi.org/10.1029/2009JG001085, 2010.

Andermann, C., Longuevergne, L., Bonnet, S., Crave, A., Davy, P., and Gloaguen, R.: Impact of transient groundwater storage on the discharge of Himalayan rivers, Nat. Geosci., 5, 127–132, https://doi.org/10.1038/Ngeo1356, 2012.

Anderson, L., Birks, J., Rover, J., and Guldager, N.: Controls on recent Alaskan lake changes identified from water isotopes and remote sensing, Geophys. Res. Lett., 40, 3413–3418, 2013.

Anderson, N. J., Saros, J. E., Bullard, J. E., Cahoon, S. M., McGowan, S., Bagshaw, E. A., Barry, C. D., Bindler, R., Burpee, B. T., and Carrivick, J. L.: The Arctic in the Twenty-First Century: Changing Biogeochemical Linkages across a Paraglacial Landscape of Greenland, Bioscience, 67, 118–133, 2017.

Barry, R. G.: The status of research on glaciers and global glacier recession: a review, Prog. Phys. Geog., 30, 285–306, 2006.

Batlle-Aguilar, J., Harrington, G. A., Leblanc, M., Welch, C., and Cook, P. G.: Chemistry of groundwater discharge inferred from longitudinal river sampling, Water Resour. Res., 50, 1550–1568, https://doi.org/10.1002/2013WR013591, 2014.

Belanger, T. V., Mikutel, D. F., and Churchill, P. A.: Groundwater seepage nutrient loading in a Florida Lake, Water Res., 19, 773–781, https://doi.org/10.1016/0043-1354(85)90126-5, 1985.

Biskop, S., Maussion, F., Krause, P., and Fink, M.: Differences in the water-balance components of four lakes in the southern-central Tibetan Plateau, Hydrol. Earth Syst. Sci., 20, 209–225, https://doi.org/10.5194/hess-20-209-2016, 2016.

Blume, T., Krause, S., Meinikmann, K., and Lewandowski, J.: Upscaling lacustrine groundwater discharge rates by fiber-optic distributed temperature sensing, Water Resour. Res., 49, 7929–7944, https://doi.org/10.1002/2012WR013215, 2013.

Böhner, J.: Säkulare Klimaschwankungen und rezente Klimatrends Zentral-und Hochasiens, Goltze, 101 Inst. of Geography, Göttingen, 1996.

Böhner, J.: General climatic controls and topoclimatic variations in Central and High Asia, Boreas, 35, 279–295, 2006.

Bolch, T., Kulkarni, A., Kääb, A., Huggel, C., Paul, F., Cogley, J. G., Frey, H., Kargel, J. S., Fujita, K., Scheel, M., Bajracharya, S., and Stoffel, M.: The State and Fate of Himalayan Glaciers, Science, 336, 310–314, 2012.

Brown, L. E., Hannah, D. M., Milner, A. M., Soulsby, C., Hodson, A. J., and Brewer, M. J.: Water source dynamics in a glacierized alpine river basin (TaillonGabiétous, French Pyrénées), Water Resour. Res., 42, W08404, https://doi.org/10.1029/2005WR004268, 2006.

Burnett, W. C., Peterson, R. N., Santos, I. R., and Hicks, R. W.: Use of automated radon measurements for rapid assessment of groundwater flow into Florida streams, J. Hydrol., 380, 298–304, 2010.

Callegary, J. B., Kikuchi, C. P., Koch, J. C., Lilly, M. R., and Leake, S. A.: Review: Groundwater in Alaska (USA), Hydrogeol. J., 21, 25–39, https://doi.org/10.1007/s10040-012-0940-5, 2013.

China Meteorological Administration: Climatological data, available at: http://www.weatherdt.com/, last access: 2 June 2016.

Cole, J., Nina, J., and Caraco, F.: Atmospheric exchange of carbon dioxide in a low-wind oligotrophic lake measured by the addition of SF_6, Limnol. Oceanogr., 43, 647–656, 1998.

Cook, P., Lamontagne, S., Berhane, D., and Clark, J.: Quantifying groundwater discharge to Cockburn River, southeastern Australia, using dissolved gas tracers ^{222}Rn and SF_6, Water Resour. Res., 42, W10411, https://doi.org/10.1029/2006WR004921, 2006.

Cook, P. G., Favreau, G., Dighton, J. C., and Tickell, S.: Determining natural groundwater influx to a tropical river using radon, chlorofluorocarbons and ionic environmental tracers, J. Hydrol., 277, 74–88, https://doi.org/10.1016/S0022-1694(03)00087-8, 2003.

Corbett, D. R., Burnett, W. C., Cable, P. H., and Clark, S. B.: Radon tracing of groundwater input into Par Pond, Savannah River site, J. Hydrol., 203, 209–227, 1997.

Cui, X., Ren, J., Qin, X., Sun, W., Yu, G., Wang, Z., and Liu, W.: Chemical characteristics and environmental records of a snow-pit at the Glacier No. 12 in the Laohugou Valley, Qilian Mountains, J. Earth Sci., 25, 379–385, 2014.

Dansgaard, W.: Stable isotopes in precipitation, Tellus A, 16, 436–468, https://doi.org/10.3402/tellusa.v16i4.8993, 1964.

Dimova, N., Paytan, A., Kessler, J. D., Sparrow, K., Garcia-Tigreros Kodovska, F., Lecher, A. L., Murray, J., and Tulaczyk, S. M.: The current magnitude and mechanisms of groundwater discharge in the Arctic: a case study from Alaska, Environ. Sci. Technol., 49, 12036–12043, 2015.

Dimova, N. T. and Burnett, W. C.: Evaluation of groundwater discharge into small lakes based on the temporal distribution of radon-222, Limnol. Oceanogr., 56, 486–494, 2011.

Dimova, N. T., Burnett, W. C., Chanton, J. P., and Corbett, J. E.: Application of radon-222 to investigate groundwater discharge into small shallow lakes, J. Hydrol., 486, 112–122, https://doi.org/10.1016/j.jhydrol.2013.01.043, 2013.

Downing, J. A. and McCauley, E.: The nitrogen: phosphorus relationship in lakes, Limnol. Oceanogr., 37, 936–945, 1992.

Dubnick, A., Wadham, J., Tranter, M., Sharp, M., Orwin, J., Barker, J., Bagshaw, E., and Fitzsimons, S.: Trickle or treat: The dynamics of nutrient export from polar glaciers, Hydrol. Process., 31, 1776–1789, https://doi.org/10.1002/hyp.11149, 2017.

Einarsdottir, K., Wallin, M. B., and Sobek, S.: High terrestrial carbon load via groundwater to a boreal lake dominated by surface water inflow, J. Geophys. Res.-Biogeo., 122, 15–29, https://doi.org/10.1002/2016JG003495, 2017.

Farinotti, D., Longuevergne, L., Moholdt, G., Duethmann, D., Molg, T., Bolch, T., Vorogushyn, S., and Guntner, A.: Substantial glacier mass loss in the Tien Shan over the past 50 years, Nat. Gesoci., 8, 716–722, https://doi.org/10.1038/ngeo2513, 2015.

Gat, J.: Stable isotopes of fresh and saline lakes, in: Physics and chemistry of lakes, Springer, Berlin, Heidelberg, 139–165, 1995.

Gat, J.: Oxygen and hydrogen isotopes in the hydrologic cycle, Annu. Rev. Earth Pl. Sc., 24, 225–262, 1996.

Genereux, D.: Quantifying uncertainty in tracerbased hydrograph separations, Water Resour. Res., 34, 915–919, 1998.

Gibson, J., Edwards, T., Bursey, G., and Prowse, T.: Estimating evaporation using stable isotopes: quantitative results and sensitivity analysis for, Nord. Hydrol., 24, 79–94, 1993.

Gibson, J. J.: Short-term evaporation and water budget comparisons in shallow Arctic lakes using non-steady isotope mass balance, J. Hydrol., 264, 242–261, 2002.

Gibson, J. J. and Edwards, T. W. D.: Regional water balance trends and evaporation-transpiration partitioning from a stable isotope survey of lakes in northern Canada, Global Biogeochem. Cy., 16, 10-1–10-14, https://doi.org/10.1029/2001GB001839, 2002.

Gibson, J. J., Birks, S. J., and Yi, Y.: Stable isotope mass balance of lakes: a contemporary perspective, Quaternary Sci. Rev., 131, 316–328, https://doi.org/10.1016/j.quascirev.2015.04.013, 2016.

Gilfedder, B., Frei, S., Hofmann, H., and Cartwright, I.: Groundwater discharge to wetlands driven by storm and flood events: Quantification using continuous Radon-222 and electrical conductivity measurements and dynamic mass-balance modelling, Geochim. Cosmochim. Ac., 165, 161–177, 2015.

Gonfiantini, R.: Environmental isotopes in lake studies, Handbook of environmental isotope geochemistry, Elsevier Scientific Publishing Company, Cornell University, 2, 113–168, 1986.

Good, S. P., Noone, D., and Bowen, G.: Hydrologic connectivity constrains partitioning of global terrestrial water fluxes, Science, 349, 175–177, https://doi.org/10.1126/science.aaa5931, 2015.

Hagerthey, S. E. and Kerfoot, W. C.: Groundwater flow influences the biomass and nutrient ratios of epibenthic algae in a north temperate seepage lake, Limnol. Oceanogr., 43, 1227–1242, 1998.

Harris, C., Arenson, L. U., Christiansen, H. H., Etzelmüller, B., Frauenfelder, R., Gruber, S., Haeberli, W., Hauck, C., Hölzle, M., and Humlum, O.: Permafrost and climate in Europe: Monitoring and modelling thermal, geomorphological and geotechnical responses, Earth-Sci. Rev., 92, 117–171, 2009.

Hasler, C. T., Midway, S. R., Jeffrey, J. D., Tix, J. A., Sullivan, C., and Suski, C. D.: Exposure to elevated pCO2 alters post-treatment diel movement patterns of largemouth bass over short time scales, Freshwater Biol., 61, 1590–1600, https://doi.org/10.1111/fwb.12805, 2016.

Hawkings, J., Wadham, J., Tranter, M., Telling, J., Bagshaw, E., Beaton, A., Simmons, S. L., Chandler, D., Tedstone, A., and Nienow, P.: The Greenland Ice Sheet as a hot spot of phosphorus weathering and export in the Arctic, Global Biogeochem. Cy., 30, 191–210, 2016.

Heckmann, T., McColl, S., and Morche, D.: Retreating ice: research in proglacial areas matters, Earth Surf. Proc. Land., 41, 271–276, https://doi.org/10.1002/esp.3858, 2016.

Hodson, A.: Phosphorus in glacial meltwaters, Glacier Science and Environmental Change, John Wiley & Sons, New York City, 81–82, 2007.

Hodson, A., Mumford, P., Kohler, J., and Wynn, P. M.: The High Arctic glacial ecosystem: new insights from nutrient budgets, Biogeochemistry, 72, 233–256, 2005.

Hood, J. L., Roy, J. W., and Hayashi, M.: Importance of groundwater in the water balance of an alpine headwater lake, Geophys. Res. Lett., 33, L13405, https://doi.org/10.1029/2006GL026611, 2006.

Hren, M. T., Bookhagen, B., Blisniuk, P. M., Booth, A. L., and Chamberlain, C. P.: δ^{18}O and δD of streamwaters across the Himalaya and Tibetan Plateau: Implications for moisture sources and paleoelevation reconstructions, Earth Planet. Sc. Lett., 288, 20–32, 2009.

Hudson, J. J., Taylor, W. D., and Schindler, D. W.: Phosphate concentrations in lakes, Nature, 406, 54–56, 2000.

Johannes, R. E.: The Ecological Significance of the Submarine Discharge of Groundwater, Mar. Ecol. Prog. Ser., 3, 365–373, 1980.

Kim, G., Burnett, W. C., Dulaiova, H., Swarzenski, P. W., and Moore, W. S.: Measurement of Ra-224 and Ra-226 activities in natural waters using a radon-in-air monitor, Environ. Sci. Technol., 35, 4680–4683, 2001.

Klaus, J. and McDonnell, J.: Hydrograph separation using stable isotopes: Review and evaluation, J. Hydrol., 505, 47–64, 2013.

Kluge, T., Ilmberger, J., von Rohden, C., and Aeschbach-Hertig, W.: Tracing and quantifying groundwater inflow into lakes using a simple method for radon-222 analysis, Hydrol. Earth Syst. Sci., 11, 1621–1631, https://doi.org/10.5194/hess-11-1621-2007, 2007.

Kluge, T., von Rohden, C., Sonntag, P., Lorenz, S., Wieser, M., Aeschbach-Hertig, W., and Ilmberger, J.: Localising and quantifying groundwater inflow into lakes using high-precision ^{222}Rn profiles, J. Hydrol., 450, 70–81, 2012.

Kraemer, T. F.: Radium isotopes in Cayuga Lake, New York: Indicators of inflow and mixing processes, Limnol. Oceanogr., 50, 158–168, 2005.

LaBaugh, J. W., Rosenberry, D. O., and Winter, T. C.: Groundwater contribution to the water and chemical budgets of Williams Lake, Minnesota, 1980–1991, Can. J. Fish. Aquat. Sci., 52, 754–767, 1995.

LaBaugh, J. W., Winter, T. C., Rosenberry, D. O., Schuster, P. F., Reddy, M. M., and Aiken, G. R.: Hydrological and chemical estimates of the water balance of a closedbasin lake in north central Minnesota, Water Resour. Res., 33, 2799–2812, 1997.

Langston, G., Hayashi, M., and Roy, J. W.: Quantifying groundwater-surface water interactions in a proglacial moraine using heat and solute tracers, Water Resour. Res., 49, 5411–5426, https://doi.org/10.1002/wrcr.20372, 2013.

Lazar, B., Weinstein, Y., Paytan, A., Magal, E., Bruce, D., and Kolodny, Y.: Ra and Th adsorption coefficients in lakes – Lake Kinneret (Sea of Galilee) "natural experiment", Geochim. Cosmochim. Ac., 72, 3446–3459, 2008.

Lazhu, Yang, K., Wang, J., Lei, Y., Chen, Y., Zhu, L., Ding, B., and Qin, J.: Quantifying evaporation and its decadal change for Lake Nam Co, central Tibetan Plateau, J. Geophys. Res.-Atmos., 121, 7578–7591, 2016.

Lecher, A. L., Kessler, J., Sparrow, K., Garcia-Tigreros Kodovska, F., Dimova, N., Murray, J., Tulaczyk, S., and Paytan, A.: Methane transport through submarine groundwater discharge to the North Pacific and Arctic Ocean at two Alaskan sites, Limnol. Oceanogr., 61, S344–355, 2015.

Lee, C. M., Jiao, J. J., Luo, X., and Moore, W. S.: Estimation of submarine groundwater discharge and associated nutrient fluxes in Tolo Harbour, Hong Kong, Sci. Total Environ., 433, 427–433, https://doi.org/10.1016/j.scitotenv.2012.06.073, 2012.

Lee, D. R.: A device for measuring seepage flux in lakes and estuaries, Limnol. Oceanogr., 22, 140–147, 1977.

Lee, D. R., Cherry, J. A., and Pickens, J. F.: Groundwater transport of a salt tracer through a sandy lakebed, Limnol. Oceanogr., 25, 45–61, 1980.

Lehmkuhl, F.: Extent and spatial distribution of Pleistocene glaciations in eastern Tibet, Quatern. Int., 45–46, 123–134, https://doi.org/10.1016/S1040-6182(97)00010-4, 1998.

Lei, Y., Yao, T., Yang, K., Sheng, Y., Kleinherenbrink, M., Yi, S., Bird, B. W., Zhang, X., Zhu, L., and Zhang, G.: Lake seasonality across the Tibetan Plateau and their varying relationship with regional mass changes and local hydrology, Geophys. Res. Lett., 44, 892–900, 2017.

Lemieux, J. M., Sudicky, E. A., Peltier, W. R., and Tarasov, L.: Dynamics of groundwater recharge and seepage over the Canadian landscape during the Wisconsinian glaciation, J. Geophys. Res.-Earth, 113, F01011, https://doi.org/10.1029/2007JF000838, 2008a.

Lemieux, J. M., Sudicky, E. A., Peltier, W. R., and Tarasov, L.: Simulating the impact of glaciations on continental groundwater flow systems: 1. Relevant processes and model formulation, J. Geophys. Res.-Earth, 113, F03017, https://doi.org/10.1029/2007JF000928, 2008b.

Lemieux, J. M., Sudicky, E. A., Peltier, W. R., and Tarasov, L.: Simulating the impact of glaciations on continental groundwater flow systems: 2. Model application to the Wisconsinian glaciation over the Canadian landscape, J. Geophys. Res.-Earth, 113, F03018, https://doi.org/10.1029/2007JF000929, 2008c.

Lerman, A., Imboden, D., and Gat, J.: Physics and chemistry of lakes, New York, Springer-Verlag, Berlin, 1995.

Levy, A., Robinson, Z., Krause, S., Waller, R., and Weatherill, J.: Longterm variability of proglacial groundwaterfed hydrological systems in an area of glacier retreat, Skeiðarársandur, Iceland, Earth Surf. Proc. Land., 40, 981–994, https://doi.org/10.1002/esp.3696, 2015.

Lewandowski, J., Meinikmann, K., Ruhtz, T., Pöschke, F., and Kirillin, G.: Localization of lacustrine groundwater discharge (LGD) by airborne measurement of thermal infrared radiation, Remote Sens. Environ., 138, 119–125, https://doi.org/10.1016/j.rse.2013.07.005, 2013.

Lewandowski, J., Meinikmann, K., Nützmann, G., and Rosenberry, D. O.: Groundwater – the disregarded component in lake water and nutrient budgets. Part 2: effects of groundwater on nutrients, Hydrol. Process., 29, 2922–2955, https://doi.org/10.1002/hyp.10384, 2015.

Li, Y. H. and Gregory, S.: Diffusion of Ions in Sea-Water and in Deep-Sea Sediments, Geochim. Cosmochim. Ac., 38, 703–714, 1974.

Liu, C., Liu, J., Wang, X. S., and Zheng, C.: Analysis of groundwater-lake interaction by distributed temperature sensing in Badain Jaran Desert, Northwest China, Hydrol. Process., 30, 1330–1341, 2015.

Liu, Y., Yao, T., Jiao, N., Tian, L., Hu, A., Yu, W., and Li, S.: Microbial diversity in the snow, a moraine lake and a stream in Himalayan glacier, Extremophiles, 15, 411–421, 2011.

Luo, X. and Jiao, J. J.: Submarine groundwater discharge and nutrient loadings in Tolo Harbor, Hong Kong using multiple geotracer-based models, and their implications of red tide outbreaks, Water Res., 102, 11–31, https://doi.org/10.1016/j.watres.2016.06.017, 2016.

Luo, X., Jiao, J. J., Moore, W., and Lee, C. M.: Submarine groundwater discharge estimation in an urbanized embayment in Hong Kong via short-lived radium isotopes and its implication of nutrient loadings and primary production, Mar. Pollut. Bull., 82, 144–154, 2014.

Luo, X., Jiao, J. J., Wang, X.-S., and Liu, K.: Temporal ^{222}Rn distributions to reveal groundwater discharge into desert lakes: implication of water balance in the Badain Jaran Desert, China, J. Hydrol., 534, 87–103, 2016.

Luo, X., Jiao, J. J., Wang, X.-s., Liu, K., Lian, E., and Yang, S.: Groundwater discharge and hydrologic partition of the lakes in desert environment: Insights from stable ^{18}O / ^{2}H and radium isotopes, J. Hydrol., 546, 189–203, 2017.

Luo, X., Jiao, J. J., Liu, Y., Zhang, X., Liang, W., and Tang, D.: Evaluation of Water Residence Time, Submarine Groundwater Discharge, and Maximum New Production Supported by Groundwater Borne Nutrients in a Coastal Upwelling Shelf System, J. Geophys. Res.-Oceans, 123, 631–655, 2018.

Ma, N., Zhang, Y., Szilagyi, J., Guo, Y., Zhai, J., and Gao, H.: Evaluating the complementary relationship of evapotranspiration in the alpine steppe of the Tibetan Plateau, Water Resour. Res., 51, 1069–1083, https://doi.org/10.1002/2014WR015493, 2015.

MacIntyre, S., Wannikhof, R., Chanton, J. P., Matson, P. A., and Hariss, R. C.: Biogenic Trace Gases: Measuring Emissions from Soil and Water, John Wiley & Sons, New York City, 52 pp., 1995.

Magnusson, J., Kobierska, F., Huxol, S., Hayashi, M., Jonas, T., and Kirchner, J. W.: Melt water driven stream and groundwater stage fluctuations on a glacier forefield (Dammagletscher, Switzerland), Hydrol. Process., 28, 823–836, 2014.

Mitamura, O., Seike, Y., Kondo, K., Goto, N., Anbutsu, K., Akatsuka, T., Kihira, M., Tsering, T. Q., and Nishimura, M.: First investigation of ultraoligotrophic alpine Lake Puma Yumco in the pre-Himalayas, China, Limnology, 4, 167–175, 2003.

Moore, W. S.: Sampling radium-228 in the deep ocean, Deep-Sea Res., 23, 647–651, 1976.

Nakayama, T. and Watanabe, M.: Missing role of groundwater in water and nutrient cycles in the shallow eutrophic Lake Kasumigaura, Japan, Hydrol. Process., 22, 1150–1172, 2008.

Paytan, A., Lecher, A. L., Dimova, N., Sparrow, K. J., Kodovska, F. G.-T., Murray, J., Tulaczyk, S., and Kessler, J. D.: Methane transport from the active layer to lakes in the Arctic using Toolik Lake, Alaska, as a case study, P. Natl. Acad. Sci. USA, 112, 3636–3640, 2015.

Qiu, J.: China: the third pole, Nature News, 454, 393–396, 2008.

Quigley, M. A. and Robbins, J. A.: Phosphorus release processes in nearshore southern Lake Michigan, Can. J. Fish. Aquat. Sci., 43, 1201–1207, 1986.

Ren, W., Yao, T., Yang, X., and Joswiak, D. R.: Implications of variations in δ^{18}O and δD in precipitation at Madoi in the eastern Tibetan Plateau, Quatern. Int., 313–314, 56–61, https://doi.org/10.1016/j.quaint.2013.05.026, 2013.

Ren, W., Yao, T., Xie, S., and He, Y.: Controls on the stable isotopes in precipitation and surface waters across the southeastern Tibetan Plateau, J. Hydrol., 545, 276–287, https://doi.org/10.1016/j.jhydrol.2016.12.034, 2017.

Rosenberry, D. O., Lewandowski, J., Meinikmann, K., and Nützmann, G.: Groundwaterthe disregarded component in lake water and nutrient budgets. Part 1: effects of groundwater on hydrology, Hydrol. Process., 29, 2895–2921, 2015.

Roy, J. W. and Hayashi, M.: Groundwater exchange with two small alpine lakes in the Canadian Rockies, Hydrol. Process., 22, 2838–2846, 2008.

Schafran, G. C. and Driscoll, C. T.: Flow pathcomposition relationships for groundwater entering an acidic lake, Water Resour. Res., 29, 145–154, 1993.

Scheidegger, J. M. and Bense, V. F.: Impacts of glacially recharged groundwater flow systems on talik evolution, J. Geophys. Res.-Earth, 119, 758–778, https://doi.org/10.1002/2013JF002894, 2014.

Schlutz, F. and Lehmkuhl, F.: Holocene climatic change and the nomadic Anthropocene in Eastern Tibet: palynological and geomorphological results from the Nianbaoyeze Mountains, Quaternary Sci. Rev., 28, 1449–1471, 2009.

Schmidt, A., Gibson, J. J., Santos, I. R., Schubert, M., Tattrie, K., and Weiss, H.: The contribution of groundwater discharge to the overall water budget of two typical Boreal lakes in Alberta/Canada estimated from a radon mass balance, Hydrol. Earth Syst. Sci., 14, 79–89, https://doi.org/10.5194/hess-14-79-2010, 2010.

Sebok, E., Duque, C., Kazmierczak, J., Engesgaard, P., Nilsson, B., Karan, S., and Frandsen, M.: Highresolution distributed temperature sensing to detect seasonal groundwater discharge into Lake Væng, Denmark, Water Resour. Res., 49, 5355–5368, 2013.

Shaw, R. D. and Prepas, E. E.: Groundwater-lake interactions: I. Accuracy of seepage meter estimates of lake seepage, J. Hydrol., 119, 105–120, https://doi.org/10.1016/0022-1694(90)90037-X, 1990.

Slaymaker, O.: Criteria to distinguish between periglacial, proglacial and paraglacial environments, Quaestiones Geographicae, 30, 85–94, 2011.

Slomp, C. P. and Van Cappellen, P.: Nutrient inputs to the coastal ocean through submarine groundwater discharge: controls and potential impact, J. Hydrol., 295, 64–86, https://doi.org/10.1016/j.jhydrol.2004.02.018, 2004.

Smerdon, B., Mendoza, C., and Devito, K.: Simulations of fully coupled lakegroundwater exchange in a subhumid climate with an integrated hydrologic model, Water Resour. Res., 43, W01416, https://doi.org/10.1029/2006WR005137, 2007.

Stets, E. G., Winter, T. C., Rosenberry, D. O., and Striegl, R. G.: Quantification of surface water and groundwater flows to open and closedbasin lakes in a headwaters watershed using a descriptive oxygen stable isotope model, Water Resour. Res., 46, W03515, https://doi.org/10.1029/2009WR007793, 2010.

Tockner, K., Malard, F., Uehlinger, U., and Ward, J.: Nutrients and organic matter in a glacial river-floodplain system (Val Roseg, Switzerland), Limnol. Oceanogr., 47, 266–277, 2002.

Valiela, I., Teal, J. M., Volkmann, S., Shafer, D., and Carpenter, E. J.: Nutrient and Particulate Fluxes in a Salt-Marsh Ecosystem – Tidal Exchanges and Inputs by Precipitation and Groundwater, Limnol. Oceanogr., 23, 798–812, 1978.

Wang, C., Dong, Z., Qin, X., Zhang, J., Du, W., and Wu, J.: Glacier meltwater runoff process analysis using δD and δ^{18}O isotope and chemistry at the remote Laohugou glacier basin in western Qilian Mountains, China, J. Geogr. Sci., 26, 722–734, 2016a.

Wang, J., Zhu, L., Wang, Y., Ju, J., Xie, M., and Daut, G.: Comparisons between the chemical compositions of lake water, inflowing river water, and lake sediment in Nam Co, central Tibetan Plateau, China and their controlling mechanisms, J. Great Lakes Res., 36, 587–595, 2010.

Wang, R., Liu, Z., Jiang, L., Yao, Z., Wang, J., and Ju, J.: Comparison of surface water chemistry and weathering effects of two lake basins in the Changtang Nature Reserve, China, J. Environ. Sci., 41, 183–194, https://doi.org/10.1016/j.jes.2015.03.016, 2016b.

Wang, S.: Frozen ground and environment in the Zoige Plateau and its surrounding mountains, Journal of Glaciology and Geocryology, 19, 39–46, 1997 (in Chinese with English abstract).

Warner, N. R., Christie, C. A., Jackson, R. B., and Vengosh, A.: Impacts of shale gas wastewater disposal on water quality in western Pennsylvania, Environ. Sci. Technol., 47, 11849–11857, 2013.

Weynell, M., Wiechert, U., and Zhang, C.: Chemical and isotopic (O, H, C) composition of surface waters in the catchment of Lake Donggi Cona (NW China) and implications for paleoenvironmental reconstructions, Chem. Geol., 435, 92–107, https://doi.org/10.1016/j.chemgeo.2016.04.012, 2016.

White, D., Lapworth, D. J., Stuart, M. E., and Williams, P. J.: Hydrochemical profiles in urban groundwater systems: New insights into contaminant sources and pathways in the subsurface from legacy and emerging contaminants, Sci. Total Environ., 562, 962–973, https://doi.org/10.1016/j.scitotenv.2016.04.054, 2016.

Wilson, J. and Rocha, C.: A combined remote sensing and multitracer approach for localising and assessing groundwater-lake interactions, Int. J. Appl. Earth Obs., 44, 195–204, 2016.

Winter, T. C.: Relation of streams, lakes, and wetlands to groundwater flow systems, Hydrogeol. J., 7, 28–45, 1999.

Wischnewski, J., Herzschuh, U., Rühland, K. M., Bräuning, A., Mischke, S., Smol, J. P., and Wang, L.: Recent ecological responses to climate variability and human impacts in the Nianbaoyeze Mountains (eastern Tibetan Plateau) inferred from pollen, diatom and tree-ring data, J. Paleolimnol., 51, 287–302, 2014.

Xin, W., Yongjian, D., Shiyin, L., Lianghong, J., Kunpeng, W., Zongli, J., and Wanqin, G.: Changes of glacial lakes and implications in Tian Shan, central Asia, based on remote sensing data from 1990 to 2010, Environ. Res. Lett., 8, 044052, https://doi.org/10.1088/1748-9326/8/4/044052, 2013.

Yang, K., Wu, H., Qin, J., Lin, C., Tang, W., and Chen, Y.: Recent climate changes over the Tibetan Plateau and their impacts on energy and water cycle: A review, Global Planet. Change, 112, 79–91, 2014.

Yao, T., Thompson, L., Yang, W., Yu, W., Gao, Y., Guo, X., Yang, X., Duan, K., Zhao, H., Xu, B., Pu, J., Lu, A., Xiang, Y., Kattel, D. B., and Joswiak, D.: Different glacier status with atmospheric circulations in Tibetan Plateau and surroundings, Nat. Clim. Change, 2, 663–667, 2012.

Yao, T., MassonDelmotte, V., Gao, J., Yu, W., Yang, X., Risi, C., Sturm, C., Werner, M., Zhao, H., He, Y., Ren, W., Tian, L., Shi, C., and Hou, S.: A review of climatic controls on $\delta^{18}O$ in precipitation over the Tibetan Plateau: Observations and simulations, Rev. Geophys., 51, 525–548, https://doi.org/10.1002/rog.20023, 2013.

Yao, Z., Wang, R., Liu, Z., Wu, S., and Jiang, L.: Spatial-temporal patterns of major ion chemistry and its controlling factors in the Manasarovar Basin, Tibet, J. Geogr. Sci., 25, 687–700, https://doi.org/10.1007/s11442-015-1196-5, 2015.

Yuan, H., Liu, E., Shen, J., Zhou, H., Geng, Q., and An, S.: Characteristics and origins of heavy metals in sediments from Ximen Co Lake during summer monsoon season, a deep lake on the eastern Tibetan Plateau, J. Geochem. Explor., 136, 76–83, https://doi.org/10.1016/j.gexplo.2013.10.008, 2014.

Zhang, B., Wu, Y., Zhu, L., Wang, J., Li, J., and Chen, D.: Estimation and trend detection of water storage at Nam Co Lake, central Tibetan Plateau, J. Hydrol., 405, 161–170, 2011.

Zhang, C. and Mischke, S.: A Lateglacial and Holocene lake record from the Nianbaoyeze Mountains and inferences of lake, glacier and climate evolution on the eastern Tibetan Plateau, Quaternary Sci. Rev., 28, 1970–1983, 2009.

Zhang, G., Yao, T., Xie, H., Kang, S., and Lei, Y.: Increased mass over the Tibetan Plateau: from lakes or glaciers?, Geophys. Res. Lett., 40, 2125–2130, 2013.

Zhang, G., Yao, T., Piao, S., Bolch, T., Xie, H., Chen, D., Gao, Y., O'Reilly, C. M., Shum, C., and Yang, K.: Extensive and drastically different alpine lake changes on Asia's high plateaus during the past four decades, Geophys. Res. Lett., 44, 252–260, 2017a.

Zhang, G., Yao, T., Shum, C., Yi, S., Yang, K., Xie, H., Feng, W., Bolch, T., Wang, L., and Behrangi, A.: Lake volume and groundwater storage variations in Tibetan Plateau's endorheic basin, Geophys. Res. Lett., 44, 5550–5560, 2017b.

Zhang, Y., Liu, C., Tang, Y., and Yang, Y.: Trends in pan evaporation and reference and actual evapotranspiration across the Tibetan Plateau, J. Geophys. Res.-Atmos., 112, D12110, https://doi.org/10.1029/2006JD008161, 2007.

Zhou, S., Kang, S., Chen, F., and Joswiak, D. R.: Water balance observations reveal significant subsurface water seepage from Lake Nam Co, south-central Tibetan Plateau, J. Hydrol., 491, 89–99, 2013.

Zlotnik, V. A., Olaguera, F., and Ong, J. B.: An approach to assessment of flow regimes of groundwater-dominated lakes in arid environments, J. Hydrol., 371, 22–30, 2009.

Zlotnik, V. A., Robinson, N. I., and Simmons, C. T.: Salinity dynamics of discharge lakes in dune environments: Conceptual model, Water Resour. Res., 46, W11548, 2010.

Zongxing, L., Qi, F., Wei, L., Tingting, W., Xiaoyan, G., Zongjie, L., Yan, G., Yanhui, P., Rui, G., and Bing, J.: The stable isotope evolution in Shiyi glacier system during the ablation period in the north of Tibetan Plateau, China, Quatern. Int., 380, 262–271, 2015.

6

Soil water stable isotopes reveal evaporation dynamics at the soil–plant–atmosphere interface of the critical zone

Matthias Sprenger, Doerthe Tetzlaff, and Chris Soulsby

Northern Rivers Institute, School of Geosciences, University of Aberdeen, Aberdeen, AB24 3UF, UK

Correspondence to: Matthias Sprenger (matthias.sprenger@abdn.ac.uk)

Abstract. Understanding the influence of vegetation on water storage and flux in the upper soil is crucial in assessing the consequences of climate and land use change. We sampled the upper 20 cm of podzolic soils at 5 cm intervals in four sites differing in their vegetation (Scots Pine (*Pinus sylvestris*) and heather (*Calluna* sp. and *Erica* Sp)) and aspect. The sites were located within the Bruntland Burn long-term experimental catchment in the Scottish Highlands, a low energy, wet environment. Sampling took place on 11 occasions between September 2015 and September 2016 to capture seasonal variability in isotope dynamics. The pore waters of soil samples were analyzed for their isotopic composition (δ^2H and δ^{18}O) with the direct-equilibration method. Our results show that the soil waters in the top soil are, despite the low potential evaporation rates in such northern latitudes, kinetically fractionated compared to the precipitation input throughout the year. This fractionation signal decreases within the upper 15 cm resulting in the top 5 cm being isotopically differentiated to the soil at 15–20 cm soil depth. There are significant differences in the fractionation signal between soils beneath heather and soils beneath Scots pine, with the latter being more pronounced. But again, this difference diminishes within the upper 15 cm of soil. The enrichment in heavy isotopes in the topsoil follows a seasonal hysteresis pattern, indicating a lag time between the fractionation signal in the soil and the increase/decrease of soil evaporation in spring/autumn. Based on the kinetic enrichment of the soil water isotopes, we estimated the soil evaporation losses to be about 5 and 10 % of the infiltrating water for soils beneath heather and Scots pine, respectively. The high sampling frequency in time (monthly) and depth (5 cm intervals) revealed high temporal and spatial variability of the isotopic composition of soil waters, which can be critical, when using stable isotopes as tracers to assess plant water uptake patterns within the critical zone or applying them to calibrate tracer-aided hydrological models either at the plot to the catchment scale.

1 Introduction

Processes in the soil–plant–atmosphere continuum exert a major influence on water partitioning into evaporation, transpiration, and recharge fluxes. Therefore, the outer part of the Earth's terrestrial surface, where the subsurface is closely coupled with the atmosphere and vegetation, is often referred to as the critical zone (Brooks et al., 2015). However, the dynamics of feedbacks between soils and vegetation remain insufficiently understood (Werner and Dubbert, 2016). Consequently, there is an increased interest in improving the conceptualization of the upper boundary of soils, as the important interface between soils–plant–atmosphere. For example, it has been shown that evapotranspiration dynamics can affect travel times of percolating water in the unsaturated zone (Sprenger et al., 2016c; Heße et al., 2017) and catchment outflows (van der Velde et al., 2015; Rinaldo et al., 2011). Additionally, understanding the age distributions of evapotranspiration water has recently gained interest in the literature (Harman, 2015; Soulsby et al., 2016a; van Huijgevoort et al., 2016; Queloz et al., 2015). However, estimates of these evapotranspiration ages and catchment travel times require a sound understanding of the storage and mixing dynamics of the subsurface and surface water pools, which form the sources of evapotranspiration within catchments. Disentangling these atmospheric losses into evaporation and transpiration is particularly challenging.

Stable isotopes of water (^{2}H and ^{18}O) are a powerful tool for the analysis of the partitioning of water (see review by Kool et al., 2014). They are often seen as ideal tracers, since they are part of the water molecule. During root water uptake, the isotopic composition of the remaining soil water is usually not altered (Wershaw et al., 1966; Dawson and Ehleringer, 1991). However, evaporation leads to isotopic fractionation, where the remaining water is generally enriched in heavy isotopes (equilibrium fractionation). Additionally, in natural open systems with a humidity of $< 100\,\%$, δ^2H is more likely to be evaporated than δ^{18}O, because of their different atomic weights, leading to kinetic non-equilibrium fractionation (Craig et al., 1963). Therefore, evaporation losses result in an isotopic signal in the residual water that is distinct from the original isotopic composition of the precipitation waters that were formed in isotopic equilibrium (Dansgaard, 1964).

Such enrichment from kinetic fractionation was found in soil water isotopes across various climatic regions, with more pronounced evaporative signals reaching deeper into the soils in arid and Mediterranean environments than in temperate regions (Sprenger et al., 2016b). However, the temporal variability of the isotopic fractionation in the field has not yet been studied, despite recent technical developments that enabled easier analysis of stable isotopes in soil water (see review by Sprenger et al., 2015a). While Rothfuss et al. (2015) sampled the soil water isotopes in a soil column undergoing evaporation in the laboratory, field studies are usually limited to a few sampling campaigns or a few days (Twining et al., 2006; Gaj et al., 2016; Volkmann et al., 2016). Only recently, Oerter and Bowen (2017) applied in situ soil water isotope measurements to cover almost 1 year, but their study was limited to one particular location. Comparison of the soil water stable isotope dynamics that occur within an entire year at a number of dominant landscape units are yet missing. Additionally, soil water isotopes have been studied much less extensively in the colder regions of the northern latitudes (but see Tetzlaff et al., 2014; Geris et al., 2015a, b). However, the knowledge of soil water isotopic composition is of fundamental importance when studying the root water uptake pattern of plants (Rothfuss and Javaux, 2017). While comparisons of the isotopic signal in soil waters with waters of plant tissues have been reported for decades (see review by Ehleringer and Dawson, 1992), recently posed research questions on which water of the soil's pore system is used by plants (Brooks et al., 2010) are enhancing ecohydrological studies on soil–plant interactions in the critical zone. While studies based on two sampling campaigns (Goldsmith et al., 2012; Evaristo et al., 2016) were supportive of the ecohydrological separation, as presented by Brooks et al. (2010), newly published work with higher temporal resolution of soil water and xylem water isotope sampling suggests that there are seasonal differences with regard to ecohydrological separation (McCutcheon et al., 2016; Hervé-Fernández et al., 2016). A comparative study by Evaristo et al. (2015) showed that the isotopic composition of plant waters – just like soil waters in the upper horizons – are usually kinetically fractionated. However, Evaristo et al. (2015) did not cover the northern latitudes in their review, since there has been only preliminary studies, looking into the root water uptake of the vegetation in this low energy region (Geris et al., 2015a, 2017). Geris et al. (2015a) found relatively little to moderate fractionation in the soil waters at -10 cm soil depth and xylem waters in the Bruntland Burn catchment in northern Scotland. However, recent findings about kinetic fractionation in the water pools and tracks of an extended drainage network in a raised bog within the Bruntland Burn showed that evaporation can have a fractionating effect on the stable isotopes of peatland waters, despite the relatively low energy available (Sprenger et al., 2017b). While such isotopic fractionation of open waters in peatlands of the northern latitude were found by others (Carrer et al., 2016; Gibson et al., 2000; Isokangas et al., 2017), the potential fractionation dynamics of the water in the upper soil layer in these cold regions have not previously been studied.

Thus, there is a particular need to better understand such ecohydrological processes in the higher latitudes (Tetzlaff et al., 2013). These environments are known to be especially sensitive to future climate change projections, since relatively little warming could cause intense changes in the water balance when snowfall and snowmelt dynamics (Mioduszewski et al., 2014) and vegetation phenology shift (Shen et al., 2014). The upper soil layers (top 30 cm), where about 90 % of the root mass is usually present in the northern temperate and boreal biomes (Jackson et al., 1996), are of special interest to better understand how vegetation influences the partitioning of soil water into evaporation, transpiration, and recharge. The marked seasonality in the northern environments and its impact on the evaporation signal in soil water stable isotopes are fertile areas for investigation.

Here, we address the following research questions in order to improve understanding of the evaporation dynamics at the soil–plant–atmosphere interface and their influences on the water storage and mixing in the critical zone:

- How do precipitation input and the soil water storage mix and affect the soil water isotope dynamics over time?

- How can one infer soil evaporation dynamics in the field from soil water isotopic fractionation?

- How do soil characteristics, vegetation cover, and aspect drive evaporation fractionation dynamics?

2 Methods and study sites

2.1 Environmental conditions

Our study was conducted in the Bruntland Burn (BB) experimental catchment ($3.2\,km^2$) in the Scottish Highlands; a subcatchment of the Girnock burn, a long-term ecohydrological research site. A detailed description of the soil and vegetation characteristics follows in Sect. 2.3. The underlying geology of the BB is characterized by granitic and metamorphic rocks (Soulsby et al., 2007). About 60 % of the catchment is covered by up to 40 m of glacial drift deposits which maintain a high groundwater storage (Soulsby et al., 2016b). The climate is temperate–boreal oceanic with mean daily air temperatures ranging between 2 °C in January and 13 °C in July. Annual precipitation (P) is about $1000\,mm\,yr^{-1}$, which is fairly evenly distributed and occurs mainly as rainfall (usually < 5 % as snow) of low intensities (50 % of rainfall at intensities of $< 10\,mm\,d^{-1}$) (Soulsby et al., 2015). The annual potential evaporation (PET) is about $400\,mm\,yr^{-1}$ and the annual runoff is around $700\,mm\,yr^{-1}$. While the runoff of the BB shows limited seasonality, though lower flows tend to be in summer, the PET estimated with the Penman–Monteith approach follows a strong seasonal dynamic with average PET rates of 0.3 to $0.7\,mm\,d^{-1}$ from November to February and 2.3 to $2.7\,mm\,d^{-1}$ from May to August (Sprenger et al., 2017b).

The seasonality of the climate is also reflected in the variability of the isotopic signal of the precipitation with depleted values being more common during winter (dropping frequently below $-80\,‰\ \delta^2H$ from November to March) and more enriched values dominant in summer. The weighted averages for the precipitation isotope signal over 5 years (2011–2016) were $P_{avg}\delta^{18}O = -8.5\,‰$ and $P_{avg}\delta^2H = -61\,‰$. The regression between $\delta^{18}O$ and δ^2H values of daily precipitation data sampled between June 2011 and September 2016 describes the local meteoric water line (LMWL):

$$\delta^2H = 7.6 \times \delta^{18}O + 4.7. \quad (1)$$

2.2 Study sites

Soil sampling focused on four different sites within the BB, where the soils are characterized as freely draining podzols (Fig. 1a). The study sites differed regarding their vegetation cover and their aspect: At two sites, Scots Pine (*Pinus sylvestris*) forest is the dominant vegetation and at the two other sites, heather (*Calluna* sp. and *Erica* sp.) shrubland is dominating (Fig. 1b). Each of the two soil–vegetation landscape units were studied at a north-facing and a south-facing slope (sites had gentle slopes), leading to the following four different sites: north-facing heather (NH), north-facing forest (NF), south-facing heather (SH), and south-facing forest (SF).

The podzols are shallow soils and frequent large clasts within the glacial drift deposit usually inhibit soil sampling below 20 to 30 cm. The soils at the four study sites were relatively similar regarding their color (Fig. 1c) and texture (Table 1). The texture was determined after ignition of the soil samples to free the soil of organic matter content. The coarse and medium sand fractions were determined by dry sieving and the fine sand, silt, and clay fractions were estimated with the hydrometer method (Gee and Bauder, 1986). The soil consist mainly of loamy sand and for NH and SH, the gravel content as well as the sand content generally increases with depth (Table 1). The bulk density of the podzols in the BB is about $0.74\,g\,cm^{-3}$ for the top 20 cm (Geris et al., 2015b).

The organic matter content of the soils was determined for each study site in replicated 5 cm depth increments ($n = 5$ per site and depth) by loss on ignition (LOI) of about 10 g soil material at 550 °C in a furnace over 2 h, according to Ball (1964). The LOI decreased linearly with soil depth at all study sites and also showed generally a strong relationship with the gravimetric water content (Table 1). We determined the gravimetric water content (GWC) of all soil samples by relating the weight loss after oven drying at 105 °C over night to the dry soil mass.

Hemispheric photos taken during the vegetation period revealed that the median of the canopy coverage (CC) of the heather was slightly lower (NH: CC = 65 %, $n = 9$; SH: CC = 62 %, $n = 9$) than for the forested sites (NF: CC = 67 %, $n = 36$; SF: CC = 69 %, $n = 46$) (Braun, 2015). The forested sites were both plantations, but with larger trees (mean diameter at breast height (DBH) = 21.8 cm) and lower tree density at SF compared to NF (mean DBH = 13.8 cm), where the tree ages were more variable (Braun, 2015). The height of the heather vegetation was limited to about 0.5 m and tree height was 12–15 m in the plantations.

Fine root (defined as 0.5–2 mm according to Zobel and Waisel, 2010) density was determined by wet sieving of the fine roots from soil cores ($100\,cm^3$) and subsequent oven drying at 70 °C. This analysis was limited to the heather sites, since the roots and boulders at the forested sites inhibited sampling of undisturbed soil cores. The roots of heather were limited to the upper 15 cm at the heather sites with almost exponential decrease at NH and a more linear decrease at SH (Table 1).

2.3 Sampling design and analysis

Soil sampling at each site was conducted at monthly intervals between September 2015 and September 2016 (except for December, February, March; $n = 11$). For each sampling campaign, soils at all four sites were sampled with a spade across five profiles in 5 cm increments down to 20 cm soil depth. For each sampling depth, five replicate samples were taken to account for the high subsurface heterogeneity (Fig. S1 in the Supplement). Each soil sample contained of

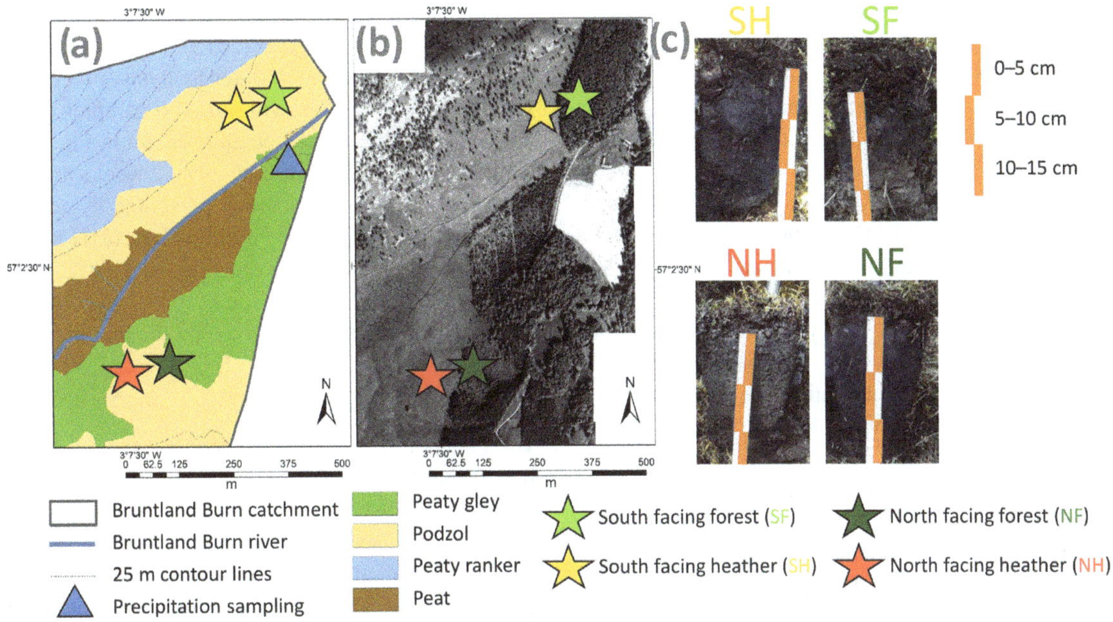

Figure 1. Location of the four sampling sites within the Bruntland Burn catchment on **(a)** a soil map and **(b)** an aerial photo. The precipitation sampling location is indicated by a blue triangle in **(a)**. **(c)** The four photos on the right show exemplary soil profiles for the four study sites.

Table 1. Vegetation and soil profile characteristics of the four study sites: percentage of coarse gravel (> 20 mm diameter) and fine gravel (20–2 mm diameter) of the soil sample, percentage of sand (S, 2–0.6 mm), silt (Si, 0.06–0.002 mm), and clay (C, < 0.002 mm) of the fine-soil matrix. Percentage of organic matter content in the soil as loss on ignition (LOI), correlation characteristics for LOI with depth and with gravimetric water content (GWC), and fine root density as percentage of total root mass. Note that for SH, NF, and SF not enough fine soil was left for soil texture analysis after ignition of organic material at 550 °C and that for NF and SF, no root density could be measured due to root thickness and stone content. ± indicate standard deviations out of five replicates for LOI values and range of two and three replicates for root density, respectively.

Site	Vegetation	Depth [cm]	Coarse gravel [% of soil]	Fine gravel [% of soil]	Sand [% of fine soil]	Silt [% of fine soil]	Clay [% of fine soil]	LOI [% of soil]	LOI vs. Depth	LOI vs. GWC	Fine root density [% of total roots]
NH	Heather	0–5	0	0	74	22	4	26 ± 15	$r = 0.66$; $p < 0.01$	$r = 0.73$; $p < 0.01$	82 ± 7
		5–10	0	5	77	19	4	11 ± 5			15 ± 5
		10–15	0	11	78	18	4	8 ± 2			3 ± 1
		15–20	2	10	80	16	4	9 ± 6			0
SH	Heather	0–5	0	0	–	–	–	92 ± 4	$r = 0.72$; $p < 0.01$	$r = 0.87$; $p < 0.01$	56 ± 18
		5–10	0	1	57	35	8	79 ± 14			24 ± 9
		10–15	0	5	78	18	4	54 ± 27			16 ± 6
		15–20	0	18	81	16	3	50 ± 39			4 ± 2
NF	Scots pine	0–5	0	1	–	–	–	28 ± 11	$r = 0.42$; $p = 0.06$	$r = 0.70$; $p < 0.01$	
		5–10	0	2	80	17	3	31 ± 20			
		10–15	0	0	77	18	5	20 ± 18			
		15–20	0	2	77	20	3	14 ± 10			
SF	Scots pine	0–5	0	1	–	–	–	95 ± 1	$r = 0.69$; $p < 0.01$	$r = 0.91$; $p < 0.01$	
		5–10	12	27	84	12	4	58 ± 36			
		10–15	15	22	85	12	3	40 ± 23			
		15–20	18	18	81	16	3	45 ± 39			

about 80 to 250 g of soil and was stored in air tight bags (Weber Packaging, Güglingen, Germany), ensuring – by manually furling the bags – that as little air as possible was inside them. The used bags ensured that there were no evaporation losses through bags, since less than 0.15 % of water was lost over 30 days in an experiment as reported by Sprenger et al. (2015a). For the sampling campaign in May, five additional sites with heather vegetation were sampled on the north- ($n = 2$) and south-facing ($n = 3$) slopes to increase the sample size for comparisons between the two slopes and get an idea of the general variability in space. In addition to the monthly sampling, two extra sampling campaigns for the two heather sites were conducted in August (4 and 9 August 2016) to investigate short time changes on both slopes and to further increase the sample number for comparison between the two slopes.

The soil samples were analyzed for their stable isotopic composition (δ^2H and δ^{18}O) according to the direct-equilibration method suggested by Wassenaar et al. (2008). The analyses were conducted within 1 week after the sampling to prevent microbial activity within the bags. The analyses were done by adding dry air to the bags that contained the soil samples, heat sealing the bags and letting the soil water equilibrate with the dry atmosphere in the bag for 2 days at constant temperature in the laboratory. The same was done in parallel with bags each filled with 10 mL of one of three different standard waters covering the range of the soil water isotopic signals: seawater ($\delta^{18}O = -0.85$ ‰ and $\delta^2H = -5.1$ ‰), Aberdeen tap water ($\delta^{18}O = -8.59$ ‰ and $\delta^2H = -57.7$), condensate of distilled tap water ($\delta^{18}O = -11.28$ ‰ and $\delta^2H = -71.8$) and for the sampling in January Krycklan snowmelt ($\delta^{18}O = -15.36$ ‰ and $\delta^2H = -114.4$). After the equilibration over 2 days, the vapor in the headspace of each bag was sampled directly with a needle connected to an off-axis integrated cavity output spectroscopy (OA-ICOS) (TWIA-45-EP, Los Gatos Research, Inc., San Jose, CA, USA). The δ^{18}O and δ^2H composition was continuously measured over 6 min, of which the last 2 min, when the water vapor pressure in the cavity was constant (standard deviation < 100 ppm), were used to calculate average values. The standard deviation for the δ^{18}O and δ^2H measurements were usually < 0.25 and < 0.55 ‰, respectively. The standards, which were treated the same way as the soil samples and measured at the beginning, the middle and the end of each sampling day, were then used for calibration to derive the isotopic composition of liquid soil waters from vapor measurements. For a detailed description of the soil water isotope analyses in the lab of the Northern Rivers Institute at the University of Aberdeen with the direct-equilibration method, we refer to Sprenger et al. (2017a). To assess the precision of the analysis, we derived the standard deviation of in total 81 measurements of the standard Aberdeen tap water sampled along with the soil samples on 27 days of laboratory analyses over 1 year. The standard deviation of the standard water analysis was 0.31 ‰ for δ^{18}O values and 1.13 ‰ for δ^2H values. Recently reported potential effects of CO_2 on the isotope analysis of vapor with wavelength-scanned cavity ring-down spectroscopy (Gralher et al., 2016) have been shown to not apply to the OA-ICSO that we used, as shown by Sprenger et al. (2017a). Potentially fractionating effects of interactions between soil water and surfaces of clay minerals (Oerter et al., 2014; Gaj et al., 2017a, b; Newberry et al., 2017) are of minor relevance for our study, since clay contents were low in the sampled soils (Table 1). We can further ensure that the sampled soil volumes always contained much more than 3 g of water as suggested by Hendry et al. (2015).

In addition to the soil water analysis, precipitation at the field site was sampled with an auto sampler on daily basis at the catchment outlet (location shown in Fig. 1a). The auto sampler was emptied at least every 2 weeks and evaporation from the sampling bottles was prevented by adding paraffin. The precipitation isotopic composition (δ^2H and δ^{18}O) was determined with the abovementioned OA-ICOS running in liquid mode with a precision of 0.4 ‰ for δ^2H values and 0.1 ‰ for δ^{18}O values, as given by the manufacturer.

2.4 Data analysis

We calculated the evaporation line (EL) as a regression line through the soil water isotope data of each sampling date in the dual-isotope space. The EL is characterized by its slope and intercept with the δ^2H axis. All regressions for EL presented here were significant at the 95 % confidence interval. For each soil water and precipitation sample, we further calculated the line-conditioned excess (lc-excess) as a function of the slope ($a = 7.6$) and the intercept ($b = +4.7$ ‰) of the LMWL (Eq. 1) as suggested by Landwehr and Coplen (2006):

$$\text{lc-excess} = \delta^2\text{H} - a \times \delta^{18}\text{O} - b. \qquad (2)$$

The lc-excess describes the deviation of the sample's δ^2H value the LMWL in the dual-isotope space (Landwehr et al., 2014), which indicates non-equilibrium kinetic fractionation processes due to evaporation after precipitation. Therefore, the lc-excess is similar to the well-established deuterium-excess (Dansgaard, 1964) that relates the deuterium composition to the global meteoric water line (GMWL). However, we found that lc-excess was advantageous over the deuterium-excess (or single isotope approaches with δ^2H or δ^{18}O) for inferring evaporation fractionation, because the lc-excess of the precipitation input is about 0 ‰ and with relatively little seasonal dynamics, while δ^2H, δ^{18}O, and d-excess can have an intense seasonal variability (Sprenger et al., 2017b). The accuracy for the liquid and soil water isotope analysis result in a precision limit for lc-excess of about 1.1 and 3.4 ‰, respectively.

To infer dynamics of potential evaporation rates, we estimated PET with the Penman–Monteith equation adjusted for the Scottish Highlands by Dunn and Mackay (1995). Note that we focus in our study on the PET dynamics and that the absolute values could vary depending on the aerodynamic and roughness parameter of different vegetation covers. We further did not partition PET into evaporation and transpiration fluxes, since PET was primarily used as a proxy for potential soil evaporation rates, and evaporation and transpiration usually show a linear relationship in temperate regions (Renner et al., 2016; Schwärzel et al., 2009). To understand the potential atmospheric drivers for the soil water isotopic composition, we investigated the effect of antecedent conditions. We calculated average values of PET over 30 days prior to each sampling campaign (PET_{30}) to account for the potential soil evaporation dynamics. Additionally, the precipitation sums and the amount weighted isotopic signal of the daily precipitation isotope samples were computed for the 7 and 30 days period prior to the sampling (P_7 and P_{30}, respectively) to assess the mixing processes between precipitation input and soil water. Weekly and monthly averages were chosen to see if the relatively young water input or the average over the last month relate differently to the observed soil water isotopic signal.

We estimated the evaporative losses f [%] based on the Craig–Gordon model (Craig and Gordon, 1965) and formulations introduced by Gonfiantini (1986) for isotope mass balance as follows:

$$f = 1 - \left[\frac{(\delta_S - \delta^*)}{(\delta_P - \delta^*)}\right]^m, \tag{3}$$

where δ_S is defined as the average weighted isotopic signal of the soil water in the upper 10 cm. The upper 10 cm were chosen, because this was the depth with the highest evaporation signal in the soil water isotopes (see results section) and where most evaporation are usually observed in laboratory experiments (Or et al., 2013). δ_P was defined as the isotopic signal of the original water source by calculating the intercept between the evaporation line of the soil water isotope data in the dual-isotope space (see Fig. S2) and the LMWL according to Javaux et al. (2016). δ^* is the limiting isotopic enrichment factor and m is the enrichment slope, both described by Gibson and Reid (2014) (Eqs. 8 and 9) therein). δ^* is a function of the air humidity h, the isotopic composition of the ambient air δ_A, and a total enrichment factor ε (Gat and Levy, 1978). For humidity, we averaged over 30 days prior to each soil water sampling date the measured humidity at a meteorological station less than 800 m away from the study sites h_{30}. δ_A was derived as function from the weighted average precipitation input of the 30 days prior to the soil sampling $\delta_{P_{30}}$ and the equilibrium isotope fractionation factor ε^+ (Gibson et al., 2008), with the latter depending on the temperature as given by Horita and Wesolowski (1994). We used air temperature data from the aforementioned meteorological station and computed values averaged over 30 days prior to soil water sampling T_{30}. The total enrichment factor ε is the sum of the equilibrium isotope fractionation factor ε^+ and the kinetic isotope fractionation factor ε_K, which is a function of h_{30}, the exponent of the diffusion coefficient ratio n and a kinetic fractionation constant, which has a value of 28.4 ‰ for $\delta^{18}O$ and 25.0 for δ^2H (Gonfiantini, 1986). Here, we define $n = 1$ in accordance to Barnes and Allison (1983), representing diffusional transport in soil pores. The enrichment slope m was calculated as a function from h_{30}, ε, and ε_k in accordance to Welhan and Fritz (1977).

f can be derived either by δ^2H or $\delta^{18}O$, indicated as f_{δ^2H} and $f_{\delta^{18}O}$. Note that $\delta_{P_{30}}$ represents in these calculations the net precipitation infiltrating the soil, which has the same isotopic composition as the measured rainfall isotope values, since no evidence for isotopic enrichment was found for the throughfall and stem flow in heather and Scots pine stands in the Bruntland Burn catchment (Braun, 2015).

Statistical analyses for the soil water isotopes (δ^2H, $\delta^{18}O$ and lc-excess) for individual sites, depths, and dates were done with non-parametric tests, since the null hypothesis that the data were drawn from a normal distribution was rejected for several sampling campaigns using the Shapiro–Wilk test for normality. We also tested whether there were significant differences between the sites (each site with $n > 200$) and sampling dates (each date with $n > 75$) or at different depths (each depth with $n > 200$) with the Kruskal–Wallis test. When significant differences were present at the 95 % confidence interval, a post hoc Dunn test with p value adjustment "Bonferroni" was applied to see which of the sampling dates or sampling depths were significantly different. For pairwise tests (e.g., aspect (north- and south-facing) or vegetation (soils beneath heather and soils beneath Scots pine; $n > 35$) for each vegetation type on each sample day and $n > 100$ for each vegetation type at each sample depth), the non-parametric Mann–Whitney–Wilcoxon test was used. We assessed if the soil water lc-excess values for different sites were significantly lower than 0 ‰ by using the Wilcoxon signed-rank test.

Mean values for each of the 11 sampling dates of δ^2H, $\delta^{18}O$, and lc-excess of soil water and P_7 or GWC, PET_{30}, P_7, and the lc-excess of P_{30} were normally distributed according to the Shapiro–Wilk test. Therefore, we used the t test for the mean of one group of samples to test if the average soil water lc-excess deviated significantly from zero for any sampling campaign. We further calculated the Pearson correlation coefficients (r) to describe linear relationships between mean values. Since δ^2H and $\delta^{18}O$ of P_{30} were not normally distributed, the Spearman rank correlation (ρ) was applied to describe relationships with these two variables. For all statistical analyses, the 95 % confidence interval was defined as significance level ($p < 0.05$). Visualizations with box plots generally show the interquartile range (IQR) as boxes, the median as line within the box, the 1.5 IQR as whiskers and data points > 1.5 IQR as points. We further make use of violin plots, where a kernel density estimation of the underlying distribution describing the data is visualized. Statistical differences derived with the post hoc Dunn test are visualized by either letters or colored markers within the box plots, where the same letter or marker color indicate that the samples are not significantly different to each other. Statistical differences between two groups derived with the Mann–Whitney–Wilcoxon test are indicated by either an asterisk or "X".

Table 2. For each soil sampling campaign: the PET on that sampling day; PET as average over the 30 days prior to the sampling day (PET_{30}); lc-excess of the precipitation input weighted averaged over 7 (P_7) and 30 (P_{30}) days, respectively; precipitation summed over 7 (P_7) and 30 (P_{30}) days prior to soil sampling; total soil sample number; minimum, median, maximum, and interquartile range (IQR) of the soil water lc-excess data; slope and intercept of the evaporation line (note that regression is significant at the 99 % confidence interval); and mean gravimetric water content (GWC).

	PET [mm d^{-1}]		P lc-excess [‰]		P [mm]		Sample [–]	Soil water lc-excess [‰]				Evaporation line Slope [–]	Evaporation line Intercept [‰]	GWC [–]
Date	PET	PET_{30}	P_7	P_{30}	P_7	P_{30}	*n*	min.	median	max.	IQR			mean
2015-09-29	2.7	1.4	−0.8	−1	1	66	91	−16.1	−5	1.1	4.2	4	−21.4	0.39
2015-10-20	1.2	1.1	−0.2	−0.7	2	60	78	−12.6	−4.1	5.3	5.2	3.6	−23.3	0.37
2015-11-26	1.1	0.8	3.4	−0.2	14	94	80	−11.1	−3	6.5	3.9	4.6	−21.1	0.37
2016-01-13	0.4	0.6	0.4	−1.6	56	434	77	−8.2	−0.8	7.5	6	8.2	10.8	0.42
2016-03-18	0.6	0.8	−5.2	−2.1	2	33	79	−11.6	−1.3	6.7	5.4	5.5	−15.2	0.37
2016-04-27	0.9	1.2	7.3	−1	21	80	79	−7.7	−1.8	5.8	6.4	5.1	−16.2	0.40
2016-05-24	1.6	2.1	0.1	0.1	31	66	80	−10	−1.4	2.9	3.7	4.9	−17.1	0.36
2016-06-18	2.6	1.9	−1.1	0.5	67	105	80	−13.2	−3.2	5.9	5.2	4.5	−21.6	0.38
2016-07-26	2.2	2.3	−2.5	−2.5	40	98	72	−14.6	−4.5	3.2	5.1	4.6	−19.1	0.37
2016-08-23	1	2.1	−3.4	−3.6	12	54	64	−16.7	−6	2	5.1	3.8	−25.4	0.35
2016-09-23	0.4	1.9	5.3	−1.6	8	51	77	−21.7	−4.1	8.9	6.2	4.3	−19.9	0.33

3 Results

3.1 Temporal dynamics in soil water isotopes

The temporal dynamic of the seasonally variable precipitation input signal between September 2015 and September 2016 was generally imprinted on the soil water (SW) isotope data at all sites. $P_{30}\delta^2H$ values usually were most similar to the lowest (most depleted) SW δ^2H values (Fig. 2b). An exception to this were the soil samples from January, because the sampling took place after a period of intense rainfall ($P_{30} = 434$ mm) that occurred at the end of December and beginning of January (Fig. 2a). The SW δ^2H (and $\delta^{18}O$) values averaged over all sites and depths for each sampling campaign correlated significantly with P_{30} δ^2H ($\rho = 0.92$, $p < 0.01$) and $P_{30}\delta^{18}O$ ($\rho = 0.97$, $p < 0.01$) and $P_7\delta^2H$ ($r = 0.74$, $p = 0.01$), and $P_7\delta^{18}O$ ($r = 0.83$, $p < 0.01$). However, the soil water averages were usually more enriched than the precipitation input (except for the January sampling). There is relatively little variability in the P lc-excess, which had a weighted average of −0.58 ‰. The average SW lc-excess was usually more negative than the P_{30} lc-excess, but the sampling campaigns on 13 January and 18 March in 2016 were an exception (Fig. 2c, Table 2). In contrast to δ^2H and $\delta^{18}O$, there was neither a relationship between the SW lc-excess with P_{30} lc-excess ($r = 0.33$, $p = 0.32$) nor with P_7 lc-excess ($r = 0.18$, $p = 0.59$). Thus, the dynamics of SW lc-excess cannot be explained by variation of the lc-excess in the input.

The seasonally variable input of P δ^2H and $\delta^{18}O$ led to significantly more depleted SW isotope values during January and March compared to the other sampling days (indicated by the letter "a" at the box plots in Fig. 3). The sampling in November and April represented a transition period, where the SW isotopic composition was significantly different to the winter and summer samples. The SW δ^2H and $\delta^{18}O$ values between May and August did not differ significantly. The soil water samples from January were the only ones that plotted along the LMWL with a slope of the EL of 8.2 (cyan dots in Fig. 3). The other soil water samples followed ELs of slopes between 3.7 and 5.4 with the lowest slopes at end of summer and beginning of autumn (Table 2). The variability of δ^2H and $\delta^{18}O$ values was highest for the sampling during winter and generally higher for $\delta^{18}O$ compared to δ^2H (note that the axes are scaled according to the GMWL in Fig. 3).

The enrichment in SW δ^2H during spring (green dots in Fig. 4a) followed an increase in PET_{30}. The highest PET_{30} in summer correspond with enriched δ^2H values. While PET_{30} decreased at the end of summer, the SW stayed enriched in ^2H until late autumn (orange dots in Fig. 4a). Thus, the SW δ^2H (and also $\delta^{18}O$, not shown) to PET_{30} relationship can be described by a linear correlation ($r = 0.65$, $p = 0.03$ for δ^2H and $r = 0.64$, $p = 0.03$ for $\delta^{18}O$). However, Fig. 4a shows that the delayed response in the SW isotopic signal to changes in PET_{30} resulted in a hysteresis pattern. The error bars in Fig. 4a, representing standard deviations of all the samples for each sampling campaign, show again that the SW δ^2H values were most variable for the sampling in January, becoming less variable during spring and lowest in autumn.

The hysteresis pattern was very pronounced for the relationship between SW lc-excess and PET_{30}, since the SW lc-excess only increased little with onset of PET_{30} in spring time (green dots in Fig. 4b). During summer, PET_{30} remained high, and SW lc-excess values became lower and stayed low, even when PET_{30} started to decrease (Fig. 4b). However, the relationship between SW lc-excess and PET_{30} could statistically be also described by a linear relationship

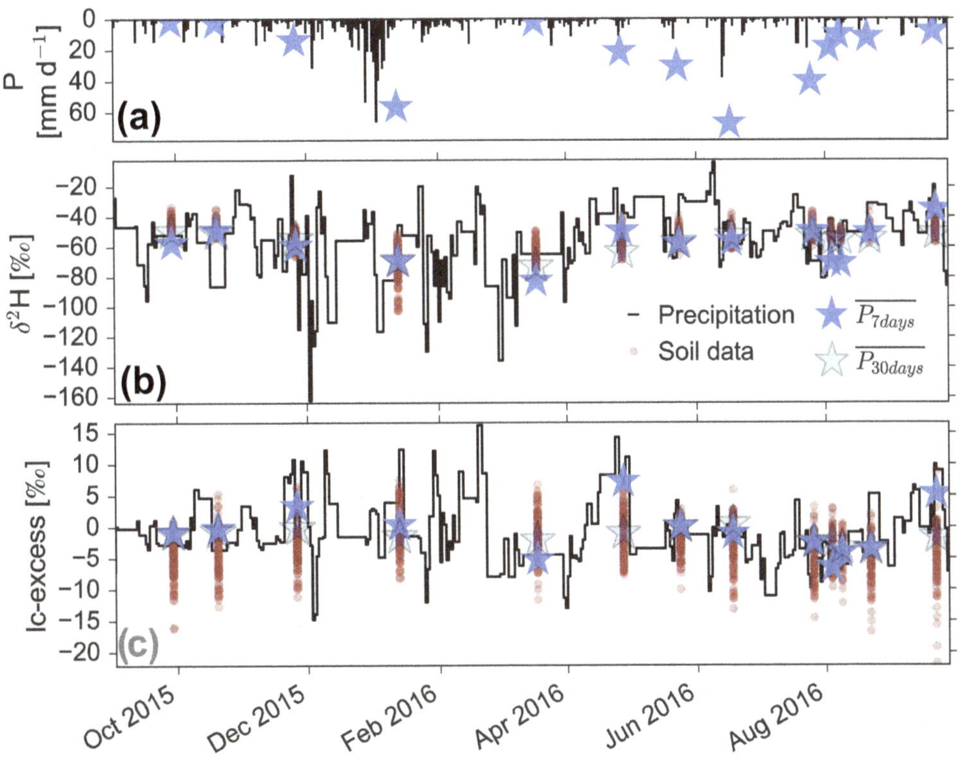

Figure 2. (a) Daily precipitation sums shown in a bar plot and precipitation sums over the 7 days prior to the soil sampling P_7 as a blue star. **(b)** Dynamics of the δ^2H and **(c)** lc-excess in the precipitation input (black line) and soil waters (half transparent brown dots). Stars indicate values as weighted averages over 7 days (P_7, blue) and 30 days (P_{30}, light blue) before the day of soil sampling.

($r=-0.64$, $p=0.04$). It is further worth noting that the average SW lc-excess in the upper 20 cm was for all sampling campaigns, apart from the January sampling, significantly <0 ‰, (t test, $p=0.63$). When limiting the analysis to samples from the upper 5 cm, there was also a hysteresis pattern in the relationship between SW lc-excess and PET_{30}.

The average GWC for each sampling campaign correlated significantly with the average SW lc-excess for that sampling day, with lower SW lc-excess when soils were drier ($r=0.67$, $p=0.02$). The GWC also correlated with SW δ^{18}O ($r=-0.75$, $p=0.01$) and δ^2H ($r=-0.74$, $p=0.01$) with isotopically more enriched values when the soil was drier.

3.2 Spatial soil water isotopes patterns

3.2.1 Differences between study sites

A comparison of the SW δ^2H and δ^{18}O values within the upper 20 cm at the sites for different sampling showed no significant differences between the sites (not shown). Differences of the SW isotope values between sites were limited to the upper 10 cm, with a significantly more depleted δ^2H signal for soil water at SH compared to NF (at 0–5 cm) and SF (at 0–10 cm), when looking at values bulked over the entire sampling period (Fig. 5a). For lc-excess, on average soil water samples at NF and SF showed lower values than NH and SH and differences were significant between NF and SH at 0–5 cm depth and for NF and SF compared to NH at 5–10 cm depth (Fig. 5b). At all four sites, the SW lc-excess averaged over the upper 20 cm was significantly <0 ‰ between May and October and not significantly <0 ‰ in January. However, the SW lc-excess for NF and SF was often lower than for NH and SH and the lc-excess or NF and SF was significantly <-5 ‰ for some sampling campaigns, while NH and SH was usually not significantly <-3 ‰. Regarding GWC (Fig. 5c), SH had significantly higher values than the other sites and NF had significantly lower values than the other sites in the upper 15 cm. The median GWCs at NH and SF were very similar, but variability was higher in the latter. The GWC was significantly linearly correlated to the organic matter content in the soil (Table 1) at all sites and this relationship explains partly the significant higher GWC at SH. This influence of organic matter content on GWC was evident during each sampling campaign: the sites with higher LOI tended to have higher average GWCs for the 11 sampling days. The coefficients of determination for the relationship between average LOI and average GWC on the sampling day was generally >0.56 (blue points in Fig. 6). Note that these coefficients of determination are limited to a sampling number of 4, given by four sampling sites, while the

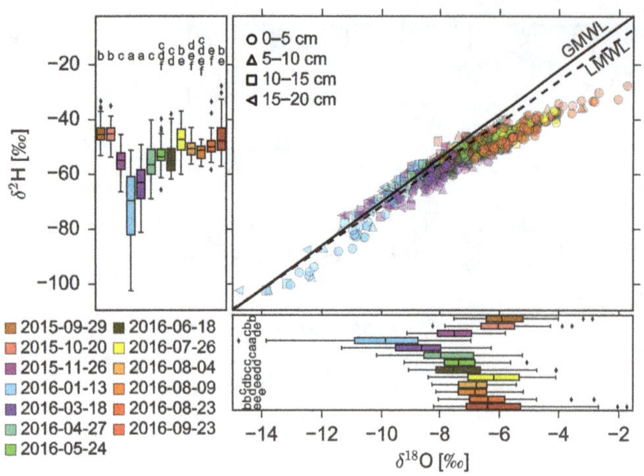

Figure 3. Dual isotope for all soil sampling campaigns and box plots for δ^{18}O and δ^2H bulked over all four sampling sites. The date of each soil sampling is indicated by colors. The depth of the soil samples is shown by different symbols in the dual-isotope plot. Sampling dates that do not have significantly different isotope data according to the post hoc Dunn test are indicated by the same letter next to the box plots. For example, the letter "a" at the box plots from sampling day 13 January 2016 and 18 March 2016, indicate that the soil water isotopes do not differ significantly from each other, but to all other sampling days. The global meteoric water line (GMWL, $\delta^2\mathrm{H} = 8 \times \delta^{18}\mathrm{O} + 10$) is shown with a solid line and local meteoric water line (LMWL, $\delta^2\mathrm{H} = 7.6 \times \delta^{18}\mathrm{O} + 4.7$) is plotted with a dashed line.

correlations given in Table 1 are based on $n = 20$ for each site. This relationship between GWC and LOI persists independently of how many dry days occurred prior to the soil sampling (Fig. 6).

While there was no effect of organic matter content on δ^2H and δ^{18}O values, average LOI at the sites showed a relationship with lc-excess on several sampling dates. These sampling campaigns, when the r^2 for the relationship between lc-excess and LOI ranged from 0.42 to 0.80, were all characterized by having at least 13 dry days during the 30-day period (40 %) prior to the sampling. For sampling campaigns with > 40 % dry days prior to the sampling, there was no relationship between lc-excess and LOI (r^2 ranged between 0.00 and 0.29, Fig. 6). Therefore, there is a significant correlation between the correlation coefficient between lc-excess and LOI and the percentage of dry days prior to the sampling ($r = -0.84$, $p < 0.01$).

3.2.2 Differences with soil depth

During most sampling dates δ^2H values became more depleted with increasing soil depth (Fig. 7a; SW δ^{18}O very similar and therefore not shown). The δ^2H values for the top 5 cm were always significantly different compared to the values at 15–20 cm. The δ^2H values at 0–10 cm were for most sampling campaigns significantly enriched compared to the soil water at 15–20 cm. However, during the November and January sampling, the upper 0–5 cm were more depleted due to the depleted precipitation δ^2H input (Fig. 1b). The highest variability of SW δ^2H within the sampling depths was found for the samples taken in January and March. Importantly, the variability of SW δ^2H generally decreased with soil depth.

The lc-excess for the upper 5 cm was always significantly more negative than at 15–20 cm (Fig. 7b). In addition, the soil signature at 5–10 cm was significantly more negative than at 15–20 cm; exceptions were November 2015 at 0–5 cm and September 2016 at 5–10 cm. The lc-excess depth profiles had a persistent pattern of steadily decreasing lc-excess values with depth, approaching SW lc-excess of 0 ‰ at 15–20 cm depth. The SW lc-excess variability usually decreased with depth, but not for the November sampling (Fig. 7b).

SW lc-excess at 15–20 cm was usually not significantly < 0 ‰, but at the top 5 cm, the SW lc-excess was significantly lower < 0 ‰ throughout the year; with only few exceptions (January and April 2016, Fig. 7).

The upper 5 cm soil had always significantly higher GWCs than the soil between 15 and 20 cm depth. This pattern of decreasing soil moisture with depth was persistent over time. The range of the GWC was always much lower below 15 cm than above (Fig. 7).

3.2.3 Differences due to vegetation cover

Comparing all soil samples bulked over depth and sampling campaigns, SW beneath Scots pine was significantly more enriched in δ^2H (Mann–Whitney–Wilcoxon Test, $p = 0.038$) and δ^{18}O ($p = 0.0062$) than the SW beneath heather sites. In addition, lc-excess was significantly lower for the SW at the forested sites compared to the heather sites ($p < 0.01$).

The temporal dynamics of the differences of δ^2H (and also SW δ^{18}O, but not shown) values for SW beneath the different vegetation sites show that the SW beneath Scots pine was (except for the sampling in March for δ^2H) always more isotopically enriched than the soil water beneath heather (Fig. 8a). However, the differences were only significantly different for 4 out of the 11 sampling campaigns. SW δ^2H values showed usually higher variability for soils beneath Scots pine for the sampling campaigns between May and October, but during winter, the variability is generally high for soils beneath both vegetation types.

The SW lc-excess was – except for the January and May sampling campaigns – more negative in the soils beneath Scots pine compared to soils beneath heather (Fig. 8b). The GWC was also always lower in the soils beneath Scots pines. These differences were most pronounced at the end of the summer and during autumn (Fig. 8c).

The differences between the SW isotopic composition under the two different vegetation types mainly stemmed from differences in the shallow soils. Despite being classed as similar podzolic soils, the SWs beneath Scots pine were significantly more enriched in the upper 10 cm in δ^2H (and also

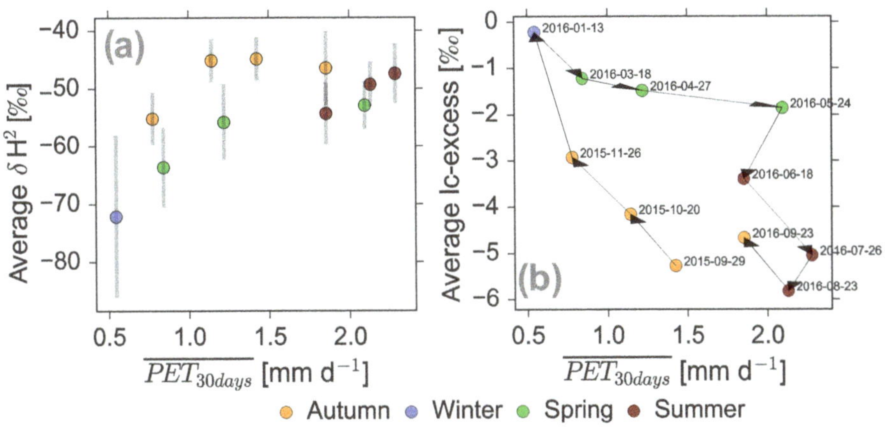

Figure 4. Variation of the **(a)** δ^2H and **(b)** lc-excess in the top 20 cm of the soil as a function of the potential evaporation averaged over the 30 days prior to the sampling day (PET_{30}). The colors of the scatter points indicate the season of the sampling. The error bars in **(a)** indicate the standard deviation of δ^2H in the soil waters. The arrows in **(b)** indicate the order of sampling and sampling dates are given.

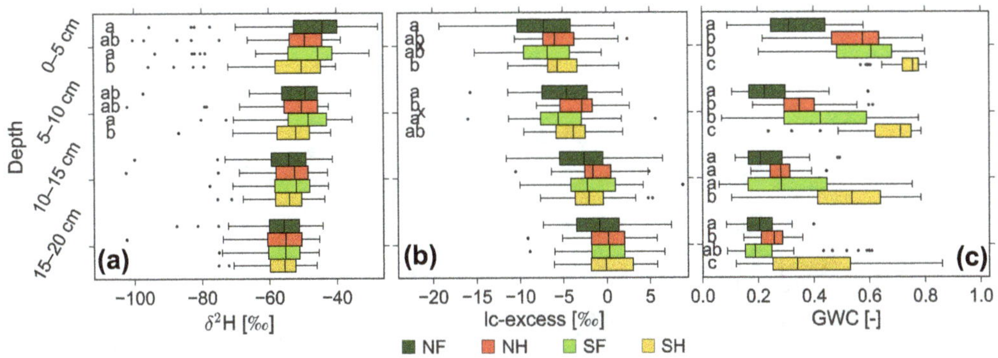

Figure 5. Site specific **(a)** δ^2H, **(b)** lc-excess, and **(c)** gravimetric water content (GWC) for the soils at the four study sites summarized for each site over all sampling dates as function of depth. For depths, where letters are shown, different letters indicate significant differences between the sites for the specific depth.

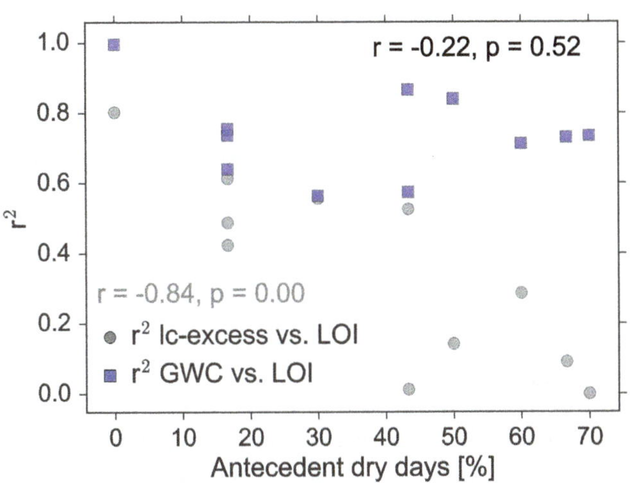

Figure 6. Relationship between coefficient of determination of the relationship between lc-excess and LOI (gray) and GWC and LOI (blue) with the percentage of dry days during the 30-day period prior to the soil sampling.

δ^{18}O, not shown) than the SW beneath heather (Fig. 9a). SW beneath Scots pine had a significantly more negative lc-excess signal for the upper 15 cm soils. The GWC of soil beneath Scots pine were over the entire soil profile significantly lower than for the soils beneath heather (Fig. 9b).

No significant differences were found when splitting the samples according to their aspect. There were neither differences when looking at all sampling depths, nor for any individual sampling depth (not shown). A comparison of the more intensive spatially distributed sampling focusing on the top 10 cm of heather soils in May 2016 at the north-facing sites ($n = 26$) and south-facing sites ($n = 33$) showed also no significant differences between aspects regarding isotopes and soil moisture. Increasing the sampling numbers for the heather sites when the two additional sampling campaigns in August were included, also did not result in significant differences between the two studied slopes. The median SW δ^2H values for the samples taken at the south- and north-facing slopes were −50.6 and −50.9 ‰, respectively. The median SW lc-excess was −3.4 and −3.3 ‰ for south- and north-

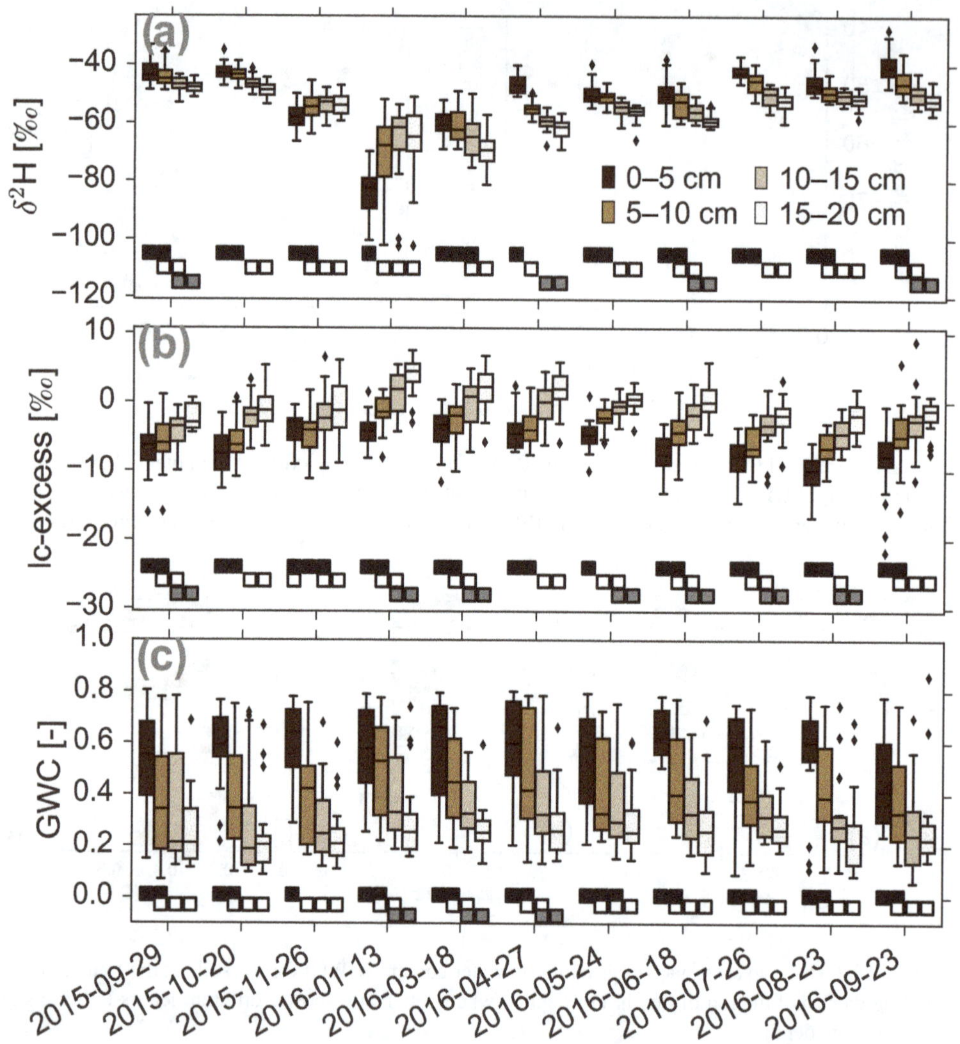

Figure 7. Differences of **(a)** δ^2H, **(b)** lc-excess, and **(c)** gravimetric water content (GWC) over depth for each sampling day. Black, white, and gray squares below the box plots indicate significant differences between the depth for each sampling day according to the Dunn test (i.e. same color reflects similarities; different colors differences). For example, for 29 September 2015, samples from 0 to 5 cm depth were significantly different from samples from the depth 10–15 cm and 15–20 cm, but not significantly different to samples from 5–10 cm depth.

facing slope, respectively. The median GWC was 0.62 and 0.54 for the samples at the south- and north-facing slope, respectively.

3.3 Evaporation estimates

Our findings clearly showed an influence of the vegetation on the evaporation signal in the soil water between April and October. Therefore, we estimated the fraction of evaporation losses in the SW beneath Scots pine and heather as described in the methods section for the nine samplings in this period. For the heather sites, the median (±SD over the period) values of $f_{\delta^2\mathrm{H}}$ and $f_{\delta^{18}\mathrm{O}}$ were 5.0 ± 4.0 and 3.7 ± 1.6 %, respectively, and for the forested sites, the values were $f_{\delta^{18}\mathrm{O}} = 9.9 \pm 3.4$ % and $f_{\delta^2\mathrm{H}} = 6.3 \pm 2.1$ %. Thus, the SW fractionation signal indicates that about 8 and 4.5 % of the originally infiltrated water (with lc-excess = 0 ‰) went back into the atmosphere by soil evaporation.

4 Discussion

4.1 Mixing of precipitation input and the critical zone water storage

Our uniquely detailed data set from 11 sampling campaigns over 1 year revealed the isotopic response of soil water to variable precipitation inputs. An understanding of the isotopic variability of the precipitation input signal is crucial for the interpretation of the soil water isotopes in terms of soil evaporation dynamics at the soil–plant–atmosphere interface. Therefore, we will first discuss how mixing of the

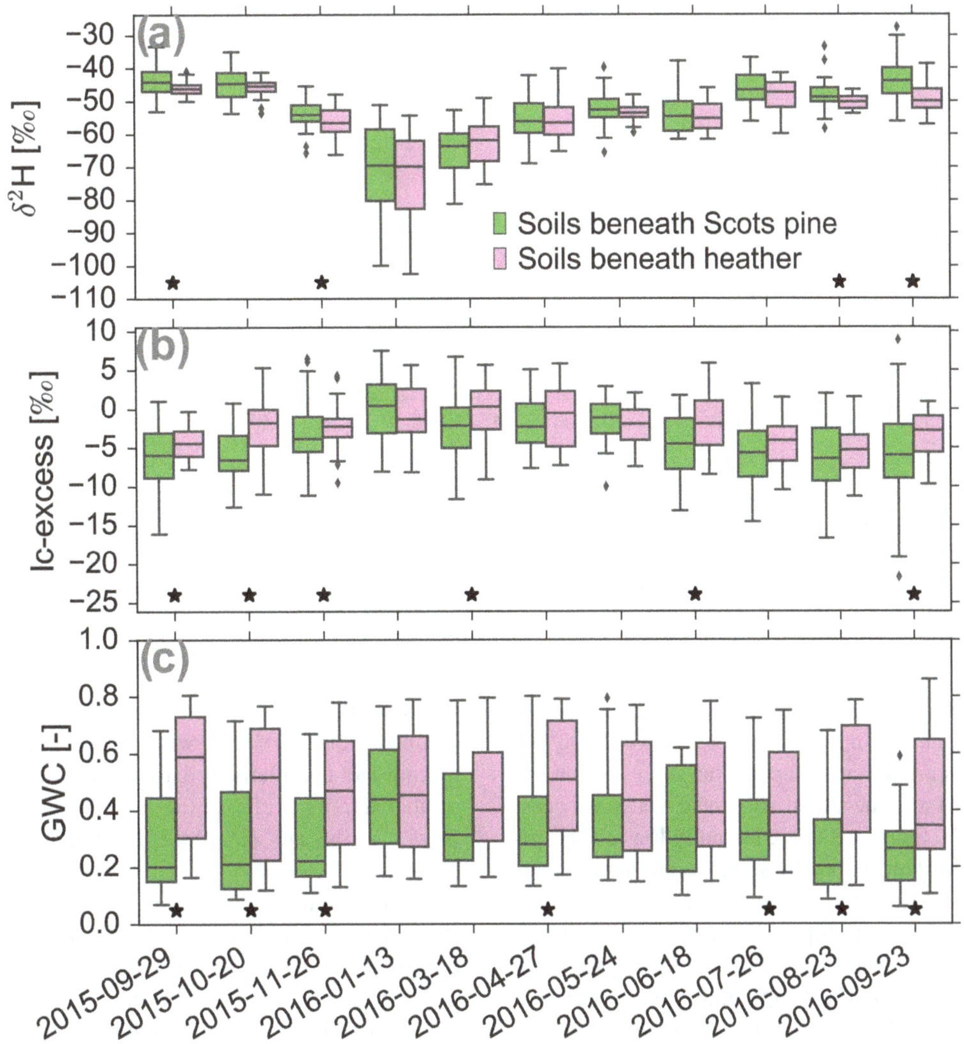

Figure 8. Differences of **(a)** δ^2H, **(b)** lc-excess, and **(c)** gravimetric water content (GWC) for soils under Scots pine (green) and soils under heather (violet) during 1 year. Significant differences estimated with the Mann–Whitney–Wilcoxon test between soils under Scots pine and soils under heather for a particular day are indicated by a star.

infiltrating precipitation within the topsoil provides the basis for the isotopic enrichment in the soil due to evaporative losses.

Given that the soil at the four sites consists of similar texture, it is not too surprising that the SW generally responded similarly to the seasonal P δ^2H and δ^{18}O input signal. However, we would expect differences in the soil physical properties due to the differences in the organic matter contents. Our data showed that higher organic matter resulted in higher GWCs at the sites. That means that the soil water storage is higher for the sites with higher organic material, which is in line with several other studies (e.g., Hudson, 1994).

While this relationship between GWC and LOI persisted independently of how many dry days occurred prior to the soil sampling, LOI showed a relationship with lc-excess depending on the dryness. This suggests that during dry periods, the external atmospheric drivers such as evaporative demand may have a high influence on the SW lc-excess. In contrast, during wet periods, the soil water storage capacities – here, mainly controlled by organic material – gain importance. With increasing new precipitation input (lc-excess close to zero), the sites with higher organic matter content have a different SW lc-excess than the sites with lower organic matter content, because there is a higher mixing volume for the sites with the high LOI compared to the soils with lower LOI (Fig. 6). Hence, future and current changes in the organic matter content of soils with high organic matter, due to, for example, land use or climate changes (Foley et al., 2005; Rees et al., 2011), would have an impact on the mixing of event and pre-event water in the podzols.

Interestingly, the highest variability of the δ^2H values in the soil water was found for the sampling after the intense

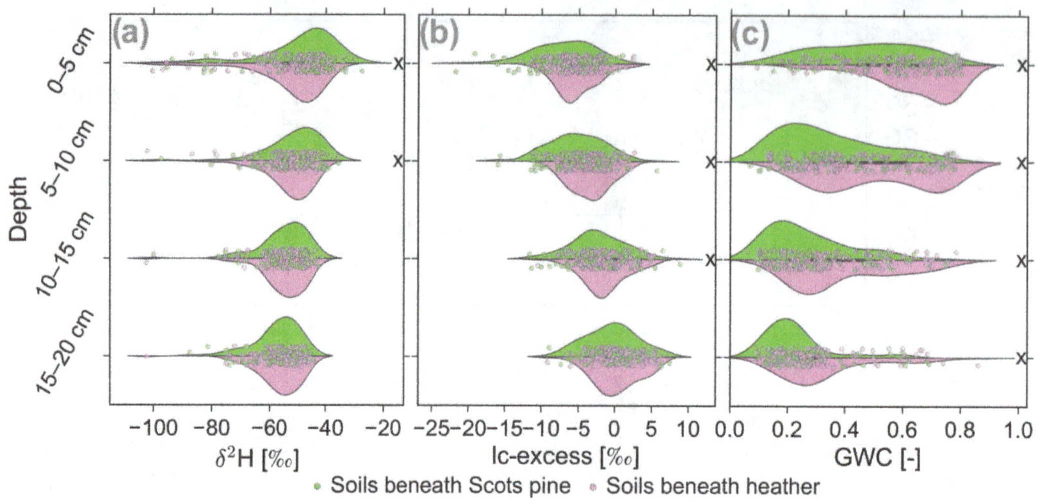

Figure 9. Violin plots showing the distribution as a kernel density estimation of the soil water **(a)** δ^2H, **(b)** lc-excess, and **(c)** gravimetric water content over all sampling campaigns with differentiation between soils beneath Scots pine (green) and soils beneath heather (purple). The symbol "X" at the depths indicates significant differences between the soils under Scots pine and soils under heather for the particular depth as estimated with the Mann–Whitney–Wilcoxon test. Dots indicate the individual sample points.

rainfalls in January 2016 (Figs. 4a, 7a), which indicates that the unsaturated zone was not homogeneously wetted by the exceptionally high precipitation input. This suggests that bypass flow occurred and the stored soil water was not necessarily well-mixed. However, the significantly lower δ^2H values in the top 5 cm indicate that, despite the potential occurrence of preferential flow, most of the depleted precipitation input was stored in the very top soil and the more enriched soil water from autumn was partly mixed with and partly replaced by the event water. The sampling in January was also special, because the GWC was about as high in the soils beneath Scots pine than in the soils beneath heather. Hence, differences in GWC between soils beneath the two vegetation types during the vegetation period as discussed below would stem from different transpiration and soil evaporation rates and not from differences in the soil water storage capacities.

The higher variability of the SW δ^2H values beneath Scots pine compared to the SW beneath heather (Fig. 9) cannot be explained by differences of the throughfall isotopic signal, since they are minor for the two vegetation types (Braun, 2015). The higher variability in the isotopic signal therefore indicates that flow paths are generally more variable in the forest soils. This is probably due to preferential flows via larger macropores formed by tree roots that also reach deeper than the shallower roots of heather. While the high variability of SW δ^2H values indicate preferential flow paths, there was also a clear signal that much of the input water percolates through the pore matrix, since we see that the depleted winter precipitation signal is still evident at −20 cm soil depth during the March sampling campaign (Fig. 7a).

The relatively high soil water storage volumes in the podzols result in a damping of the δ^2H signal in the SW compared to the precipitation input. There was also a more damped isotopic signal with soil depth due to increased mixing and cumulatively larger soil water storage. Because of this damping effect, the differences between the sites (Figs. 5a, 9a) and sampling times (Fig. 7) decreased with depth.

Our data support the findings of Geris et al. (2015b), who showed – with soil water samples extracted with suction lysimeters – a more damped signal in deeper layers of the soils beneath heather than beneath Scots pine. However, we did not see a more delayed response beneath Scots pine compared to podzols beneath heather as observed by Geris et al. (2015b). This discrepancy could arise from the fortnightly sampling frequency by Geris et al. (2015b) and monthly sampling frequency in the current study, as well the effect of an unusually dry summer in the former study.

In contrast to Geris et al. (2015b), who sampled the soil–vegetation units with duplicates, our sampling design with five replicates allowed for a clearer assessment of the spatial heterogeneity of the subsurface. For example, the standard deviation of the SW δ^2H and δ^{18}O values for the 5 cm depth increments at each site was – for the entirely sampled upper 20 cm of soil – always higher than the measurement accuracy of 1.13 and 0.31 ‰, respectively. At Bruntland Burn – as in most northern temperate and boreal biomes (Jackson et al., 1996) – topsoils contain almost all the root biomass. For the interpretation of potential sources of root water uptake this means that the uncertainty of the potential water source signal – caused by the heterogeneous isotopic composition at particular depths within the rooting zone – is higher than the measurement errors. Our field measurements underline, therefore, the need for an improved spatial resolution of soil water sampling when studying root water uptake patterns with stable isotopes, as recently called for by Berry et al. (2017). Further, this high variability will potentially impact the application of soil water isotopes for the calibration of soil physical models and the resulting interpretation (Sprenger et al., 2015b).

The observed mixing processes have implications for modeling of water movement within the critical zone. The soil water isotope response seems to follow a replacement of the pore waters by a mixture of newly infiltrated precipitation and older water, which could probably be described by the advection dispersion equation as frequently applied for isotope modeling in the unsaturated zone (Adomako et al., 2010; Stumpp et al., 2012; Mueller et al., 2014; Sprenger et al., 2015b). However, during intense rainfall, the precipitation input seems to partly bypass the matrix leading to rapid changes in the tracer signal in the subsoil and a generally high variability of the soil water isotopes within the entire soil profile. Such preferential flow processes might be better represented by soil physical models that take different mobility of water within the pore space into account, like the dual-permeability (Gerke and van Genuchten, 1993) or dual-porosity (van Genuchten and Wierenga, 1976) approaches. For example, Gouet-Kaplan et al. (2012) showed with column experiments that a significant amount of pre-event (or "old") water stayed in a sandy pore system. They found that these processes could be described by a dual-porosity (mobile–immobile) model. Such incomplete mixing within the unsaturated zone could have an impact on catchment transit time estimates (van der Velde et al., 2015). Given the high storage volumes of the periglacial deposits (Soulsby et al., 2016b) and the high water-mixing volume in the riparian zone (Tetzlaff et al., 2014) in the Bruntland Burn, it is yet unclear, which role the relatively slow moving and partly mixed soil waters in the soil matrix at the hillslopes could play regarding long tails of transit times in the Scottish Highlands, as reported by Kirchner et al. (2010).

4.2 Evaporation dynamics within the soil–plant–atmosphere interface

The correlation between the seasonally variable P and SW isotopic signals inhibited an assessment of soil evaporation processes from either δ^2H or δ^{18}O. Instead, the SW lc-excess values were independent from the P input and were therefore indicative of kinetic fractionation due to soil evaporation. The monthly soil water sampling revealed soil evaporation dynamics (in terms of lc-excess) during times of the highest potential evaporation for the Scottish Highlands. The evaporation fluxes were not expected to be high given the relatively low energy environment at the study site. The soil waters showed, nevertheless, a clear fractionation signal in terms of their lc-excess (25th percentile < -6 ‰ between June and October) and the evaporation line (3.6 and 4.5 between June and October, Table 2). This fractionation signal in the soil water was of the same magnitude as for surface waters in the peatland drainage network of a raised bog in the Bruntland Burn, where the lc-excess reached values < -5 ‰ and the EL slope ranged from 3.9 to 4.9 between May and September (Sprenger et al., 2017b). In comparison to arid and Mediterranean environments, where SW lc-excess can fall below -20 ‰ (McCutcheon et al., 2016, and reviewed in Sprenger et al., 2016b), the SW lc-excess in the Scottish Highlands remained relatively high. While the lc-excess was usually not significantly different from zero at 15–20 cm soil depth in the studied podzols of the study site, soils studied by McCutcheon et al. (2016) in a much drier environment (Dry Creek, Idaho) showed that the SW lc-excess from the surface down to -70 cm was significantly lower than zero.

The high soil water storage contributes to the SW lc-excess dynamics shown in the hysteresis pattern for the relationship between SW lc-excess and PET_{30}, which revealed that there was a delayed response of the SW lc-excess to the onset and offset of soil evaporation in spring and autumn, respectively. We explain the hysteresis by generally low soil evaporation fluxes, relatively high humidity in the Scottish Highlands, and high soil water storage. The high soil water storage means that relatively high evaporation losses are needed in spring to change the ^{2}H to ^{18}O ratios by kinetic fractionation. The variability of SW lc-excess values from July to September was relatively small, indicating that a steady state between soil water fractionation by soil evaporation and input of unfractionated precipitation was reached. In autumn, when the soil evaporation ceased, relatively high unfractionated precipitation input was needed to dilute the evaporation fractionation signal of the soil water again (Fig. 4).

In contrast to our findings of a pronounced evaporation fractionation in the soil water, Geris et al. (2015b), who sampled the mobile soil water for its isotopic composition with suction lysimeters, saw little to no fractionation. The sampling by Geris et al. (2015b) took place at the same/similar sites (podzols beneath Scots pine and heather within the Bruntland Burn), but during different years. However, the sampling period of Geris et al. (2015b) covered summer 2013, which was exceptional dry ($\sim$ 10-year return period) and would have been therefore very likely to induce evaporation fractionation in soil waters. We conclude that the differences between the two studies most likely relate to the different sampling methods. Suction lysimeters, as applied by Geris et al. (2015b), are known to be limited to sample the mobile water within the pore space, but the direct-equilibration method applied in our study also samples more tightly bound waters (Sprenger et al., 2015a). Consequently, we can infer that the mobile waters sampled with suction lysimeters will be relatively young waters that percolated without experiencing pronounced water losses due to soil evaporation. In contrast, the soil water data presented in our study represent a bulk pore water sample, where more tightly bound waters will be integrating older ages and, therefore, affected by kinetic fractionation during periods of atmospheric evaporative demand. However, comparisons between mobile and bulk soil water isotopic signals showed that tightly bound water at soil depth deeper than 10 cm was usually more depleted than the mobile waters during spring or summer, due to filling of fine pores of a relatively dry soil with depleted precipitation several months earlier (Brooks et al., 2010; Geris et al., 2015a; Oerter and Bowen, 2017). Our study shows that, in contrast to old isotopically distinct infiltration water, also the legacy of evaporation losses over previous months allows for separation between pools of different water mobility in the soil–plant–atmosphere interface. This means that old (more tightly bound) water might have not only a distinct δ^2H or δ^{18}O signal compared to mobile water due to seasonally variable precipitation inputs, but also an evaporative enrichment signal from periods of high soil evaporation. In conclusion, when relating isotope values of xylem water to soil water to study root water uptake patterns, an evaporation signal (that is lc-excess of xylem water <0) would not be paradoxical, but simply represent the range of available soil water in the subsurface.

So far, mostly cryogenic extraction was used to analyze the isotopic composition of bulk (mobile and tightly bound) soil waters and relate these data to, for example, root water uptake pattern. However, recent experimental studies indicated potential for fractionation during the cryogenic extraction (Orlowski et al., 2016), probably induced by water–mineral interactions (Oerter et al., 2014; Gaj et al., 2017a, b). The direct-equilibration method – as a potential alternative analysis method – we applied in our study was so far usually limited to study percolation processes (e.g., Garvelmann et al., 2012; Mueller et al., 2014; Sprenger et al., 2016a). To our knowledge, only Bertrand et al. (2012) used the direct-equilibration method to study the water use by plant water use and the influence of evaporation fractionation, but they bulked the soil data of the upper 20 cm.

However, our results emphasize the importance of sampling the upper most soil layer when studying plant–water interactions and soil evaporation dynamics. The evaporation was predominantly taking place at the interface to the atmosphere, where the lowest SW lc-excess values were found throughout the sampling period. Due to the relatively wet conditions of the studied soils, the evaporation signal dropped sharply within the first 15 cm soil depth and was always significantly lower at the top 5 cm compared to the SW at 15–20 cm depth. Geris et al. (2015a) sampled with cryogenic extraction the isotopic composition of the bulk soil water at −10 cm soil depth at the site NF during the summer of 2013. They found soil water isotopes of the same magnitude at −10 cm (δ^2H between −40 and −55 ‰), but their sample size was too little to assess evaporation fractionation patterns.

The pronounced differences of the soil water samples in the dual-isotope space over little soil depth is crucial when assessing potential water sources of the vegetation. The lc-excess signal could therefore provide additional information as an end-member when applying mixing models to derive root water uptake depths and times, since δ^2H and lc-excess do not necessarily correlate. So far, usually either δ^2H or δ^{18}O, but not lc-excess are being considered to delineate water sources of the vegetation (Rothfuss and Javaux, 2017). In line with the call for a higher temporal resolution of soil water isotope sampling (Berry et al., 2017), the highly dynamic isotopic signal during the transition between the dormant and growing seasons underlines the importance of not limiting the soil water isotope sampling to a few sampling campaigns (usually $n \leq 3$ in Brooks et al., 2010; Evaristo et al., 2016; Goldsmith et al., 2012), when investigating root water uptake patterns. In the light of recent studies dealing with potential water sources of vegetation that showed how variable the plant water isotopic signal can be over a year (Hervé-Fernández et al., 2016; McCutcheon et al., 2016), a proper understanding of the temporal variability of the potential water sources (i.e., bulk soil water, groundwater, stream water) appears to be critical.

4.3 Vegetation affects critical zone evaporation losses

The vegetation was the main reason for differences in the soil water evaporation signal between the four study sites. While we limited the study to one soil type, we still saw differences between the soils in terms of their organic matter content leading to different soil water storage capacity. However, these differences in water storage capacity seemed to only influence the evaporation signal during periods of high precipitation input, when evaporation is already likely to be low (Fig. 6). The aspect and exposition of the slopes did not affect the soil water evaporation signal, which probably can be explained by the low angle (4°) for the north- and south-facing slopes at our site. Differences in net radiation for the two studied slopes were found to be small with the north-facing slope receiving about 4 % less radiation than the south-facing slope between April and October (Ala-Aho et al., 2017).

The significant differences of the SW lc-excess in soils beneath Scots pine and heather are mainly due to the differences of the vegetation structure. The heather forms a dense soil cover just about 20 cm above the ground with an understory of mosses and lichens. Thus, the vegetation shades most of the soil and generates a microclimate with little direct exchange between soil vapor and atmospheric vapor. The soil climatic conditions beneath heather are therefore characterized by higher relative humidity and less radiation input than for the soils beneath Scots pine. In addition, the soils beneath the Scots pine are not densely vegetated by understory, which allows for more open exchange between soil vapor and atmospheric vapor at the forested sites. A similar conclusion of microclimatic conditions driving the differences in the evaporation signal was drawn by Midwood et al. (1998), who also reported a higher isotopic enrichment of soil surface soil waters under groves and woody clusters than under grassland.

The fractionation signal was shown to be more pronounced for evaporation losses from drier soils compared to wetter soils (Allison et al., 1983; Barnes and Allison, 1988).

Higher canopy storage, higher interception losses and reduced net precipitation in the forested stands, where throughfall is about 47 % of precipitation, compared to the heather with throughfall being about 35 % of precipitation (Braun, 2015) will influence the GWC of the soils beneath the two vegetation types. The higher transpiration rates of the trees (Wang et al., 2017) compared to the heather further influences the soil moisture dynamics, leading to significantly lower GWC in the forested sites. In consequence, soil evaporation taking place from the drier soils beneath Scots pine will result in higher evaporation fractionation than for soil evaporation from soils beneath heather.

The evaporation estimates are limited to the period of the highest evaporation fluxes in the Bruntland Burn and are likely to be lower than what we present, since we relate the isotopic enrichment to the long-term average input signal in our calculations. However, the assumption of relating the enrichment to the average input seems to be valid, since we have seen the steady state balance of unfractionated input and kinetic fractionation due soil evaporation during the summer months. Nevertheless, detailed estimates of the evaporative fluxes require transient modeling of all water fluxes (i.e., precipitation, transpiration, evaporation, recharge) and their isotopic composition, which will be subject of future work. Potential transient numerical modeling approaches that account for isotopic fractionation of the soil water isotopes are available (Braud et al., 2005; Rothfuss et al., 2012; Mueller et al., 2014).

The frequently measured soil water isotope data we have presented and the inherent evaporation signal can help to calibrate or benchmark the representation of soil–vegetation–atmosphere interactions in tracer-aided hydrological modeling from the plot (Rothfuss et al., 2012; Sprenger et al., 2016b) to the catchment scale (Soulsby et al., 2015; van Huijgevoort et al., 2016). So far, the isotopic composition of mobile water has been used to better constrain semi-distributed models (Birkel et al., 2014; van Huijgevoort et al., 2016). Including the isotopes of the bulk soil water, could allow for an improved conceptualization of the soil–plant–atmosphere interface, which is crucial for an adequate representation of evaporation and transpiration. Especially the marked differences of the evaporation signal within the first few centimeters of the soil depth will significantly affect the age distribution of transpiration (Sprenger et al., 2016c), evaporation (Soulsby et al., 2016a) or evapotranspiration (Harman, 2015; Queloz et al., 2015; van Huijgevoort et al., 2016).

5 Conclusions

Our study provides a unique insight into the soil water stable isotope dynamics in podzolic soils under different vegetation types at a northern latitude site. We showed that, despite a relatively low energy environment in the Scottish Highlands, the temporal variability of soil water isotopic enrichment was driven by changes of soil evaporation over the year. The monthly frequency of the soil water isotope sampling corroborates the importance of covering transition periods in a climate with seasonally variable isotopic precipitation input signals and evaporative output fluxes. Missing sampling these periods of higher temporal variability in spring and autumn could pose problems when referring plant water isotopes to potential soil water sources or when using soil water isotopic information for calibrating hydrological models. Especially the delayed response of the soil water lc-excess to evaporation (hysteresis effect) provides valuable insight into how unfractionated precipitation input and kinetic fractionation due to soil evaporation both affect mixing processes in the upper layer of the critical zone. The fact that the evaporation signal generally disappears within the first 15 cm of the soil profile emphasizes the importance and the spatial scale of the processes taking place at the soil–vegetation–atmosphere interface of the critical zone.

The vegetation type played a significant role for the evaporation losses, with a generally higher soil evaporation signal in the soil water isotopes beneath Scots pine than beneath heather. Notably, these differences – as indicated in the soil water lc-excess – remain limited to the top 15 cm of soil. The vegetation cover directly affected the evaporation losses with soil evaporation being twice as high as beneath Scots pine (10 % of infiltrating water) compared to soils beneath heather (5 % of the infiltrating water) during the growing season.

The presented soil water isotope data covering the seasonal dynamics at high spatial resolution (5 cm increments at four locations) will allow us to test efficiencies of soil physical models in simulating the water flow and transport in the critical zone (as called by for Vereecken, 2016). The data can further provide a basis to benchmark hydrological models for a more realistic representation of the celerities and velocities when simulating water fluxes and their ages within catchments (McDonnell and Beven, 2014).

Author contributions. DT, CS, and MS established the sampling design together, MS conducted the field and laboratory work and statistical analysis, all authors were involved in the data interpretation, MS prepared the manuscript with contributions from both co-authors.

Competing interests. The authors declare that they have no conflict of interest.

Acknowledgements. We are thankful for the support by Audrey Innes during all laboratory work. We further thank Jonathan Dick for running the isotope analysis of precipitation samples and

Annette C. Raffan for her support in the soil texture analysis. We would also like to thank the European Research Council (ERC, project GA 335910 VeWa) for funding.

Edited by: Hilary McMillan

References

Adomako, D., Maloszewski, P., Stumpp, C., Osae, S., and Akiti, T. T.: Estimating groundwater recharge from water isotope (δ^2H, δ^{18}O) depth profiles in the Densu River basin, Ghana, Hydrolog. Sci. J., 55, 1405–1416, https://doi.org/10.1080/02626667.2010.527847, 2010.

Ala-aho, P., Tetzlaff, D., McNamara, J. P., Laudon, H., and Soulsby, C.: Using isotopes to constrain water flux and age estimates in snow-influenced catchments using the STARR (Spatially distributed Tracer-Aided Rainfall-Runoff) model, Hydrol. Earth Syst. Sci. Discuss., https://doi.org/10.5194/hess-2017-106, in review, 2017.

Allison, G., Barnes, C., and Hughes, M.: The distribution of deuterium and ^{18}O in dry soils 2. Experimental, J. Hydrol., 64, 377–397, https://doi.org/10.1016/0022-1694(83)90078-1, 1983.

Ball, D. F.: Loss-on-ignition as an estimate of organic matter and organic carbon in non-calcareous soils, J. Soil Sci., 15, 84–92, https://doi.org/10.1111/j.1365-2389.1964.tb00247.x, 1964.

Barnes, C. and Allison, G.: The distribution of deuterium and ^{18}O in dry soils: 1. Theory, J. Hydrol., 60, 141–156, https://doi.org/10.1016/0022-1694(83)90018-5, 1983.

Barnes, C. and Allison, G.: Tracing of water movement in the unsaturated zone using stable isotopes of hydrogen and oxygen, J. Hydrol., 100, 143–176, https://doi.org/10.1016/0022-1694(88)90184-9, 1988.

Berry, Z. C., Evaristo, J., Moore, G., Poca, M., Steppe, K., Verrot, L., Asbjornsen, H., Borma, L. S., Bretfeld, M., Hervé-Fernández, P., Seyfried, M., Schwendenmann, L., Sinacore, K., de Wispelaere, L., and McDonnell, J.: The two water worlds hypothesis: Addressing multiple working hypotheses and proposing a way forward, Ecohydrol., https://doi.org/10.1002/eco.1843, online first, 2017.

Bertrand, G., Masini, J., Goldscheider, N., Meeks, J., Lavastre, V., Celle-Jeanton, H., Gobat, J.-M., and Hunkeler, D.: Determination of spatiotemporal variability of tree water uptake using stable isotopes (δ^{18}O, δ^2H) in an alluvial system supplied by a high-altitude watershed, Pfyn forest, Switzerland, Ecohydrol., 7, 319–333, https://doi.org/10.1002/eco.1347, 2012.

Birkel, C., Soulsby, C., and Tetzlaff, D.: Developing a consistent process-based conceptualization of catchment functioning using measurements of internal state variables, Water Resour. Res., 50, 3481–3501, https://doi.org/10.1002/2013WR014925, 2014.

Braud, I., Bariac, T., Gaudet, J. P., and Vauclin, M.: SiSPAT-Isotope, a coupled heat, water and stable isotope (HDO and H2^{18}O) transport model for bare soil. Part I. Model description and first verifications, J. Hydrol., 309, 277–300, https://doi.org/10.1016/j.jhydrol.2004.12.013, 2005.

Braun, H.: Influence of vegetation on precipitation partitioning and isotopic composition in northern upland catchments, Master's Thesis, Professur für Hydrologie, Albert-Ludwigs-Universität Freiburg i. Br., 106 pp., 2015.

Brooks, J. R., Barnard, H. R., Coulombe, R., and McDonnell, J. J.: Ecohydrologic separation of water between trees and streams in a Mediterranean climate, Nat. Geosci., 3, 100–104, https://doi.org/10.1038/NGEO722, 2010.

Brooks, P. D., Chorover, J., Fan, Y., Godsey, S. E., Maxwell, R. M., McNamara, J. P., and Tague, C.: Hydrological partitioning in the critical zone: Recent advances and opportunities for developing transferrable understanding of water cycle dynamics, Water Resour. Res., 51, 6973–6987, https://doi.org/10.1002/2015WR017039, 2015.

Carrer, G. E., Rousseau, A. N., Jutras, S., and Fossey, M.: Assessment of the impact of pools on the water isotopic signature of a boreal patterned peatland, Hydrol. Process., 30, 1292–1307, https://doi.org/10.1002/hyp.10715, 2016.

Craig, H. and Gordon, L. I.: Deuterium and oxygen 18 variations in the ocean and the marine atmosphere, in: Stable Isotopes in Oceanographic Studies and Paleotemperatures, edited by: Tongiorgi, E., Stable Isotopes in Oceanographic Studies and Paleotemperatures, Spoleto, Italy, Laboratorio di Geologica Nucleare, Pisa, Italy, 9–130, 1965.

Craig, H., Gordon, L. I., and Horibe, Y.: Isotopic exchange effects in the evaporation of water: 1. Low-temperature experimental results, J. Geophys. Res., 68, 5079–5087, https://doi.org/10.1029/JZ068i017p05079, 1963.

Dansgaard, W.: Stable isotopes in precipitation, Tellus, 16, 436–468, https://doi.org/10.1111/j.2153-3490.1964.tb00181.x, 1964.

Dawson, T. E. and Ehleringer, J. R.: Streamside trees that do not use stream water, Nature, 350, 335–337, https://doi.org/10.1038/350335a0, 1991.

Dunn, S. M. and Mackay, R.: Spatial variation in evapotranspiration and the influence of land use on catchment hydrology, J. Hydrol., 171, 49–73, https://doi.org/10.1016/0022-1694(95)02733-6, 1995.

Ehleringer, J. R. and Dawson, T. E.: Water uptake by plants: perspectives from stable isotope composition, Plant Cell Environ., 15, 1073–1082, https://doi.org/10.1111/j.1365-3040.1992.tb01657.x, 1992.

Evaristo, J., Jasechko, S., and McDonnell, J. J.: Global separation of plant transpiration from groundwater and streamflow, Nature, 525, 91–94, https://doi.org/10.1038/nature14983, 2015.

Evaristo, J., McDonnell, J. J., Scholl, M. A., Bruijnzeel, L. A., and Chun, K. P.: Insights into plant water uptake from xylem-water isotope measurements in two tropical catchments with contrasting moisture conditions, Hydrol. Process., 30, 3210–3227, https://doi.org/10.1002/hyp.10841, 2016.

Foley, J. A., Defries, R., Asner, G. P., Barford, C., Bonan, G., Carpenter, S. R., Chapin, F. S., Coe, M. T., Daily, G. C., Gibbs, H. K., Helkowski, J. H., Holloway, T., Howard, E. A., Kucharik, C. J., Monfreda, C., Patz, J. A., Prentice, I. C., Ramankutty, N., and Snyder, P. K.: Global consequences of land use, Science, 309, 570–574, https://doi.org/10.1126/science.1111772, 2005.

Gaj, M., Beyer, M., Koeniger, P., Wanke, H., Hamutoko, J., and Himmelsbach, T.: In situ unsaturated zone water stable isotope (^{2}H and ^{18}O) measurements in semi-arid environments:

a soil water balance, Hydrol. Earth Syst. Sci., 20, 715–731, https://doi.org/10.5194/hess-20-715-2016, 2016.

Gaj, M., Kaufhold, S., Koeniger, P., Beyer, M., Weiler, M., and Himmelsbach, T.: Mineral mediated isotope fractionation of soil water, Rapid Commun. Mass Sp., 31, 269–280, https://doi.org/10.1002/rcm.7787, 2017a.

Gaj, M., Kaufhold, S., and McDonnell, J. J.: Potential limitation of cryogenic vacuum extractions and spiked experiments, Rapid Commun. Mass Sp., 31, 821–823, https://doi.org/10.1002/rcm.7850, 2017b.

Garvelmann, J., Külls, C., and Weiler, M.: A porewater-based stable isotope approach for the investigation of subsurface hydrological processes, Hydrol. Earth Syst. Sci., 16, 631–640, https://doi.org/10.5194/hess-16-631-2012, 2012.

Gat, J. R. and Levy, Y.: Isotope hydrology of inland sabkhas in the Bardawil area, Sinai, Limnol. Oceanogr., 23, 841–850, https://doi.org/10.4319/lo.1978.23.5.0841, 1978.

Gee, G. W. and Bauder, J. W.: Particle-size Analysis, in: Methods of Soil Analysis: Part 1– Physical and Mineralogical Methods, edited by: Klute, A., SSSA Book Series, American Society of Agronomy, Madison, WI, 383–411, 1986.

Geris, J., Tetzlaff, D., McDonnell, J., Anderson, J., Paton, G., and Soulsby, C.: Ecohydrological separation in wet, low energy northern environments?: A preliminary assessment using different soil water extraction techniques, Hydrol. Process., 29, 5139–5152, https://doi.org/10.1002/hyp.10603, 2015a.

Geris, J., Tetzlaff, D., McDonnell, J., and Soulsby, C.: The relative role of soil type and tree cover on water storage and transmission in northern headwater catchments, Hydrol. Process., 29, 1844–1860, https://doi.org/10.1002/hyp.10289, 2015b.

Geris, J., Tetzlaff, D., McDonnell, J. J., and Soulsby, C.: Spatial and temporal patterns of soil water storage and vegetation water use in humid northern catchments, Sci. Total Environ., 595, 486–493, https://doi.org/10.1016/j.scitotenv.2017.03.275, 2017.

Gerke, H. H. and van Genuchten, M. T.: A dual-porosity model for simulating the preferential movement of water and solutes in structured porous media, Water Resour. Res., 29, 305–319, https://doi.org/10.1029/92WR02339, 1993.

Gibson, J. J. and Reid, R.: Water balance along a chain of tundra lakes: A 20-year isotopic perspective, J. Hydrol., 519, 2148–2164, https://doi.org/10.1016/j.jhydrol.2014.10.011, 2014.

Gibson, J. J., Price, J. S., Aravena, R., Fitzgerald, D. F., and Maloney, D.: Runoff generation in a hypermaritime bog-forest upland, Hydrol. Process., 14, 2711–2730, https://doi.org/10.1002/1099-1085(20001030)14:15<2711:AID-HYP88>3.0.CO;2-2, 2000.

Gibson, J. J., Birks, S. J., and Edwards, T. W. D.: Global prediction of δA and δ2H-δ18O evaporation slopes for lakes and soil water accounting for seasonality, Global Biogeochem. Cy., 22, GB2031, https://doi.org/10.1029/2007GB002997, 2008.

Goldsmith, G. R., Muñoz-Villers, L. E., Holwerda, F., McDonnell, J. J., Asbjornsen, H., and Dawson, T. E.: Stable isotopes reveal linkages among ecohydrological processes in a seasonally dry tropical montane cloud forest, Ecohydrol., 5, 779–790, https://doi.org/10.1002/eco.268, 2012.

Gonfiantini, R.: Environmental isotopes in lake studies, in: The Terrestrial Environment, B: Handbook of Environmental Isotope Geochemistry, edited by: Fritz, P. and Fontes, J. C., Elsevier, Amsterdam, 113–168, 1986.

Gouet-Kaplan, M., Arye, G., and Berkowitz, B.: Interplay between resident and infiltrating water: Estimates from transient water flow and solute transport, J. Hydrol., 458–459, 40–50, https://doi.org/10.1016/j.jhydrol.2012.06.026, 2012.

Gralher, B., Herbstritt, B., Weiler, M., Wassenaar, L. I., and Stumpp, C.: Correcting laser-based water stable isotope readings biased by carrier gas changes, Environ. Sci. Technol., 50, 7074–7081, https://doi.org/10.1021/acs.est.6b01124, 2016.

Harman, C. J.: Time-variable transit time distributions and transport: Theory and application to storage-dependent transport of chloride in a watershed, Water Resour. Res., 51, 1–30, https://doi.org/10.1002/2014WR015707, 2015.

Hendry, M. J., Schmeling, E., Wassenaar, L. I., Barbour, S. L., and Pratt, D.: Determining the stable isotope composition of pore water from saturated and unsaturated zone core: improvements to the direct vapour equilibration laser spectrometry method, Hydrol. Earth Syst. Sci., 19, 4427–4440, https://doi.org/10.5194/hess-19-4427-2015, 2015.

Hervé-Fernández, P., Oyarzún, C., Brumbt, C., Huygens, D., Bodé, S., Verhoest, N. E. C., and Boeckx, P.: Assessing the "two water worlds" hypothesis and water sources for native and exotic evergreen species in south-central Chile, Hydrol. Process., 30, 4227–4241, https://doi.org/10.1002/hyp.10984, 2016.

Heße, F., Zink, M., Kumar, R., Samaniego, L., and Attinger, S.: Spatially distributed characterization of soil-moisture dynamics using travel-time distributions, Hydrol. Earth Syst. Sci., 21, 549–570, https://doi.org/10.5194/hess-21-549-2017, 2017.

Horita, J. and Wesolowski, D. J.: Liquid-vapor fractionation of oxygen and hydrogen isotopes of water from the freezing to the critical temperature, Geochim. Cosmochim. Ac., 58, 3425–3437, https://doi.org/10.1016/0016-7037(94)90096-5, 1994.

Hudson, B. D.: Soil organic matter and available water capacity, J. Soil Water Conserv., 49, 189–194, 1994.

Isokangas, E., Rossi, P. M., Ronkanen, A.-K., Marttila, H., Rozanski, K., and Kløve, B.: Quantifying spatial groundwater dependence in peatlands through a distributed isotope mass balance approach, Water Resour. Res., 53, 2524–2541, https://doi.org/10.1002/2016WR019661, 2017.

Jackson, R. B., Canadell, J., Ehleringer, J. R., Mooney, H. A., Sala, O. E., and Schulze, E. D.: A global analysis of root distributions for terrestrial biomes, Oecologia, 108, 389–411, https://doi.org/10.1007/BF00333714, 1996.

Javaux, M., Rothfuss, Y., Vanderborght, J., Vereecken, H., and Bruggemann, N.: Isotopic composition of plant water sources, Nature, 536, E1-3, https://doi.org/10.1038/nature18946, 2016.

Kirchner, J. W., Tetzlaff, D., and Soulsby, C.: Comparing chloride and water isotopes as hydrological tracers in two Scottish catchments, Hydrol. Process., 24, 1631–1645, https://doi.org/10.1002/Hyp.7676, 2010.

Kool, D., Agam, N., Lazarovitch, N., Heitman, J., Sauer, T., and Ben-Gal, A.: A review of approaches for evapotranspiration partitioning, Agr. Forest Meteorol., 184, 56–70, https://doi.org/10.1016/j.agrformet.2013.09.003, 2014.

Landwehr, J. M. and Coplen, T. B.: Line-conditioned excess: a new method for characterizing stable hydrogen and oxygen isotope ratios in hydrologic systems, in: International Conference on Isotopes in Environmental Studies, Aquatic Forum, Monte-Carlo, Monaco, 25–29 October 2004, IAEA, Vienna, 132–135, 2006.

Landwehr, J. M., Coplen, T. B., and Stewart, D. W.: Spatial, seasonal, and source variability in the stable oxygen and hydrogen isotopic composition of tap waters throughout the USA, Hydrol. Process., 28, 5382–5422, https://doi.org/10.1002/hyp.10004, 2014.

McCutcheon, R. J., McNamara, J. P., Kohn, M. J., and Evans, S. L.: An Evaluation of the Ecohydrological Separation Hypothesis in a Semiarid Catchment, Hydrol. Process., 31, 783–799, https://doi.org/10.1002/hyp.11052, 2016.

McDonnell, J. J. and Beven, K.: Debates-The future of hydrological sciences: A (common) path forward? A call to action aimed at understanding velocities, celerities and residence time distributions of the headwater hydrograph, Water Resour. Res., 50, 5342–5350, https://doi.org/10.1002/2013WR015141, 2014.

Midwood, A. J., Boutton, T. W., Archer, S. R., and Watts, S. E.: Water use by woody plants on contrasting soils in a savanna parkland: assessment with δ^2H and δ^{18}O, Plant Soil, 205, 13–24, https://doi.org/10.1023/A:1004355423241, 1998.

Mioduszewski, J. R., Rennermalm, A. K., Robinson, D. A., and Mote, T. L.: Attribution of snowmelt onset in Northern Canada, J. Geophys. Res.-Atmos., 119, 9638–9653, https://doi.org/10.1002/2013JD021024, 2014.

Mueller, M. H., Alaoui, A., Kuells, C., Leistert, H., Meusburger, K., Stumpp, C., Weiler, M., and Alewell, C.: Tracking water pathways in steep hillslopes by δ^{18}O depth profiles of soil water, J. Hydrol., 519, 340–352, https://doi.org/10.1016/j.jhydrol.2014.07.031, 2014.

Newberry, S. L., Prechsl, U. E., Pace, M., and Kahmen, A.: Tightly bound soil water introduces isotopic memory effects on mobile and extractable soil water pools, Isot. Environ. Healt. S., 1–14, https://doi.org/10.1080/10256016.2017.1302446, 2017.

Oerter, E. and Bowen, G.: In situ monitoring of H and O stable isotopes in soil water reveals ecohydrologic dynamics in managed soil systems, Ecohydrol., 10, e1841, https://doi.org/10.1002/eco.1841, 2017.

Oerter, E., Finstad, K., Schaefer, J., Goldsmith, G. R., Dawson, T., and Amundson, R.: Oxygen isotope fractionation effects in soil water via interaction with cations (Mg, Ca, K, Na) adsorbed to phyllosilicate clay minerals, J. Hydrol., 515, 1–9, https://doi.org/10.1016/j.jhydrol.2014.04.029, 2014.

Or, D., Lehmann, P., Shahraeeni, E., and Shokri, N.: Advances in Soil Evaporation Physics – A Review, Vadose Zone J., 12, 1–16, https://doi.org/10.2136/vzj2012.0163, 2013.

Orlowski, N., Breuer, L., and McDonnell, J. J.: Critical issues with cryogenic extraction of soil water for stable isotope analysis, Ecohydrol., 9, 3–10, https://doi.org/10.1002/eco.1722, 2016.

Queloz, P., Carraro, L., Benettin, P., Botter, G., Rinaldo, A., and Bertuzzo, E.: Transport of fluorobenzoate tracers in a vegetated hydrologic control volume: 2. Theoretical inferences and modeling, Water Resour. Res., 51, 2793–2806, https://doi.org/10.1002/2014WR016508, 2015.

Rees, R. M., Black, H., Chapman, S. J., Clayden, H., Edwards, A. C., and Waldron, S.: Loss of soil organic matter, in: The State of Scotland's Soil. Natural Scotland, edited by: Dobbie, K. E., Bruneau, P., and Towers, W., available at: www.sepa.org.uk/land/land_publications.aspx (last access: 24 July 2017), 23–34, 2011.

Renner, M., Hassler, S. K., Blume, T., Weiler, M., Hildebrandt, A., Guderle, M., Schymanski, S. J., and Kleidon, A.: Dominant controls of transpiration along a hillslope transect inferred from ecohydrological measurements and thermodynamic limits, Hydrol. Earth Syst. Sci., 20, 2063–2083, https://doi.org/10.5194/hess-20-2063-2016, 2016.

Rinaldo, A., Beven, K. J., Bertuzzo, E., Nicotina, L., Davies, J., Fiori, A., Russo, D., and Botter, G.: Catchment travel time distributions and water flow in soils, Water Resour. Res., 47, W07537, https://doi.org/10.1029/2011wr010478, 2011.

Rothfuss, Y. and Javaux, M.: Reviews and syntheses: Isotopic approaches to quantify root water uptake: a review and comparison of methods, Biogeosciences, 14, 2199–2224, https://doi.org/10.5194/bg-14-2199-2017, 2017.

Rothfuss, Y., Braud, I., Le Moine, N., Biron, P., Durand, J.-L., Vauclin, M., and Bariac, T.: Factors controlling the isotopic partitioning between soil evaporation and plant transpiration: Assessment using a multi-objective calibration of SiSPAT-Isotope under controlled conditions, J. Hydrol., 442–443, 75–88, https://doi.org/10.1016/j.jhydrol.2012.03.041, 2012.

Rothfuss, Y., Merz, S., Vanderborght, J., Hermes, N., Weuthen, A., Pohlmeier, A., Vereecken, H., and Brüggemann, N.: Long-term and high-frequency non-destructive monitoring of water stable isotope profiles in an evaporating soil column, Hydrol. Earth Syst. Sci., 19, 4067–4080, https://doi.org/10.5194/hess-19-4067-2015, 2015.

Schwärzel, K., Menzer, A., Clausnitzer, F., Spank, U., Häntzschel, J., Grünwald, T., Köstner, B., Bernhofer, C., and Feger, K.-H.: Soil water content measurements deliver reliable estimates of water fluxes: A comparative study in a beech and a spruce stand in the Tharandt forest (Saxony, Germany), Agr. Forest Meteorol., 149, 1994–2006, https://doi.org/10.1016/j.agrformet.2009.07.006, 2009.

Shen, M., Tang, Y., Chen, J., Yang, X., Wang, C., Cui, X., Yang, Y., Han, L., Le Li, Du, J., Zhang, G., and Cong, N.: Earlier-season vegetation has greater temperature sensitivity of spring phenology in northern hemisphere, PloS one, 9, e88178, https://doi.org/10.1371/journal.pone.0088178, 2014.

Soulsby, C., Tetzlaff, D., van den Bedem, N., Malcolm, I. A., Bacon, P. J., and Youngson, A. F.: Inferring groundwater influences on surface water in montane catchments from hydrochemical surveys of springs and streamwaters, J. Hydrol., 333, 199–213, https://doi.org/10.1016/j.jhydrol.2006.08.016, 2007.

Soulsby, C., Birkel, C., Geris, J., Dick, J., Tunaley, C., and Tetzlaff, D.: Stream water age distributions controlled by storage dynamics and nonlinear hydrologic connectivity: Modeling with high-resolution isotope data, Water Resour. Res., 51, 7759–7776, https://doi.org/10.1002/2015WR017888, 2015.

Soulsby, C., Birkel, C., and Tetzlaff, D.: Characterizing the age distribution of catchment evaporative losses, Hydrol. Process., 30, 1308–1312, https://doi.org/10.1002/hyp.10751, 2016a.

Soulsby, C., Bradford, J., Dick, J., McNamara, J. P., Geris, J., Lessels, J., Blumstock, M., and Tetzlaff, D.: Using geophysical surveys to test tracer-based storage estimates in headwater catchments, Hydrol. Process., 30, 4434–4445, https://doi.org/10.1002/hyp.10889, 2016b.

Sprenger, M., Herbstritt, B., and Weiler, M.: Established methods and new opportunities for pore water stable isotope analysis, Hydrol. Process., 29, 5174–5192, https://doi.org/10.1002/hyp.10643, 2015a.

Sprenger, M., Volkmann, T. H. M., Blume, T., and Weiler, M.: Estimating flow and transport parameters in the unsaturated zone with pore water stable isotopes, Hydrol. Earth Syst. Sci., 19, 2617–2635, https://doi.org/10.5194/hess-19-2617-2015, 2015b.

Sprenger, M., Erhardt, M., Riedel, M., and Weiler, M.: Historical tracking of nitrate in contrasting vineyards using water isotopes and nitrate depth profiles, Agr. Ecosyst. Environ., 222, 185–192, https://doi.org/10.1016/j.agee.2016.02.014, 2016a.

Sprenger, M., Leistert, H., Gimbel, K., and Weiler, M.: Illuminating hydrological processes at the soil-vegetation-atmosphere interface with water stable isotopes, Rev. Geophys., 54, 674–704, https://doi.org/10.1002/2015RG000515, 2016b.

Sprenger, M., Seeger, S., Blume, T., and Weiler, M.: Travel times in the vadose zone: variability in space and time, Water Resour. Res., 52, 5727–5754, https://doi.org/10.1002/2015WR018077, 2016c.

Sprenger, M., Tetzlaff, D., and Soulsby, C.: No influence of CO_2 on stable isotope analyses of soil waters with OA-ICOS, Rapid Commun. Mass Sp., 31, 430–436, https://doi.org/10.1002/rcm.7815, 2017a.

Sprenger, M., Tetzlaff, D., Tunaley, C., Dick, J., and Soulsby, C.: Evaporation fractionation in a peatland drainage network affects stream water isotope composition, Water Resour. Res., 53, 851–866, https://doi.org/10.1002/2016WR019258, 2017b.

Stumpp, C., Stichler, W., Kandolf, M., and Šimůnek, J.: Effects of Land Cover and Fertilization Method on Water Flow and Solute Transport in Five Lysimeters: A Long-Term Study Using Stable Water Isotopes, Vadose Zone J., 11, vzj2011.0075, https://doi.org/10.2136/vzj2011.0075, 2012.

Tetzlaff, D., Carey, S., Soulsby, C., Tetzlaff, D., Soulsby, C., Buttle, J., Capell, R., Carey, S. K., Laudon, H., McDonnell, J., McGuire, K., Seibert, J., and Shanley, J.: Catchments on the cusp? Structural and functional change in northern ecohydrology, Hydrol. Process., 27, 766–774, https://doi.org/10.1002/hyp.9700, 2013.

Tetzlaff, D., Birkel, C., Dick, J., Geris, J., and Soulsby, C.: Storage dynamics in hydropedological units control hillslope connectivity, runoff generation and the evolution of catchment transit time distributions, Water Resour. Res., 50, 969–985, https://doi.org/10.1002/2013WR014147, 2014.

Twining, J., Stone, D., Tadros, C., Henderson-Sellers, A., and Williams, A.: Moisture Isotopes in the Biosphere and Atmosphere (MIBA) in Australia: A priori estimates and preliminary observations of stable water isotopes in soil, plant and vapour for the Tumbarumba Field Campaign, Global Planet. Change, 51, 59–72, https://doi.org/10.1016/j.gloplacha.2005.12.005, 2006.

van der Velde, Y., Heidbüchel, I., Lyon, S. W., Nyberg, L., Rodhe, A., Bishop, K., and Troch, P. A.: Consequences of mixing assumptions for time-variable travel time distributions, Hydrol. Process., 29, 3460–3474, https://doi.org/10.1002/hyp.10372, 2015.

van Genuchten, M. T. and Wierenga, P. J.: Mass transfer studies in sorbing porous media I. Analytical solutions, Soil Sci. Soc. Am. J., 40, 473–480, 1976.

van Huijgevoort, M. H. J., Tetzlaff, D., Sutanudjaja, E. H., and Soulsby, C.: Using high resolution tracer data to constrain water storage, flux and age estimates in a spatially distributed rainfall-runoff model, Hydrol. Process., 30, 4761–4778, https://doi.org/10.1002/hyp.10902, 2016.

Vereecken, H.: Modeling Soil Processes: Review, Key challenges and New Perspectives, Vadose Zone J., 15, 1–57, 2016.

Volkmann, T. H. M., Haberer, K., Gessler, A., and Weiler, M.: High-resolution isotope measurements resolve rapid ecohydrological dynamics at the soil-plant interface, New Phytol., 210, 839–849, https://doi.org/10.1111/nph.13868, 2016.

Wang, H., Tetzlaff, D., and Soulsby, C.: Assessing the environmental controls on Scots pine transpiration and the implications for water partitioning in a boreal headwater catchment, Agr. Forest Meteorol., 240–241, 58–66, https://doi.org/10.1016/j.agrformet.2017.04.002, 2017.

Wassenaar, L., Hendry, M., Chostner, V., and Lis, G.: High Resolution Pore Water δ^2H and $\delta^{18}O$ Measurements by H_2O (liquid) -H_2O (vapor) Equilibration Laser Spectroscopy, Environ. Sci. Technol., 42, 9262–9267, https://doi.org/10.1021/es802065s, 2008.

Welhan, J. and Fritz, P.: Evaporation pan isotopic behavior as an index of isotopic evaporation conditions, Geochim. Cosmochim. Ac., 41, 682–686, https://doi.org/10.1016/0016-7037(77)90306-4, 1977.

Werner, C. and Dubbert, M.: Resolving rapid dynamics of soil-plant-atmosphere interactions, New Phytol., 210, 767–769, https://doi.org/10.1111/nph.13936, 2016.

Wershaw, R. L., Friedman, I., and Heller, S. J.: Hydrogen isotope fractionation of water passing through trees, in: Advances in Organic Geochemistry, edited by: Hobson, F. and Speers, M., Pergamon Press, New York, 55–67, 1966.

Zobel, R. W. and Waisel, Y.: A plant root system architectural taxonomy: A framework for root nomenclature, Plant Biosyst., 144, 507–512, https://doi.org/10.1080/11263501003764483, 2010.

7

Application of the pore water stable isotope method and hydrogeological approaches to characterise a wetland system

Katarina David[1], **Wendy Timms**[2], **Catherine E. Hughes**[3], **Jagoda Crawford**[3], **and Dayna McGeeney**[4]

[1]School of Minerals and Energy Resource Engineering, and Connected Waters Initiative, University of New South Wales, Sydney, Australia
[2]School of Engineering, Deakin University, Waurn Ponds, Australia
[3]Australian Nuclear Science and Technology Organisation, Sydney, Australia
[4]Australian Museum, Sydney, Australia

Correspondence: Katarina David (k.david@unsw.edu.au)

Abstract. Three naturally intact wetland systems (swamps) were characterised based on sediment cores, analysis of surface water, swamp groundwater, regional groundwater and pore water stable isotopes. These swamps are classified as temperate highland peat swamps on sandstone (THPSS) and in Australia they are listed as threatened endangered ecological communities under state and federal legislation.

This study applies the stable isotope direct vapour equilibration method in a wetland, aiming at quantification of the contributions of evaporation, rainfall and groundwater to swamp water balance. This technique potentially enables understanding of the depth of evaporative losses and the relative importance of groundwater flow within the swamp environment without the need for intrusive piezometer installation at multiple locations and depths. Additional advantages of the stable isotope direct vapour equilibration technique include detailed spatial and vertical depth profiles of $\delta^{18}O$ and $\delta^{2}H$, with good accuracy comparable to other physical and chemical extraction methods.

Depletion of $\delta^{18}O$ and $\delta^{2}H$ in pore water with increasing depth (to around 40–60 cm depth) was observed in two swamps but remained uniform with depth in the third swamp. Within the upper surficial zone, the measurements respond to seasonal trends and are subject to evaporation in the capillary zone. Below this depth the pore water $\delta^{18}O$ and $\delta^{2}H$ signature approaches that of regional groundwater, indicating lateral groundwater contribution. Significant differences were found in stable pore water isotope samples collected after the dry weather period compared to wet periods where recharge of depleted rainfall (with low $\delta^{18}O$ and $\delta^{2}H$ values) was apparent.

The organic-rich soil in the upper 40 to 60 cm retains significant saturation following precipitation events and maintains moisture necessary for ecosystem functioning. An important finding for wetland and ecosystem response to changing swamp groundwater conditions (and potential ground movement) is that basal sands are observed to underlay these swamps, allowing relatively rapid drainage at the base of the swamp and lateral groundwater contribution.

Based on the novel stable isotope direct vapour equilibration analysis of swamp sediment, our study identified the following important processes: rapid infiltration of rainfall to the water table with longer retention of moisture in the upper 40–60 cm and lateral groundwater flow contribution at the base. This study also found that evaporation estimated using the stable isotope direct vapour equilibration method is more realistic compared to reference evapotranspiration (ET). Importantly, if swamp discharge data were available in combination with pore water isotope profiles, an appropriate transpiration rate could be determined for these swamps. Based on the results, the groundwater contribution to the swamp is a significant and perhaps dominant component of the water balance. Our methods could complement other monitoring studies and numerical water balance models to improve prediction of the hydrological response of the swamp to changes in water conditions due to natural or anthropogenic influences.

1 Introduction

Stable isotopes of water (δ^{18}O and δ^{2}H) have been widely used to understand groundwater and surface water interaction and recharge processes in aquifer systems (Barnes and Allison, 1988; Cuthbert et al., 2014). Although less common than liquid water isotope studies, pore water (vapour) stable isotope techniques have been applied to investigate groundwater flux and interpret the paleoenvironment (Hendry et al., 2013; Harrington et al., 2013), determine slope runoff contribution to groundwater (Garvelman et al., 2012) and characterise multi-layered sedimentary sequences (David et al., 2015). Pore water stable isotope analysis was less common due to sampling difficulties (Soderberg et al., 2012) and high cost (Harrington et al., 2013), until advances in laser spectroscopy improved the speed and accuracy of the analysis (Hendry and Wassenaar, 2009; Hendry et al., 2015).

Examples of wetland research indicate that it remains challenging to quantify some components of the water balance (Bijoor et al., 2011), since only a few studies have investigated the groundwater contribution (Hunt et al., 1998). The significance of groundwater in maintaining the swamp ecosystem function has been discussed in the literature (Chang et al., 2009; Kaller et al., 2015). In Australia, swamp studies have evaluated geomorphology (Fryirs et al., 2016; Cowley et al., 2016), management (Kohlhagen et al., 2013), the relationship between vegetation and groundwater (Hose et al., 2014), processes that result in denudation and sedimentation in the headwaters of the swamps (Prosser et al., 1994), natural and anthropogenic vegetation change in swamps (Bickford and Gell, 2005) and the impact of mining subsidence (CoA, 2014b). However, there is limited literature on the importance of groundwater storage, flow and which water source contributes to maintaining moisture in swamp systems.

The temperate highland peat swamps on sandstone (THPSS) swamps in eastern Australia are endangered ecological communities with endemic flora and fauna that are dependent on water balance. The direct influences on the water regime of these swamps are changes in weather patterns, natural storm activity (Smith et al., 2001), fire (Middleton and Kleinebecker, 2012; CoA, 2014a) and the effects of mining subsidence (CoA, 2014b). Ecologically, the THPSS swamps are sensitive to changing swamp moisture content (CoA, 2014a; Young, 2017), and the importance of groundwater in these systems has been discussed by Eamus and Froend (2006), Fryirs et al. (2014) and Hose et al. (2014).

There is direct and indirect evidence that the THPSS swamps (Newnes Plateau shrub swamp) are mostly saturated, including vegetation patterns, specific species of plants, piezometer records (Benson and Baird, 2012) and spring discharge (Johnson, 2007). Maintaining groundwater levels is necessary for the health for such a wetland system (Clifton and Evans, 2001).

Despite the awareness of these factors, there is limited research to predict hydrogeological changes in swamps and ecological response under changing water conditions (Mitsch and Gosselink, 2000; Cowley et al., 2016). Furthermore, studies describing the impact of environmental changes on swamp ecology are rare and there is insufficient understanding of natural variation in swamp ecology over time (CoA, 2014a). Although groundwater is often assumed to be important for the sustainability of wetlands and swamp ecosystems, very little is known on how these systems would respond to changing swamp groundwater and regional groundwater conditions. The existing literature recognises that natural variation in swamp ecology is not well understood given the complexity and interaction of swamp groundwater, groundwater and surface water.

Long drought periods in Australia result in temporary drying of water bodies, pooling of water and reduction in baseflow contribution to wetlands (Lake, 2003). Following such long dry periods, the rainfall may not be sufficient for a swamp to recover its original condition (Bond et al., 2008; Middleton and Kleinebecker, 2012). For example, Smith et al. (2001) report loss of swamps in Africa and Australia as a result of climate change. Vegetation removal, drainage of swamp and undermining are known to be critical human impacts (Kohlhagen et al., 2013; Valentin et al., 2005). As such, mining and urbanisation have degraded and considerably damaged the THPSS swamps (CoA, 2014b), although the actual impact often cannot be quantified due to limited baseline and monitoring data (Paterson, 2004). However, it is generally recognised that rock fracturing, changes in elevation gradient and catchment conditions can compromise the stability and integrity of these swamps (CoA, 2014b).

The objective of this research was to improve understanding of intact swamps under natural conditions by characterising the sediments, waters and organic materials and developing the conceptual model for the swamp system. We hypothesise that groundwater is an important contributor to the swamp water balance and is connected to the regional groundwater system. Therefore, we investigate, for the first time using the direct equilibration method, the vertical profiles of stable δ^{18}O and δ^{2}H isotopes of pore water within the swamp. We then compare those to stable isotopes of regional groundwater, rainfall and surface water as endpoint members. Supported by sediment lithology logs and organic and carbon content of sediments, these stable isotope results enabled the development of a conceptual model of the swamp water cycle.

2 Site description

The research site is located west of Sydney, NSW, Australia, in the World Heritage listed Blue Mountains on the Newnes Plateau (Fig. 1) between Lithgow and Blue Mountains local

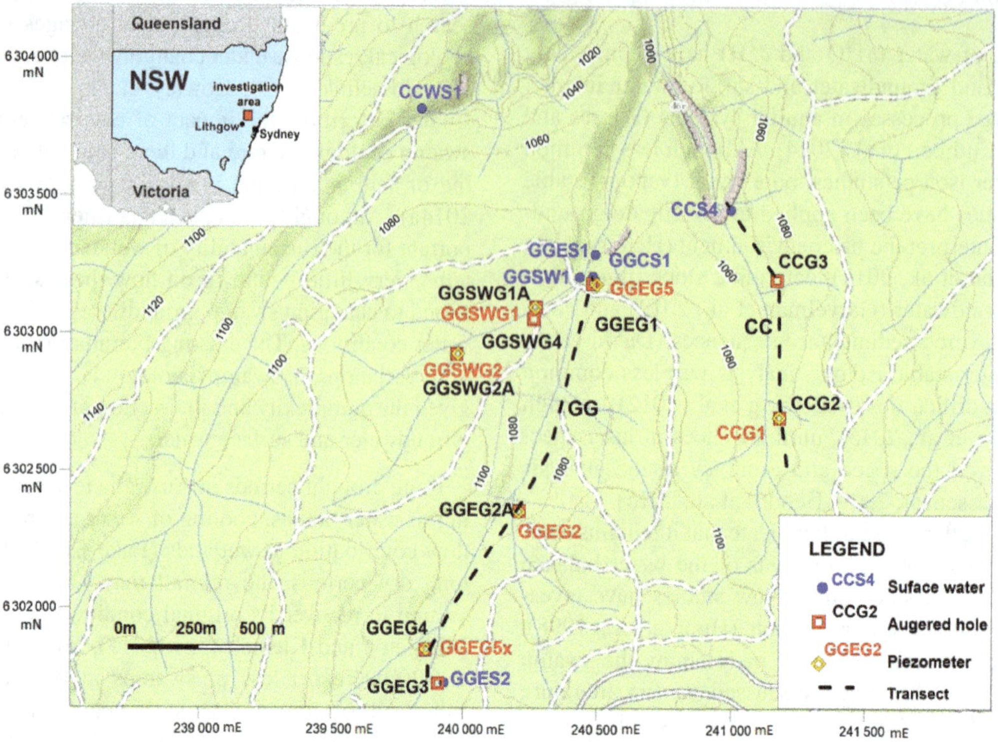

Figure 1. Map of selected Newnes Plateau swamps with locations of samples and transects.

government areas. The elevation of the plateau ranges from 1000 to 1200 m Australian Height Datum (mAHD).

The Triassic Narrabeen Sandstone outcrops over most of the study area. It comprises mainly quartzose sandstone and minor claystone and shale (Yoo et al., 2001). The swamps on the Newnes Plateau are classified as shrub swamps (OEH, 2017) based on the dominant shrub ecological community, and they occur at the highest elevation of any sandstone-based swamp in Australia. These swamps occur in low slope headwaters of the Newnes Plateau as narrow and elongated sites with impeded drainage (OEH, 2017) and are also classified as THPSS belonging to both headwater and valley infill types (CoA, 2014a). Mapping by Keith and Benson (1988) and Benson and Keith (1990) indicates that the shrub swamps cover 650 ha of land on the Newnes Plateau, with the largest swamp being 40 ha and average size less than 6 ha. Keith and Myerscough (1993) relate the swamps to other upland swamps in the Sydney Basin in terms of biogeography. However, the difference from other TPHSS is the presence of a permanent water table (Benson and Baird, 2012).

The three swamps selected, identified as CC (swamp area 7 ha, catchment area 150 ha), GG (swamp area 11 ha, catchment area 190 ha) and GGSW (swamp area 5 ha, catchment area 57 ha), are in the upper Carne Creek catchment (Fig. 1). Carne Creek is a tributary of the Wolgan River (catchment area 5310 ha) that ultimately flows to the Hawkesbury River and Pacific Ocean. The three swamps selected thus have a total area approximately 0.043 % of the Wolgan River catchment. Except for the headwaters including these swamps, the Wolgan River has been designated part of the Colo Wild River area recognising substantially unmodified conditions and high conservation value (NSW Government, 2008). The swamps in this study are located to the east of current underground mining operations, with coal extraction currently occurring on the western side of the GGSW and GG swamps, though not directly below these swamps. The swamps are elongated with a gentle gradient and typically terminate with a sandstone rockbar. The swamp groundwater level responds rapidly to rainfall recharge (Centennial Coal, 2016) and there is an indication that swamp systems are fed by lateral groundwater inflow (Benson and Baird, 2012). Further indirect evidence of regional groundwater interaction with swamp sediments and long-term saturation are the consistently stable swamp groundwater levels over time (Centennial Coal, 2016). Chalson and Martin (2009) undertook radiocarbon dating on pollen from a swamp on the Newnes Plateau and found that the calibrated ages were 11 000 to 7500 years (sampling depth 55–90 cm) and decreasing to 1800 years at 40 cm depth. These ages support the existence of the swamps during the early wetter and warmer Holocene, through to seasonally variable climate in the period from the mid to late Holocene (Allen and Lindesay, 1998). Given the seasonality of rainfall events, groundwater interaction must have existed through the mid and late Holocene to enable the swamp survival during dry periods.

Two sedimentary formations underlay the Newnes Plateau swamps: the Burralow Formation and the Banks Wall Sandstone. The Burralow Formation underlies the upper headwater part of the swamps and comprises an interbedded coarse-grained sandstone sequence with frequent fine-grained, clay-rich sandstone, siltstones, claystones and shales. The thickness of the Burralow Formation ranges from around 40 m in the upgradient part of the swamps but is absent in downgradient parts of the swamps. Banks Wall Sandstone typically forms the base of the lower parts of most swamps (McHugh, 2014).

Information on the natural groundwater regime in the swamps is very limited (CoA, 2014). It is generally considered that the sandstone underlying the swamps provides the barrier to water loss due to its relatively low permeability (CoA, 2014). However, in some cases, joints and bedding planes within sandstone can provide recharge to swamps (Coffey, 2008). NSW DP (2008) considers that headwater swamps, such as the ones described in this study, are likely to be perched above the regional water table. CoA (2014a) indicates that regional groundwater in those swamps can interact with the swamp system, but where this occurs the connection is ephemeral as it is dependent on the perched aquifer. The groundwater residence time is short, and the water is fresh. Information available to date suggests that the dominant source of water to Sydney Basin (headwater) swamps is rainfall and run-off recharge (NSW PAC, 2009).

The climate on the Newnes Plateau is temperate with higher rainfall in November to March and lower rainfall from April to October. Average yearly rainfall at the closest long-term meteorological station in Lidsdale (Bureau of Meteorology (BoM, 2017) Station SN63132, 12 km west of the study area) is 765 mm (890 mAHD) and 1270 mm at Mt Wilson (BoM Station 63246 21 km south-east (SE) of the study area) (1010 mAHD). The temperature varies from an average of 19.6 °C in summer to 5.8 °C in winter (Lidsdale).

3 Methods

3.1 Fieldwork and sampling

Fieldwork was undertaken during 2016 and 2017 with the swamps in a natural state and recovered from earlier wildfire in 2013. The first sampling event 24 to 25 May 2016 occurred following an extremely dry weather period of four months below the long-term average rainfall (BoM Lithgow Station SN63226, 900 mAHD, 13 km SW of the study area with 139 years of data records) for February to May (46.6, 36.8, 6.6 and 20.8 mm for each of the months). A total of 34 pore water samples and 5 surface and groundwater samples were collected. A repeat sampling on 25 to 26 October 2016 occurred after four months of above average rainfall from June to September (170.2, 102, 61.8 and 92 mm for each of the months). During October 2016 sampling event 14 pore water samples and 13 surface and groundwater samples were collected. Sampling on 30 May 2017 occurred under different climate conditions with both above and below average rainfall trend in the months preceding the sampling event. A total of 27 pore water samples, 3 surface and 3 groundwater samples were collected in May 2017. The spatial depth resolution varied from 10 to 20 cm depending on the penetration of the corer. Figure 2 shows the variation in monthly long-term rainfall (139 years) and comparison with rainfall in 2016 and 2017.

In total seven sediment cores were obtained by coring using a Russian D hand corer (40 mm diameter) to rock refusal (between 0.45 and 1.4 m), and three transects (CC, GG and GGSW) were prepared along the length of the swamps (Figs. 3, 4 and 5). Samples were geologically logged after extraction, by noting the lithology, grain size and roundness, matrix and colour. The hand-cored holes were restored by returning soil material to the hole immediately after sampling. This was undertaken to ensure no change occurred to the endangered and protected ecological system as a result of sampling. The coring on the CC transect was repeated in October 2016 at a distance of less than 0.5 m from the original hole. The coring locations were selected to represent swamp stratigraphy from upstream to downstream and to provide a spatial coverage across the three swamps. In addition, three cored locations were selected such that they were adjacent to an existing piezometer (CCG1 on transect CC and GGEG2A and GGEG4 on the GG transect). The purpose of this, in addition to determining the stable isotope profiles, was to enable comparison with the swamp groundwater measurements and to collect regional groundwater from the underlying sandstone aquifer where possible.

Sediment cores were divided into subsamples of 10–20 cm length, packed into Ziploc bags and kept in cool storage for later analysis of moisture content and organic matter content. The samples for pore water analysis were temporarily double packed in Ziploc bags by minimising the airspace in the bag, stored in the cooled ice box in accordance with the sampling protocol developed by Wassenaar et al. (2008) and further improved by Hendry et al. (2015). The use of clear Ziplock bags for storage of samples for pore water analysis has been found (Hendry et al., 2015) to result in evaporation loss and isotopic fractionation only after 10–15 days after sample collection. The same afternoon after collection, samples were packed in tough high-grade food storage plastic bags with air extracted, double sealed, separately stored in an additional plastic bag and were kept at a 4 °C to prevent evaporation. Vacuum packing was required to minimise atmospheric moisture contamination. All isotopic field controls during sampling and analysis were implemented; this included: quick storage in tough plastic bags, immediate double bagging during collection and vacuum packing the same afternoon. Storage time for samples after collection

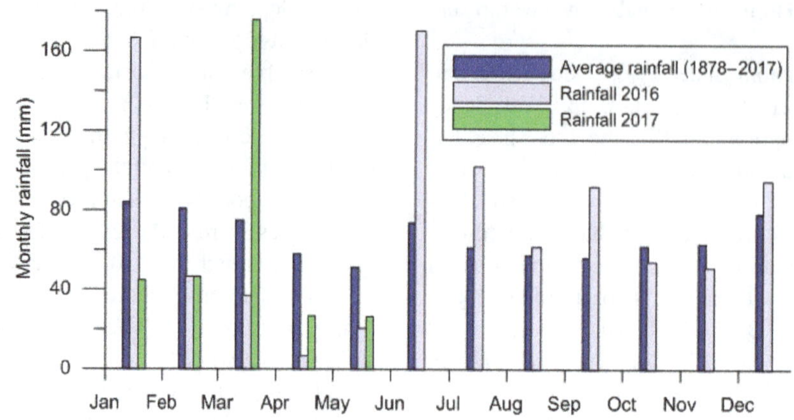

Figure 2. Long-term average monthly rainfall at Lithgow station (Bureau of Meteorology Station No. 063226) compared to rainfall during 2016 and 2017.

was 3 days in the cool environment (4 °C) before they were analysed.

Swamp groundwater was sampled directly from the cored hole and field parameters were measured immediately (pH, electrical conductivity (EC), dissolved oxygen (DO), temperature). This was repeated for all three sampling events; however, some bores were dry and some not accessible. Swamp groundwater and regional groundwater from existing piezometers (CCG1, GGEG2, GGEG5x, GGEG5 and GGSWG1) was gauged and sampled by bailing three volumes and then the same procedure was followed as the cored holes. Swamp and sandstone piezometers were installed by the mining company prior to our research study. Swamp piezometers were installed to the base of the swamp, where auger refusal did not allow further progress. The typical installation depth is around 1 to 1.3 m. To minimise disturbance of the swamp, all swamp piezometers were installed by manual coring an 80 mm diameter hole to refusal and pushing the slotted 50 mm diameter PVC tube into the hole. A full PVC casing was attached to the top of the pipe. The sandstone piezometer is 8.5 m depth with 50 mm diameter PVC casing that includes a 3 m length of screen at the bottom of the hole. The piezometer installation was extended with casing to the top. The top was sealed by grout, and a steel monument constructed for protection. Surface water samples were collected at the downgradient end of the swamp but also at one upgradient location (GGES2) where this was possible.

For this study the Australian Nuclear Science and Technology Organisation (ANSTO) provided event-based $\delta^{18}O$ and δ^2H data for precipitation from Mt Werong for the period covered in this research. Mt Werong (Hughes and Crawford, 2013) is located around 70 km south of this research site, however, within the same climatic environment and at a similar elevation to the investigated swamps.

3.2 Sample analysis

The swamp sediment samples were analysed for $\delta^{18}O$ and δ^2H by $H_2O_{(water)}$–$H_2O_{(vapour)}$ pore water equilibration (Wassenaar et al., 2008) and off-axis ICOS. The Los Gatos (LGR) water vapour analyser (WVIA RMT-EP model 911-0004) located at the University of NSW (UNSW), Australia, was used for sample analysis. All samples and standards have been stored at 4 °C prior to the analysis, and all have been allowed the same time on the laboratory bench in the temperature-controlled laboratory during preparation and have followed the same treatment. Samples ($n = 34$, 14 and 27 for each of the sampling events) were prepared in the lab by transferring the samples to a tough Ziploc bag. The 1 L sample bags were inflated with dry air and left on the laboratory bench within the controlled temperature for a period of between 17 and 24 h to allow vapour equilibration. Timing of vapour equilibration is dependent on compactness of the core sample, whether it is broken into pieces and whether it is unconsolidated (Wassenaar et al., 2008). The timing varies for different geologic materials and must be determined experimentally (Hendry et al., 2015) for each material. Work by Wassenaar et al. (2008) and David et al. (2015) indicates that for compact, low-permeability, consolidated materials around 3 days is required for core sample equilibration. The samples in this research are broken down, unconsolidated, saturated and high-permeability; therefore, a shorter equilibration time is considered justified. In addition, the optimal equilibration time in this research is considered to be achieved when a headspace water content of 23 000 to 28 000 ppm H_2O was measured in the bag. This headspace water content is important for accurate sampling (Hendry et al., 2015). Once the sample has reached complete isotopic equilibrium, the vapour was collected by perforating the bag containing the sample with a sharp needle and transferring it directly from the bag to the LGR vapour analyser. The connection between the needle and the LGR inlet fitting was via

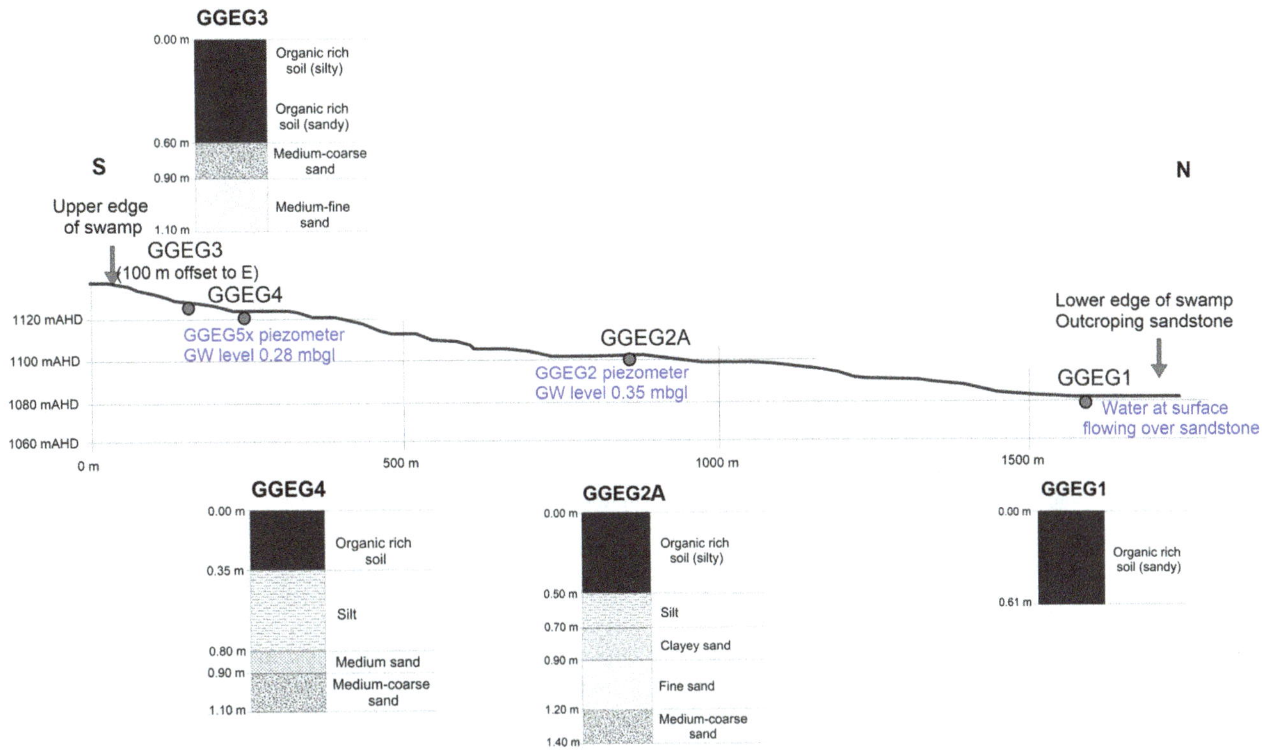

Figure 3. Interpreted long section of swamp GG with swamp groundwater levels as measured in May 2016.

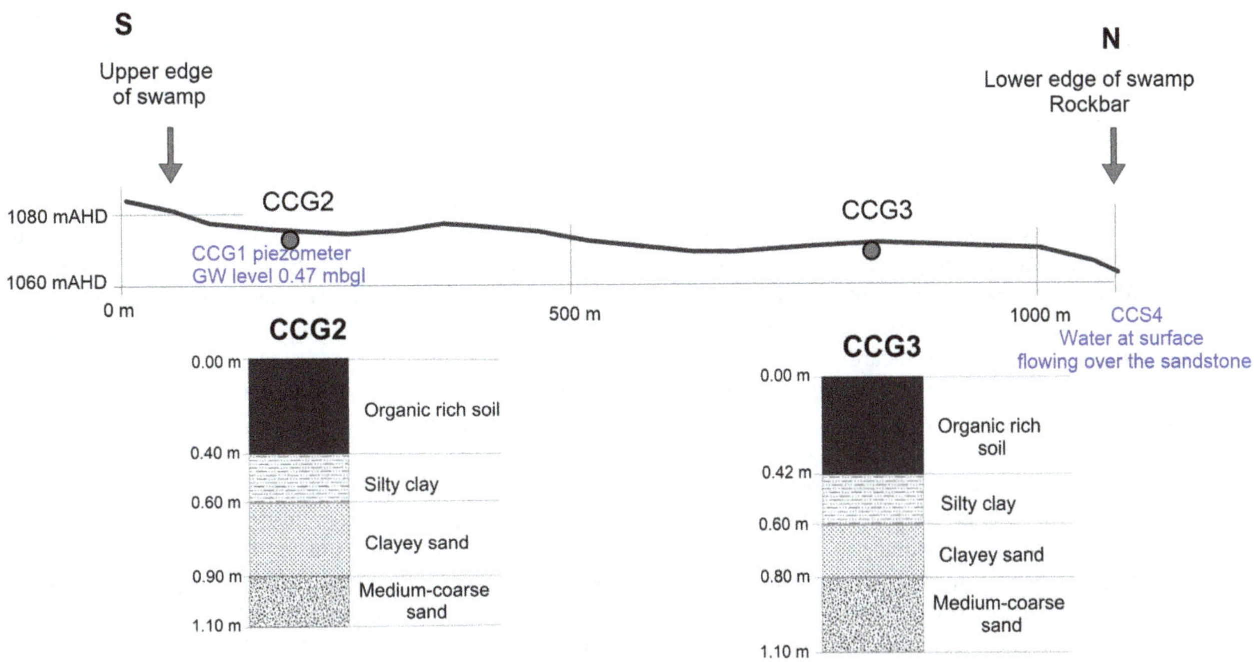

Figure 4. Interpreted long section of swamp CC with swamp groundwater levels as measured in May 2016.

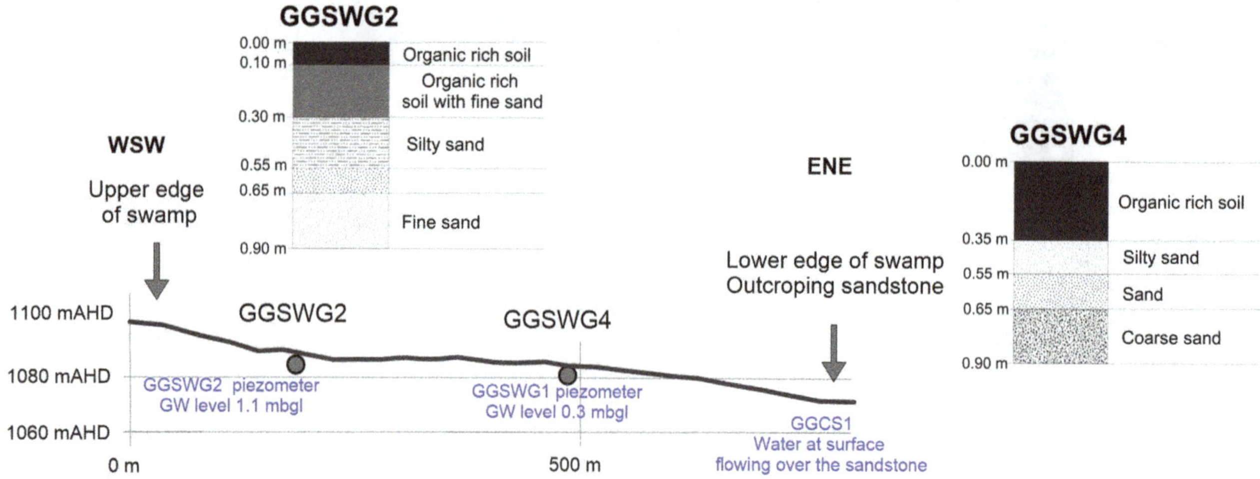

Figure 5. Interpreted long section of swamp GGSW with swamp groundwater levels as measured in October 2016.

a flexible, thick-walled, soft plastic tube, fitted tightly with fittings on both sides. The tight fitting was required to limit the atmospheric air ingress into the LGR. The contamination by atmospheric air during sampling is considered negligible. This is based on the measurement of ambient air moisture of around 14 000 to 15 000 ppm, while the headspace for samples had a range of 23 000 to 28 000 ppm H_2O.

Analysis of the vapour sample was undertaken along with the standards (1 mL) prepared in the similar manner to the core samples. The equilibration time for standards was around 20 min based on literature and air moisture (Wassenaar et al., 2008). A new set of three standards (one primary and two secondary) were run after every third sample. It is not possible to sample the headspace repeatedly using this technique, as 1 L headspace only allows sampling once (60–90 s). Repeated inflating of the same sample with dry air results in incorrect readings. Following each set of samples and standards, the analysis was suspended for a period of around 10–15 min, to allow the LGR to reach the stable atmospheric air readings and reduce any memory effect. Linear regression for $\delta^{18}O$ and δ^2H was established between the liquid values for standards and raw headspace vapour (fractionation factor at 25 °C) readings for the same standards. Regression was used to calibrate the vapour results for samples. Calibration was undertaken with two secondary $\delta^{18}O$ and δ^2H standards (Los Gatos 2A −16.14 ‰ $\delta^{18}O$ and −123.6 ‰ δ^2H and 5A −2.80 ‰ $\delta^{18}O$ and 9.5 ‰ δ^2H) and normalised with one primary VSMOW/VSMOW2 standard run during the analysis. LGR standards were stored in accordance with the protocol, at 4 °C, and ampules were fully sealed to prevent any exchange with the atmosphere. The use of LRG standards as secondary standards has been used in other studies such as Penna et al. (2010) on reproducibility and repeatability of the laser absorption spectroscopy measurements and was found that LGR standards performed satisfactorily.

Replicate sample analyses using the direct vapour equilibration method (mean difference of six samples) indicate reproducibility of results in our research within an uncertainty of 0.68 ‰ for δ^2H and 0.04 ‰ for $\delta^{18}O$. Reported instrument precision of 0.5 ‰ δ^2H and 0.15 ‰ $\delta^{18}O$ over 10 s and drift of 0.75 ‰ δ^2H and 0.3 ‰ $\delta^{18}O$ over 15 min was minimised by correcting the readings. The dataset for each sample was corrected for drift by back correction using standards within each set and then applying the same regression analysis to the relevant samples. For each sample the standard deviation and instrument drift error were calculated. Following the standard operating procedures, the precision in this research was 0.6 ‰ for δ^2H and 0.23 ‰ for $\delta^{18}O$ over 60 s.

Hendry et al. (2015) report the analytical precision of the vapour equilibration method (±0.40 ‰ for $\delta^{18}O$ and ±2.1 ‰ for δ^2H) to be comparable to or better than physical extraction from cores using high-speed centrifugation, cryogenic micro-distillation, azeotropic and microwave distillation or isotope ratio mass spectrometry (IRMS) based direct equilibration methods as discussed in Kelln et al. (2001). Based on work by Allison and Hughes (1983) and Revesz and Woods (1990), the direct vapour equilibration method precision is also better than for methods obtained by chemical water extractions (Hendry et al., 2015). This is achieved by limiting fractionation losses by short storage time, single procedure once the samples are in the laboratory and use of standards and water isotopic data as a cross check. Water samples (surface water, swamp groundwater and regional groundwater, $n = 21$) were analysed for $\delta^{18}O$ and δ^2H by the off axis-integrated cavity output spectrometry (OC-ICOS) technique using an LGR analyser located at UNSW Australia. Two secondary standards and a VSMOW/VSMOW2 standard were used to calibrate and normalise the samples.

Gravimetric water content (ASTM D2974-14, 2014, and ASTM D2216-10, 2010) was measured by weighing the sediment samples ($n = 70$), drying at 100 °C for 24 h and reweighing (Reynolds, 1970); 100 % gravimetric water content relates to water holding capacity and organic content of the material. The analysis was undertaken at the School of Mining Engineering, UNSW Australia. Organic matter content was measured by the loss on ignition method (LOI), by weighing (following initial drying at 100 °C) and by drying in a furnace oven at 550 °C (Heiri et al., 2001). The analysis was conducted at the Water Research Laboratory, UNSW Australia.

The vapour equilibration method is sensitive to the presence of volatile organic compounds (VOCs) in samples (Millar et al., 2018) and may require spectral correction post analysis; however, usually the presence of these hydrocarbons is not known (Hendry et al., 2015). There are limited studies related to quantification of VOCs in peat (Mezhibor and Bonn, 2014) and impact of organic matter on isotopic composition (Orlowski et al., 2016). For water samples, the LGR's post analysis software automatically applied a check for spectral interference. A similar approach was reported by Millar et al. (2018), Schultz et al. (2011) and Orlowski et al. (2018). There was no evidence of spectral contamination. However, this analysis was not possible for vapour analysis of soil samples due to the different processing method. The similarity between the isotopic composition of water and vapour collected from the same horizon suggests a strong interaction between groundwater and pore water; as the groundwater analyses show no evidence of VOC contamination, there is no reason to suspect that the pore water samples would contain concerning levels of VOCs despite the high peat organic content. Furthermore, in one of the rare VOC quantification studies on peat, Mezhibor and Bonn (2014) found that peat had a mean concentration of isoprene and acetaldehyde (VOCs characteristic of natural plant organic emission) up to 0.26 ppb. The ecosystem in their study is similar to this one and for such low-volume % concentrations the spectral corrections are not considered to be necessary. Precipitation samples were analysed at the ANSTO Environmental Isotope Laboratory using a cavity ring-down spectroscopy method on a Picarro L2120-I Water Analyser (reported accuracy of ± 1.0, ± 0.2 ‰ for δ^2H and δ^{18}O respectively). The lab runs a minimum of two in-house standards calibrated against VSMOW/VSMOW2 and SLAP/SLAP2 with samples in each batch.

For simple statistical analysis of moisture content, precipitation and organic matter content, an XLStat software package (XLStat, 2017) was used. The Barnes and Allison (1988) model was implemented for this project using R (R core team, 2013), to investigate the evaporative losses based on isotopic composition of water. For the Barnes and Allison (1988) model volumetric water content was calculated from the measured gravimetric water content and bulk density. Bulk density was obtained from known lithology and measured data (Cowley et al., 2016) and porosity data from a swamp study by Walczak et al. (2002). To estimate effective liquid diffusivity of isotopes, particle size and tortuosity values were obtained from the literature (Maidment, 1993; Shackelford and Daniel, 1991; Barnes and Allison, 1988).

4 Results

4.1 Stratigraphy, organic matter and moisture content

Four stratigraphic units are recognised along all three Newnes Plateau swamp transects CC, GG and GGSW (Figs. 3 to 5), similar to a general classification derived by Fryirs et al. (2014) for THPSS in the Blue Mountains and Southern Highlands regions. The cross sections presented in Figs. 3 to 5 were prepared on the basis of logged cores extracted as part of this research. These units are typically from the base upward medium to coarse sand, medium sand to clayey sand, silt to sandy clay and organic-rich soil (sandy) at the top.

The base of the swamp is comprised of quartz sandstone, the Banks Wall Sandstone of the Narrabeen Group. The alluvial sands (with sub-angular quartz grains) overlying the sandstone are off-white opaque to transparent, medium- to coarse-grained with occasional quartz grains up to 2.5 mm in diameter. The term "sub-angular" defines the roundness of quartz or any other sediment grain. This is important as it points to material transport information; angular grains have been subject to limited transport.

These sands are overlain by medium sand grading to fine sand in the GG transect, with 15 % organic matter and a minor clay component. However, in the CC transect this layer is missing and sand transitions upwards to clayey sand with iron staining. The total thickness of these two sandy units varies from 10 to 50 cm, increasing in the downgradient direction. At the most downgradient site on the GG transect the sand layer is absent. Typically, the basal sand is overlain by a silt and silty clay that is thickest in the middle of the swamp (20–45 cm). The silt is dark grey in colour and contains approximately 40 % organic matter with a strong organic smell. Organic smell relates to a high percentage of organic matter (peat). The uppermost unit is an organic-rich soil or peat (20–60 cm thick), occasionally silty with abundant roots.

The swamp groundwater level is shallow, and it varied in piezometers (installed to 1.5 m depth) in May 2016 from 0.35 m below ground level (b.g.l.) in GGEG2 to 0.47 m b.g.l. in CCG1 (Figs. 3 and 4). The swamp groundwater level in cored holes was similar to that in shallow piezometers; however, there was a significant difference represented by a rise of up to 0.4 m at all measured locations following the wetter period. The initial rise is mainly attributed to rainfall. During this wetter period, swamp groundwater levels recorded at GGSWG1 and GGEG2 were 0.05 and 0.09 m b.g.l. respectively. No overland flow was observed at any time, and the swamps did not have a formed channel. The only surface water observed in the swamps was at the lower edge of the swamp and flowing over the rockbar.

The swamp sediments are variably saturated, with gravimetric water content measurements exceeding 100 % weight (dry mass basis) in the top 30 cm. This is typical for a high organic matter proportion (GG samples) (Fig. 6). Within the same vertical profile, the organic matter content varied with depth and decreased from 60 % to 10 %. At a depth from 60 to 120 cm the gravimetric water content decreased to an average of 17 % for CC and 32 % for the GG swamp during both the May and October 2016 sampling periods. The average organic matter decreased to 3.7 % for all swamp locations below 80 cm depth.

During May 2016, following the dry period, upgradient and downgradient samples in CC swamp had similar gravimetric water content. A clear distinction was observed after wet weather period between the upgradient CCG2, having overall lower gravimetric water content, and downgradient CCG3, with higher gravimetric water content. A trend with an increase in moisture content downstream has been observed in all three swamps. However, at GGEG, the undulating topographic gradient means that changing moisture conditions exist along the length of the swamp. An overall increase in moisture content to around 80 cm depth in CCG3, was also recorded following the wet weather period although the increase was not statistically significant ($p>0.05$).

4.2 Stable isotopes of water and pore water

The relationship between surface water, swamp groundwater, regional groundwater and swamp pore water δ^{18}O and δ^{2}H data is presented in Fig. 7. This figure also shows the local meteoric water line (LMWL) for Lithgow ($\delta^{2}\text{H} = 7.99\,\delta^{18}\text{O} + 16.6$; Hughes and Crawford, 2013) and weighted rainfall average for Mt Werong which is based on the past 12 years of data ($\delta^{18}\text{O} = -6.87$ ‰, $\delta^{2}\text{H} = -37.3$ ‰). The δ^{18}O of rainfall varies seasonally with higher values in summer and lower in winter.

Stable isotope data from precipitation events at Mt Werong are plotted (excluding the rainfall below 5 mm) for three periods (January to May 2016, May to October 2016, and January to May 2017). The stable isotope data for these events plot on or close to the previously defined LMWL for Lithgow (note that the LMWL for Mt Werong of δ^{2}H is $8.08\,\delta^{18}\text{O} + 16.6$; Hughes and Crawford, 2013) and has a similar slope but higher intercept than that for Lithgow.

For May 2016 with dry and warm antecedent conditions, pore water stable isotope ranges are −7.20 to 3.10 ‰ δ^{18}O and −45.7 to −22.3 ‰ δ^{2}H (Fig. 7a). Pore water samples at CC and GG in May 2016 were clearly evaporated, lying along lines with slopes of 4.2 and 4.6 respectively, even though no single initial value for pore water evaporation was discernible. Two major rainfall periods (27 mm 2 weeks prior to May 2016 sampling and 153.5 mm in January 2016) had no noticeable influence on the swamp pore water isotope composition. The intersection points of the regressed trend lines of pore water and LMWL plot within the lower δ^{18}O and δ^{2}H rainfall range.

Stable isotopes for swamp pore water collected in October 2016 (Fig. 7b) range from −4.50 to −7.50 ‰ δ^{18}O and −25.0 to −47.0 ‰ δ^{2}H. Pore water stable isotope values from samples collected in the wet and cool antecedent conditions plot along the LMWL very close to the weighted rainfall average. This is consistent with a winter rainfall signature.

The pore water samples collected in May 2017 from GGSW swamp lie along a slope of 6 which aligns with a wetter period in early 2017 compared to 2016. Samples from CC swamp collected in May 2017 are more enriched in ^{2}H (i.e. have a higher *D*-excess (d), defined as $d = \delta^{2}\text{H} - 8\delta^{18}\text{O}$, Dansgaard, 1964) than previously collected samples indicating greater evaporative influence. Rainfall samples for bigger rainfall events in the period from December 2016 to May 2017 plot along the LMWL, except events in the April prior to the 2017 sampling which have a significantly higher *D*-excess ($d = 24.5$ ‰, rainfall of 68 mm). The pore water returned to the LMWL between May and October 2016 and shifted to the left of the LMWL for the May 2017 sampling.

Swamp groundwater samples collected in October 2016 and May 2017 are enriched in ^{18}O and ^{2}H relative to the rainfall weighted average for Mt Werong (2005–2017). Surface water samples collected mainly at the downstream point of the swamp plot close to the LMWL and are lower in δ^{18}O and δ^{2}H relative to pore water samples, and relative to large rainfall events preceding the sampling event.

Surface water samples (−6.50 to −7.70 ‰ δ^{18}O and −37.0 and −44.4 ‰ δ^{2}H) plot within the range of δ^{18}O and δ^{2}H for swamp groundwater samples (−6.2 to −7.9 ‰ δ^{18}O and −32.4 and −44.7 ‰ δ^{2}H). The statistical significance of the difference between the isotopic composition of surface water and swamp groundwater on both GG and GGSW transects was analysed by comparing the means of δ^{18}O and δ^{2}H (October 2016 and May 2017) for these two datasets using a t-test. Based on the mean, we test the hypothesis that there is no statistical difference between the datasets (surface water and swamp water). The calculated p-value was significantly more than 0.05 (for $\delta^{18}\text{O}$ $p = 0.34$ and for $\delta^{2}\text{H}p = 0.27$; ($n = 20$), indicating that the null hypothesis cannot be rejected and there is no significant difference between these two datasets.

The δ^{18}O and δ^{2}H data for pore water are plotted with depth along with surface water and groundwater from the GG and GGSW swamps (Fig. 8a, b). Seasonal pore water and swamp groundwater variations (May and October 2016 sampling) for the CC swamp are compared to the rainfall isotopic signature collected at Lithgow (Hughes and Crawford, 2013) (Figs. 8c and 7d). The δ^{18}O values of pore water (May 2016) in the GG and GGSW swamps (Fig. 8) show a tendency towards depletion with depth with greater variability at a depth of 40–65 cm. Below 100 cm depth, the δ^{18}O

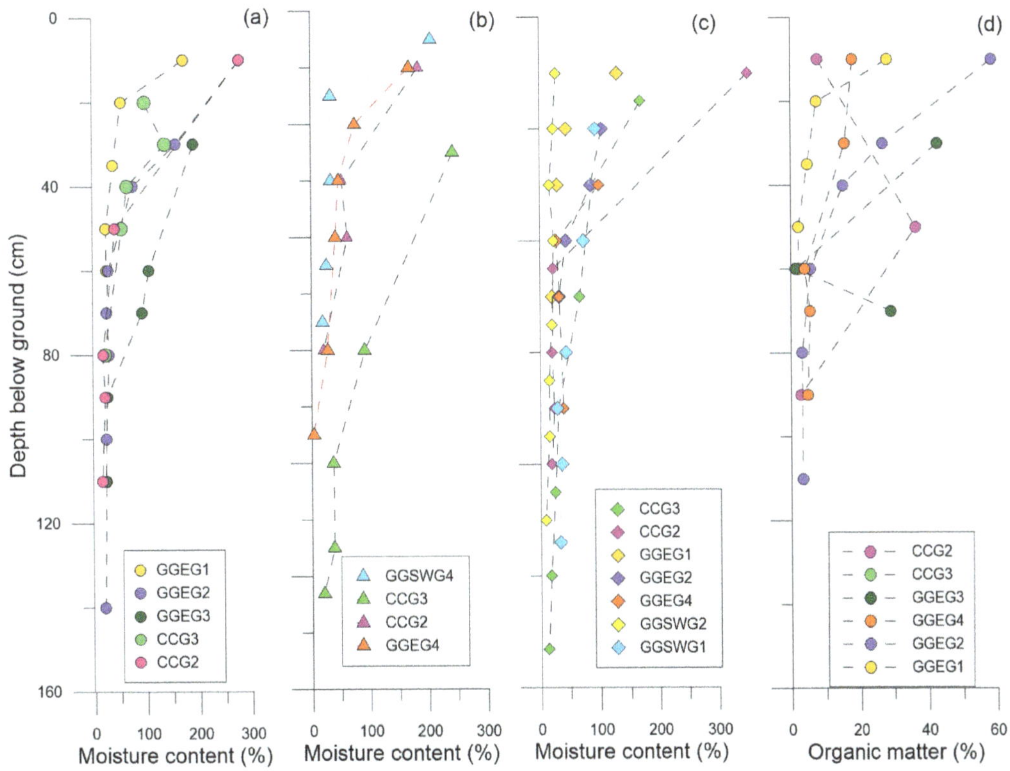

Figure 6. Gravimetric water content (% weight) for May 2016 **(a)**, October 16 **(b)** and May 17 **(c)** and organic matter content in the CC and GG swamps **(d)** shown with depth.

values of pore water approach the swamp groundwater and regional groundwater signature.

It can be observed that pore water samples from the CC swamp from both upstream (location CCG2) and downstream (location CCG3) have lower $\delta^{18}O$ after longer wet and cool antecedent conditions with a $\delta^{18}O$ shift of around 1 ‰–3 ‰ (Fig. 8c and d). $\delta^{18}O$ and $\delta^{2}H$ for pore water at CCG2 during May and October 2016 show a statistically significant difference between the wet and dry periods ($\delta^{18}O$ ($p = 0.003$) and $\delta^{2}H$ ($p = 0.02$)), similar to $\delta^{18}O$ of pore water at CCG3 ($p = 0.01$). The CC samples collected in May 2017 have lower $\delta^{18}O$ values of pore water compared to October 2016 samples and are similar to significant rainfall in March 2017.

Swamp groundwater samples collected from piezometers screened across both top of sandstone and the bases of swamp sediments (CCG1 and GGEG2) have a similar $\delta^{18}O$ signature to pore water at a depth below 110 cm. Surface water $\delta^{2}H$ for October 2016 is more negative than the pore water value (−37.7 ‰ $\delta^{2}H$) in the upper 70 cm and is similar to the typical winter rainfall signature.

4.3 Water balance

During dry periods swamp pore water is subject to evaporation and becomes enriched in ^{18}O and ^{2}H. Therefore, the fractional loss of water through evaporation can be quantified if other water loss processes do not isotopically fractionate (Gonfiantini, 1986) or/and if the stable isotope composition of inflow and outflow and site weather data is known (Lawrence et al., 2007). To evaluate the evaporative losses based on isotopic composition of water, we used the Barnes and Allison (1988) analytical model to represent the change in isotopic profile in unsaturated soils due to evaporation. This model, based on deterministic approach, was selected because the stable isotopes diffusivities vary slowly with water content and a relatively good agreement is reported with experimental results (Barnes and Allison, 1988; Shanafield et al., 2015). The disadvantage of using the soil profile to estimate evaporation is that an assumption of steady state is needed and there is some uncertainty in dispersivity and tortuosity values (Shanafield et al., 2015). The support for the selection of the Barnes and Allison (1988) model for the vegetated wetland environment is shown in the recent work undertaken by Piayda et al. (2017). They found that regardless of the presence of vegetation or bare soil, the total evapotranspirative water loss of soil and understorey remains un-

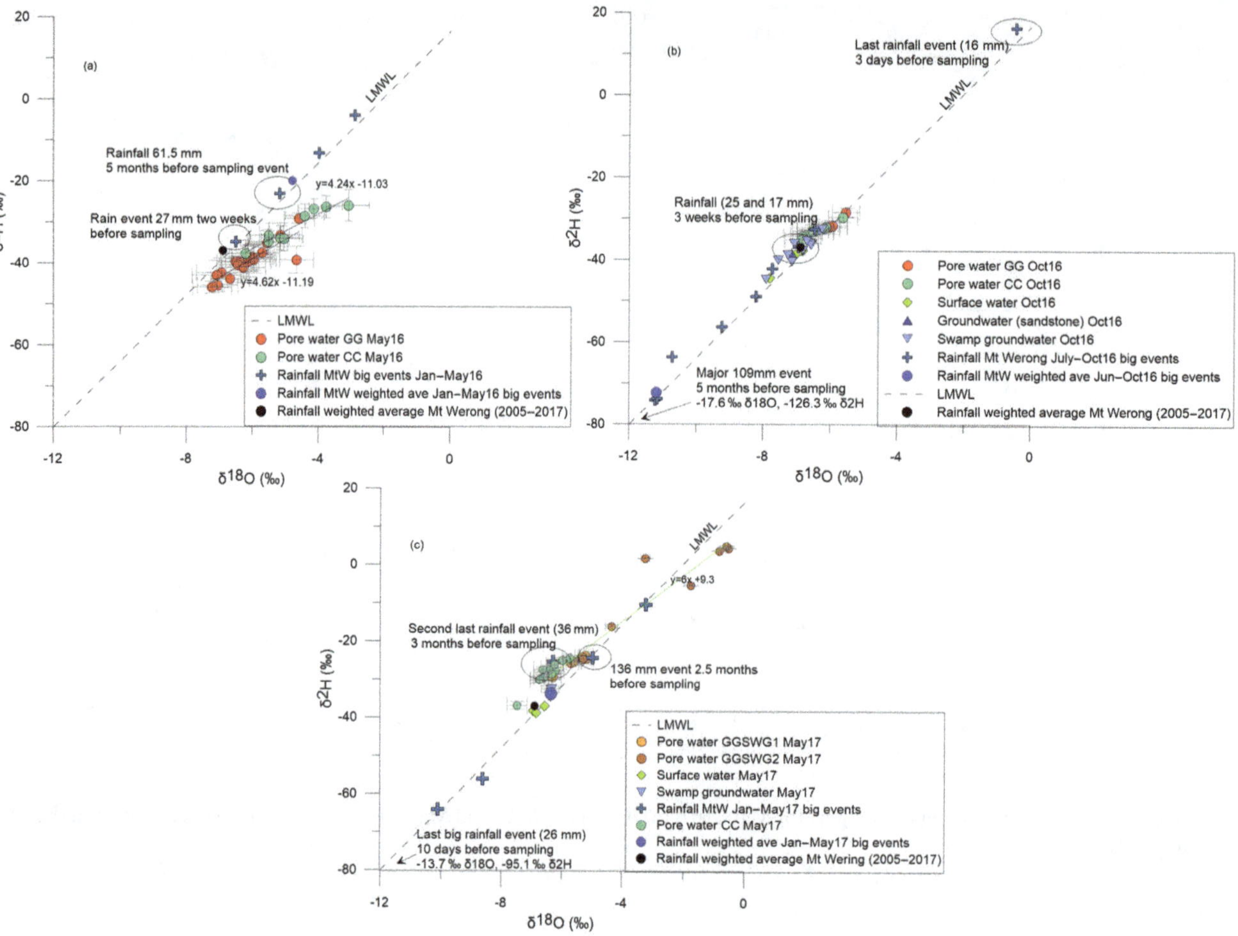

Figure 7. Stable δ^{18}O and δ^{2}H composition of surface water, swamp groundwater, regional groundwater, swamp pore water, weighted rainfall average for Mt Werong (2005–2017) and LMWL for Lithgow (Hughes and Crawford, 2013) May 2016 **(a)**, October (2016) **(b)** and May 2017 **(c)**.

changed. Furthermore, the modelling is considered to be applicable by focusing on vertical flow. Modelling included the samples from the base of the swamp (not sides) where vertical flow is dominant due to high permeability of the peat.

We applied the model to pore water data from all three sampling periods considering realistic input variables into the model as given in Table 1. The model ran with the evaporation factor adjusted such that it matched the observed data; all other parameters remain constant. A linear relationship was identified between particle size and tortuosity, and the final estimated tortuosity values are given in the Supplement (Table S1).

The results for unsaturated soil modelling at all sampled depth points based on δ^{18}O and δ^{2}H indicate an evaporative loss in the unsaturated zone of 4 to 9 mm day^{-1} in May 2016 (dry) period, and <1 mm day^{-1} for the wetter and cooler period between May and October 2016. Evaporation of less than 1 mm was estimated in CC swamp in both wet and dry periods, and at the upstream point on GGSW swamp.

The model was not sensitive to temperature; modelling at both 21.9 and 10 °C resulted in only minor differences in evaporation (<0.04 mm day^{-1}). Model results for drier and wetter periods are presented in the Supplement (Fig. S1). The data for the May 2016 period (dry) show a clear evaporative enrichment profile towards the surface (upper 0.4 to 0.6 m) and uniform δ^{2}H with depth (Fig. S1a and b). No changes in isotopic composition are observed below a depth of 0.6 m.

The water balance was prepared such that it incorporates the following parameters: rainfall, runoff from each of the swamps, and evaporation. The deficit in the water balance is attributed to groundwater contribution. Two options are considered in the water balance with respect to evaporation: evaporation based on unsaturated soil model results (E) and

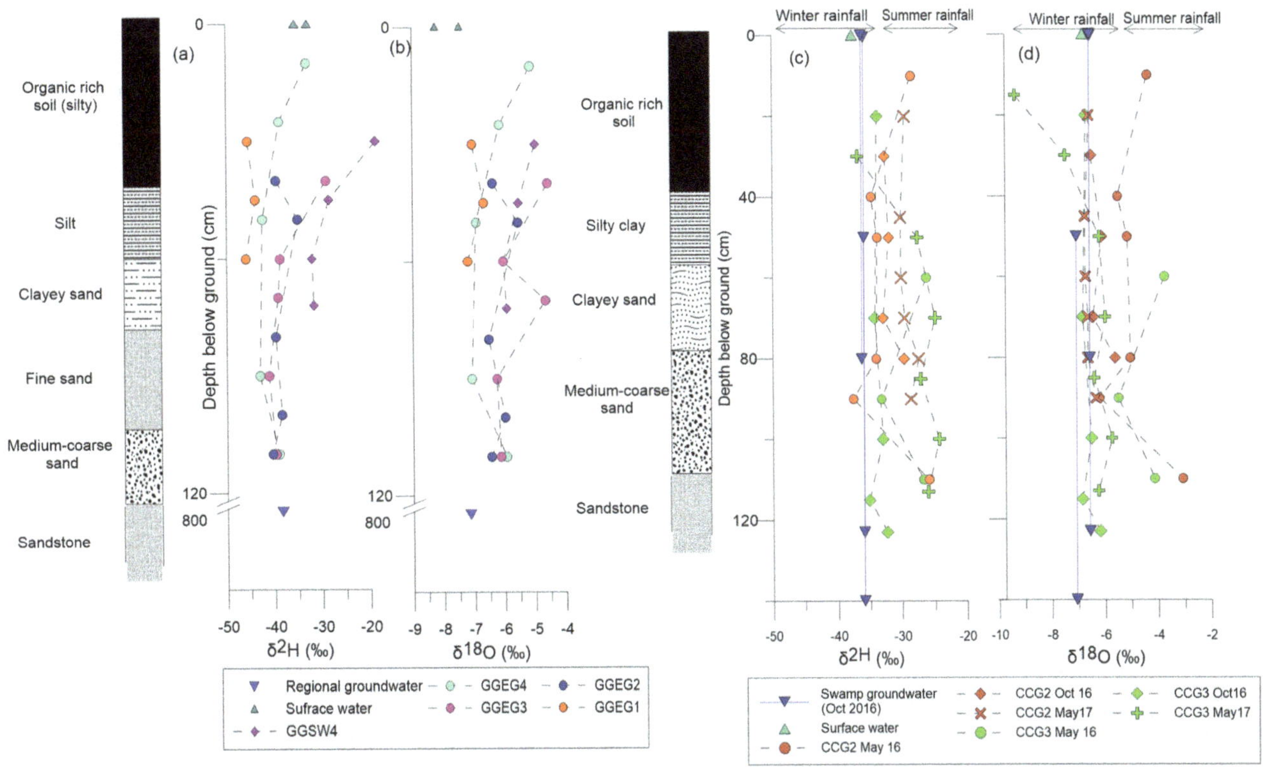

Figure 8. $\delta^{18}O$ and $\delta^{2}H$ variation with depth in GG and GGSW swamps (May 2016) with typical lithology log. Regional groundwater sample was collected at the downstream point of the GG swamp **(a, b)**. $\delta^{18}O$ and $\delta^{2}H$ variation with season and depth in CC swamp (May and October 2016 and May 2017) with typical lithology log. Swamp groundwater represents cumulative water through the swamp within shallow piezometers and cored holes **(c, d)**. Swamp groundwater samples were not collected at all locations in May 2016 due to dry conditions. Depth of augured holes was not exactly the same in all sampling events.

reference data (ET_c). The rainfall data from Lithgow BoM Station 63132 and reference evapotranspiration (ET) data from Nullo Mountain BoM Station 62100 (94 km north of the study site in the same mountain range and similar elevation and climate) indicate that in the dry period (February to May 2016) the ET significantly exceeded the rainfall (Table 1). The ET represents evapotranspiration computed from the reference surface (grass) using meteorological data (Allen et al., 1998). Crop evapotranspiration (ET_c) calculation incorporates the ground cover, canopy properties and aerodynamic resistance for the specific crop into the calculation. In our case the ET_C is applied to a wetland system.

Figure 9 shows the water deficit and estimated regional groundwater contribution to each of the swamps for the ET_C and *E* methods. The relative regional groundwater contribution is dominant in the dry weather period when it exceeds total rainfall. This regional groundwater contribution range represents the minimum and conservative value given that discharge from the swamp is not included in the water balance and that the estimates based on the *E* method do not include transpiration losses.

5 Discussion

5.1 Swamp stratigraphy, geomorphology and groundwater condition

Swamp sediments are thin (less than 1.5 m) and are deposited directly on the sandstone basement. Typically, the organic soil or peat is 40–60 cm thick, underlain by unconsolidated alluvial sand and sandy silt with organic-rich thin bands. The geomorphology of the Newnes swamps is consistent with the intact swamp classification as reported by Fryirs et al. (2016), and with moisture and organic matter content as reported in Blue Mountain swamps by Cowley et al. (2016). The lithology indicates that the sediment transport is alluvial; however, it is limited and occurring over relatively short distances (length of the swamp).

An important finding of this research is that no evidence was observed for a clay-rich layer with sealing properties at the base of these swamps. A conceptual model of swamp sediments that are hydraulically connected with the underlying sandstone is proposed (Fig. 10). However, there is likely to be a decrease in permeability at this interface. A degree of hydraulic connection between the regional groundwater and

Table 1. Water/mass balance components: measured rainfall and ET data (Lithgow and Nullo Mountain), runoff, measured E and estimated balance deficit (negative values are groundwater contribution).

	February 2016	March 2016	April 2016	May 2016
Total monthly rainfall Lithgow station (SN63132) (mm month^{-1})	28.8	61.2	6.2	26
Reference ET Nullo Mountain (SN62100) (mm month^{-1})	119.9	92	76.6	51
Evaporation (pore water stable isotope profiles) mm month^{-1}	117–267	123–273	120–270	123–273
Runoff estimate				
CC	31	65.6	6.6	28
GG	25	53	5.4	22.5
GGSW	16.4	35	3.5	15
Balance deficit (groundwater component) (ETc)				
CC	−60.2	34.8	−63.8	2.9
GG	−66.2	22.1	−65.0	−2.5
GGSW	−74.7	4.1	−66.9	−10.2
Balance deficit (groundwater component) (4 mm)				
CC	−56.3	2.8	−107.2	−70.1
GG	−62.3	−9.9	−108.4	−75.5
GGSW	−70.8	−27.9	−110.3	−83.2
Balance deficit (groundwater component) (9 mm)				
CC	−201.3	−152.2	−257.2	−225.1
GG	−207.3	−164.9	−258.4	−230.5
GGSW	−215.8	−182.9	−260.3	−238.2

these elongated gentle gradient shrub swamps (50 mm m^{-1} average; Cardno, 2014) is further supported by gravimetric water content results. The stable gravimetric water content below 0.4 m depth in CC and 0.6 m depth in GG and GGSW swamps indicates stable saturated conditions likely supported by lateral groundwater inflow.

Groundwater levels in the swamps were observed to be similar to regional groundwater level within the underlying sandstone (monitoring screen at a depth of around 10 m b.g.l.) at the downstream end of GG swamp indicating that these two units could be hydraulically connected. Typically, the swamp groundwater levels in THPSS (CC swamp) rise and decline in response to rainfall recharge (Centennial Coal, 2016) with very little lag time. Rapid infiltration and discharge in the swamp groundwater system is indicated by low swamp groundwater salinity (measured in this study) (David et al., 2018). Given high moisture and organic matter content and evidence of seasonal precipitation in $\delta^{18}O$ and δ^2H profiles ($p<0.05$) in the upper swamp horizons, we conclude that in this zone the high water holding capacity increases residence time following the initial infiltration (vertical swamp groundwater flow). The $\delta^{18}O$ and δ^2H of pore water in this variably saturated zone exhibits summer evaporation trends and a winter rainfall signature. The lateral groundwater discharge to the swamp is characterised by longer residence time compared to water exchange through the swamp based on lower $\delta^{18}O$ values and minor change between the sampling events. The similarity in EC and pH values between surface and swamp groundwater (David et al., 2018) further supports relatively rapid infiltration and possibility of both lateral and upward local groundwater inflow that provides baseflow to the swamp. However, local differences in swamp strata do exist: e.g. the difference in gravimetric water content in the CC swamp between the upgradient and downgradient location. This difference can be explained by higher permeability in the upgradient part of the swamp resulting in quicker drainage, increased groundwater contribution in the lower part of the swamp and/or lateral throughflow.

To validate this conceptual model, a simple water balance was completed based on the evaporative losses estimated by the analytical model (Barnes and Allison, 1988). Using the results from the dry weather period February to May 2016, we obtain evaporation estimates ranging from 1 to 9 mm day^{-1}. The evaporation occurs in the top 0.4 m of the vertical profile, with an absence of fractionation below this depth where pore water isotope values are similar to swamp water and regional groundwater. These evaporation rates (1 to 9 mm day^{-1}) suggest high evaporation compared to rainfall in the same time period (Table 1). During the wet period (Fig. S1c, d) we observe the lower δ^2H at the surface; this is related to a big rainfall event, 10 days before sampling (Fig. 7c).

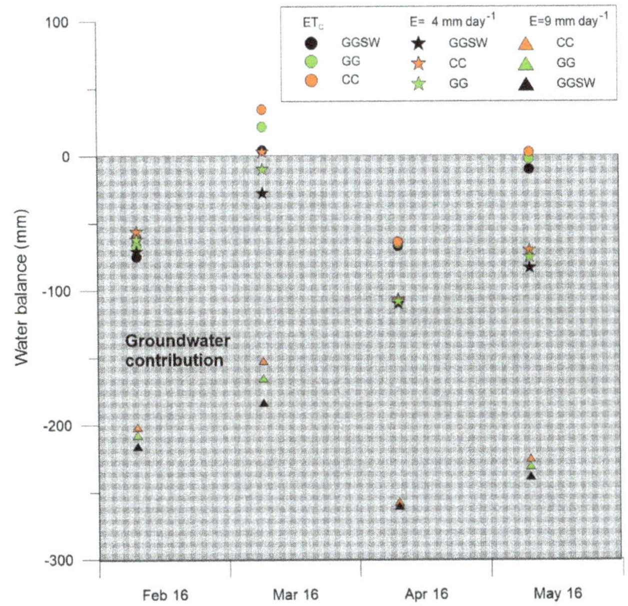

Figure 9. Water balance during the dry period estimated using ETc and E for each of the swamps.

With an ET_c ranging from 1.7 to 4.4 mm day^{-1} and E ranging mainly from 4 to 9 mm day^{-1}, the ET_c / E ratio for these swamps would be 0.7 to 0.3. This ratio is at the lower end of measured ET_c / E ratio for typical wetlands indicating that reference ET could underestimate that based on realistic evaporation rates obtained by matching the modelled to observed data. The ET_c for typical wetland vegetation (sedge) in temperate climates ranges from 0.8 to 1.2 (Allen et al., 1998; Mohamed et al., 2012) and 0.7 was reported in a swamp in the Murrumbidgee, Australia (Linacre et al., 1967). The ET_c in our case is less than the estimated E based on stable isotope data. As transpiration does not fractionate, the actual evapotranspiration in the dry and warm period would have to be greater than the estimated evaporation. This would result in higher water balance losses, requiring more water be supplied from other sources.

Runoff represents only a small component of the water budget for several reasons. Firstly, the 10 % slope gradient of the ridges, 3 % slope gradient along the swamp floor and densely vegetated sides and base of the swamp minimise the runoff significantly. Secondly, the upper soil layer is peat with significant water holding capacity compared to other soil types, and as indicated by the gravimetric water content measured in CC and GG swamps.

A simple mass balance comprising the rainfall (input), runoff (input) from the catchment considered two different approaches in dry period using E or ET_c. When ET_c (output), was used, March had excess water with a deficit in February, April and May of between 10 and 60 mm. However, the same mass balance calculated with E using 4 mm day^{-1}, has water deficit of between 10 and 113 mm month^{-1} for any month. If E of 9 mm day^{-1} is used in the water balance, water deficit occurs in every month in the range from 10 to 260 mm month^{-1} (Table 1). Either way, two important output components are not considered in this mass balance: transpiration and discharge at the rockbar downgradient of the swamp. The estimation of these two components is uncertain, but inclusion in the water balance would increase the water deficit further. Importantly, if swamp discharge data were available in combination with pore water isotope profiles, an appropriate crop transpiration could be determined for these swamps, a factor that is typically a large unknown in water balance studies.

It is evident that given the water deficit, even without two output components, an additional water source must have maintained the swamp groundwater levels. We therefore conclude that groundwater is a significant contributor to swamp water balance, particularly during dry periods. For example, in the GG swamp the swamp groundwater levels are in the range from 0.28 to 0.38 m b.g.l., and if groundwater inflow were not occurring under the same evaporation conditions, the depth to water in the swamp would be greater.

Furthermore, measured loss of moisture as shown in Fig. 6 indicates that significant loss occurs in such a dry weather period in the top 40 cm (up to 150 % by weight), while the lower parts of the swamp remain saturated. The estimate of groundwater contribution in the drier period (February to May 2016) ranges from 10 to over 113 mm month^{-1} if calculated using 4 mm day^{-1} of evaporation, and up to −260 mm day^{-1} if E of 9 mm day^{-1} is used. The water balance was undertaken for the dry period only as evaporation from the soil profile using stable isotopes is considered to be most accurate during that period. Thus, even in these months when the water balance is positive, groundwater contribution is likely, as evident from discharge at the rockbar observed at the end of the dry period. Although there is a compelling explanation for significant groundwater contribution to the swamp water balance, the actual volume of groundwater cannot be estimated without knowledge of swamp groundwater and regional groundwater recession rate and/or measurement of discharge from the swamp.

It is clear that the water balance in swamps can be obtained if all the components are known (rainfall, runoff, groundwater contribution, evaporation, evapotranspiration, discharge); however, due to the THPSS swamps being difficult to access and being protected under state and federal legislation, it is not possible to undertake intrusive drilling to obtain all the hydrogeological information. The application of stable isotopes has enabled estimation of evapotranspiration from the swamp and assisted in development of the conceptual model.

5.2 Swamp groundwater and regional groundwater movement within the swamp system

The vertical depth profiles of pore water $\delta^{18}O$ and $\delta^{2}H$ can provide time series information by tracing the influence of

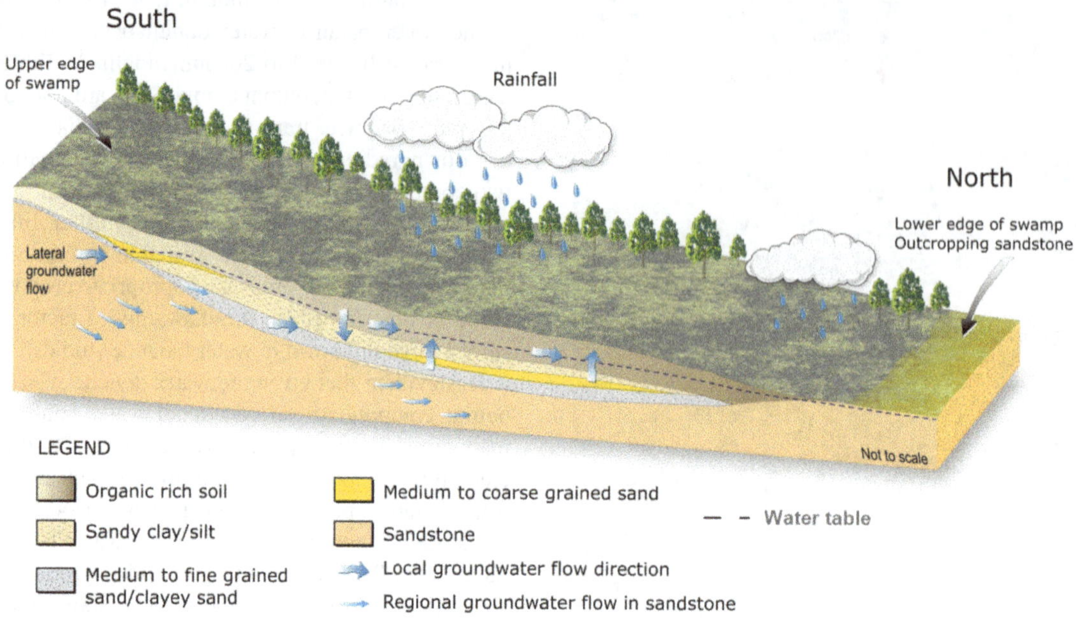

Figure 10. Conceptual representation of water dynamics in the swamp system.

the rainfall isotopic signature in recharging water. The pore water direct vapour equilibration method is used in a swamp environment and results compared with end-members which included surface water, rainfall and groundwater. Although stable isotope data in precipitation change in the short term, this end-member is well constrained based on the good-quality dataset for precipitation. Constraining the interpretation of isotope results with these end-members enabled groundwater inputs to be identified.

The evaporation response in the upper 40 cm is consistent with depth of penetration dependent on evaporation rate, soil type and time between rainfall events (Mathieu and Bariac, 1996; Melayah et al., 1996; dePaolo et al., 2004). As evaporation proceeds, capillary rise of swamp groundwater reduces the $\delta^{18}O$ enrichment closer to the surface. Moisture content data reveal variability at 30–70 cm depth, which is also observed in $\delta^{18}O$ and $\delta^{2}H$ profiles and is related to interlayering of fine- and coarser-grained material, consistent with other studies (dePaolo et al., 2004). The pore water regression line intercepts the LMWL at a lower $\delta^{18}O$ and $\delta^{2}H$ value than weighted average rainfall. The isotope signature in the partially saturated zone (variable from 0.05 to 0.4 m b.g.l. in the swamp) in the summer period (May 2016 sampling event) is a result of evaporation as observed from depth profiles and moisture content. This agrees with numerical experiments conducted by Benettin et al. (2018) where the soil water samples' trend lines were found to be products of seasonality of evaporative fractionation. Swamp pore water in May 2016 has lower δ values than rainfall and is therefore likely to be from bigger, more isotopically depleted events in the autumn and winter of the prior year which are lower in $\delta^{18}O$ and $\delta^{2}H$ (including 230 mm in April 2015: $-7.60‰\,\delta^{18}O$, $-39.2‰\,\delta^{2}H$; 108 mm in August 2015: $-9.8‰\,\delta^{18}O$, $-61.6‰\,\delta^{2}H$; and two smaller but highly ^{18}O and ^{2}H depleted events in June and July). This agrees with annual weighted averages at Mt Werong of -34.9 and $-46.5‰\,\delta^{2}H$ in 2015 and 2016 respectively. A major rainfall event in June 2016 (92.8 mm at Lithgow and 109 mm at Mt Werong, $-17.70‰\,\delta^{18}O$, $-126.3‰\,\delta^{2}H$) had not obviously affected swamp pore water.

Although the same rainfall events generally affect both Mt Werong and Newnes, and occur at the same time, the amount of rainfall at Newnes is typically smaller than at Mt Werong. Whilst we would expect that larger rainfall events would lead to the most significant infiltration and recharge of swamp groundwater and regional groundwater, and therefore influence the pore water signature more, the data seem to suggest that small recent rainfall events are very important in October 2016 and May 2017, following the wetter conditions experienced in the second half of 2016 and early 2017. The importance of smaller rainfall events for recharge is also consistent with gravimetric water content data which remained stable throughout the wetter and drier periods at depths below 0.8 m in CC and 0.6 m in GG and GGSW swamps. Another contributing factor may be that groundwater provides a moderating effect, particularly during wetter periods, reducing the effects that evaporation has on pore water isotope composition.

Statistically there is no difference between the mean of the surface water and swamp groundwater stable isotope samples for GG and GGSW swamp. The reason for similarity of surface and swamp groundwater samples is assumed to be short infiltration time to the water table and/or mixing with lateral regional groundwater, with surface water sample points being located largely in the groundwater discharge zone. There is a difference in $\delta^{18}O$ and $\delta^{2}H$ values ($p<0.05$, $n=18$) between samples collected after dry and warm versus wet and cool antecedent conditions. The October 2016 (cool weather) samples from CC swamp are typically lower in $\delta^{18}O$ and $\delta^{2}H$ and we conclude that these values are within the range of winter rainfall isotope values. Below 100 cm depth the pore water values of $\delta^{18}O$ remain uniform and consistent with the regional groundwater value but also with surface water. We infer this to represent swamp groundwater derived from vertical infiltration and laterally from sandstone respectively. We therefore consider the main processes to be rapid infiltration through the swamp sediments to the water table but at the same time high water retention in the upper horizons, and slow lateral exchange of pore water below the vadose zone.

The vertical topographic difference from swamp headwaters to the downstream end of the swamp (typically a sandstone rockbar) is around 40 m. This elevation difference is too small to result in any difference in isotopic signature of precipitation, therefore, given the spatial response and assuming a homogeneous environment with vertical flow, pore water $\delta^{18}O$ and $\delta^{2}H$ should be similar (Garvelman et al., 2012). However, observed variation in profiles is not uniform, and is caused by vertical rainfall infiltration in the upper part of the profile and lateral flow at the base. The lateral flow within the swamp sediments is further enhanced by regional groundwater flow contribution from the valley sides. Such lateral flow is reported in these swamps where sandstone is underlain by a claystone layer (Corbett et al., 2014).

Factors such as fine-grained content of lithological units, reported by other studies (dePaolo et al., 2012), have been found to result in a bigger shift to lower $\delta^{18}O$ and $\delta^{2}H$ values and variation in isotope signature with depth. The reason for this is related to hydraulic conductivity of the unconsolidated soil. For example, the biggest variation in $\delta^{18}O$ and $\delta^{2}H$ was observed in silt and clayey sand units (Fig. 8a and b) which contain a higher percentage of particles <2 µm. Contrary to observations by Garvelman et al. (2012), we did not find the variability in $\delta^{18}O$ and $\delta^{2}H$ to be a result of soil saturation and depth of the vadose zone only, but also as a function of lithology and different grain size material (peat, organic soil with sand and silt). Variations in particle size, porosity and permeability would then influence groundwater flow and storage.

6 Conclusions

The hydrogeological and isotopic characterisation of these swamp environments provides a baseline understanding for future comparison of any hydrological changes due to natural or human activities. This study applies the vapour equilibration method for determining stable isotopes of pore water in a wetland system. This unique pore water isotope approach combined with other data and information has significantly improved a conceptual model of wetland hydrology. As found by Wassenaar et al. (2008), the pore water stable isotope method allows efficient sample collection without permanent disturbance, collection of vertically discretised data at any practicable frequency and without the need for more complex methods of water extraction.

This study found, for several upland peat swamps, that swamp groundwater is a dominant component of the water balance, its contribution being larger than rainfall during dry weather periods. This finding is consistent with environmental tracer studies suggesting that 19 %–80 % of water in Blue Mountains swamps is from groundwater, particularly in steeper and rounder catchments (Young, 2017). Furthermore, these swamp groundwater systems appeared to be in hydraulic connection with the underlying sandstone regional groundwater, given similar groundwater levels and the lack of a clayey layer at the base of the swamp. Although rainfall infiltration to the water table occurs rapidly, the high water holding capacity of upper organic-rich layers maintains the moisture for long periods. These processes are confirmed by the results of the water balance, in particular during dry periods. The majority of flow through the swamp system is via lateral groundwater flow where flow rate depends on heterogeneity within this layer and hydraulic conditions. Under natural intact conditions, upward or downward flow between the swamp system and underlying rock is controlled by groundwater heads, the slope, and the hydraulic conductivity contrast at the interface.

The conceptual model presented here provides a valuable benchmark from which to evaluate potential changes in swamps following underground mining and forestry activity. The improved understanding in the water balance in these swamps also has implications in other areas of the Blue Mountains where urbanisation has a significant impact on upland swamps. The role that catchments have on the health of a swamp is important in supporting its flora and fauna, with groundwater likely to be a primary factor that contributes to the long-term survival of the ecosystem (Gorissen et al., 2017). The protection of this ecological community is therefore dependent on maintenance of catchment stability and groundwater baseflow contribution if forestry activity and ground movement or deformation due to mining occur in the swamp catchment.

Measurement of pore water stable isotopes of peat and sediment within the swamp ecosystem provides direct information on the depth at which the evaporation occurs and understanding of the water cycle. Evaporation obtained from the stable isotope direct equilibration method was found to be more realistic than reference evapotranspiration. In particular, based on current research of the water balance in wetland and swamp systems and ecology around the world, the application of this method could be beneficial to define water availability for flora and fauna in swamps where a thick organic soil/peat and sedimentary layer exists.

Competing interests. The authors declare that they have no conflict of interest.

Acknowledgements. This research did not receive any grants from funding agencies in the public, commercial, or not-for-profit sectors. The authors would like to acknowledge the support of the Centre for Water Initiative and School of Biological and Earth Sciences for assistance with sample analysis. Rainfall isotope analysis was funded independently by ANSTO. We thank Andy Baker and anonymous referees for providing constructive suggestions to improve this paper. We acknowledge Bob Cullen for collecting rainfall samples at Mt Werong, Barbara Gallagher, Jennifer van Holst (ANSTO) and Fang Bian (UNSW) for analysis of rainfall samples, the Sydney Catchment Authority for providing rainfall data, and Karina Meredith (ANSTO) for advice on evaporation modelling.

Edited by: Christine Stumpp

References

Allen, R. and Lindesay, J.: Past climates of Australasia, in: Climates of the Southern Continents, edited by: Hobbs, J. E., Lindesay, J. A., and Bridgman, H. A., 208–247, Wiley & Sons, Chichester, 1998.

Allen, R. G., Pereira, L. S., Raes, D., and Smith, M.: Crop evapotranspiration – Guidelines for computing crop water requirements – FAO Irrigation and drainage paper 56, Fao Rome 300, D05109, 1998.

Allison, G. B. and Hughes, M. W.: The use of natural tracers as indicators of soil-water movement in a temperate semi-arid region, J. Hydrol, 60, 157–173, https://doi.org/10.1016/0022-1694(83)90019-7, 1983.

Barnes, C. J. and Allison, G. B.: Tracing of water movement in the unsaturated zone using stable isotopes of hydrogen and oxygen, J. Hydrol, 100, 143–176, 1988.

Benettin, P., Volkmann, T. H. M., von Freyberg, J., Frentress, J., Penna, D., Dawson, T. E., and Kirchner, J. W.: Effects of climatic seasonality on the isotopic composition of evaporating soil waters, Hydrol. Earth Syst. Sci., 22, 2881–2890, https://doi.org/10.5194/hess-22-2881-2018, 2018.

Benson, D. and Baird, I. R. C.: Vegetation, fauna and groundwater interrelations in low nutrient temperate montane peat swamps in the upper Blue Mountains, New South Wales, Cunninghanmia, 12, 267–307, 2012.

Benson, D. H. and Keith, D. A.: The natural vegetation of the Wallerawang 1 : 100000 map sheet, Cunninghamia, 2, 305–335, 1990.

Bickford, S. and Gell, P.: Holocene vegetations change, Aboriginal wetland use and the impact of European settlement on the Fleurieu Peninsula, South Australia, Holocene, 15, 200–215, 2005.

Bijoor, N. S., Pataki, D. E., Rocha, A. V., and Goulden, M. L.: The application of $\delta 18O$ and δD for understanding water pools and fluxes in a Typha marsh, Plant. Cell Environ., 34, 1761–1775, 2011.

Bond, N. R., Lake, P. S., and Arthington, A. H.: The impacts of drought on freshwater ecosystems: an Australian perspective, Hydrobiologia, 600, 3–16, 2008.

Bureau of Meteorology (BoM): Climate Data Online, Canberra, available at: http://www.bom.gov.au/climate/data/index.shtml?bookmark=_136, last access: 14 April 2017.

Cardno: Aquatic Ecology and Stygofauna Assessment, Appendix G to Environmental Impact Statement for the Springvale Mine Extension Project, Report no. 49913131, 2014.

Centennial Coal: Subsidence management status report LW411-418, Springvale, available at: http://majorprojects.planning.nsw.gov.au (last access: 25 April 2017), 2016.

Chang, C. C. Y., McCormick, P. V., Newman, S., and Elliott, E. M.: Isotopic indicators of environmental change in a subtropical wetland, Ecol. Indic., 9, 825–836, 2009.

Chalson, J. M. and Martin, H. A.: Holocene history of the vegetation of the Blue Mountains, NSW South Wales, P. Linn. Soc. N. S. W., 130, 77–109, 2009.

Clifton, C. and Evans, R.: Environmental Water requirements of groundwater dependent ecosystem, Environmental flows initiative technical report number 2, Commonwealth of Australia, Canberra, 2001.

Coffey: Transient groundwater modelling study, Upper Nepean (Kangaloon) borefield, Sydney Catchment Authority Southern Highlands, NSW, 2008.

Commonwealth of Australia (CoA): Temperate Highland Peat Swamps on Sandstone ecological characteristics, sensitivities to change, and monitoring and reporting techniques, Knowledge report prepared by Jacobs SKM for the Department of the Environment, Commonwealth of Australia, Canberra, 2014a.

Commonwealth of Australia (CoA): Temperate Highland Peat Swamps on Sandstone longwall mining engineering design-subsidence prediction, buffer distances and mine design options, Knowledge report prepared by Coffey Geotechnics for the Department of the Environment, Commonwealth of Australia, Canberra, 2014b.

Corbett, P., White, E., and Kirsch, B.: Hydrogeological characterisation of temperate highland peat swamps on sandstone on the Newnes Plateau, Proceedings of the 9th Triennial Conference on Mine Subsidence, Pokolbin, 11–13 May 2014.

Cowley, K. L., Fryirs, K. A., and Hose, G. C.: Identifying key sedimentary indicators of geomorphic structure and function of upland swamps in the Blue Mountains for use in condition assessment and monitoring, Catena, 147, 564–577, 2016.

Cuthbert, M. O., Baker, A., Jax, C. N., Graham, P. W., Treble, P. C., Anderson, M. S., and Acworth, R. I.: Drip water isotopes in semi-arid karst: Implication for speleothem paleoclimatology, Earth Planet. Sc. Lett., 395, 194–204, 2014.

Dansgaard, W.: Stable isotopes in precipitation, Tellus, 16, 436–468, 1964.

David, K., Timms, W., and Baker, A.: Direct stable isotope porewater equilibration and identification of groundwater processes in heterogeneous sedimentary rock, Sci. Total Environ., 538, 1010–1023, 2015.

David, K., Timms, W., Baker, A., and McGeeney, D.: Hydrogeochemical characterisation of swamps near underground mining development, 11th International Mine Water Association conference proceedings, Pretoria, South Africa, 2018.

dePaolo, D. J., Conrad, M. E., Maher, K., and Gee, G. W.: Evaporation effects on oxygen and hydrogen isotopes in deep vadose zone pore fluids at Hanford, Washington, Vadose Zone J., 3, 220–232, 2004.

Eamus, D. and Froend, R. H.: Groundwater dependent ecosystems: the where, what and why of GDEs, Aust. J. Bot., 54, 91–96, 2006.

Fryirs, K., Freidman, B., Williams, R., and Jacobsen, G.: Peatlands in eastern Australia? Sedimentology and age structure of temperate highland peat swamps on sandstone (THPSS) in the Southern Highlands and Blue Mountains of NSW, Australia, Holocene, 24, 1527–1538, 2014.

Fryirs, K. A., Cowley, K., and Hose, G. C.: Intrinsic and extrinsic controls on the geomorphic condition of upland swamps in eastern NSW, Catena, 137, 100–112, 2016.

Garvelmann, J., Külls, C., and Weiler, M.: A porewater-based stable isotope approach for the investigation of subsurface hydrological processes, Hydrol. Earth Syst. Sci., 16, 631–640, https://doi.org/10.5194/hess-16-631-2012, 2012.

Gonfiantini, R.: Environmental isotopes in lake studies, in: Handbook of Environmental Isotope Geochemistry, edited by: Fritz, P. and Fontes, J. C., 113–168, Elsevier, New York, NY, USA, 1986.

Gorissen, S., Greenlees, M., and Shine, R.: Habitat and Fauna of an Endangered Swamp Ecosystem in Australia's Eastern Highlands, Wetlands, 37, 269–276, 2017.

Harrington, G. A., Gardner, W. P., Smerdon, B. D., and Hendry, M. J.: Palaeohydrogeological insights from natural tracer profiles in aquitard porewater, Great Artesian Basin, Australia, Water Resour. Res., 49, 4054–4070, 2013.

Heiri, O., Lotter, A. F., and Lemcke, G.: Loss on ignition as a method for estimating organic and carbonate content in sediments: reproducibility and comparability of results, J. Paleolimnol., 25, 101–110, 2001.

Hendry, M. J. and Wassenaar, L. I.: Inferring heterogeneity in aquitards using high resolution $\delta^{18}O$ and $\delta^{2}H$ profiles, Ground Water, 47, 639–645, 2009.

Hendry, M. J., Barbour, S. L., Novakowski, K., and Wassenaar, L. I.: Paleohydrogeology of the Cretaceous sediments of the Williston Basin using stable isotopes of water, Water Resour. Res., 49, 4580–4592, 2013.

Hendry, M. J., Schmeling, E., Wassenaar, L. I., Barbour, S. L., and Pratt, D.: Determining the stable isotope composition of pore water from saturated and unsaturated zone core: improvements to the direct vapour equilibration laser spectrometry method, Hydrol. Earth Syst. Sci., 19, 4427–4440, https://doi.org/10.5194/hess-19-4427-2015, 2015.

Hose, G. C., Bailey, J., Stumpp, C., and Fryirs, K.: Groundwater depth and topography correlate with vegetation structure of an upland peat swamp, Budderoo Plateau, NSW, Australia, Ecohydrology, 7, 1392–1402, 2014.

Hughes, C. E. and Crawford, J.: Spatial and temporal variation in precipitation isotopes in the Sydney Basin, Australia, J. Hydrol., 489, 42–55, 2013.

Hunt, R. J., Bullen, T. D., Krabbenhoft, D. P., and Kendall, C.: Using stable isotopes of water and strontium to investigate the hydrology of a natural and a constructed wetland, Ground Water, 36, 434–443, 1998.

Johnson, D.: Sacred waters: the story of the Blue Mountains gully traditional owners, Halstead Press, Broadway, N.S.W., 237 pp., 2007.

Kaller, M. D., Keim, R. F., Edwards, B. L., Raynie, H. A., Pasco, T. E., Kelso, W. E., and Allen, R. D.: Aquatic vegetation mediates the relationship between hydrologic connectivity and water quality in a managed floodplain, Hydrobiologia, 760, 29–41, 2015.

Keith, D. A. and Benson, D. H.: The natural vegetation of the Katoomba 1 : 100000 map sheet, Cunninghamia, 2, 107–143, 1988.

Keith, D. A. and Myerscough, P. J.: Floristics and soil relations of upland swamp vegetation near Sydney, Aust. J. Ecol., 18, 325–344, 1993.

Kelln, C. J., Wassenaar, L. I., and Hendry, M. J.: Stable isotopes ($\delta^{18}O$, $\delta^{2}H$) of pore waters in clay-rich aquitards: A comparison and evaluation of measurement techniques, Ground Water Monit. R., 21, 108–116, 2001.

Kohlhagen, T., Fryirs, K., and Semple, A. L.: Highlighting the need and potential for use of interdisciplinary science in adaptive environmental management: The case of Endangered upland swamps in the Blue Mountains, NSW, Australia, Geogr. Res., 51, 439–453, 2013.

Lake, P. S.: Ecological effects of perturbation by drought in flowing waters, Freshwater Biol., 48, 1161–1172, 2003.

Lawrence, D. M., Thornton, P. E., Oleson, K. W., and Bonan, G. B.: Partitioning of evaporation into transpiration, soil evaporation, and canopy evaporation in aGCM: impacts on land-atmosphere interaction, J. Hydrometeorol., 8, 862–880, 2007.

Linacre, E. T.: Climate and evaporation from crops, J. Irr. Drain. Div.-ASCE, 93, 61–79, 1967.

Maidment, D.: Handbook of Hydrology, David R. Maidment, editor in chief, McGraw-Hill Inc., 1993.

Mathieu, R. and Bariac, T.: An isotopic study (^{2}H and ^{18}O) of water movements in clayey soils under a semiarid climate, Water Resour. Res., 32, 779–789, 1996.

McHugh, E.: The geology of the shrub swamps within Angus Place, Springvale and the Springvale mine extension project areas, Appendix 18, available at: http://majorprojects.planning.nsw.gov.au/index.pl?action=_view_job&job_id=_5594 (last access: 4 January 2018), 2014.

Melayah, A., Bruckler, L., and Bariac, T.: Modeling the transport of water stable isotopes in unsaturated soils under natural conditions. 1. Theory, Water Resour. Res., 32, 2047–2065, 1996.

Mezhibor, A. M. and Bonn, B.: Detection of volatile organic compounds in upland peat by means of proton-transfer-reaction mass spectrometry, Procedia Chem., 10, 203–208, 2014.

Middleton, B. A. and Kleinebecker, T.: The Effects of Climate-Change-Induced Drought and Freshwater Wetlands in: Global

Change and the Function and Distribution of Wetlands, edited by: Middleton, B. A., Springer, the Netherlands, 2012.

Millar, C., Pratt, D., Schneider, D., and McDonnell, J. J.: A comparison of extraciotn systems for pant water stable isotope analysis, Rapid Commun. Mass Sp., 32, 1031–1044, 2018.

Mitsch, W. J. and Gosselink, J. G.: Wetlands, 3rd edn., Wiley, New York, 2000.

Mohamed, Y. A., Bastiaanssen, W. G. M., Savenije, H. H. G., van den Hurk, B. J. J. M., and Finlayson, C. M.: Wetland versus open water evaporation: An analysis and literature review, Phys. Chem. Earth, 47–48, 114–121, 2012.

NSW DP: Impacts of underground coal mining on natural features in the Southern Coalfield: strategic review, report prepared by Hebblewhite, I., Galvin, J., Mackie, C., West, R., and Collins, D., NSW Government, Department of Planning, Sydney, 2008.

NSW Government: Colo River, Wollemi and Blue Mountains National Parks, Wild River Assessment 2008, Department of Environment, Climate Change and Water NSW, Sydney South, NSW, 2008.

NSW PAC: The Metropolitan Coal Project review report, NSW Planning Assessment Commission, Sydney, 2009.

Office of Environment and Heritage (OEH): Newnes Plateau Shrub Swamp in the Sydney Basin Bioregion – endangered ecological community listing, available at: http://www.environment.nsw.gov.au/determinations/NewnesPlateauShrubSwampEndSpListing.htm, last access: 10 April 2017.

Orlowski, N., Breuer, L., and McDonnell, J. J.: Critical issues with cryogenic extraction of soil water for stable isotope analysis, Ecohydrology, 9, 1–5, 2016.

Orlowski, N., Breuer, L., Angeli, N., Boeckx, P., Brumbt, C., Cook, C. S., Dubbert, M., Dyckmans, J., Gallagher, B., Gralher, B., Herbstritt, B., Hervé-Fernández, P., Hissler, C., Koeniger, P., Legout, A., Macdonald, C. J., Oyarzún, C., Redelstein, R., Seidler, C., Siegwolf, R., Stumpp, C., Thomsen, S., Weiler, M., Werner, C., and McDonnell, J. J.: Inter-laboratory comparison of cryogenic water extraction systems for stable isotope analysis of soil water, Hydrol. Earth Syst. Sci., 22, 3619–3637, https://doi.org/10.5194/hess-22-3619-2018, 2018.

Paterson, P. W.: Upland swamp monitoring and assessment programme, MSc thesis, National Center for Groundwater Management, University of Technology, Sydney, 2004.

Penna, D., Stenni, B., Šanda, M., Wrede, S., Bogaard, T. A., Gobbi, A., Borga, M., Fischer, B. M. C., Bonazza, M., and Chárová, Z.: On the reproducibility and repeatability of laser absorption spectroscopy measurements for δ^2H and δ^{18}O isotopic analysis, Hydrol. Earth Syst. Sci., 14, 1551–1566, https://doi.org/10.5194/hess-14-1551-2010, 2010.

Piayda, A., Dubbert, M., Siegwolf, R., Cuntz, M., and Werner, C.: Quantification of dynamic soil-vegetation feedbacks following an isotopically labelled precipitation pulse, Biogeosciences, 14, 2293–2306, https://doi.org/10.5194/bg-14-2293-2017, 2017.

Prosser, P. I., Chappell, J., and Gillespie, R.: Holocene valley aggradation and gully erosion in headwater catchments, South-eastern highlands of Australia, Earth Surf. Proc. Land., 19, 465–480, 1994.

Revesz, K. and Woods, P. H.: A method to extract soil water for stable isotope analysis, J. Hydrol., 115, 397–406, 1990.

R Core Team: R: A language and environment for statistical computing, R Foundation for Statistical Computing, Vienna, Austria, available at: http://www.R-project.org/ (last access: 15 April 2018), 2013.

Reynolds, S. G.: The gravimetric method of soil moisture determination, part I. A study of equipment, and methodological problems, J. Hydrol., 11, 258–273, 1970.

Schultz, N. M., Griffis, T. J., Lee, X., and Baker, J. M.: Identification and correction of spectral contamination in 2H / 1H and 18O / 16O measured in leaf, stem, and soil water, Rapid Commun. Mass Sp., 25, 3360–3368, 2011.

Shackelford, C. D. and Daniel, D. E.: Diffusion in Saturated Soil. I: Background, J. Geotech. Eng.-ASCE, 117, 467–484, 1991.

Shanafield, M., Cool, P. G., Gutierrez-Jurado, H. A., Faux, R., Cleverly, J., and Eamus, D.: Field comparison of methods for estimating groundwater discharge by evaporation and evapotranspiration in an arid-zone playa, J. Hydrol., 527, 1073–1083, 2015.

Smith, J. B., Schellnhuber, H. J., and Mirza, M. M. Q.: Vulnerability to climate change and reasons for concern: a synthesis, in: Climate change 2001: impacts: adaptation and vulnerability, edited by: McCarthy, J. J., White, K. S., Canziani, O., Learly, N., and Dokken, D. J., Cambridge University Press, Cambridge, 913–970, 2001.

Soderberg, K., Good, S. P., Wang, L., and Caylor, K.: Stable isotopes of water vapour I the vadose zone: A review of measurement and modeling techniques, Vadose Zone J., 11, https://doi.org/10.2136/vzj2011.0165, 2012.

Valentin, C., Poesen, J., and Li, Y.: Gully erosion: Impacts, factors and control, Catena, 63, 132–153, 2005.

Walczak, R., Rovdan, E., and Witkowska-Walczak, B.: Water retention characteristics of peat and sand mixtures, Int. Agrophys., 16, 161–165, 2002.

Wassenaar, L. I., Hendry, M. J., Chostner, V. L., and Lis, G. P.: High resolution pore water _2H and _18O measurements by H2O (liquid) – H2O (vapor) equilibration laser spectroscopy, Environ. Sci. Technol., 42, 9262–9267, 2008.

XLSTAT 2017: Data Analysis and Statistical Solution for Microsoft Excel, Addinsoft, Paris, France, 2017.

Yoo, E. K., Tadros, N. Z., and Bayly, K. W.: A compilation of the geology of the Western Coalfield. Geological Survey of New South Wales, Report GS2001/204 (unpublished), 2001.

Young, A. R. M.: Upland swamps in the Sydney region, Thirroul, NSW Dr Ann Young, 2017.

8

High-frequency NO_3^- isotope ($\delta^{15}N$, $\delta^{18}O$) patterns in groundwater recharge reveal that short-term changes in land use and precipitation influence nitrate contamination trends

Martin Suchy[1], **Leonard I. Wassenaar**[2], **Gwyn Graham**[1], **and Bernie Zebarth**[3]

[1]Environment and Climate Change Canada, 201-401 Burrard St., Vancouver, BC, Canada
[2]International Atomic Energy Agency, Vienna International Centre, 1400, Vienna, Austria
[3]Fredericton Research and Development Centre, Agriculture and Agri-Food Canada, 850 Lincoln Rd., P.O. Box 20280 Fredericton, NB, E3B 4Z7, Canada

Correspondence: Leonard I. Wassenaar (l.wassenaar@iaea.org)

Abstract. Poultry manure is the primary cause of nitrate (NO_3^-) exceedances in the transboundary Abbotsford–Sumas aquifer (ASA; Canada–USA) based on synoptic surveys two decades apart, but questions remained about seasonal and spatial aspects of agricultural nitrate fluxes to the aquifer to help better focus remediation efforts. We conducted over 700 monthly $\delta^{15}N$ and $\delta^{18}O$ of nitrate assays, focusing on shallow groundwater (<5 years old) over a 5-year period to gain new insight on spatio-temporal sources and controls of groundwater nitrate contamination. NO_3^- concentrations in these wells ranged from 1.3 to 99 mg N L^{-1} ($n = 1041$) with a mean of 16.2±0.4 mg N L^{-1}. The high-frequency ^{15}N and ^{18}O isotope data allowed us to identify three distinctive NO_3^- source patterns: (i) primarily from synthetic fertilizer, (ii) dynamic changes in nitrate due to changes in land use, and (iii) from a mix of poultry manure and fertilizer. A key finding was that the source(s) of nitrate in recharge could be quickly influenced by short-term near-field management practices and stochastic precipitation events, which ultimately impact long-term nitrate contamination trends. Overall, the isotope data affirmed a subtle decadal-scale shift in agricultural practices from manure increasingly towards fertilizer nitrate sources; nevertheless, poultry-derived N remains a predominant source of nitrate contamination. Because the aquifer does not generally support denitrification, remediation of the Abbotsford–Sumas aquifer is possible only if agricultural N sources are seriously curtailed, a difficult proposition due to longstanding high-value intensive poultry and raspberry and blueberry operations over the aquifer.

1 Introduction

The global widespread use and over-application of synthetic and manure N nutrients in agriculture have caused widespread groundwater nitrate (NO_3^-) contamination in surficial aquifers around the world (Haslauer et al., 2005; Hamilton and Helsel, 1995; Spalding and Exner, 1993). Furthermore, with global trends towards increased agricultural intensification, threats to groundwater quality are correspondingly heightened (Vorosmarty et al., 2000; Böhlke, 2002). In agricultural settings, elevated groundwater NO_3^- concentrations typically reflect a supply of N from manure, synthetic fertilizer, or soil organic matter which exceeds, or is poorly synchronized with, crop N requirements (Canter, 1997). The risk of NO_3^- contamination is especially high in phreatic aquifers comprised of coarse-grained permeable soils with minimal propensity for natural attenuation and remediation processes such as microbial denitrification. Studies have used stable isotopes of nitrate ($\delta^{15}N$, $\delta^{18}O$) to differentiate different source inputs (Mitchell et al., 2003; Wassenaar et al., 2006; Xue et al., 2009; Pastén-Zapata et al., 2014), while others have used $\delta^{15}N$ and $\delta^{18}O$ to examine the anaerobic remediation of nitrate by microbial denitrification (Böhlke and Denver, 1995; Kellman and Hillaire-Marcel, 2003).

Others used isotopes of nitrate to assess soil N transformations (Savard et al., 2010) or temporal variations in agricultural leachate to groundwater (Ostrom et al., 1998; Loo et al., 2017; Savard et al., 2007).

Concentrations of non-agricultural NO_3^- in aquifers that are low (<1 mg N L^{-1}) and below drinking water standards are usually attributed to atmospheric N deposition, organic N from plant decomposition or land breakage, and geological sources that are mobilized due to disruptions in water recharge fluxes such as introduction of irrigation (Canter, 1997). Choi et al. (2003) suggests low groundwater NO_3^- concentrations consistently below 3 mg N L^{-1} with $\delta^{15}N$ values between +5 ‰ and +8 ‰ are likely derived from natural soil or organic N sources (average $\delta^{15}N$ +5 ‰). Loo et al. (2017) reported non-agricultural soil $\delta^{15}N$ nitrate ranges between +3.7 ‰ and +4.9 ‰ in our study area (Table 1).

Sources of nitrate from animal waste arise from dispersed agricultural field applications and/or point-source manure storage facilities (liquid and solid). Under aerobic soil conditions, NO_3^- quickly forms from oxidation of NH_4^+ after manure application (Aravena et al, 1993). Due to preferential volatilization of ^{14}N in gaseous NH_3 from NH_4^+ during wet storage and/or application of manure, manure-derived NO_3^- is accordingly enriched in ^{15}N (Kendall, 1998). Nitrate derived from manure or septic waste can have $\delta^{15}N$ values between +10 ‰ and +25 ‰, (Wassenaar, 1995; Kreitler, 1975; Heaton, 1986; Aravena and Robertson, 1998), generally revealing little distinctive ^{15}N isotopic resolution between these two organic N waste sources. Poultry manure solids have average $\delta^{15}N$ values of approximately +7.9 ‰ in the study area (Loo et al., 2017; Wassenaar, 1995). In North America, urea ($CO(NH_2)_2$) (nitrogen, phosphorus, potassium (N-P-K): 46-0-0) is one of the most common forms of synthetic fertilizers used (Overdahl et al., 2007). Other forms of synthetic fertilizers include ammonium nitrate (NH_4-NO_3) (34-0-0) and ammonium sulfate (NH_4-SO_4) (22-0-0), as shown in Table 1. Each of these is manufactured by fixation of atmospheric N ($\delta^{15}N = \sim 0$ ‰), resulting in $\delta^{15}N$ values from −2.8 ‰ to +0.3 ‰ (Loo et al., 2017). In the Abbotsford–Sumas aquifer (ASA) area, berry-specific fertilizer blends are commonplace (Table 1), where N is derived from one of the above sources (Loo et al., 2017; Wassenaar, 1995).

The $\delta^{18}O$ values of synthetic fertilizer-derived NO_3^- typically range between +18 ‰ and +22 ‰, because the oxygen in nitrate originates from air ($\delta^{18}O = +23.5$ ‰) and ^{18}O-depleted H_2O (Amberger and Schmidt, 1987). Nitrate derived from NH_4-NO_3 fertilizers, where 50 % of the oxygen is from nitrification of NH_4 fertilizer and 50 % is from synthetic NO_3^- fertilizer, has reported $\delta^{18}O$ values around +13 ‰ (Aravena et al., 1993). In this study, the $\delta^{18}O$ of groundwater is relatively uniform (-11.1 ± 0.4 ‰; Wassenaar, 1995), which would result in $\delta^{18}O$ values around zero ‰ for NO_3^- derived only from the nitrification of soil or organic N or ammonium sources.

In the phreatic transboundary Abbotsford–Sumas aquifer (Canada–USA, Fig. 1), long-term nitrate contamination trends and isotopic studies have been conducted over several decades. The isotopic apportionment of NO_3^- sources in the aquifer was based on two summertime synoptic isotopic studies spanning a full decade, which revealed that poultry manure was the predominant source of NO_3^-, with localized long-term shifts towards inorganic fertilizer sources (Wassenaar, 1995; Wassenaar et al., 2006) due to changes in agricultural practices (Zebarth et al., 2015). These agricultural changes include decadal shifts from beef and dairy feed production to lower N uptake berry production, changes in irrigation methods from sprinklers to low-pressure drip fertigation, planting of cover crop for berry alley management, and a slow transition from raspberries to blueberries. One critique of the previous synoptic isotope studies was that sampling (and hence interpretations) was prejudiced to summer snapshots and thereby could be biased, especially for the numerous shallow and highly responsive water table wells spanning the aquifer and winter-biased recharge (Environment Canada, 2014). The seasonal dynamics of NO_3^- sources and fluxes and the potential for isotopic changes due to soil and unsaturated zone NO_3^- cycling were not fully evaluated and need to be considered to improve surface nutrient applications and agricultural management practices.

To address this knowledge gap, we conducted high-frequency (monthly) NO_3^- concentration and isotope analysis of the ASA over a 5-year period, with a focus on shallow wells having groundwater residence times of <5 years as determined by ^3H-He age dating. Our aim was to determine whether high-frequency (monthly) nitrate and isotope ($\delta^{15}N$, $\delta^{18}O$) sampling improved upon previous interpretations of N sources and processes and whether important seasonal changes in the proportion of NO_3^- sources recharging to groundwater were overlooked by synoptic snapshots. Our goal was to gain improved insight on the spatio-temporal sources and controls of groundwater–nitrate dynamics and thereby to help better inform agricultural nutrient management practices and potential NO_3^- remediation efforts in the aquifer.

2 Materials and methods

2.1 Study area and hydrogeologic setting

The Abbotsford–Sumas aquifer is a shallow phreatic transboundary aquifer located in southwestern British Columbia, Canada, and northwestern Washington state, USA (Fig. 1). The ASA is the most intensively studied nitrate-contaminated aquifer in Canada (Zebarth et al., 1998, 2015) and covers an area of about 200 km^2, with approximately 40 % of the total surface area in Canada (Cox and Kahle, 1999). Our study area encompassed approximately 40 km^2 on the Canadian side of the aquifer, between the Abbotsford

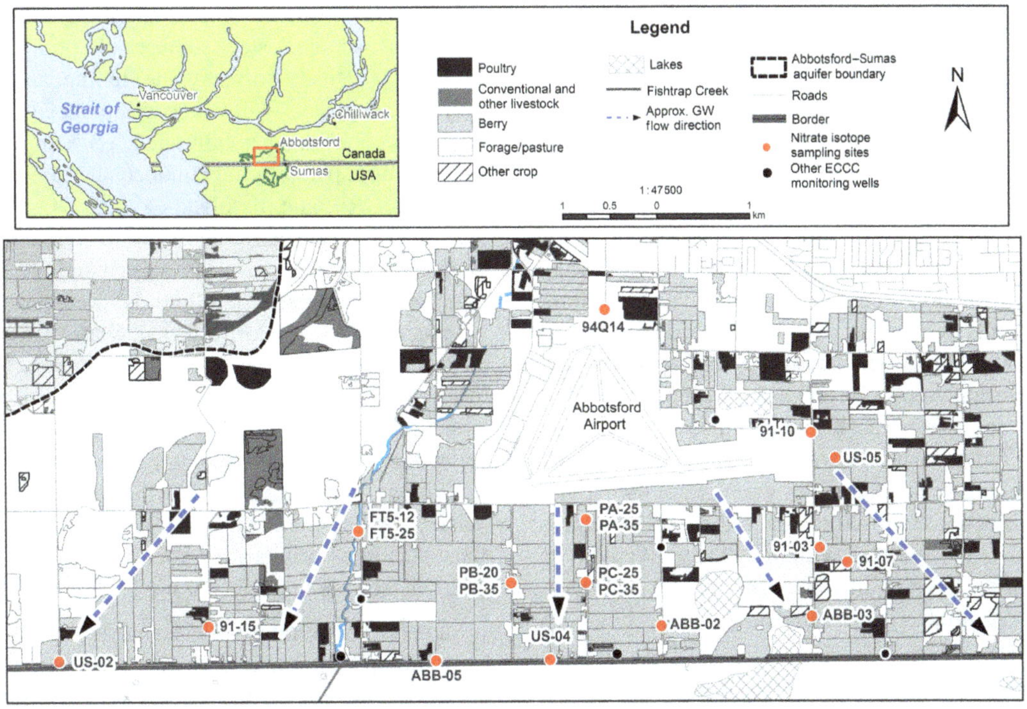

Figure 1. Location of the Abbotsford–Sumas aquifer (ASA), southwestern BC, Canada, and northwestern Washington state, USA, along with simplified agricultural land use and sampling locations with groundwater mean residence times (MRT) of <5 years. Arrows show the approximate groundwater flow direction.

Table 1. Local synthetic fertilizer, poultry manure, soil N, and leachate δ^{15}N values used in the Abbotsford area.

Source	δ^{15}N (AIR, ‰)	Reference
Poultry manure (total N)	+7.9	Loo et al. (2017)
Poultry manure (total N)	+8.1	Wassenaar (1995)
Poultry manure (total N)	+7.9	Wassenaar (1995)
Urea (total N)	−0.7	Loo et al. (2017)
NH_4-NO_3 (total N)	−2.8	Loo et al. (2017)
NH_4-SO_4 (total N)	+0.3	Loo et al. (2017)
Urea (total N)	−0.6	Wassenaar (1995)
NH_4-SO_4 (total N)	−0.9	Wassenaar (1995)
Soil N (total N)	+3.8 to +4.6	Loo et al. (2017)
Soil N (total N)	+3.7 to +4.1	Wassenaar (1995)
Irrigation water – average (NO_3-N)	+9.0	Loo et al. (2017)
Weighted fertilizer treatment leachate (NO_3-N)	$+3.2 \pm 2.3$	Loo et al. (2017)
Weighted manure leachate (NO_3-N)	$+7.3 \pm 1.2$	Loo et al. (2017)

International Airport and the Canada–USA border (Fig. 1). Land use on the aquifer is predominantly commercial raspberry and blueberry production, mixed with intensive commercial poultry barn operations (Fig. 1) and is <5 % rural residential; (BC Ministry of Agriculture, unpublished data).

The aquifer is typically 10–25 m thick but reaches 70 m thickness towards the southeast portion (Cox and Kahle, 1999). The aquifer comprises coarse glacio-fluvial sand and gravel with minor till and clayey silt lenses (Armstrong et al., 1965), with glacio-marine clays confining the aquifer below (Halstead, 1965). The high sand and gravel content results in a high transmittance of water, with mean hydraulic conductivities (K) of 1.6×10^{-3} m s^{-1} (Chesnaux and Allen, 2007) to 9.5×10^{-4} m s^{-1} (Cox and Kahle, 1999). The thin organic-poor soils (0–70 cm) are medium-textured aeolian deposits, moderately well to well drained, and are classified as Orthic Humo-Ferric Podzols (Luttmerding, 1980).

Average annual precipitation across the aquifer (1981–2010) is 1538 mm, of which 70 % falls between October and March (Environment Canada, 2014). Annual recharge estimates range from 850 to 1100 mm (Zebarth et al., 2015), and water table depths typically vary widely between 2 and 20 m below surface depending on the location and season. Annual water table fluctuations average ~ 3.6 m (Scibek and Allen, 2006). The overall flow direction in the aquifer is south (Fig. 1), southeast, and southwest at linear velocities of up to $450\,\mathrm{m\,yr^{-1}}$ (Liebscher et al., 1992; Cox and Kahle, 1999).

The aquifer is highly vulnerable to surface-derived NO_3^- and other contamination because of (i) intensive agricultural activity; (ii) highly permeable soil, coarse sand, and gravel lithology; and (iii) high precipitation amounts in the fall and winter when nutrient uptake by crops is lowest and the NO_3^- leaching potential is greatest (Kohut et al., 1989; Liebscher et al., 1992). Elevated groundwater–nitrate concentrations exceeding drinking water guidelines are observed since the 1970s (Zebarth et al., 2015). Mitchell et al. (2003) and others (Wassenaar et al., 1995) showed vertical stratification of nitrate in groundwater was linked to agricultural practices, with the greatest nitrate concentrations ($>20\,\mathrm{mg\,N\,L^{-1}}$) occurring in shallow water table regions (<10 m b.g.l., metres below ground level), while average groundwater–nitrate concentrations ($<10\,\mathrm{mg\,N\,L^{-1}}$) in deep wells ($>10$ m screen depth below average static water level) were lower and relatively stable over time. Based on Environment and Climate Change Canada (ECCC) monitoring, the highest seasonal and temporal variations in NO_3^- are found in wells screened near the water table. Both seasonal and long-term temporal variations in groundwater–nitrate over decadal timeframes are well documented (Liebscher et al., 1992; Graham et al., 2015). The aquifer has little widespread intrinsic capacity to sustain microbial denitrification (self-remediation) because of largely aerobic conditions and the low organic content of the aquifer materials (Wassenaar, 1995), but it can occur in localized pockets around riparian discharge zones.

2.2 Sample collection and analysis

Monthly groundwater samples ($n = 56$ per well) were collected from 19 selected monitoring wells from September 2008 to March 2013. These wells were selected based on the following criteria: (1) groundwater having a <5-year residence time based on ^{3}H-He age dating (Wassenaar et al., 2006), (2) representative spatial coverage within the monitoring network, and (3) wells representing aerobic groundwater where denitrification does not occur (Tesoriero, 2000; Wassenaar et al., 2006). These criteria helped to ensure that high-frequency nitrate and isotopic patterns stem from short-term nitrate responses unaffected by historical or subsurface biogeochemical processes or mixing with deeper water and could therefore be more explicitly linked to contemporary landscape and agricultural activities and practices happening roughly within a 5-year timeframe.

Static water level measurements were taken prior to pumping and were reported in metres above mean sea level (m a.s.l.). Groundwater was sampled from the wells using a Grundfos® stainless-steel submersible pump, Teflon® lined LDPE (low-density polyethylene) tubing, and stainless-steel fittings and valves. Well water was pumped through a flow-through cell housing a calibrated YSI® multi-probe sonde (temperature; pH; specific conductance; oxidation reduction potential, ORP; and dissolved oxygen, DO). General chemistry and NO_3^- isotope water samples were collected after at least three well volumes were purged and the YSI® field parameters were stabilized. All bottles were rinsed three times with sample water prior to filling. Water samples for major ion and nutrient concentrations were taken in 1 L LDPE bottles, filtered through 0.45 µm cellulose acetate membrane filters, stored at 5 °C, and analyzed within 5 days for nitrate using standard ion chromatography techniques. Nitrate concentrations were determined at the Pacific and Yukon Laboratory for Environmental Testing in North Vancouver, BC, Canada. Nitrate results are reported as $\mathrm{mg\,N\,L^{-1}}$ with a minimum detectable limit of $0.02\,\mathrm{mg\,L^{-1}}$.

Samples for nitrate isotope analyses (δ^{15}N, δ^{18}O) were field filtered through 0.45 µm cellulose acetate membrane filters and frozen (−40 °C) in 125 mL HDPE (high-density polyethylene) bottles. Nitrate isotope assays were conducted by the University of Calgary Stable Isotope Laboratory, using a microbial reduction to N_2O described elsewhere (Casciotti et al., 2002; Sigman et al., 2001). All δ^{15}N values are reported relative to the atmospheric air reference (Mariotti, 1983) and normalized by analyzing reference materials IAEA-N3 ($\delta^{15}N_{AIR} = +4.7$ ‰), USGS32 ($\delta^{15}N_{AIR} = +180$ ‰), USGS34 ($\delta^{15}N_{AIR} = -1.8$ ‰), and USGS35 ($\delta^{15}N_{AIR} = +2.7$ ‰) along with samples. The analytical uncertainty for δ^{15}N was ±0.5 ‰. The δ^{18}O values were reported relative to the VSMOW reference (Coplen, 1994) and determined by analyzing reference materials IAEA-N3 ($\delta^{18}O_{VSMOW} = +25.6$ ‰), USGS32 ($\delta^{18}O_{VSMOW} = +25.7$ ‰), USGS34 ($\delta^{18}O_{VSMOW} = -27.9$ ‰), and USGS35 ($\delta^{18}O_{VSMOW} = +57.5$ ‰). The analytical uncertainty for δ^{18}O was ±1.0 ‰.

Nitrate and chloride concentrations were log-transformed prior to analysis to ensure normal distributions and were evaluated using principal component analysis (PCA) and factor analysis. Statistical analyses (at the 95 % confidence level), including multivariate time series analyses, were conducted using the Kruskal–Wallis methods for determining seasonality, log-normal transformations, Mann–Kendall trend analyses, and Gaussian mixture and Bayesian clustering models using WQHydro®, ProUCL 5®, and XLSTAT® (Lettenmaier, 1988; Thas et al., 1998). Seasonal Mann–Kendall trend analyses were deemed inappropriate for evaluating nitrate concentration seasonality as the repeating periods were correlated to precipitation patterns instead of calendar month and because peak nitrate concentration timings varied from year to year, resulting in a determination of non-seasonality.

3 Results and discussion

3.1 Groundwater nitrate concentrations

Results of monthly nitrate concentrations in the water table wells in the aquifer over the 5-year sampling period ranged from 1.3 to 99.0 $mg\,N\,L^{-1}$ ($n = 1041$), having a mean concentration ($\pm$SE) of 16.2$\pm$0.1 $mg\,N\,L^{-1}$. Approximately 76 % of the shallow groundwater locations (16 of 19 sites) exceeded the maximum allowable concentration (MAC) of 10 $mg\,N\,L^{-1}$ in the Canadian Drinking Water Guidelines (Health Canada, 2013). These nitrate exceedances were consistent with previous observations of high nitrate concentrations in shallow wells in the aquifer (Hii et al., 1999). Previous studies reported NO_3^- concentrations exceeding the MAC in 58 %, 69 %, and 59 % of wells (Wassenaar, 1995; Zebarth et al., 1998; Wassenaar et al., 2006), respectively. The current study only had ~ 50 % monitoring wells in common with previous investigations because early studies also sampled deeper monitoring wells containing older groundwater.

A time-series analysis showed that overall NO_3^- concentrations steadily increased from 14.1 to 18.4 $mg\,L^{-1}$ in the targeted shallow wells over our 5-year study period, which contrasted with long-term declines observed for a wider depth variety of wells in the Canadian portion of the aquifer (Zebarth et al., 2015). Graham et al. (2015) identified several key drivers causing the short-term (intra- and inter-year) nitrate trends (increases or declines) that contrasted with the long-term (inter-decadal) declines. These key drivers were primarily stochastic rainfall patterns (wet vs. dry years, with first major rainfalls occurring in October) and short-term land use change factors. The overall increasing nitrate trend in the 19 wells could be attributed to the marked increases in NO_3^- concentrations in three of the wells occurring in the second half of our study, without which the average NO_3^- concentration in the remaining 16 wells (15.5 $mg\,L^{-1}$) did not change. These nitrate increases were attributed to (i) clearing of an adjacent woodlot, and the subsequent application of large quantities of poultry manure as a soil amendment up-gradient of PC-25 and PC-35 in 2011; and (ii) a raspberry field up-gradient of US-02 that underwent a renovation cycle (described in Zebarth et al., 2015), which likely also included soil organic amendments. Wells 94Q-14, PA-25, and PA-35 did not exceed the nitrate MAC because these sites were located up-gradient of the most intensive agricultural production areas.

Almost half the 19 shallow monitoring wells (47 %) showed NO_3^- seasonality, with maximum concentrations usually occurring in the springtime. Nitrate accumulates in the soil and root zones over the summer, and a large proportion of nitrate flushing to the water table happens with the first major recharge events in the fall and winter rainy season (Kowalenko, 2000). Subsequent recharge typically has lower nitrate concentrations as the availability of dissolved soil nitrate drops. Previous evidence of NO_3^- flushing in the fall is shown by Wassenaar (1995) and Zebarth et al. (1998), when precipitation, recharge rates, and soil-NO_3^- are at their peak. Coupled with vadose zone infiltration lag times of several months, which do not vary significantly between sites (Herod et al., 2015), accordingly peak NO_3^- concentrations reaching the water table are usually observed in the springtime.

All wells have aerobic groundwater (mean = 8.9 $mg\,O_2\,L^{-1}$; Table S1 in the Supplement); however, two sites (ABB-03 and US-02) showed short intervals of lowered DO levels (< 1 $mg\,L^{-1}$) in the winter months, coinciding with higher water tables. Chloride levels were on average 8.7 $\pm$ 3.0 $mg\,L^{-1}$. At six sites (91-10, 91-15, PA-25, PA-35, US-02, and US-05), NO_3^- and Cl concentrations exhibited a covariance (Pearson's R correlation coefficients > 0.5), suggesting similar sources. Three sites (PC-25, FT5-12, and FT5-25) exhibited an offset between NO_3^- and Cl; however, the Cl peaks usually lagged behind NO_3^- peaks by 1–3 months, which is inconsistent with the conservative properties of Cl^-. This was also observed by Malekani (2012) and may be caused by differences in timing of NO_3^- and Cl availability in combination with growing season leaching under drip irrigation or variable transit times though the unsaturated zone. The remaining sites exhibited limited seasonal nitrate and chloride variability or showed no correlation between these variables, suggesting more mixing within local groundwater and possibly longer transit times in the unsaturated zone.

3.2 Nitrate N and O isotopes

Overall, the mean ($\pm$SE) groundwater nitrate $\delta^{15}N$ value from the 19 study wells was +7.9 $\pm$ 0.1 ‰ ($n = 717$), which was consistent with positive $\delta^{15}N$ values of local poultry manure solid sources (Wassenaar, 1995; Loo et al., 2017) as summarized in Table 1. Higher $\delta^{15}N$ source values were observed in nitrate extracted from soil below former manure stockpiles and ranging up to +14 ‰, due to ammonia volatilization (Wassenaar, 1995). Mean nitrate $\delta^{18}O$ was −1.7$\pm$0.1 ‰ ($n = 717$), which was typical of values derived during the nitrification of manure or synthetic fertilizers (Xue et al., 2009). Previously measured groundwater $\delta^{18}O_{H_2O}$ in the aquifer ranged between −10 ‰ and −12 ‰ and, coupled with O derived from air (+23.5 ‰), the current nitrate $\delta^{18}O$ values were comparable with earlier $\delta^{18}O$ values (Wassenaar, 1995) of −1.0 $\pm$ 0.3 ‰ ($n = 16$) and +0.5 $\pm$ 0.8 ‰ ($n = 40$) a decade later (Wassenaar et al., 2006).

Table 2. Bayesian clustering model of NO_3^-N and $\delta^{15}N$ means by class.

Class	1	2	3	4	5
Mean (NO_3^-)	4.4	13.2	13.5	22.9	55.2
Mean ($\delta^{15}N$)	3.7	5.4	7.9	10.7	13.2

To further assess sources and seasonality of groundwater nitrate from these 19 shallow wells, the results were evaluated using nitrate concentrations and isotopic compositions. A Keeling plot of $1/NO_3^-$ vs. $\delta^{15}N$ (Fig. 2a), supported by a Gaussian mixing model, suggests three main nitrate groupings with the following proportions and interpretation: (i) a historical mixing (47 %) trend between high NO_3^- and high $\delta^{15}N$ (manure-derived) and intermediate NO_3^- and low $\delta^{15}N$ values (fertilizer-derived); (ii) fertilizer and soil N dominant (47 %) low-intermediate NO3 and low $\delta^{15}N$ (+2 ‰ to +4 ‰); (iii) intermediate NO_3^- and intermediate $\delta^{15}N$ (+8 ‰), with a mixed source of manure or soil and N fertilizer (6 %). A Bayesian VVV clustering model (which considers the volume, shape, and orientation of data points) using $\delta\ ^{15}N$ and NO_3^- suggested five possible groupings (Fig. 2b), with means shown in Table 2. These findings altogether suggest that field-scale agricultural management practices upgradient of the monitoring wells resulted in four quantifiably distinctive nitrate isotopic clusters (Table 3 – source grouping). The first two Bayesian groups were amalgamated as they suggested the same isotopic source, but with partial isotopic enrichment.

Another clustering approach, based on $\delta^{15}N$ trends and seasonality in the 19 wells over the course of the study was also evaluated. In this case, sites were separated into four clusters (Table 3 – trend grouping) as follows: (a) no trend with stable $\delta^{15}N$ values (SD < ±1.0 ‰), (b) no trend with variable $\delta^{15}N$ values (SD > ±1.0 ‰), (c) ^{15}N enrichment trends, and (d) ^{15}N depletion trends.

3.2.1 Nitrate isotopic variations

Considering the Bayesian and Gaussian clustering approaches altogether, we separated the nitrate and isotope data into four distinctive groups (Fig. 3) based on their isotopic values (three primary groups and one subgroup), both in relation to each other and to well-known NO_3^- sources.

Group 1a groundwater was impacted by synthetic fertilizer and/or organic N and showed little isotopic variability, while group 1b was similar but impacted by clear short-term spikes in $\delta^{15}N$ and NO_3^-. Group 2 was dominated by poultry manure with some influence of ^{15}N-depleted sources, while group 3 was dominated solely by poultry manure N.

The four wells categorized into group 1a, with $\delta^{15}N$ values of +3 ‰ to +8 ‰ representing 21 % of the 19 sites (PA-25, PA-35, 91-07, and US-04), had a mean $\delta^{15}N$ value of +5.0 ‰. The N isotope distribution of these samples suggests they were dominated by synthetic fertilizers and natural (background) soil N sources ($\delta^{15}N$ of −1.0 ‰ and +4.0 ‰, respectively). Loo et al. (2017) reported that weighted $\delta^{15}N$ of fertilizer treatment leachate in the ASA is +3.2 ± 2.3 ‰. Sampling wells in this group did not exhibit large seasonal swings in NO_3^- concentration or $\delta^{15}N$ values, although strong seasonality was found for NO_3^- in wells PA-25 and PA-35. These isotope data suggest a combination of annual synthetic fertilizer applications with occasional poultry manure application as a soil amendment, which is a common agricultural practice in this area, particularly with blueberry crops.

The group 1b wells were distinctive because the mean nitrate $\delta\ ^{15}N$ value was more depleted than poultry manure (+6.7 ‰), but spanned a wider $\delta^{15}N$ range from +2 ‰ to +16 ‰ and represented only two wells (PC-25 and PC-35). In addition, both exhibited nitrate ^{18}O enrichment, coupled with increasing $\delta^{15}N$ values (Fig. 3a) and NO_3^- concentrations. Well PC-25 was likely subjected to localized and temporal root zone denitrification since some $\delta\ ^{18}O$ values increased above +5 ‰; however, groundwater DO values were never below 8.8 mg L^{-1}, suggesting groundwater microbial denitrification processes were unlikely in the aquifer. The positive $\delta^{15}N$ values coupled with elevated NO_3^- (Fig. 3b) concentrations were more likely the result of soil amendment practices whereby poultry manure is applied to fields during crop replacement cycles to augment soil carbon and nitrogen content (Zebarth et al., 2015). As previously indicated, this site may also have been affected by recent adjacent woodlot clearing and poultry manure application following planting of a new blueberry crop in 2011–2012. If the elevated $\delta\ ^{15}N$ values after January 2012 are omitted from these two wells, the mean $\delta^{15}N$ value drops to +4.2 ‰, which corresponds to group 1a. Furthermore, most of the group 1a and 1b wells fall along the same groundwater flow path (Fig. 1).

Wells categorized as group 2 had a mean $\delta^{15}N$ of +7.8 ‰, which corresponded to both leachate from manure-treated fields (+7.3 ± 1.2 ‰; Loo et al., 2017) and poultry manure in general (∼ +7.8 ‰). More ^{15}N-depleted samples were likely influenced by synthetic fertilizers or residual soil N, while ^{15}N-enriched samples represented temporal soil zone denitrification. Group 2 wells include 91-03, 91-15, 94Q-14, ABB-02, ABB-03, ABB-05, FT5-12, FT5-25, PB-20, and PB-35. Wells in this group were in the majority, representing 53 % of the sites, and as with group 1 did not exhibit large seasonal or inter-annual swings in NO_3^- concentrations or their $\delta^{15}N$ values, other than NO_3^- concentrations and $\delta^{15}N$ values that were elevated compared to group 1. Based on these results, it appeared that poultry manure applications or excess residual soil N from historical poultry manure applications influenced the nitrate contamination level in these wells.

The group 3 wells (91-10, US-02, and US-05) had a mean $\delta^{15}N$ value of +12.6 ‰ and a $\delta^{15}N$ range of +9 ‰ to +16 ‰,

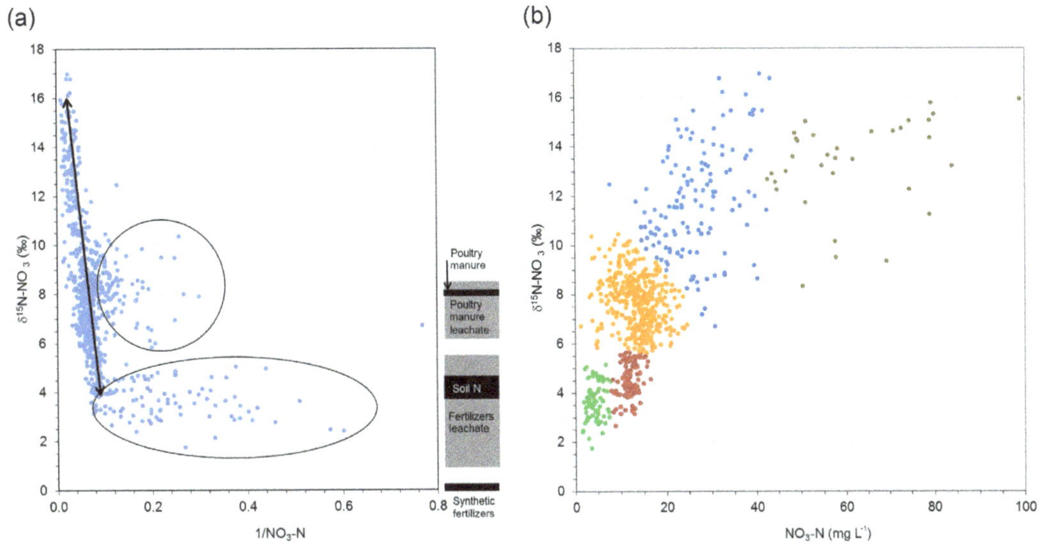

Figure 2. **(a)** Keeling plot of $1/NO_3$ (x axis) vs. $\delta^{15}N$ (y axis). Three distinct groups are described as follows: (i) arrow represents the mixing line between fertilizer and enriched manure endmembers, (ii) (bottom) wide range NO_3 (mineral fertilizer leachate and soil N), and (iii) (middle group) manure–fertilizer mixture. $\delta^{15}N$ reference sources (Wassenaar, 1995; Loo et al., 2017). **(b)** $\delta^{15}N$ vs. nitrate Bayesian clustering model suggests five distinct groupings.

Table 3. Nitrate isotopic distribution and trend grouping classification.

	Source grouping		$\delta^{15}N$ trend grouping
1a	$\delta^{15}N$ range (+3 ‰ to +8 ‰), $\delta^{18}O$ range (−5 ‰ to +2 ‰)	A	No trend – stable (SD < ±1.0 ‰)
1b	$\delta^{15}N$ range (+2 ‰ to +16 ‰), $\delta^{18}O$ (−7 ‰ to +7 ‰)	B	No trend – variable (SD > ±1.0 ‰)
2	$\delta^{15}N$ range (+6 ‰ to +10 ‰), $\delta^{18}O$ range (−5 ‰ to +2 ‰)	C	Enrichment
3	$\delta^{15}N$ range (+9 ‰ to +16 ‰), $\delta^{18}O$ range (−5 ‰ to +2 ‰)	D	Depletion

which was more ^{15}N enriched than local poultry manure or manure leachates (Table 1). These ^{15}N-enriched results likely resulted from ammonia volatilization of the source poultry manure and temporal soil zone denitrification. Ammonia volatilization occurs in poultry manure piles and during field application. The mineralized residual ammonium can have $\delta^{15}N$ values up to +25 ‰ but is dependent on pH, temperature, humidity, and other environmental factors (Kendall, 1998). Group 3 sites are all located down-gradient of current and former poultry barns or known locations of on-field poultry storage piles, which was shown by Wassenaar (1995) to result in ^{15}N-enriched values in soil N from +7.5 ‰ to +13.6 ‰ that are transported to the aquifer.

3.2.2 5-year isotopic trends

The 19 monitoring wells were evaluated based on their nitrate $\delta^{15}N$ and $\delta^{18}O$ isotopic trends over the 5-year study period (Table 4). The trend evaluation was conducted using Mann–Kendall (monthly data) and Seasonal Kendall (bimonthly data) non-parametric tests for detection of upward or downward trends in a time series at the $p > 0.05$ level of significance. For individual wells, if there was insufficient evidence to detect a trend, individual well results were grouped as being stable or variable, depending on whether the $\delta^{15}N$ standard deviation was < or > 1.0 ‰, respectively. Wells exhibiting seasonality were identified as group B. The analysis showed no statistically significant temporal trend in $\delta^{15}N$ during the study period; however, if results from the three nitrate "spiking" sites (US-02, PC-25, and PC-35) were removed, a statistically significant $\delta^{15}N$ depletion trend was observed over the 5-year period, with $\delta^{15}N$ declining from +9.1 ‰ to +7.4 ‰. This finding corresponded to the previously reported finding of a decadal-scale nitrate ^{15}N depletion trend in the aquifer, which was attributed to a long-term shift from manure to fertilizer use (Wassenaar et al., 2006).

Four wells (91-15, ABB-02, ABB-05, and FT5-12) were classified into trend group A, where analyses did not support a significant upward or downward $\delta^{15}N$ trend and ±SD ≤ 1.0 ‰ (Fig. 5a). All four wells (21 %) were from distribution group 2, where $\delta^{15}N$ values were +6 ‰ to +10 ‰. Interestingly, all group A sites exhibited appreciable NO_3^- variability, but only FT5-12 depicted any seasonality, with peak nitrate concentrations occurring in winter, likely the re-

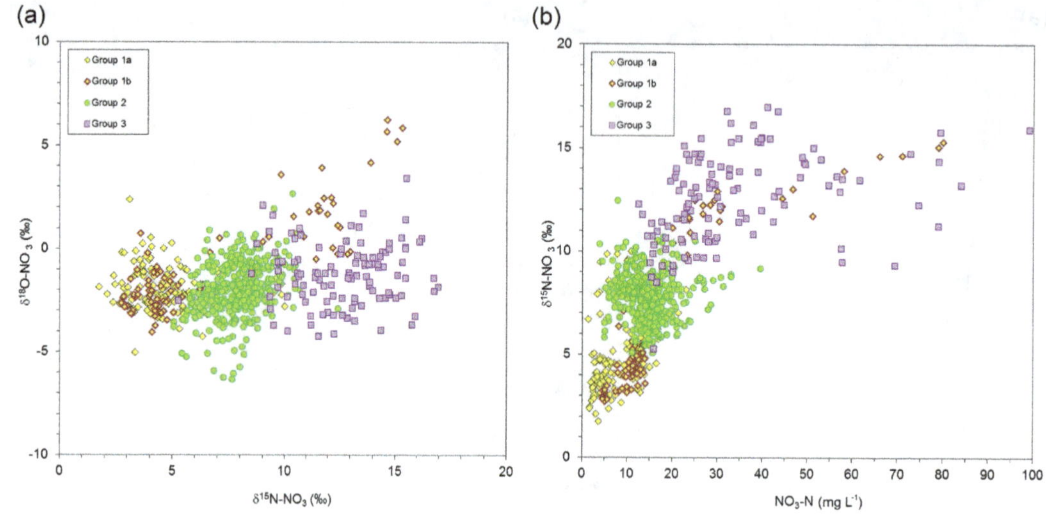

Figure 3. **(a)** Nitrate δ^{18}O vs. δ^{15}N cross plot. **(b)** Distribution of 19 well sites grouped by δ^{15}N range and δ^{18}O. Group 1a: δ^{15}N range (3 ‰ to 8 ‰). Group 1b: δ^{15}N range (+2 ‰ to +16‰). Group 2: δ^{15}N range (+6 ‰ to +10 ‰). Group 3: δ^{15}N range (+9 ‰ to +16 ‰).

Table 4. Results summary with ^{3}H / ^{3}He groundwater ages in years (Wassenaar et al., 2006); average water column height (metres; mid-screen depth below average static water level); isotopic distribution and trend groupings; NO_3-N, δ^{15}N, and δ^{18}O values (mean, standard deviation, and confidence intervals ($\alpha = 0.05$)). NA means not available.

Site ID	^{3}H / ^{3}He age (yr)	Average water column height (m)	Source group	Trend group	NO_3^- mg N L^{-1}			$\delta^{15}N_{AIR}$ ‰			$\delta^{18}O_{VSMOW}$ ‰		
					Mean	SD	CI	Mean	SD	CI	Mean	SD	CI
91-03	3.5	2	2	C	17.2	3.8	1.0	7.0	0.6	0.2	−2.0	1.0	0.3
91-07	2.7	1.8	1a	C	13.2	3.3	0.9	5.4	1.5	0.5	−2.6	0.8	0.3
91-10	3.2	3	3	B	33.2	11.6	3.0	13.1	1.6	0.5	−0.9	1.4	0.4
91-15	5.94	7.2	2	A	12.1	3.7	1.0	8.9	1.0	0.3	−1.4	0.8	0.3
94Q-14	4.2	6.3	2	D	7.7	1.9	0.5	8.0	0.8	0.3	−0.9	0.8	0.3
ABB-02	5.5	5	2	A	14.0	3.4	0.9	8.0	0.5	0.2	−3.5	1.4	0.4
ABB-03	0.9	5.2	2	D	12.4	3.9	1.0	7.5	1.2	0.4	−1.3	1.6	0.5
ABB-05	4.3	6.7	2	A	16.2	2.3	0.6	6.4	0.6	0.2	−2.6	0.9	0.3
FT5-12	NA	2	2	A	16.1	6.4	1.7	8.4	0.9	0.3	−1.9	1.0	0.3
FT5-25	NA	5.6	2	C	13.1	2.6	0.7	8.8	0.9	0.3	−1.5	1.0	0.3
PA-25	4.2	2.9	1a	B	5.8	3.4	0.9	4.1	1.8	0.6	−1.7	1.1	0.4
PA-35	4.7	6.7	1a	B	4.6	2.2	0.6	5.1	2.2	0.7	−1.4	0.8	0.3
PB-20	1.3	2.4	2	B	18.9	5.1	1.3	8.0	1.1	0.4	−2.1	1.2	0.4
PB-35	4.8	6.7	2	D	17.0	3.7	1.0	7.4	0.7	0.2	−2.1	0.9	0.3
PC-25	1.5	2.2	1b	C	17.5	19.6	5.1	6.8	4.5	1.4	−0.3	3.0	0.9
PC-35	4.4	6.3	1b	C	14.9	7.0	1.8	6.6	3.3	1.0	−1.4	1.6	0.5
US-02	1	4.6	3	B	38.6	22.8	6.0	11.4	2.3	0.7	−1.7	1.7	0.5
US-04	5	6.9	1a	C	13.2	2.0	0.5	5.4	0.5	0.2	−2.2	0.7	0.2
US-05	<1	1	3	B	27.3	10.1	3.2	13.4	2.2	0.8	−1.3	1.2	0.4

sult of organic soil N mobilization following higher precipitation periods. Average NO_3^- concentrations were 16.1 ± 6.4 mg N L^{-1}. The decoupling of δ^{15}N from NO_3^- suggested a consistent isotopic NO_3^- source, with limited residual build-up of enriched ^{15}N, where concentrations were likely driven by seasonal periods of enhanced recharge.

Trend group B comprised six wells (91-10, PA-25, PA-35, PB-20, US-02, and US-05) with no significant δ^{15}N trend over the study period (Fig. 5b), but exhibiting high δ^{15}N variability around the mean (±SD ≥ 1.0 ‰). The degree of δ^{15}N and NO_3^- variability differed for most wells in this group; however, all sites exhibited strong δ^{15}N and NO_3^- coupling, with at least a 5 ‰ change in δ^{15}N and 15 mg N L^{-1} fluc-

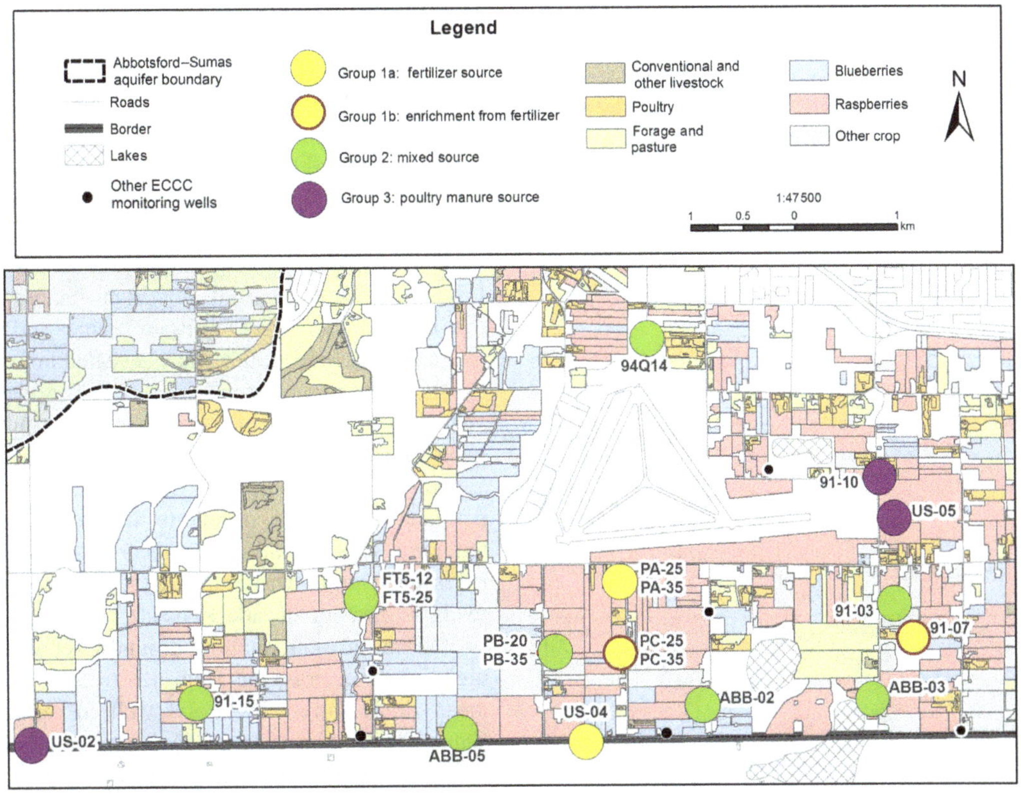

Figure 4. Spatial distribution of δ^{15}N source groupings, along with agricultural land use.

tuation in NO_3^- concentrations. Decreasing DO concentrations in US-02 were associated with increasing δ^{15}N values; however, in this case NO_3^- and Cl concentrations (Table S1) were synchronous, suggesting the same source. In fact, the up-gradient field of this well had undergone a renovation cycle in the preceding months, where old raspberry plants were removed followed by a poultry manure soil amendment prior to replanting. Sites 91-10 and US-05 showed similar δ^{15}N and NO_3^- fluctuations, albeit smaller in magnitude, with corresponding increases in chloride and elevated DO concentrations. Sites 91-10 and US-05 are close to each another (< 200 m apart) along a similar groundwater flow path, suggesting these variations are linked. No other sites in this group were spatially proximal. Sites PB-20, PA-25, and PA-35 exhibited varying degree of coupled δ^{15}N and NO_3^- seasonality, suggesting nitrate leaching was the primary driver of NO_3^- variability. For PA-25, increasing NO_3^- concentrations with ^{15}N enrichment (although variable in degree) were systematically observed each winter, suggesting nitrate mobilization occurred during peak winter rainfall periods.

Six sites were identified as trend group C, with increasing δ^{15}N trends (91-03, 91-07, FT5-25, PC-25, PC-35, and US-04). These sites were evenly distributed between distribution groups 1a (3), 1b (2), and 2 (1), suggesting one driver controlling local NO_3^- concentrations and δ^{15}N values. Enriching ^{15}N trends (often along a flow path) are usually associated with progressive microbial denitrification; however, all sites had high DO aerobic concentrations (> 8 mg L^{-1}). Sites PC-25 and PC-35, which exhibited some degree of coupled δ^{15}N and δ^{18}O enrichment at a 2 : 1 ratio, also showed increasing NO_3^- concentrations, suggesting heavy loading of poultry manure. Prior to the marked increase in NO_3^- and δ^{15}N in the spring of 2012, PC-25 and PC-35 exhibited a significant, albeit gradual, increasing ^{15}N trend (Fig. 5c). This revealed a second subtle driver – the increased precipitation that occurred between 2008 and 2011 and its effect on groundwater nitrate concentrations, as shown by Graham et al. (2015). Wells 91-03, FT5-25, and US-04 did not undergo any up-gradient crop replacement or soil amendments and exhibited various degrees of NO_3^- and δ^{15}N seasonality, further strengthening the precipitation link as a potential driver. The δ^{15}N trend could be linked to the enhanced mobilization and infiltration of ^{15}N-depleted organic soil N.

Group D exhibited a ^{15}N depletion trend (Fig. 5d); consisted of monitoring wells 94Q-14, PB-35, and ABB-03; and had a negative δ^{15}N shift of 1 ‰–3 ‰ and δ^{15}N values between +6 ‰ and +10 ‰ (group 2). Well 94Q-14 showed δ^{15}N seasonality, but not in NO_3^-, with concentrations mostly below the MAC. PB-35 showed small seasonality in NO_3^- concentrations but none in δ^{15}N, indicating possible mixing and dilution due to a shift in nitrogen sources. Wassenaar et al. (2006) suggested that a negative δ^{15}N shift may

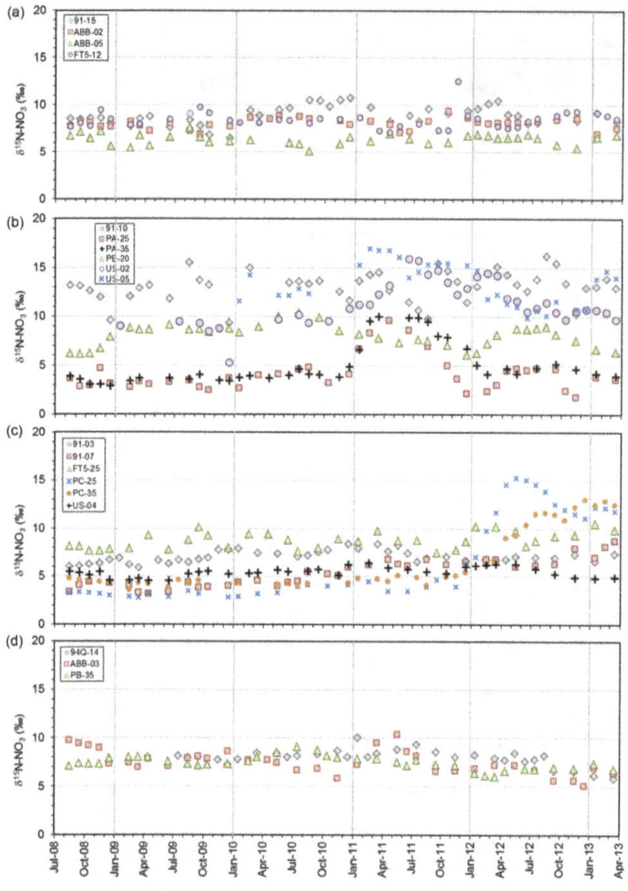

Figure 5. $\delta^{15}N\text{-}NO_3$ time series plots: **(a)** no trend – stable (SD < ±1.0), **(b)** no trend – variable (SD > ±1.0), **(c)** enrichment trend, and **(d)** depletion trend.

be attributed to the longer-term change in nitrogen sources used from poultry manure to synthetic fertilizers. Lastly, ABB-03 showed no significant trend in NO_3^- concentrations or in $\delta^{18}O$ of nitrate; however, $\delta^{15}N$ and $\delta^{18}O$ were correlated, while $\delta^{15}N$ and NO_3^- were inversely correlated. Furthermore, ABB-03 exhibited 3–4-month intervals of lower-than-usual DO conditions ($\Delta \geq 4\,mg\,L^{-1}$ compared to other seasons; Table S1), which corresponds to periods of ^{15}N enrichment and decreasing NO_3, suggesting localized denitrification, that were repeatable to various degrees on a seasonal basis but were most prominent in 2011. These findings suggest localized and temporally limited denitrification may be occurring in the soil root zone in some areas, contributing to ^{15}N enrichment and variability of NO_3^- concentrations. Site ABB-03 was not near Fishtrap Creek (Fig. 1), which Tesoriero (2000) and Wassenaar et al. (2006) identified as a localized riparian denitrification hot spot. Enrichment in ^{15}N of nitrate at these sites appeared to be from temporal drivers that could be overlooked in one-time synoptic sampling (Wassenaar, 1995).

4 Conclusions and outlook

This study represents an unprecedented high-frequency 5-year seasonal spatio-temporal study of the water table well with over 700 nitrate isotopic assays, revealing the dynamics of nitrate recharging the transboundary Abbotsford–Sumas aquifer. The high (monthly) temporal frequency of nitrate and isotopic data aimed to address concerns that infrequent nitrate isotopic or concentration synoptic samplings of shallow groundwater overlook important factors of seasonality that may be key drivers of nitrate sources and fluxes to shallow aquifers. Indeed, our study revealed new important scientific information not previously seen in the synoptic surveys that will help managers better tackle nutrient management strategies to help reduce groundwater pollution.

Overall, and unsurprisingly, we found the predominant perennial source of nitrate to the aquifer at all spatio-temporal scales within the 5-year intensive sampling period was animal waste (poultry) sources, which was already known for decades. Nitrate concentrations in young (< 5 years old) and shallow groundwater were persistently high in nitrate, ranging from 1.3 to 99 mg-N L^{-1}, with a mean of 16.2 mg-N L^{-1}, and well in exceedance of the Canadian drinking water MAC of 10 mg-N L^{-1} for 76 % of the wells. The study also verified a postulated and subtle decadal-scale shift towards ^{15}N-depleted nitrate sources, likely reflecting systematic changes in agriculture practices from the early days of indiscriminate manure disposal towards more targeted use of synthetic fertilizers or from changes in crop types and associated nutrient practices, as evidenced by the mean $\delta^{15}N$ value for nitrate of $+7.9 \pm 3.0$ ‰ compared to $+10.2 \pm 4.0$ ‰ in the 1990s. Synthetic fertilizer and soil N represent a comparatively higher N loading in the central portions of the ASA but are flanked on both sides by higher poultry-manure-dominated N loadings (Fig. 4). The high nitrate concentrations in contemporary recharging groundwater and lack of targeted nitrogen reduction measures means widespread nitrate contamination of the aquifer is likely to persist into the foreseeable future, and our data affirm little evidence for persistent or widespread natural attenuation of nitrate by subsurface denitrification processes at any time of the year. Nitrate remediation of the aquifer will only be possible if agricultural N sources are dramatically reduced or eliminated.

In some wells we found that localized agricultural practices (i.e. nearby organic N soil amendment) had a nearly immediate multi-year negative impact, mainly exhibited by marked increases in poultry-derived N, and lasting for several years and across seasons. This common practice at field renovation and post-woodlot clearing resulted in spatial clustering and differing short-term trends for water table nitrate and its isotopes across the aquifer (Figs. 4 and 6), further revealing that infiltrating NO_3^- and its isotopic composition can change quickly in direct response to contemporary near-field practices. Conversely, this suggests N source cutoff as

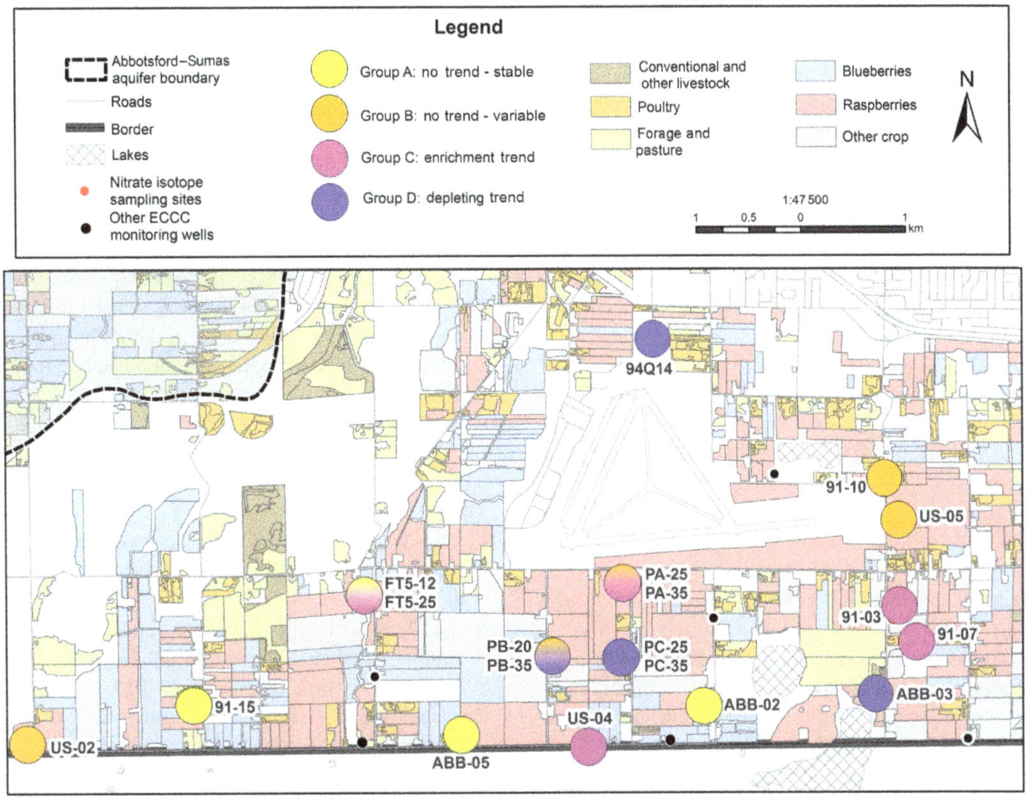

Figure 6. Spatial distribution of $\delta^{15}N$ trend groupings, along with agricultural land use.

a groundwater remediation effort could be similarly as effective. Despite 53 % of shallow wells showing no isotopic trends, 47 % showed an ^{15}N enrichment or depletion trend in nitrate, and about half of the wells exhibited nitrate seasonality in NO_3^- concentrations and/or $\delta^{15}N$ values controlled by temporal infiltration of residual mineralized N or weak, short-term denitrification.

Due to the rapid shift in NO_3^- and isotopic values of recharging groundwater immediately following field renovation and soil amendment practices, this study reinforces the importance of designing and conducting appropriate spatio-temporal nitrate sampling to reduce the risk of misinterpreting nitrate concentrations and its isotopic data through the more common practice of occasional synoptic surveys. The dynamics of nitrate in younger (< 5 years old) water table wells, however, also imply it would be prudent to monitor deeper, older groundwater which smoothes out short-term fluctuations and hence record longer-term and aquifer-wide trends.

For the ASA agricultural area specifically, measuring the impact of changes in nutrient management practices associated with the switch from raspberry to blueberry crops or field renovation is required to determine its impacts on groundwater nitrate dynamics. Decisions on future groundwater aquifer nitrate management need to take into consideration (i) permanent or cyclical changes in the planned crop types and (ii) the associated nutrient management practices involved with them. Subtle shifts in nitrate in the ASA may be unexpectedly influenced by the recent replacement of raspberries with blueberries, which may to be less reliant of cyclical poultry manure soil amendments.

Author contributions. LIW and BZ prepared the funding proposal. LIW, GG, and MS established the site-selection criteria and selected sites, while MS undertook field activities. MS and LIW analyzed data statistically and conducted the data interpretation. MS prepared the manuscript with the help of LIW, and all co-authors were asked to review the manuscript.

Competing interests. The authors declare that they have no conflict of interest.

Acknowledgements. Funding for this project was provided by Environment Canada (LIW, MS, GG), Agriculture and Agri-Food Canada (BZ), and the Canadian Water Network. The authors would also like to thank Steven Taylor from the Isotope Lab at the University of Calgary. Geoff Koehler assisted with sampling. Tommy Diep assisted with GIS mapping and field activities (Environment and Climate Change Canada).

Edited by: Christine Stumpp

References

Amberger, A. and Schmidt, H. L.: The natural isotope content of nitrate as an indicator of its origin, Geochim. Cosmochim. Ac., 51, 2699–2705, 1987.

Aravena, R. and Robertson, W.: Use of multiple Isotope Tracers to Evaluate Denitrification in Groundwater, Study of Nitrate from a Large-Flux Septic System Plume, Ground Water, 36, 975–982, 1998.

Aravena, R., Evans, M. L., and Cherry, J. A.: Stable isotopes of oxygen and nitrogen in source identification of NO_3 – from septic systems, Ground Water, 31, 180–186, 1993.

Armstrong, J. E., Crandell, D. R., Easterbrook, D. J., and Noble, J. B.: Late Pleistocence stratigraphy and chronology of southwestern British Columbia and northwestern Washington, Geol. Soc. Am. Bull.,76, 321–330, 1965.

Böhlke, J. K.: Groundwater recharge and Agricultural Contamination, Hydrogeol. J., 10, 153–179, 2002.

Böhlke, J. K. and Denver, J. M.: Combined Use of Groundwater Dating, Chemical, and Isotopic Analyses to Resolve the History and Fate of Nitrate Contamination in Two Agricultural Watersheds, Atlantic Coast Plain, Maryland, Water Resour. Res., 9, 2319–2339, 1995.

Canter, L. W.: Nitrates in Groundwater: CRC Press, Lewis Publishing, NY, 1997.

Casciotti, K. L., Sigman, D. M., Hastings, M. G., Bohlke, J. K., and Hilkert, A.: Measurement of the oxygen isotopic composition of nitrate in seawater and freshwater using the denitrifier method, Anal. Chem., 74, 4905–4912, 2002.

Chesnaux, R. and Allen, D. M.: Simulating Nitrate Leaching Profiles in a Highly Permeable Vadose Zone, Environ. Model. Assess., 13, 527–539, https://doi.org/10.1007/s10666-007-9116-4, 2007.

Choi, W. J., Lee, S. M., and Ro, H. M.: Evaluation of Contaminant Sources of Groundwater NO_3^- using Nitrogen Isotope Data: A Review, Geosci. J., 7, 81–87, 2003.

Coplen, T. B.: Reporting of Stable Hydrogen, Carbon, and Oxygen Isotopic Abundances, Pure Appl. Chem., 66, 273–276, 1994.

Cox, S. E. and Kahle, S. C.: Hydrogeology, groundwater quality and sources of nitrate in lowland glacial aquifers of Whatcom county, Washington and British Columbia, Canada, USGS Water Resources Investigation Report, 98–4195, 1999.

Environment Canada: Abbotsford A., Canadian climate normal 1981–2010 station 1100030, available at: http://climate.weather.gc.ca/climate_normals/results_1981_2010_e.html?stnID=702&autofwd=1, last access: 12 October 2014.

Graham, G., Allen, D. M., and Finkbeiner, B.: Climate controls on nitrate concentration variability in the Abbotsford-Sumas aquifer, British Columbia, Canada, Environ. Earth Sci., 73, 2895–2907, 2015.

Halstead, E. C.: Ground Water Supply-Fraser Lowland, British Columbia, Environment Canada, National Hydrology Research Paper, 26, 80 pp., 1965.

Hamilton, P. A. and Helsel, D. R.: Effects of agriculture on groundwater quality in 5 regions of the United States, Ground Water, 33, 217–226, 1995.

Haslauer, C. P., Rudolph, D. L., and Thomson, N. R.: Impacts of changing agricultural land-use practices on municipal groundwater quality: Woodstock, Ontario: Bringing Groundwater Quality Research to the Watershed Scale, 4th International Groundwater Quality Conference, IAHS Publications, 297, 223–233, 2005.

Health Canada: Guidelines for Canadian Drinking Water Quality: Guideline Technical Document – Nitrate and Nitrite, 2013.

Heaton, T. H. E.: Isotopic studies of nitrogen pollution in the hydrosphere and Atmosphere – A Review, Chem. Geol., 59, 87–102, 1986.

Herod, M. N., Suchy, M., Cornett, R. J., Kieser, W. E. Clark, I. D., and Graham, G.: The atmospheric transport of iodine-129 from Fukushima to British Columbia, Canada and its deposition and transport into groundwater, Water Resour. Res., 51, 9628–9645, https://doi.org/10.1002/2015WR017325, 2015.

Hii, B., Liebscher, H., Mazalek, M., and Tuominen, T.: Ground water quality and flow rates in the Abbotsford Aquifer, British Columbia, Environment Canada, Pacific and Yukon Region, 1999.

Kellman, L. M. and Hillaire-Marcel, C.: Evaluation of nitrogen isotopes as indicators of nitrate contamination sources in an agricultural watershed, Agr. Ecosyst. Environ., 95, 87–102, 2003.

Kendall, C.: Tracing nitrogen sources and cycling in catchments, in: Isotope tracers in catchment hydrology, 1st ed., edited by: Kendall C. and McDonnell, J. J., Elsevier Science B.V., Amsterdam, 506, 1998.

Kohut, A. P., Sather, S., Kwong, J., and Chowijka, K.: Nitrate contamination of the Abbotsford aquifer, B.C. Ministry of Water, Land and Air Protection, 18 pp., 1989.

Kowalenko, C. G.: Nitrogen pools and processes in agricultural systems of Coastal British Columbia – A review of publish research, Can. J. Plant Sci. 80, 1–10, 2000.

Kreitler, C. W.: Determining the Source of Nitrate in Groundwater by Nitrogen Isotope Studies: Report of Investigation 83, Bureau of Economic Geology, University of Texas at Austin, Austin Texas, 99, 57, 1975.

Lettenmaier, D. P.: Multivariate Nonparametric-Tests for Trend in Water-Quality, Water Resour. Bull., 24, 505–512, 1988.

Liebscher, H., Hii, B., and McNaughton, D.: Nitrates and pesticides in the Abbotsford aquifer, Southwestern British Columbia, Environment Canada, 83 pp., 1992.

Loo, S. E., Ryan, M. C., Zebarth, B. J., Kuchta, S. H., Neilsen, D., and Mayer, B.: Use of $\delta^{15}N$ and $\delta^{15}O$ Values for Nitrate source Identification under Irrigated Crops: A Cautionary Vadose Tale, J. Environ. Qual., 46, 528–536, 2017.

Luttmerding, H. A.: Soils of the Langley-Vancouver map area, BC Soil Survey Report No. 15, 1980.

Malekani, F.: A new approach to estimate monthly groundwater nitrate loading from an individual field, MS, University of Calgary, Calgary, Alberta, 2012.

Mariotti, A.: Atmospheric nitrogen is a reliable standard for natural 15N abundance measurements, Nature, 303, 685–687, 1983.

Mitchell, R. J., Babcock, R. S., Gelinas, S., Nanus, L., and Stasney, D. E.: Nitrate Distributions and Source Identification in the Abbotsford-Sumas Aquifer, Northwestern Washington State, J. Environ. Qual., 32, 789–800, 2003.

Ostrom, N. E., Knoke, K. E., Hedin, L. O., Robertson, G. P., and Smucker, A. J. M.: Temporal trends in nitrogen isotope values of nitrate leaching from an agricultural soil, Chem. Geol., 146, 219–227, 1998.

Overdahl, C. J., Rehm, G. W., and Meredith, H. L.: Fertilizer Urea, http://www.extension.umn.edu/distribution/cropsystems/dc0636.html (last access: 10 September 2014), 2007.

Pastén-Zapata, E., Ledesma-Ruiz, R., Harter T., Ramírez, A. I., and Mahlknecht, J.: Assessment of sources and fate of nitrate in shallow groundwater of an agricultural area by using a multi-tracer approach, Sci. Total Environ., 470–471, 855–864, doi.org/10.1016/j.scitotenv.2013.10.043, 2014.

Savard, M. M., Paradis, D., Somers, G., Liao, S., and van Bochove, E.: Winter nitrification contributes to excess NO_3^- in groundwater of an agricultural region, Water Resour. Res., 43, W06422, https://doi.org/10.1029/2006WR005469, 2007.

Savard, M. M., Somers, G., Smirnoff, A., Paradis, D., van Bochove, E., and Liao, S.: Nitrate isotopes unveil distinct seasonal N-sources and the critical role of crop residues in groundwater contamination, J. Hydrol., 381, 134–141, 2010.

Scibek, J. and Allen, D. M.: Modeled Impacts of Predicted Climate Change on Recharge and Groundwater Levels, Water Resour. Res., 42, W11405, https://doi.org/10.1029/2005WR004742, 2006.

Sigman, D. M., Casciotti, K. L., Andreani, M., Barford, C., Galanter, M., and Bohlke, J. K.: A bacterial method for the nitrogen isotopic analysis of nitrate in seawater and freshwater, Anal. Chem., 73, 4145–4153, 2001.

Spalding, R. F. and Exner, M. E.: Occurrence of nitrate in groundwater – a review, J. Environ. Qual., 22, 392–402, 1993.

Tesoriero, A. J., Liebscher, H., and Cox, S. E.: Mechanism and rate of denitrification in an agricultural watershed: Electron and mass balance along groundwater flow paths, Water Resour. Res., 36, 1545–1559, 2000.

Thas, O., Van Vooren, L., and Ottoy, J. P.: Nonparametric test performance for trends in water quality with sampling design applications, J. Am. Water Resour. As., 34, 347–357, 1998.

Vorosmarty, C. J., Green, P., Salisbury, J., and Lammers, R. B.: Global water resources: vulnerability from climate change and population growth, Science, 289, 284–288, 2000.

Wassenaar, L. I.: Evaluation of the Origin and Fate of Nitrate in the Abbotsford Aquifer Using the Isotopes of ^{15}N and ^{18}O in NO_3, Appl. Geochem., 10, 391–405, 1995.

Wassenaar, L. I., M. Hendry, J. M., and Harrington, N.: Decadal Geochemical and Isotopic Trends for Nitrate in the Transboundary Abbotsford-Sumas Aquifer and Implications for Agricultural Beneficial Management Practices, Environ. Sci. Technol, 40, 4626–4632, 2006.

Xue, D., Botte, J., De Baets, B., Accoe, F., Nestler, A., Taylor, P., Van Cleemput, O., Berglund, M., and Boeckx, P.: Present limitations and future prospects of stable isotope methods for nitrate source identification in surface- and groundwater, Water Res., 43, 1159–1170, 2009.

Zebarth, B. J., Hii, B., Liebscher, H., Chipperfield, K., Paul, J. W., Grove, G., and Szeto, S. Y.: Agricultural land use practices and nitrate concentration in the Abbotsford-Sumas Aquifer, British Columbia, Canada, Agr. Ecosyst. Environ., 69, 99–112, 1998.

Zebarth, B. J., Ryan, M. C, Graham, G., Forge, T. A., and Neilsen, D.: Groundwater Monitoring to Support Development of BMPs for Groundwater Protection: The Abbotsford-Sumas Aquifer Case Study, Ground Water Monit. R., 35, 82–96, https://doi.org/10.1111/gwmr.12092, 2015.

9

Isotopic reconnaissance of urban water supply system dynamics

Yusuf Jameel[1], **Simon Brewer**[2], **Richard P. Fiorella**[1], **Brett J. Tipple**[3,4], **Shazelle Terry**[5], **and Gabriel J. Bowen**[1,3]

[1]Geology and Geophysics, University of Utah, 115 S 1460 E, Salt Lake City, Utah 84112, USA
[2]Department of Geography, University of Utah, 332 S 1400 E, Salt Lake City, Utah 84112, USA
[3]Global Change and Sustainability Center, University of Utah, 115 S 1460 E, Salt Lake City, Utah 84112, USA
[4]Department of Biology, University of Utah, 257 S 1400 E, Salt Lake City, Utah 84112, USA
[5]Jordan Valley Water Conservancy District, 8215 S 1300 W, West Jordan, Utah 84088, USA

Correspondence: Yusuf Jameel (yusuf.jameel@utah.edu)

Abstract. Public water supply systems (PWSS) are critical infrastructure that is vulnerable to contamination and physical disruption. Exploring susceptibility of PWSS to such perturbations requires detailed knowledge of supply system structure and operation. The physical structure of the distribution system (i.e., pipeline connections) and basic information on sources are documented for most industrialized metropolises. Yet, most information on PWSS function comes from hydrodynamic models that are seldom validated using observational data. In developing regions, the issue may be exasperated as information regarding the physical structure of the PWSS may be incorrect, incomplete, undocumented, or difficult to obtain in many cities. Here, we present a novel application of stable isotopes in water (SIW) to quantify the contribution of different water sources, identify static and dynamic regions (e.g., regions supplied chiefly by one source vs. those experiencing active mixing between multiple sources), and reconstruct basic flow patterns in a large and complex PWSS. Our analysis, based on a Bayesian mixing model framework, uses basic information on the SIW and production volumes of sources but requires no information on pipeline connections in the system. Our work highlights the ability of stable isotopes in water to analyze PWSS and document aspects of supply system structure and operation that can otherwise be challenging to observe. This method could allow water managers to document spatiotemporal variation in flow patterns within PWSS, validate hydrodynamic model results, track pathways of contaminant propagation, optimize water supply operation, and help monitor and enforce water rights.

1 Introduction

Public water supply systems (PWSS) are an important component of the critical infrastructures supporting human development across the globe. The complexity of PWSS can vary widely, ranging from linear, single-source distribution systems to branched distribution networks using multiple water sources. To understand the stability of water supplies, conduct risk evaluation, and develop effective and efficient responses for particular threats (supply contamination, infrastructure failure, etc.), it is critical to understand the physical and spatial structure of the distribution network, connectivity within the system, and the links between the point-of-use and environmental water sources.

The physical structure of the distribution system and basic information on water sources are generally well documented for most first-world metropolises. In these settings, water managers traditionally rely on network analyses to study different aspects of water distribution systems, including pressure gradients, flow rates, water losses from the supply system, identification of vulnerable sections, and tracking of disinfectants and contaminants (Boryczko and Tchórzewska-Cieślak, 2014; Pietrucha-Urbanik, 2015; Yoo et al., 2015). These analyses are generally robust; however, they are seldom validated using observational data and can suffer from shortcomings including the absence of unique solutions in underdetermined systems, assumption of invariant flow rates, uncomprehensive or non-inclusiveness of uncertainty in the analysis (Waldrip et al., 2016), and outdated and/or incorrect information on infrastructure (Liggett and Chen, 1994). Beyond statistical and computational issues, hydrodynamic modeling requires extensive and detailed information about the PWSS, including node elevation, pipe length and diameter, and pump operation data. For many communities in the developing world, where distribution networks are commonly unregulated and decentralized, even basic information on supply system structure and source contributions may be incorrect, incomplete, undocumented, or difficult to obtain. Hydrodynamic modeling of PWSS in such cases can be challenging and prone to significant errors.

With growing water security challenges due to climate change (Vörösmarty et al., 2010), expanding complexity and dynamicity of urban water systems, and increasing detrimental effects of aging water infrastructure in many countries (Hanna-Attisha et al., 2016; Kaushal, 2016; Larsen et al., 2016; Schnoor, 2016), it is important to develop new techniques and methods to study PWSS that require minimal information on the physical structure and connectivity within the supply system. In this regard, new techniques and methods are being developed to (1) understand failure in the water distribution system with limited, imprecise, and ambiguous information on the supply structure (Najjaran et al., 2005; Ismail et al., 2011; Bolar et al., 2013; Kabir et al., 2015) and (2) analyze the water distribution system in a probabilistic framework (Waldrip et al., 2016, 2018).

Stable isotopes in water (SIW) can serve as an important tool to study water management within complex PWSS. SIW are naturally occurring tracers of the terrestrial hydrological cycle and significant isotopic differences between water sources can exists at catchment, regional, and global scales due to seasonal biases in recharge, differences in meteoric water composition, altitude, and meteorological factors such as temperature, humidity, and wind speed (Dansgaard, 1964). Varying isotopic signatures among the water sources (precipitation, river, lakes, reservoirs, shallow and deep groundwater, etc.) make SIW an effective tracer to understand and investigate natural and human–natural coupled systems (for examples please refer to Dawson and Ehleringer, 1991; Rozanski et al., 1992; Gat, 1995; von Grafenstein et al., 1999; Kennedy et al., 2011; Klaus and McDonnell, 2013; Gabor et al., 2017; Matheny et al., 2017; and Jameel et al., 2018).

In urban settings, stable isotopes and other geochemical tracers have been used successfully to understand effects of storm water control measures on urban streams (Jefferson et al., 2015), detect infiltration rates in urban sewers (De Bénédittis et al., 2005; Kracht et al., 2007), partition waste water and groundwater in urban sewers (De Bondt et al., 2018), and determine the age of drinking water in PWSS (Waples et al., 2015). Recent studies have also shown that stable isotopes of tap water in urban areas can be used to characterize active water management practices, identify linkages between socioeconomic factors and water management practices, and quantify the effects of climate variability on water resources (Ehleringer et al., 2016; Jameel et al., 2016; Tipple et al., 2017; Zhao et al., 2017, Wang et al., 2018).

Here, we collaborated with the Jordan Valley Water Conservancy District (JVD, also referred to as Jordan Valley District) to conduct an isotopic survey of waters from their service area within the Salt Lake Valley metropolitan area (SLV) of northern Utah, USA (Fig. 1). JVD is a multi-source public water distribution network (Fig. 1) and we attempt to understand mixing between water sources at various sites (subsequently referred to as distribution sites) distributed on the transmission lines using SIW. This work extends the earlier work of Jameel et al. (2016) and Tipple et al. (2017) beyond identifying broad water management patterns of a PWSS and explores the capacity of SIW to provide quantitative, spatially and temporally resolved estimates of source contributions within a well-defined and characterized PWSS.

We conducted our study during a 6-month period (May–October 2015), and using information on the production volume from the different sources, we analyze the stable isotope data at a monthly resolution within a Bayesian framework to generate quantitative estimates (with uncertainty) of the contribution of individual sources at the distribution sites. These analyses reveal basic information on supply and transport dynamics within the system, reflecting the physical structure of the supply system and the geographic distribution of sources. Finally, we combine the monthly analyses to characterize the spatial structure of the system in terms of contribution areas for the different sources across the supply network. Our results suggest that SIW-based Bayesian isotope mixing models (BIMM) could be a powerful and useful tool to interrogate PWSS, provide observational validation to hydrodynamic models, track contaminants and disinfectants within the supply system, and monitor water rights in PWSS managed by or for multiple stakeholders. This technique can be particularly useful in understanding water management practices of urban centers in the developing world that are undergoing rapid expansions and are generally decentralized, which makes conventional hydrodynamic techniques difficult to apply.

2 Methods

2.1 Site description

The Jordan Valley District is a wholesale supplier to 17 water districts in the Salt Lake Valley (USA) and retails directly to several locations in SLV located primarily on the northeastern part of the valley (known as the Jordan Valley retail area, Fig. 1). As a wholesaler, Jordan Valley District sells water to these 17 districts from fixed locations on their distribution line and is not responsible for managing and distributing water in these districts beyond the transfer point.

In general, JVD relies on 3–5 sources at any given time to supply water to its service area; however, during the summer season (June–August), an additional 5–7 sources are often used to meet increased municipal water demand (per-

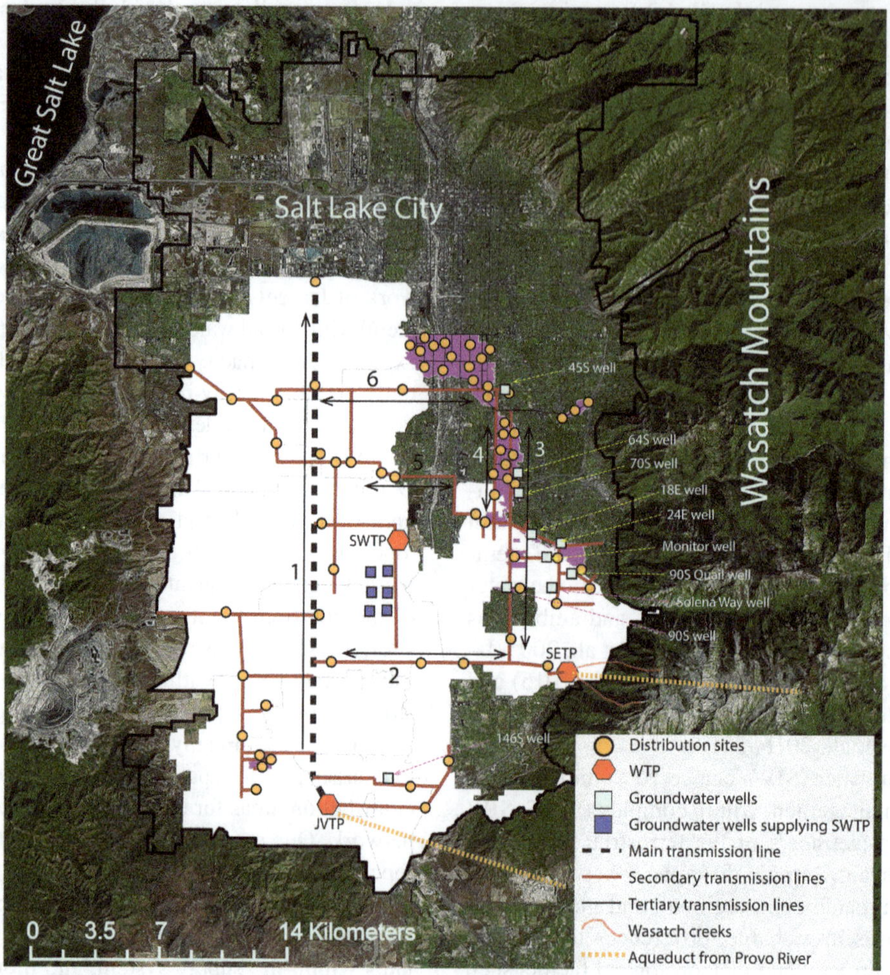

Figure 1. Jordan Valley Water Conservancy District (JVD) wholesale area (white) and Jordan Valley retail area (purple) within the Salt Lake metropolitan valley (black border). Water from wells shown in light blue color is pumped directly into the transmission lines on which they are located. The aqueducts from the Provo River and the Wasatch creeks are shown for illustrative purposes only. WTP: water treatment plant, JVTP: Jordan Valley Water Treatment Plant, SETP: Southeast Water Treatment Plant, SWTP: Southwest Water Treatment Plant. Source of base map: ESRI digital media.

sonal communication, JVD operations manager). Water is sourced primarily through the Provo River system (>75 % of total water supplied annually) and is supplemented with water from Wasatch creeks and groundwater wells depending on demand. The Wasatch creek sources carry runoff from snowmelt in the Wasatch Mountains (Fig. 1) and are used only in spring and early summer. There are approximately 25 active groundwater wells managed by the Jordan Valley District. Not all wells operate simultaneously, rather only two to five wells operate at any given time and the operating wells are rotated every few months due to contractual obligations and also to avoid overexploitation.

Jordan Valley District operates three water treatment plants (WTP). The Jordan Valley Water Treatment Plant (JVTP, also referred to as Jordan Valley Treatment Plant in the text) is the largest water treatment plant and is situated at the southern end of the valley (Fig. 1). It has a maximum operational capacity of 681 374 m^3 per day (80 million gallons per day) and treats water from the Provo River. The Southeast Water Treatment Plant (SETP, also referred to as Southeast Treatment Plant in the text) is a significantly smaller WTP with a maximum operational capacity of 75 708 m^3 per day (20 million gallons per day) situated on the southeastern side of the valley. It also treats water from the Provo River, but during spring and early summer (Mid-April to June) most of the water treated at SETP is from the Wasatch creeks. The Southwest Water Treatment Plant (SWTP) has a maximum operational capacity of 26 497 m^3 per day (7 million gallons per day), is located in the middle of the valley, and treats water from groundwater wells located near the treatment plant. Groundwater wells supplying the SWTP (shown as dark blue squares in Fig. 1) have a high salt concentration and require extensive purification before being pumped into the distribution system. In contrast, groundwater wells located on the eastern side of the valley (shown as light blue squares in Fig. 1) have lower concentrations of dissolved salts and do not require additional treatment before entering the distribution system (personal communication, JVD operations manager).

The Jordan Valley District water distribution system consists of one primary (Fig. 1), several secondary (line 2 through 6, Fig. 1), and numerous tertiary transmission lines. Water can move in either direction in all the transmission lines; however, in transmission line 1, water primarily moves from south to north. Water from Jordan Valley Treatment Plant is pumped directly into transmission line 1 and water from Southeast Treatment Plant is pumped into transmission lines 2 and 3 (Fig. 1). Water from Southwest Treatment Plant is supplied mainly to residential areas in the vicinity of the WTP (these supply connections are not shown in Fig. 1), though some water from SWTP is also pumped directly into transmission lines 5 and 6 (bypassing line 1). Water from wells in the eastern side of the valley is pumped directly into the transmission lines on which the respective wells are located. Most of the secondary transmission lines are interconnected via tertiary and quaternary lines (not shown in Fig. 1 except for the tertiary connections in the Jordan Valley retail area).

2.2 Sample acquisition and processing

Each month, from May to October 2015, we sampled water at sources contributing to the Jordan Valley District service area and at numerous locations (distribution sites or simply sites, Fig. 1) on the Jordan Valley District transmission lines. Source water samples were collected as effluent from the WTPs and directly from the groundwater wells, while distribution site samples were collected from monitoring taps on the transmission lines. For source water samples obtained from water treatment plants, we measured the pre- and post-treatment (i.e., influent to the WTP and effluent from the WTP) isotopic composition of the water. We did not observe any significant isotopic difference between pre- and post-treatment samples (differences in δ^2H and δ^{18}O less than 0.7 ‰ and 0.2 ‰, respectively). Distribution sites and surface water sources (Provo River and Wasatch creeks) were sampled one to three times per month. Groundwater wells were sampled one to five times, respectively, during the entire study period (May to October 2015). When a given well was not sampled in its month of operation, the mean value observed for the same well during other month(s) of our study period was used to characterize water supplied from that well. This substitution was justified given that previous work showed little temporal variability in the isotope values of water supplied from Salt Lake Valley groundwater wells (Jameel et al., 2016). The distribution sites are routinely monitored by Jordan Valley District for water quality analysis and are located across the supply network based on Jordan Valley District's monitoring program. As such, the distribution sites are more densely distributed in the Jordan Valley retail area because Jordan Valley District is responsible for water quality monitoring within this area. In other districts, where Jordan Valley District wholesales water, samples were collected only from the primary and secondary transmission lines.

Southeast Treatment Plant water was sourced mostly from the Wasatch creeks in May and June 2015, and from the Provo River from July to October 2015. Jordan Valley Treatment Plant water was sourced from the Provo River for the entire period of analysis. Therefore, we considered SETP and JVTP as separate sources in May and June and as a single source from July to October. Isotope ratios for effluent from SETP and JVTP were not statistically different between July and October (Hotelling multivariate t test, $p>0.05$, $n=9$). Additionally, groundwater wells situated close to each other and having similar isotope values (differences in δ^2H and δ^{18}O less than 0.5 ‰ and 0.1 ‰, respectively) were also combined together for our analyses (such as wells 64S and 70S in June, July, and August 2015).

2.3 Isotope analysis

Samples were collected in 4 mL clean glass vials and stored in a refrigerator at 4 °C prior to analysis. The samples were analyzed within a few weeks of collection at the Stable Isotope Ratio Facility for Environmental Research (SIRFER) facility, University of Utah, on a cavity ring-down spectrometer (L2130-i; Picarro, Inc., Santa Clara, CA) following protocols described in Good et al. (2014), following van Geldern and Barth (2012). Values are reported in δ notation: $\delta = (R_{\text{sample}}/R_{\text{standard}} - 1)$, where R_{sample} and R_{standard} are the ^{2}H / ^{1}H or ^{18}O / ^{16}O ratios for the sample and standard, respectively, and the VSMOW standard is referenced (Coplen, 1988). Accuracy and precision were checked using a secondary laboratory reference water, and the analytical precision for these analyses was ± 0.3 ‰ for δ^2H and ± 0.03 ‰ for δ^{18}O (± 1 SD).

2.4 Bayesian mixing model and statistical analyses

We estimated the fractional contribution of the different sources at the distribution sites for each month using a Bayesian isotope mixing model. The advantages of a Bayesian approach include (1) simultaneous analysis of both isotope ratios (δ^2H and δ^{18}O), (2) inclusion of prior information into the statistical analysis (such as those from previous studies, ancillary data, and subjective knowledge), (3) explicit incorporation of analytical and sampling uncertainties into the model, and (4) robust estimates of uncertainty and quantification of most likely solutions in an underdetermined system (number of sources greater than number of isotopes plus 1).

The Bayesian mixing model described here is similar to those used in other studies involving stable isotope data

(Ogle and Barber, 2008; Parnell et al., 2010; Cable et al., 2011; Mailloux et al., 2014). For our analysis, we first define the likelihood of the source isotope data. For this, we assumed that the different isotopic observations of each source (J) for a given month are coming from a bivariate normal distribution with a mean vector $\left[\mu\delta^2\mathbf{H}_J,\ \mu\delta^{18}\mathbf{O}_J\right]$ and a precision matrix ($\boldsymbol{\Omega}_J$) inverse of a covariance matrix) that reflects the temporal variability in the source isotope values. The bivariate normal distribution accounts for the potential correlation between δ^2H and δ^{18}O. By using a joint distribution, and its associated variance–covariance matrix, we capture both the individual variability and the co-variation. Thus,

$$\begin{bmatrix} \delta^2\mathrm{H}_{1J} & \delta^{18}\mathrm{O}_{1J} \\ \vdots & \vdots \\ \vdots & \vdots \\ \delta^2\mathrm{H}_{NJ} & \delta^{18}\mathrm{O}_{NJ} \end{bmatrix} \sim \mathrm{Mnormal}\left(\left[\mu\delta^2\mathbf{H}_J,\ \mu\delta\delta^{18}\mathbf{O}_J\right],\ \boldsymbol{\Omega}_J\right), \tag{1}$$

where $\delta^2\mathrm{H}_{1J}\ldots\delta^2\mathrm{H}_{NJ}$ and $\delta^{18}\mathrm{O}_{1J}\ldots\delta^{18}\mathrm{O}_{NJ}$ are the N observations of δ^2H and δ^{18}O of source J for that month, $\left[\mu\delta^2\mathbf{H}_J,\ \mu\delta^{18}\mathbf{O}_J\right]$ is the mean vector, and $\boldsymbol{\Omega}_J$ is the precision matrix. Similar to the source model, we assumed that for a supply site (I), the monthly averaged isotope values $[\delta^2\mathrm{H}_I,\ \delta^{18}\mathrm{O}_I]$ follow a bivariate normal distribution with mean vector $\left[\mu\delta^2\mathbf{H}_I,\ \mu\delta^{18}\mathbf{O}_I\right]$ and a precision matrix ($\boldsymbol{\Omega}_I$). Thus, for a supply site (I),

$$\left[\delta^2\mathrm{H}_I,\ \delta^{18}\mathrm{O}_I\right] \sim \mathrm{Mnormal}\left(\left[\mu\delta^2\mathbf{H}_I,\ \mu\delta^{18}\mathbf{O}_I\right],\ \boldsymbol{\Omega}_I\right). \tag{2}$$

The mean stable isotope values of the supply site can also be expressed as a mixing model, where the mean value for supply site I $\left(\mu\delta^2\mathbf{H}_I,\ \mu\delta^{18}\mathbf{O}_I\right)$ is the sum of the mean values of the sources weighted by their fractional contributions. Therefore, if K sources were used in a given month, $\left(\mu\delta^2\mathbf{H}_I,\ \mu\delta^{18}\mathbf{O}_I\right)$ for each supply site (I) is denoted by

$$\left[\mu\delta^2\mathbf{H}_I,\ \delta\mu^{18}\mathbf{O}_I\right] = \sum_{J=1}^{J=K} (f_J)\left[\mu\delta^2\mathbf{H}_J,\ \delta\mu^{18}\mathbf{O}_J\right], \tag{3}$$

where f_J is the proportional contribution from a given source J at supply site I and is our parameter of interest, i.e., the parameter to be estimated by the BIMM. Values of f were described using the Dirichlet distribution, a multivariate generalization of the beta distribution that follows the mass-balance constraint, i.e., $0 \le f_J \le 1$ and $\sum_{J=1}^{J=K}(f_J) = 1$. The Dirichlet distribution is characterized by parameter vector $\boldsymbol{\alpha} = \{\alpha_1, \alpha_2, \alpha_3, \ldots, \alpha_K\}$, such that the mean value associated with each f is $f_J = \alpha_J / \sum\{\alpha_1, \alpha_2, \alpha_3, \ldots, \alpha_K\}$.

In Bayesian analysis, the parameters to be estimated are initially assigned values (or distributions) that are believed to be the best estimate of the parameters. These initial values are referred to as prior values. After observing the data, the initial (or prior) values of the parameters are updated, and posterior values estimates of the parameter values are obtained. Please refer to Parnell et al. (2010) and Hoff (2009) for more detailed information on Bayesian isotope mixing model and Bayesian statistical methods, respectively. In our analysis, the parameters to be estimated were the fractional contribution of the different sources ($f's$) at the distribution sites. We therefore assign prior values to $f's$ that were then updated using the observed isotope ratios data at the distribution sites $\left(\mu\delta^2\mathbf{H}_I,\ \delta\mu^{18}\mathbf{O}_I\right)$ to obtain the posterior $f's$.

In general, if no information exists about the parameters, they are assigned a non-informative prior value. The default non-informative prior assigned to the Dirichlet distribution is the Jeffreys prior, where each element of the vector $\boldsymbol{\alpha}$ is assigned a value of K^{-1} (with K being the number of sources) (Parnell et al., 2010). However, if preexisting information about the parameters exists, they are assigned informed prior values. In our case, we had existing information on the volumetric contribution of each source to the water system as a whole (obtained from JVD, Table 1) as well as the distance between the sources and the distribution sites, each of which, we assert, should affect the probability that a given source supplied water to a given site. Therefore, we assigned prior values for each supply site based on this information. First, we assumed that the probability of a source supplying a given distribution site was proportional to the volume of water that source supplies to the JVD distribution system (Table 1). Thus, sources contributing more water to the JVD system have a higher probability of supplying water to any given site than do lower-volume sources. Second, we assumed that the probability of a source supplying water to a given site was inversely proportional to the distance between the source (e.g., water treatment plant or well location) and the distribution site. Thus, sources closer to a distribution site have a higher probability of supplying water at that site. We combined both pieces of prior information to obtain a normalized prior estimate, as described below.

In the first step, we calculated prior weights for the Dirichlet parameters for each source (J) based upon the proportional volume of water produced (V) by that source:

$$\alpha_{J\_\mathrm{volume}} = \frac{V_J}{\sum_{J=1}^{J=K} V_J}. \tag{4}$$

Second, we distance weighted each source's prior inversely based upon its distance (D) from supply site I:

$$\alpha_{JI\_\mathrm{distance}} = \frac{1}{\frac{D_{JI}}{\sum_{J=1}^{J=K} D_{JI}}}. \tag{5}$$

We then combined the volume and distance weighted priors to obtain a prior estimate of the mean contribution from source J at supply site I:

$$\alpha_{JI_{\mathrm{prior}}} = \frac{\alpha_{J\_\mathrm{volume}} \cdot \alpha_{JI\_\mathrm{distance}}}{\sum_{J=1}^{J=K} \alpha_{J\_\mathrm{volume}} \cdot \alpha_{JI\_\mathrm{distance}}}. \tag{6}$$

Table 1. Volumetric contribution of each source and their proportional fractional contribution (V) for each month from May to October 2015. Monthly supply data were provided by Jordan Valley Water Conservancy District. Values were rounded to 1 decimal place except for 45S Well in May 2015 and 90S Quail Well in October 2015.

May 2015			June 2015			July 2015		
Source	Volume of water supplied (in m^3)	Percent of total volume supplied (V)	Source	Volume of water supplied (in m^3)	Percent of total volume supplied (V)	Source	Volume of water supplied (in m^3)	Percent of total volume supplied (V)
Provo River (JVTP)	3.7×10^6	60.6	Provo River (JVTP)	13×10^6	80.5	Provo River (JVTP and SETP)	16.8×10^6	86.2
Wasatch creeks (SETP)	1.5×10^6	24.3	Wasatch creeks (SETP)	1.5×10^6	9.2	SWTP	0.5×10^6	2.6
SWTP	0.4×10^6	6.5	SWTP	0.6×10^6	3.8	64S/70S Well	0.5×10^6	2.8
Solena Way Well	0.3×10^6	4.6	64S/70S Well	0.4×10^6	2.4	Siesta Well	0.4×10^6	1.9
64S Well	0.2×10^6	3.0	Solena Way Well	0.2×10^6	1.0	18E Well	0.3×10^6	1.6
45S Well	0.06×10^6	1.0	Minor sources	0.5×10^6	3.1	Monitor Well	0.3×10^6	1.5
						Minor sources	0.6×10^6	3.3

August 2015			September 2015			October 2015		
Source	Volume of water supplied (in m^3)	Percent of total volume supplied (V)	Source	Volume of water supplied (in m^3)	Percent of total volume supplied (V)	Source	Volume of water supplied (in m^3)	Percent of total volume supplied (V)
Provo River (JVTP and SETP)	10.6×10^6	86.2	Provo River (JVTP and SETP)	10.6×10^6	92.1	Provo River (JVTP and SETP)	6.4×10^6	90.0
SWTP	0.7×10^6	5.4	SWTP	0.3×10^6	2.7	SWTP	0.4×10^6	6.1
64S/70S Well	0.5×10^6	3.9	64S/70S Well	0.3×10^6	2.8	64S/70S Well	0.2×10^6	3.2
90S Well	0.3×10^6	2.3	90S Well	0.2×10^6	1.4	90S Quail Well	0.05×10^6	0.7
Minor sources	0.3×10^6	2.2	90S Quail Well	0.1×10^6	0.9			

For example, if there were three sources supplying 3000, 1500, and 1500 m^3 of water, respectively, to the JVD system that were located 4, 1, and 10 km away from supply site I, then the Dirichlet prior vector would be {0.3125, 0.625, 0.0625} for this site I.

Based upon the above-described method, the prior contributions of selected sources at the distribution sites for June 2015, in spatial and isotope space, are shown in Figs. 2 and 3, respectively. The prior values were then updated using the observed isotope values at the distribution sites $\left(\mu\delta^2\mathbf{H}_I, \delta\mu^{18}\mathbf{O}_I\right)$ to obtain the posterior values of $f's$. We estimated the posterior fractional (f) contributions to each site I from each source J using the JAGS software package (Plummer, 2003), which can be integrated in the R statistical language using different R packages (Plummer, 2013; Denwood, 2016). We ran three parallel Markov chain Monte Carlo (MCMC) simulations for 300 000 iterations per chain, which were thinned every 50 steps. The first 40 000 iterations were discarded as burn-ins, providing us with 5200 samples for calculating the posterior statistics. We checked the convergence using the coda package (Plummer et al., 2006) and Gelman diagnostics (Gelman et al., 2014). All statistical analysis was performed in R (R core team, 2018).

2.5 Model results interpretation and cross-validation

For qualitative interpretation and to identify spatiotemporal variations in the association between sources and distribution sites, we considered any distribution site that our mixing analysis suggested was receiving more than 70 % (mean contribution) of its water from a single source to be supplied predominantly by that source. Sites where the analysis suggested less than 70 % water came from a single source were considered to receive water from multiple sources.

For each month, we compared the fractional production volume of each source with the fraction of the service area that our analysis suggested was served by the source. We calculated the areal contribution of the different sources for each month as a cross-check of the results obtained by BIMM. As a first-order approximation, we expected strong agreement between volumetric and areal contribution of a source as the area supplied by a source should be proportional to its volumetric supply. To calculate the areal coverage of a given source, we first calculated the area of influence (A_I) of each site on the transmission line, defined as the area of the Thiessen polygon associated with the site. For each source J, values of A_I were multiplied by the mean fractional contribution from that source ($f_{\mathrm{I},J}$). The resulting values were summed across all distribution sites ($\sum A_\mathrm{I} X f_{\mathrm{I},J}$) and divided by the total area of Jordan Valley District supply region to obtain the areal coverage (A) of that source.

3 Results and discussion

3.1 Sources and distribution sites' isotope ratios

Source water isotope values, measured across all months, ranged from -16.6 ‰ to -14.9 ‰ for δ^{18}O and -122.5 ‰ to -114.1 ‰ for δ^2H. Four sources (JVTP, SETP, SWTP, and well 64S) operated for the entire sampling period, and other sources operated intermittently. For each month, approximately 90 % (or more) of the water was supplied by the three WTPs (JVTP, SETP and SWTP), with the majority being supplied by JVTP (Table 1). Groundwater wells situated on the eastern side of the valley contributed approximately 10 % of the total water supplied each month, with well 64S supplying 1–3 % of the total volume each month.

The isotope values of SWTP and well 64S were distinct from each other and from those of JVTP and SETP for all months (Fig. 4). Isotope ratios of JVTP and SETP water were distinct (Hotelling multivariate t test, $p<0.05$, $n=6$) for May and June 2015 only. From July 2015 onwards, water from the Provo River was used by both of these WTPs; therefore, similar isotope ratios were expected. Well 64S had the lowest isotope ratios measured for any source and exhibited the highest deuterium excess values ($\sim$ 10 ‰) among all the wells, where deuterium excess is defined as δ^2H–8δ^{18}O (Dansgaard, 1964). Values from SWTP, in contrast, showed evidence for evaporative isotope effects, with high δ^{18}O values and relatively low deuterium excess ($\sim$ 4.2 ‰). JVTP isotope ratios increased from June to October 2015, as did SETP isotope ratios from July to October 2015, which can be due to evaporative enrichment of the heavy isotopes in upstream reservoirs of the Provo River system from spring to fall (mean d excess for JVTP decreased from 5.1 ‰ in June 2015 to 3.9 ‰ in October 2015; see Table S1 in the Supplement for monthly deuterium excess values).

The most and least negative isotope values of water from distribution sites were similar to the values observed for well 64S/70S and SETP, respectively. With the exception of a few sites in the May 2015 survey, distribution site isotope ratios fell within the convex hull defined by the source waters (Fig. 4). For each month, a number of distribution sites exhibited values similar to JVTP (Fig. 4). Clustering of supply site values was also observed near well 64S and SETP source values. At no point during the study did we observe any distribution sites with isotope values similar to those of SWTP source water (Fig. 4). For all months except October, approximately 20 %–30 % of the supply site values did not cluster near any major source, but rather were situated between sources. This pattern is consistent with expectations for mixing of water from multiple sources within this PWSS.

3.2 Source contributions at the distribution sites

We first illustrate the implementation of the BIMM for June 2015 (Fig. 5). Our model builds upon the work of

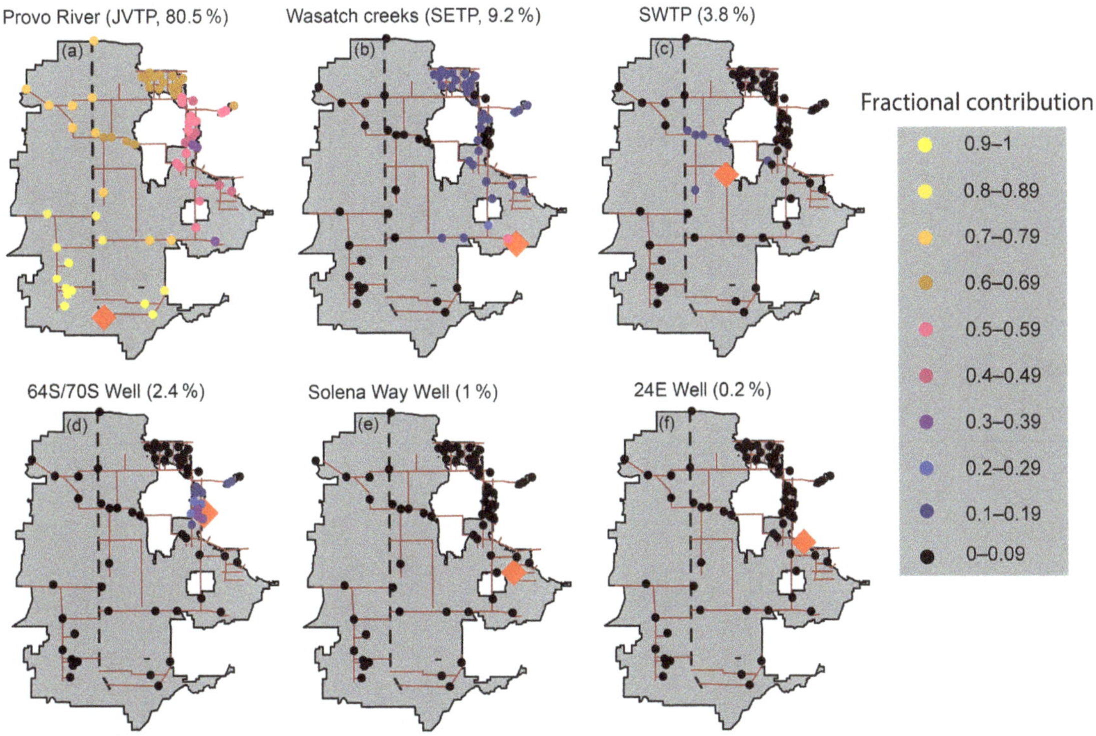

Figure 2. Mean prior contribution of selected sources at the distribution sites for June 2015 based upon Eq. (6) described in Sect. 2.4. Distribution sites are shown as circles, and the color reflects the assigned prior contribution from the different sources. The source location is shown as a red diamond in each panel. The name of each source and its percent volumetric contribution (V) is shown above each panel.

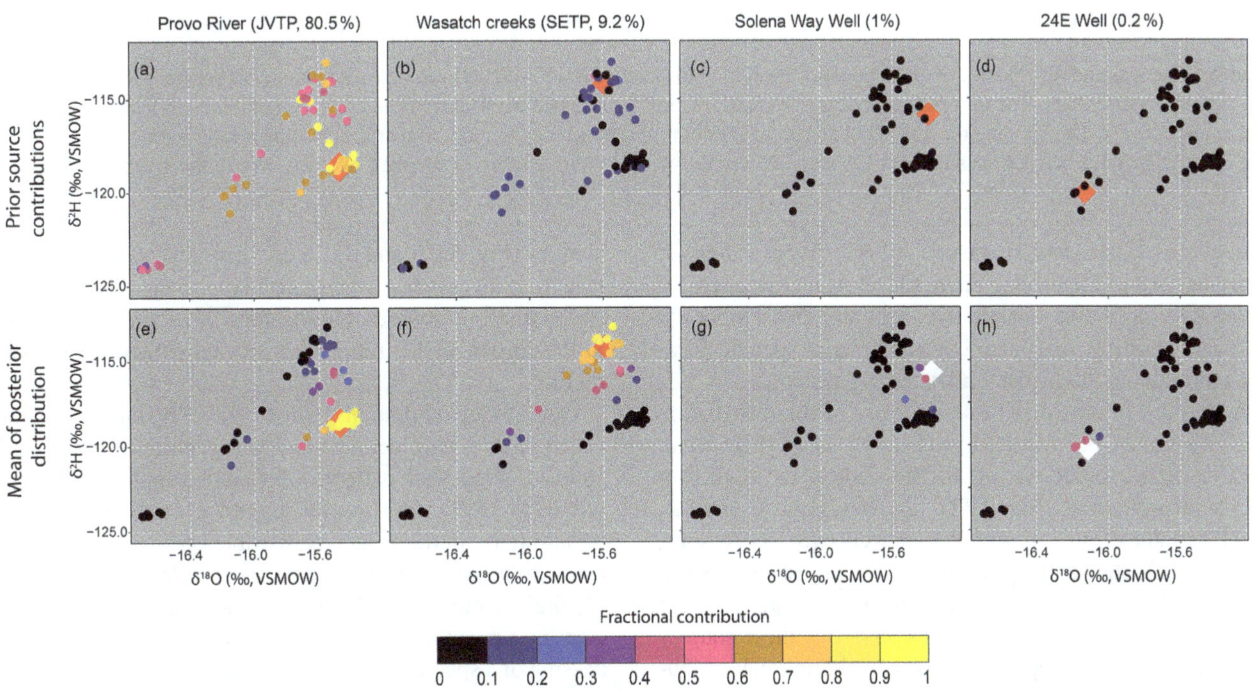

Figure 3. Prior contribution of selected sources at the distribution sites **(a–d)** and mean posterior contribution of selected sources at distribution sites **(e–h)** in isotope space for June 2015. Red diamonds represent sources and the circles represent distribution sites. For clarity, diamonds in **(a)** through **(h)** have been enlarged and in **(g)** and **(h)** are shown in white. The name of each source and its percent volumetric contribution (V) is shown above each panel.

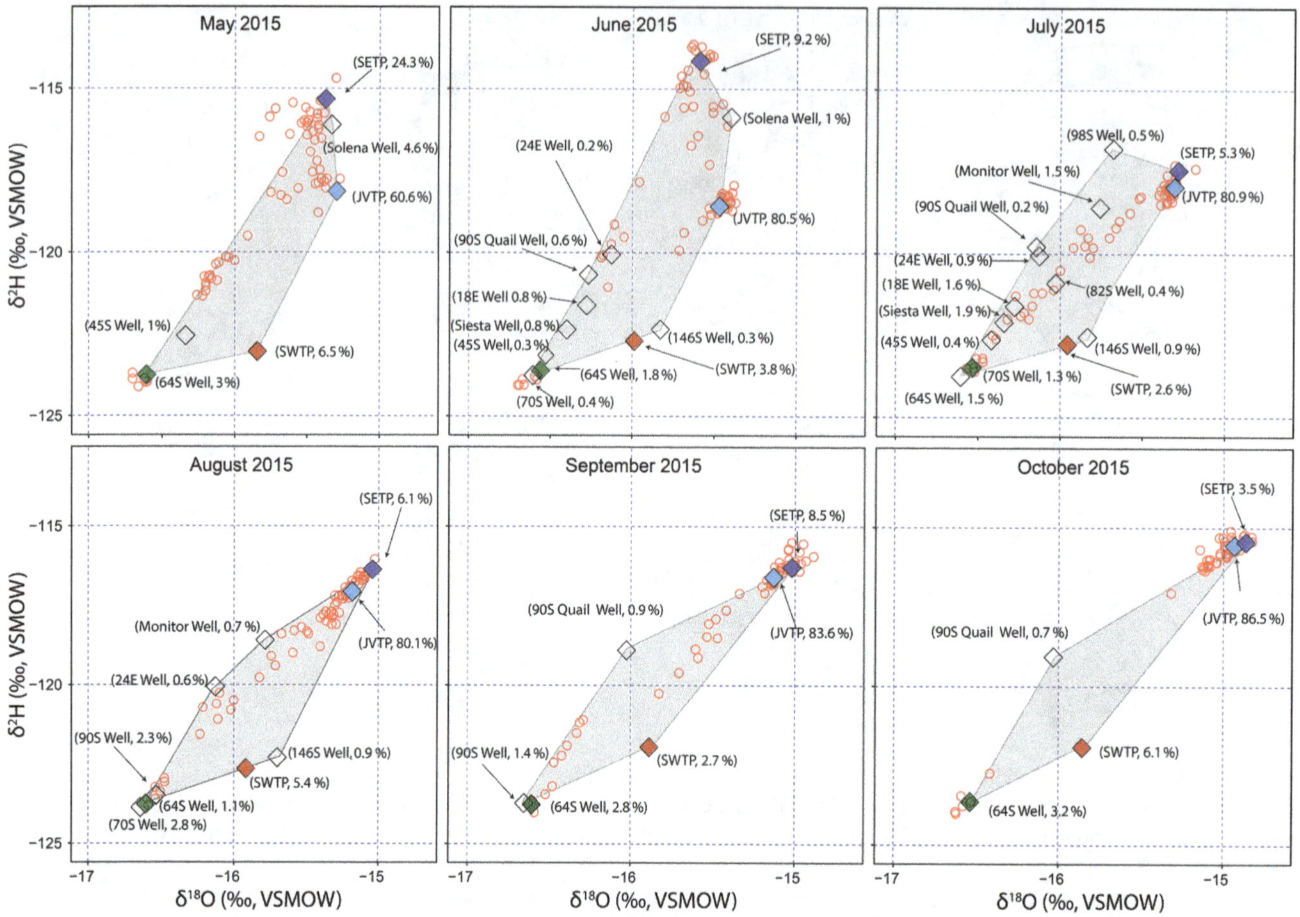

Figure 4. Sources and distribution site isotope ratios from May to October 2015. Red hollow circles and diamonds represent distribution sites and sources, respectively. The four major sources (JVTP, SETP, SWTP, and 64S Well) have been colored light blue, dark blue, orange, and green, respectively. The grey region is the convex hull of the sources (defined as the minimum area enclosing all the source isotope values). The percentage value besides the name of the source represents the total volume of water supplied by the source in the given month.

Jameel et al. (2016) and Tipple et al. (2017), but goes beyond their analyses of identifying district level water management patterns by providing quantitative, spatially and temporally resolved estimates of source contributions at specific locations throughout the Jordan Valley supply system.

According to our model, the majority of the distribution sites (45 out of 65) received most (>70 %) of their water from a single source. At all of these sites, the dominant source identified was either JVTP (24 sites, Figs. 3e and 5a), SETP (15 sites, Figs. 3f and 5b), or well 64S/70S (6 sites, Fig. 5d). This shows that three of the four largest sources operating at the time dominated the supplies of a large number of sites and that the number of sites served by these sources was approximately proportional to the volumetric contribution from each source. Our analysis suggests that the remaining 20 sites did not receive water predominantly from a single source, but had contributions from multiple sources.

Most of the sites receiving large proportional contributions from JVTP, SETP, and well 64S/70S were located on the transmission lines known to be directly connected to these sources (Fig. 5a, b and d). In contrast, distribution sites distant from all sources were more likely to exhibit mixing between multiple sources. During June 2015, all but three sites in the Jordan Valley retail area showed evidence for source mixing.

Our model output, in context with known physical infrastructure (i.e., pipelines) and geographic locations of the sources, suggested patterns of source–supply connectivity within the JVD. Our results suggest (1) subtle differences in mixing proportions among distribution sites receiving water mainly from the two largest sources (JVTP and SETP), (2) limited mixing at distribution sites located on transmission lines receiving water from multiple sources, and (3) bypassing of a specific transmission line during water transport. Below, we discuss each of these observations in more detail.

For sites on the western portion of the Jordan Valley District, the model-inferred mean JVTP contributions were uniformly large (>90 %), suggesting that JVTP was likely the dominant source supplying water to these sites (Fig. 5a). This was expected, as most of these sites have limited connectiv-

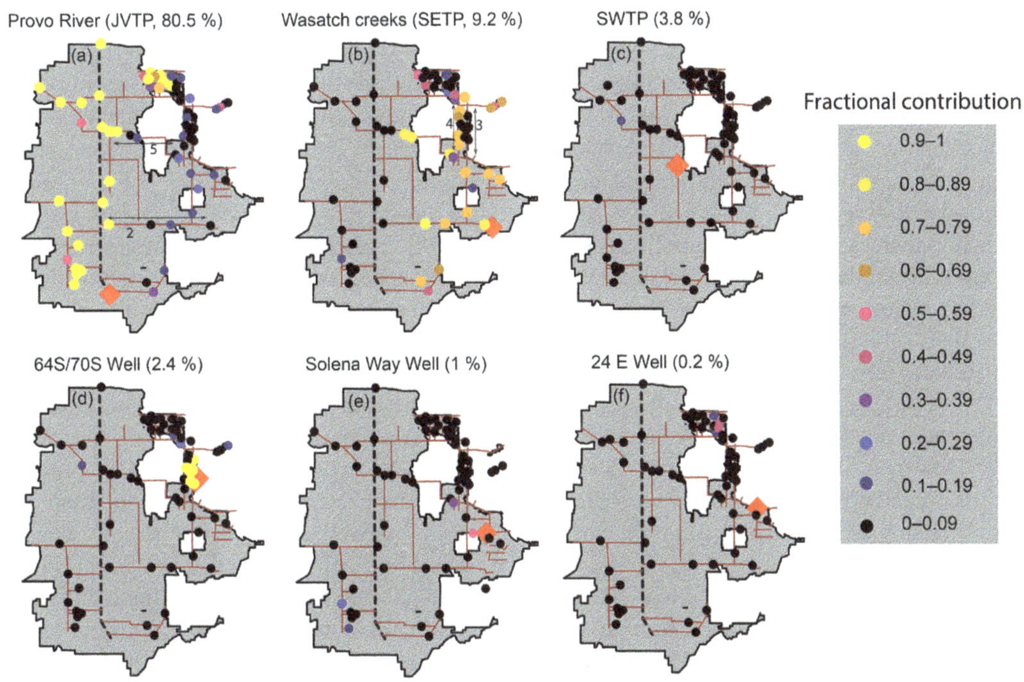

Figure 5. Mean of the posterior contribution of selected sources at the distribution sites for June 2015. Distribution sites are shown as circles and the color reflects the mean of the posterior contribution from the respective source at that site. The source in each panel is shown as a red diamond. The name of the source and its percent volumetric contribution (V) is shown above each panel. Transmission lines 2 and 5 are shown in **(a)** and lines 3 and 4 are shown in **(b)**.

ity to other sources apart from the SWTP. In contrast, our model suggested that most distribution sites receiving water predominantly from the SETP had mean contributions of SETP waters of 70 %–90 % (Fig. 5b). This likely reflects minor contributions of water from JVTP and several smaller sources in close proximity to SETP (Fig. 5a, e and f), implying that, although these sites are served chiefly by a single source, they also receive a significant fraction of water from other sources and thus are exposed to any supply or contamination issues associated with those minor sources. According to our model, sites receiving more than 60 % of water from SETP had an average contribution of 12 % from the minor sources with some sites receiving as much as 27 % of water from the minor sources.

Our model suggested limited mixing between JVTP, SETP, and other minor sources for distribution sites located on transmission lines 2 and 5 (Fig. 5a) that could receive water from all of these sources. Distribution sites on these lines mainly received water either from JVTP or SETP (more than 70 %), with contributions from other sources generally less than 30 % (Fig. 5a and b). Considering that JVTP supplied more than 80 % of all water in June 2015, we expected mixing and a large contribution from the JVTP at sites along these transmission lines (lines 2 and 5). One factor that may have caused limited mixing between these sources within the supply lines is the higher elevation of SETP (1532 m) and other minor sources on the eastern side of the valley compared to JVTP (1424 m). We suggest that the higher gravitational potential energy of water introduced from SETP and minor sources may create a pressure differential that limits mixing between these two sources; however this remains a hypothesis to be tested.

Our model suggested negligible presence of SETP water in transmission line 3 (<15 %), whereas the mean contribution of SETP in a closely running parallel transmission line 4 was high (>60 %, Fig. 5b). This result implies that water moving northward from SETP bypassed line 3. This is most likely due to the presence of well 64S/70S (Fig. 5d) on line 3, which our results suggested was the principal source to all the sites on line 3. This highlights the ability of the isotope mixing model to capture small-scale interactions between sources and supply connections.

The BIMM presented here is blind to the actual physical connections in the Jordan Valley District service area. Nonetheless, our results closely match the specific linkages between sources and distribution sites along known transmission lines. The ability of BIMM to identify patterns of source–supply connectivity within this system suggests the potential to use similar isotope-based methods to obtain information from less documented public water supply systems.

3.3 Assessment of uncertainties and model limitations

In addition to providing point estimates of source water contributions, the BIMM also provides estimates of uncertainty. To analyze the uncertainties, we divided the isotope values of the distribution sites measured in June 2015 into three groups. Group 1 consisted of sites with isotope values similar to one of the major sources (Fig. 6a), group 2 consisted of sites with isotope values in-between the SETP and JVTP endmembers (Fig. 6b), and group 3 consisted of distribution sites with isotope values similar to one of the minor sources but significantly different from any of the major sources (Fig. 6c).

According to our model, all sites in group 1 had large contributions (>70 %) from one of the major sources (JVTP, SETP or 64S/70S wells), and we observed narrow 95 % credible intervals (CIs, ranging mostly from 0.6 to 1) for the proportional contributions from these sources (Fig. 7a and b). At these sites the CIs for other sources were also narrow and ranged from 0.0 to 0.3 (Fig. 7c–h), indicating high levels of confidence that other sources were minor contributors to these sites. The effectiveness of BIMM in providing tight and robust posterior distributions for group 1 sites is due to the strong similarity between source and distribution site isotope values in this group and the distinct isotope ratios of water from these sources relative to all others.

For group 2 sites, the model predicted mixing primarily between water from the JVTP and the SETP (Fig. 7a and b), with contributions of 40 % to 60 % from each of these sources. According to our model, the contribution from the Solena Way Well was minimal at all group 2 sites except for a single site situated very close to the well (Fig. 5e). Given that the Solena Way Well contributed only 1 % of the total water to the system, dominance of JVTP and SETP water at most group 2 sites is reasonable. Further, most of the group 2 sites were situated in the Jordan Valley retail area, far from the Solena Way Well (>5 km, Fig. 5a, b and e). The CIs associated with different sources at these sites were larger than group 1 sites. Most of these sites had CIs ranging from 0.0 to 0.6 for JVTP (Fig. 7a), from 0.3 to 0.6 for SETP (Fig. 7b), and from 0.0 to 0.6 for the Solena Way Well (Fig. 7g). The tighter CIs of contributions from SETP compared to JVTP and Solena Way Well at these sites suggests that our model is more confident about the contribution from SETP (i.e., between 30 % and 60 %) compared to contributions from JVTP and Solena Way Well. As observed for group 1, the CIs associated with other sources were small, in general, ranging from 0.0 to 0.2. The advantages of including distance and volume effects in our model were reflected in this group, as our model preferred mixing between water from the JVTP and the SETP over possible mixtures between water from the Solena Way Well and other minor sources.

Group 3 exhibited isotope values that were distinct from all major sources and were similar to one or more minor sources. According to the model, no source (major and minor included) contributed more than 50 % (mean) at these sites. In general, for a given supply site in this group, our model assigned the highest mean contribution to the minor source with isotope ratios most similar to those of the supply site water (for example see Fig. 5e). The CIs associated with proportional contributions from the different sources were large, however, and for some sources ranged from 0.0 to 0.9 (Fig. 7h). This suggests that more than one possible source or multi-source mixture was consistent with the isotopic and prior constraints for these sites, resulting in identifiability issues that are commonly observed in isotope mixing models (Cable et al., 2011; Erhardt and Bedrick, 2013; Parnell et al., 2013). In our case, non-unique assignments for group 3 sites arose due to the presence of multiple sources with comparable isotope values near the distribution sites and also due to several probable potential mixing solutions between SETP, the 64S/70S Well, and these minor wells. The issue was compounded further by similar and low prior probabilities associated with the minor sources, making it difficult for the model to identify one distinct source as a major contributor.

Our results highlight the robustness as well as the limitations of our model. Both the use of informative priors and the comprehensive assessment and interpretation of uncertainty are likely to improve the quality of inferences drawn from our method. A key outcome of the priors specified here is that volumetrically minor sources were not identified as a major contributor to distribution sites, even though in many cases they had similar isotope values, except in cases where proximity provided additional evidence suggesting that they were likely sources. This result was also observed in July and August 2015. Consideration of credible intervals estimated in the analysis shows substantial and interpretable variation in the confidence of source water estimates among different sites. Even in cases where relatively high mean source contributions were assigned to a given site, robustness in the model solutions can be recognized through review of credible intervals and used to more accurately interpret these results.

3.4 Spatiotemporal variations in source contributions

We extended our analysis to all months from May to October 2015, to assess changes in the patterns of water distribution as water demand, source types, and production volumes changed through the sampling period (Table 1).

Mixing between sources was high in May, with only 25 % of the distribution sites receiving more than 70 % of their water from a single source (Fig. 8a). For May, most of the distribution site values were intermediate to the source water values (Fig. 4), clearly indicating substantial mixing across most parts of the distribution system. A handful of supply site samples in May also fell outside of the convex hull defined by the sources, suggesting that our sampling may not have captured all contributing sources, but the conclusion of pervasive mixing is not likely to be affected by this omission. In contrast, our model suggests that almost 70 % of the sites were sup-

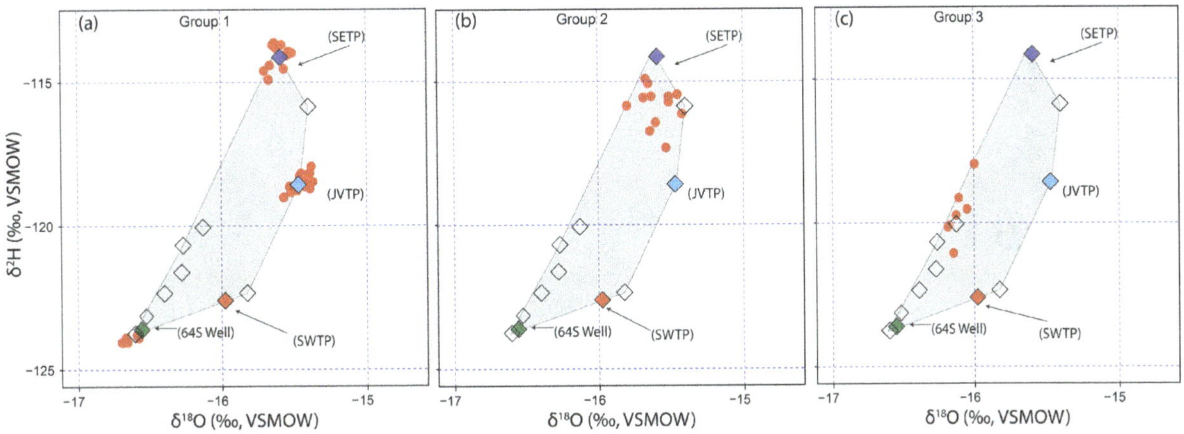

Figure 6. Distribution sites and sources for June 2015 shown as red circles and diamonds. The four major sources (JVTP, SETP, SWTP, and 64S Well) have been colored light blue, dark blue, orange, and green, respectively, and are labeled. Minor sources are shown as hollow diamonds. **(a)** group 1, **(b)** group 2, and **(c)** group 3 distribution sites.

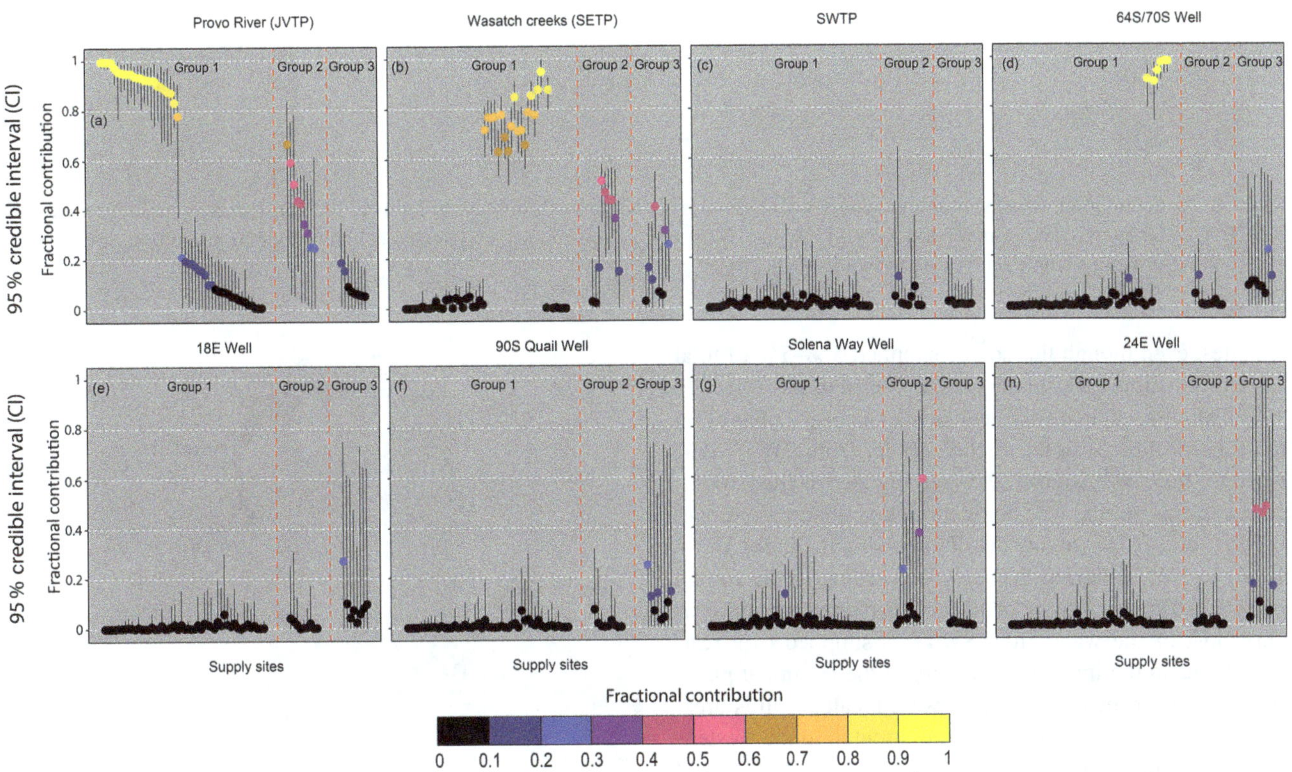

Figure 7. Mean (circle) and 95 % credible interval (vertical black lines) associated with the source contributions at the distribution sites for the different groups for June 2015. Sites in **(a)** have been sorted with decreasing contribution from JVTP for each group. The same sorting order is maintained for **(b–h)**.

plied chiefly by a single source in June and July, with this value increasing to more than 75 % in August and September (Fig. 8b–e). By October, the supply system had transitioned to a single, major source, and our results showed no significant mixing between sources for that month (Fig. 8f). Except for May 2015, where we observed large-scale mixing between different sources throughout the Jordan Valley District, distribution sites receiving water from multiple sources were limited mostly to the Jordan Valley retail area. Since this area is distant from all major sources and is surrounded by multiple transmission lines, mixing observed at the distribution sites is not surprising.

Perhaps the most surprising part of our analysis was our inability to detect contributions from SWTP at the distribu-

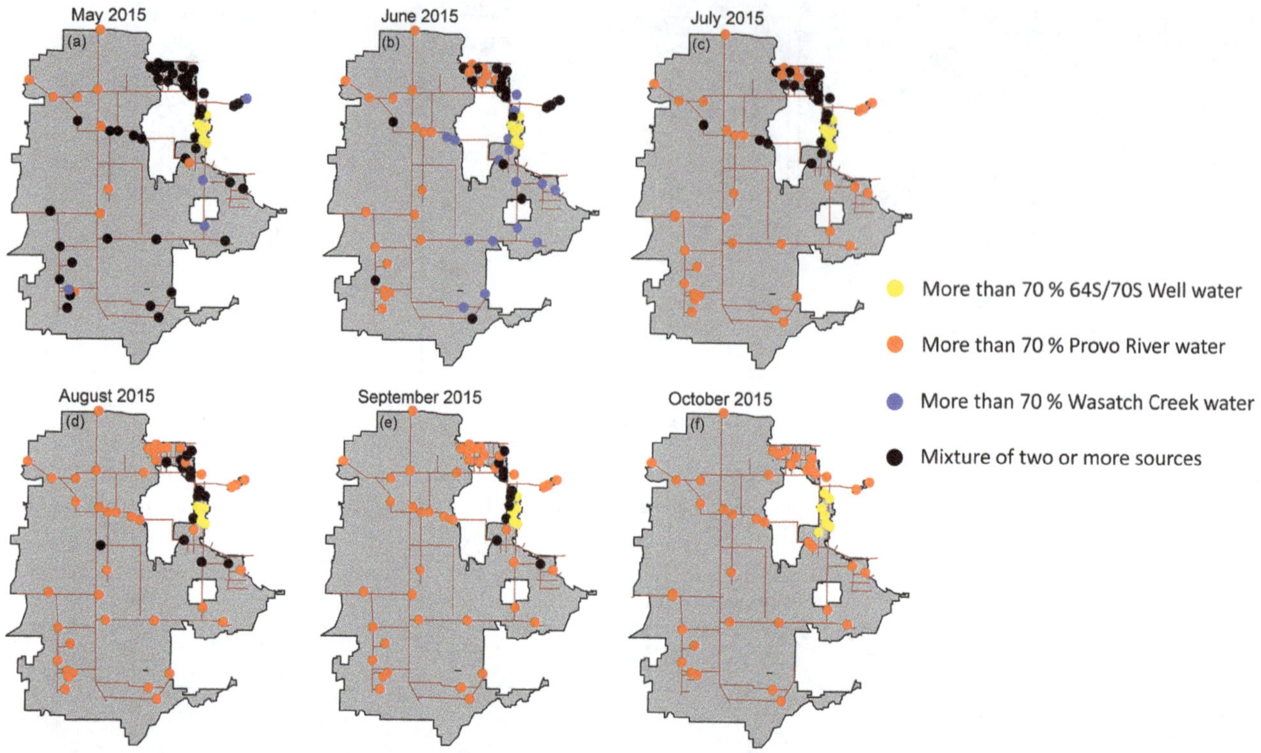

Figure 8. Spatiotemporal variation in sources and distribution sites connectivity from May to October 2015. Distribution sites receiving more than 70 % water from a single source are shown in orange, blue, and yellow circles, and sites receiving water from multiple sources (less than 70 % water from a single source) are shown in black circle.

tion sites, even though this source supplied 3 %–5 % of total water production each month. Small contributions (10 % to 20 %) from this source were indicated at a couple of sites on transmission line 5, situated relatively far from SWTP, during June, July, and August 2015 (see Fig. 5c for June 2015). However, this source was not identified as the predominant source (i.e., >70 %) at any distribution site, including those closest to SWTP, during the study. According to the Jordan Valley District operations managers (personal communication) most of the water from SWTP is supplied to a residential area in the immediate vicinity of the treatment plant, and none of our distribution sites were located in this area. A small fraction of the SWTP water is routed to the western part of the Jordan Valley District, which is possibly reflected in our results suggesting minor contributions from this source to distribution sites along distribution line 5.

We combined our model output for different months to highlight variability and quantify the mean contribution for each source at the different distribution sites from May to October 2015 (Fig. 9). Our result suggests that most of the sites received water from multiple sources or switched sources during our analysis period, with the exception of a few sites receiving Provo River and 64S/70S Well water for all the 6 months. Our results show significant changes throughout the sampling period, highlighting the complex and dynamic operation of the distribution system. We have developed the monthly (Fig. 8) and 6-month averaged (Fig. 9) contribution from the different sources at the distribution sites based upon one to three samples collected each month; however, such maps can be developed at varying spatiotemporal scales depending upon the purpose and application of the method.

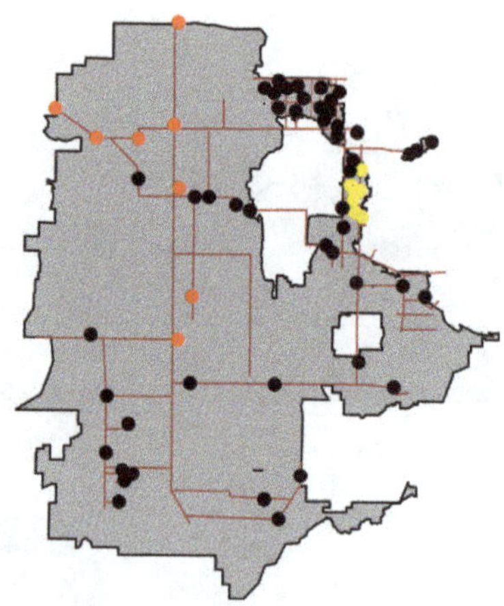

Figure 9. Mean contribution of different sources at the distribution sites during May to October 2015. Sites in orange and yellow circles received water primarily (>70 %) from the Provo River and 64S/70S Well, respectively, throughout the entire sampling period. Sites in black circles received water from multiple sources or switched sources at least once during the sampling period.

Table 2. Comparison between volumetric (V) and areal (A) contributions of the different sources from May to October 2015. V for each source was calculated from the monthly supply data of each source (see Table 1). A was calculated using methods described in Sect. 2.5. JVTP and SETP are considered as separate sources in May and June 2015 and combined sources from July to October 2015. Sources contributing less than 1 % of the total volume have been grouped together as "minor sources" for June, July, and August 2015. All values are in percent.

May 2015			June 2015			July 2015		
Source	V	A	Source	V	A	Source	V	A
Provo River (JVTP)	60.6	47.7	Provo River (JVTP)	80.5	63.9	Provo River (JVTP and SETP)	86.2	87.1
Wasatch creeks (SETP)	24.3	30.2	Wasatch creeks (SETP)	9.2	27.3	SWTP	2.6	1.7
SWTP	6.5	5.7	SWTP	3.8	2.7	64S/70S Well	2.8	2.9
Solena Way Well	4.6	6.4	64S/70S Well	2.4	2.4	Siesta Well	1.9	1.8
64S Well	3.0	7.1	Solena Way Well	1.0	0.8	18E Well	1.6	1.4
45S Well	1.0	2.8	Minor sources	3.1	2.9	Monitor Well	1.5	1.6
						Minor sources	3.3	3.2
August 2015			**September 2015**			**October 2015**		
Source	V	A	Source	V	A	Source	V	A
Provo River (JVTP and SETP)	86.2	86.9	Provo River (JVTP and SETP)	92.1	91.3	Provo River (JVTP and SETP)	90.0	94.3
SWTP	5.4	4.5	SWTP	2.7	2.3	SWTP	6.1	1.7
64S/70S Well	3.9	2.8	64S/70S Well	2.8	3.4	64S/70S Well	3.2	3.1
90S Well	2.3	2.2	90S Well	1.4	1.9	90S Quail Well	0.7	0.8
Minor sources	2.2	3.4	90S Quail Well	0.9	1.1			

3.5 Model cross-validation

To validate the results obtained by BIMM we compared the volumetric contribution of the sources obtained from the Jordan Valley District (Table 1), with their areal contribution estimated using the BIMM. Volumetric and areal contributions were strongly and systematically correlated across all sources (Table 2). However, our model predicted that the Wasatch creeks supplied a larger fraction of the area than suggested by their volumetric contribution and that the Provo River sources supplied a smaller area than implied by volumetric production numbers in May and June (Table 2). This discrepancy could reflect differences in water demand across the service area, although most of the area our analysis suggests was served by the Wasatch creeks source is heavily populated, and the corridor served by the Provo River water source includes more industrial development that may overconsume water per unit area. Nonetheless, the overall similarity between the areal coverage estimated here and reported volumetric production numbers provides an additional line of evidence supporting the robustness of the BIMM.

3.6 Model improvements and future application of BIMM in other urban water systems

We have shown here that BIMM provides robust estimates of the contribution from different sources to distribution sites within a PWSS. In our analysis, the isotopic compositions of major sources were distinct, allowing our model to quantify the contribution from the major sources at the distribution sites with robust estimates of uncertainty across the supply system. However, for distribution sites with isotope values intermediate to candidate sources (group 3, Fig. 6), the specificity of our result was limited by non-unique solutions. These challenges and limitations could be addressed with the inclusion of other conservative tracers such as chloride, calcium, and strontium (and its stable isotopes) that might vary significantly between the different sources, thus providing additional constraints and improved model predictions. Additional system data, such as pressure, elevation gradients, and flow velocity within the system, might also be included within the model to improve accuracy.

A key prerequisite for future successful implementation of the BIMM in other PWSSs is that all sources in the PWSS be characterized and have significantly distinct isotopic and/or geochemical signatures. In PWSSs with negligible isotopic and geochemical variability between the sources, the capacity of the BIMM to characterize the system would likely be limited, and it will provide results with limited practical applicability. Finally, the BIMM approach is sample based and an appropriate sampling design would be required to accurately connect sources and distribution sites and extract meaningful information from the analysis. The sampling design should consider factors such as source compositions, system operations, water residence time, water demand, and population density within the PWSS to develop a robust sampling strategy for implementing the BIMM. It is essential to capture temporal variations, especially for surface water sources or other sources with rapid water transit time, to establish accurate association between the sources and distribution sites. In our analysis, our monthly sampling protocol captured the successive isotope enrichment of the Provo River source that was vital to the success of our model.

The framework applied here can be useful in establishing source water footprints, pathways, and interactions of water sources within PWSS. In cities across the developed world that use hydrodynamic models (such as EPANET and WaterCAD) (Rossman, 2002) to predict water quality and contaminant concentration across their supply systems, the accuracy of these predictions can be evaluated by comparing the observed and predicted stable isotopes in water (or other conservative geochemical tracers) at several distribution sites using the hydrodynamic model. In many developing and rapidly growing cities across the world where applying hydrodynamic models is challenging and difficult a framework similar to that shown here can be used to develop GIS products such as (1) service maps of the different sources, (2) regions within the PWSS undergoing seasonal source switching, and (3) regions serviced by surface or groundwater, respectively. These products can be helpful in moderating water rights issues, tracking of source- and WTP-related contaminants, evaluating the susceptibility to climatic variations, and investigating long- and short-term effects of source water quality on public health. Considering that stable isotope analysis of most water samples is now rapid (minutes) and inexpensive, geochemically based BIMMs offer an attractive tool for studying and monitoring PWSS in support of management and water security.

4 Conclusions

Water isotopes have been used extensively to monitor and understand the natural component of the water cycle (Gat, 1996; Aggarwal et al., 2005; Bowen and Good, 2015); however their application in urban water systems has been limited. Recent work has shown the capacity of water isotopes to record information about water management and quantify effects of climatic variability on water resources (Jameel et al., 2016; Tipple et al., 2017). Moving beyond the coarse resolution of these studies, our work has highlighted the ability of water isotopes to provide information about PWSS operation at a much finer scale. Here, we have shown the ability of water isotopes to provide estimates of the contributions of multiple water sources across a large metropolitan PWSS and inform our understanding of the physical structure and operation of the system. The method developed here does not rely on independent information about pipe networks, flow velocities, pressure gradients, or other details of the PWSS that are integral to hydrodynamic models, and thus it can be used to interrogate PWSS where this information is lacking or to independently validate hydrodynamic model results. Our application used only two isotope (δ^2H and δ^{18}O) measurements, supplemented with information on source volumes and geographic locations. Future applications could improve upon our work by including additional geochemical tracers and flow rates, adding additional information on distribution system structure (where available), collecting samples with higher spatiotemporal resolution, and refining the statistical model.

Author contributions. YJ and GJB designed the project. YJ, SB, RPF and GJB developed the model. All authors contributed to interpreting the model results. YJ wrote the first draft of the paper. All authors contributed substantially to revisions and gave final approval for publication.

Competing interests. The authors declare that they have no conflict of interest.

Acknowledgements. We thank JVD for collaborating on this project and Heidi Nilsson for collecting the water samples. This work was supported by U.S. National Science Foundation grant 1208732.

Edited by: Laurent Pfister

References

Aggarwal, P. K., Froehlich, K. F., and Gat, J. R.: Isotopes in the water cycle, Springer, the Netherlands, 2005.

Bolar, A., Tesfamariam, S., and Sadiq, R.: Condition assessment for bridges: a hierarchical evidential reasoning (HER) framework, Struct. Infrastruct. Eng., 9, 648–666, 2013.

Boryczko, K. and Tchórzewska-Cieślak, B.: Analysis of risk of failure in water main pipe network and of delivering poor quality water, Environ. Prot. Eng., 40, 77–92, 2014.

Bowen, G. J. and Good, S. P.: Incorporating water isoscapes in hydrological and water resource investigations, WIREs Water, 2, 107–119, 2015.

Cable, J., Ogle, K., and Williams, D.: Contribution of glacier meltwater to streamflow in the Wind River Range, Wyoming, inferred via a Bayesian mixing model applied to isotopic measurements, Hydrol. Process., 25, 2228–2236, 2011.

Coplen, T. B.: Normalization of oxygen and hydrogen isotope data, Chem. Geol., 72, 293–297, 1988.

Dansgaard, W.: Stable isotopes in precipitation, Tellus A, 16, 436–468, https://doi.org/10.3402/tellusa.v16i4.8993, 1964.

Dawson, T. E. and Ehleringer, J. R.: Streamside trees that do not use stream water, Nature, 350, 335–337, 1991.

De Bénédittis, J. and Bertrand-Krajewski, J. L.: Measurement of infiltration rates in urban sewer systems by use of oxygen isotopes, Water Sci. Technol., 52, 229–237, 2005.

De Bondt, K., Seveno, F., Petrucci, G., Rodriguez, F., Joannis, C., and Claeys, P.: Potential and limits of stable isotopes ($\delta^{18}O$ and δD) to detect parasitic water in sewers of oceanic climate cities, J. Hydrol., 18, 119–142, 2018.

Denwood, M. J.: Runjags: An R package providing interface utilities, model templates, parallel computing methods and additional distributions for MCMC models in JAGS, J. Stat. Soft., 71, 1–25, 2016.

Ehleringer, J. R., Barnette, J. E., Jameel, Y., Tipple, B. J., and Bowen, G. J.: Urban water – A new frontier in isotope hydrology, Iso. Env. Health Stud., 52, 477–486, 2016.

Erhardt, E. B. and Bedrick, E. J.: A Bayesian framework for stable isotope mixing models, Environ. Ecol. Stat., 20, 1–21, 2013.

Gabor, R. S., Hall, S. J., Eiriksson, D. P., Jameel, Y., Millington, M., Stout, T., Barnes, M. B., Gelderloos, A., Tennant, H., Bowen, G. J., Neilson, B. T., and Brooks, P. D.: Persistent Urban Influence on Surface Water Quality via Impacted Groundwater, Environ. Sci. Technol., 51, 9477–9487, 2017.

Gat, J.: Stable isotopes of fresh and saline lakes, in: Physics and chemistry of lakes, Springer, Heidelberg, 139–165, 1995.

Gat, J.: Oxygen and hydrogen isotopes in the hydrologic cycle, Annu. Rev. Earth Planet. Sci., 24, 225–262, 1996.

Gelman, A., Carlin, J. B., Stern, H. S., Dunson, D. B., Vehtari, A., and Rubin, D. B.: Bayesian data analysis, CRC press, Boca Raton, 2014.

Good, S. P., Mallia, D. V., Lin, J. C., and Bowen, G. J.: Stable isotope analysis of precipitation samples obtained via crowdsourcing reveals the spatiotemporal evolution of superstorm sandy, PloS One, 9, e91117, https://doi.org/10.1371/journal.pone.0091117, 2014.

Hanna-Attisha, M., LaChance, J., Sadler, R. C., and Schnepp, A.: Elevated blood lead levels in children associated with the Flint drinking water crisis: a spatial analysis of risk and public health response, Amer. J. Pub. Health, 106, 283–290, 2016.

Hoff, P. D.: A first course in Bayesian statistical methods, Springer, New York, 2009.

Hydroshare: https://www.hydroshare.org/, Search: First Author: Yusuf Jameel, last access: 20 April 2018.

Ismail, M. A., Sadiq, R., Soleymani, H. R., and Tesfamariam, S.: Developing a road performance index using a Bayesian belief network model, J. Frankl. I., 348, 2539–2555, 2011.

Jameel, Y., Brewer, S., Good, S. P., Tipple, B. J., Ehleringer, J. R., and Bowen, G. J.: Tap water isotope ratios reflect urban water system structure and dynamics across a semiarid metropolitan area, Water Resour. Res., 52, 5891–5910, 2016.

Jameel, Y., Stein, S., Grimm, E., Roswell, C., Wilson, A. E., Troy, C., Höök, T. O., and Bowen, G. J.: Physicochemical characteristics of a southern Lake Michigan river plume, J. Great Lakes Res., 4, 209–218, 2018.

Jefferson, A. J., Bell, C. D., Clinton, S. M., and McMillan, S. K.: Application of isotope hydrograph separation to understand contributions of stormwater control measures to urban headwater streams, Hydrol. Process., 29, 5290–5306, 2015.

Kabir, G., Tesfamariam, S., Francisque, A., and Sadiq, R.: Evaluating risk of water mains failure using a Bayesian belief network model, Euro. J. Oper. Res., 240, 220–234, 2015.

Kaushal, S. S.: Increased salinization decreases safe drinking water, Environ. Sci. Tech., 50, 2765–2766, 2016.

Kennedy, C. D., Bowen, G. J., and Ehleringer, J. R.: Temporal variation of oxygen isotope ratios ($\delta^{18}O$) in drinking water: implications for specifying location of origin with human scalp hair, Foren. Sci. Int., 208, 156–166, 2011.

Klaus, J. and McDonnell, J.: Hydrograph separation using stable isotopes: review and evaluation, J. Hydrol., 505, 47–64, 2013.

Kracht, O., Gresch, M., and Gujer, W.: A Stable Isotope Approach for the Quantification of Sewer Infiltration, Environ. Sci. Technol., 41, 5839–5845, 2007.

Larsen, T. A., Hoffmann, S., Lüthi, C., Truffer, B., and Maurer, M.: Emerging solutions to the water challenges of an urbanizing world, Science, 352, 928–933, 2016.

Liggett, J. A. and Chen, L. C.: Inverse transient analysis in pipe networks, J. Hydraul. Eng., 120, 934–955, 1994.

Mailloux, J. M., Ogle, K., and Frost, C. D.: Application of a Bayesian model to infer the contribution of coalbed natural gas produced water to the Powder River, Wyoming and Montana, Hydrol. Process., 28, 2361–2381, 2014.

Matheny, A. M., Fiorella, R. P., Bohrer, G., Poulsen, C. J., Morin, T. H., Wunderlich, A., Vogel, C. S., and Curtis, P. S.: Contrasting strategies of hydraulic control in two codominant temperate tree species, Ecohydrology, 10, https://doi.org/10.1002/eco.1815, 2017.

Najjaran, H., Sadiq, R., and Rajani, B.: Condition assessment of water mains using fuzzy evidential reasoning, paper presented at Systems, Man and Cybernetics, 2005 IEEE Intern. Conf., Waikoloa, Hi, USA, 3466–3471, https://doi.org/10.1109/ICSMC.2005.1571684, 2005.

Ogle, K. and Barber, J. J.: Bayesian data – model integration in plant physiological and ecosystem ecology, Progress in Botany, Springer, Heidelberg, https://doi.org/10.1007/978-3-540-72954-9_12, 281–311, 2008.

Parnell, A. C., Inger, R., Bearhop, S., and Jackson, A. L.: Source partitioning using stable isotopes: coping with too much variation, PloS One, 5, e9672, https://doi.org/10.1371/journal.pone.0009672, 2010.

Parnell, A. C., Phillips, D. L., Bearhop, S., Semmens, B. X., Ward, E. J., Moore, J. W., Jackson, A. L., Grey, J., Kelly, D. J., and Inger, R.: Bayesian stable isotope mixing models, Environmetrics, 24, 387–399, 2013.

Pietrucha-Urbanik, K.: Failure analysis and assessment on the exemplary water supply network, Eng. Fail. Anal., 57, 137–142, 2015.

Plummer, M.: JAGS: A program for analysis of Bayesian graphical models using Gibbs sampling, in Proceedings of the 3rd international workshop on distributed statistical computing, edited, p. 125, Vienna, Austria, 2003.

Plummer, M.: Rjags: Bayesian graphical models using MCMC, R package version, 3, 2013.

Plummer, M., Best, N., Cowles, K., and Vines, K.: CODA: convergence diagnosis and output analysis for MCMC, R news, 6, 7–11, 2006.

R Core Team.: R: A Language and Environment for Statistical Computing, R Found. for Stat. Comput., Vienna, Austria, 2018.

Rossman, L. A.: EPANET 2, User's Manual, Water supply and Water Resources Division, Nat. Risk Manag. Res. Lab., Cincinnati, USA, 2001.

Rozanski, K., Araguas-Araguas, L., and Gonfiantini, R.: Relation between long-term trends of oxygen-18 isotope composition of precipitation and climate, Science, 258, 981–985, 1992.

Schnoor, J. L.: Recognizing Drinking Water Pipes as Community Health Hazards, J. Chem. Educ., 93, 581–582, https://doi.org/10.1021/acs.jchemed.6b00218, 2016.

Tipple, B. J., Jameel, Y., Chau, T. H., Mancuso, C. J., Bowen, G. J., Dufour, A., Chesson, L. A., and Ehleringer, J. R.: Stable hydrogen and oxygen isotopes of tap water reveal structure of the San Francisco Bay Area's water system and adjustments during a major drought, Water Res., 119, 212–224, 2017.

van Geldern, R. and Barth, J. A.: Optimization of instrument setup and post-run corrections for oxygen and hydrogen stable isotope measurements of water by isotope ratio infrared spectroscopy (IRIS), Limnol. Oceanogr.-Meth., 10, 1024–1036, 2011.

von Grafenstein, U., Erlenkeuser, H., Brauer, A., Jouzel, J., and Johnsen, S. J.: A mid-European decadal isotope-climate record from 15 500 to 5000 years BP, Science, 284, 1654–1657, 1999.

Vörösmarty, C. J., McIntyre, P., Gessner, M. O., Dudgeon, D., Prusevich, A., Green, P., Glidden, S., Bunn, S. E., Sullivan, C. A., and Liermann, C. R.: Global threats to human water security and river biodiversity, Nature, 467, 555–561, 2010.

Waldrip, S. H., Niven, R. K., Abel, M., and Schlegel, M.: Maximum Entropy Analysis of Hydraulic Pipe Flow Networks, J. Hydraul. Eng., 142, 1–23, https://doi.org/10.1061/(ASCE)HY.1943-7900.0001126, 2016.

Waldrip, S. H., Niven, R. K., Abel, M., and Schlegel, M.: Reduced-Parameter Method for Maximum Entropy Analysis of Hydraulic Pipe Flow Networks, J. Hydraul. Eng., 144, 2333–2344, 04017060, 2018.

Wang, S., Zhang, M., Bowen, G. J., Liu, X., Du, M., Chen, F., Qiu, X., Wang, L., Che, Y., and Zhao, G.: Water source signatures in the spatial and seasonal isotope variation of Chinese tap waters, Water Resour. Res., 54, https://doi.org/10.1029/2018WR023091, 2018.

Waples, J. T., Bordewyk, J. K., Knesting, K. M., and Orlandini, K. A.: Using Naturally Occurring Radionuclides To Determine Drinking Water Age in a Community Water System, Envir. Sci. Tech., 49, 9850–9857, https://doi.org/10.1021/acs.est.5b03227, 2015.

Waterisotopes Database: http://waterisotopes.org, Query: Project ID: 00065, last access: 22 April 2018.

Yoo, D. G., Chung, G., Sadollah, A., and Kim, J. H.: Applications of network analysis and multi-objective genetic algorithm for selecting optimal water quality sensor locations in water distribution networks, KSCE J. Civ. Eng., 19, 2333–2344, 2015.

Zhao, S., Hu, H., Tian, F., Tie, Q., Wang, L., Liu, Y., and Shi, C.: Divergence of stable isotopes in tap water across China, Sci. Rep., 7, 43653, https://doi.org/10.1038/srep43653, 2017.

10

Effects of climatic seasonality on the isotopic composition of evaporating soil waters

Paolo Benettin[1], **Till H. M. Volkmann**[2], **Jana von Freyberg**[3,4], **Jay Frentress**[5], **Daniele Penna**[6], **Todd E. Dawson**[7], **and James W. Kirchner**[3,4,8]

[1]Laboratory of Ecohydrology ENAC/IIE/ECHO, École Polytechinque Fédérale de Lausanne (EPFL), Lausanne, Switzerland
[2]Biosphere 2, University of Arizona, Tucson, AZ, USA
[3]Department of Environmental Systems Science, ETH Zurich, Zurich, Switzerland
[4]Research Unit Mountain Hydrology and Mass Movements, Swiss Federal Institute for Forest, Snow and Landscape Research (WSL), Birmensdorf, Switzerland
[5]Faculty of Science and Technology, Free University of Bolzano, Bolzano, Italy
[6]Department of Agricultural, Food and Forestry Systems, University of Florence, Florence, Italy
[7]Department of Integrative Biology, University of California, Berkeley, CA, USA
[8]Department of Earth and Planetary Science, University of California, Berkeley, CA, USA

Correspondence: Paolo Benettin (paolo.benettin@epfl.ch)

Abstract. Stable water isotopes are widely used in ecohydrology to trace the transport, storage, and mixing of water on its journey through landscapes and ecosystems. Evaporation leaves a characteristic signature on the isotopic composition of the water that is left behind, such that in dual-isotope space, evaporated waters plot below the local meteoric water line (LMWL) that characterizes precipitation. Soil and xylem water samples can often plot below the LMWL as well, suggesting that they have also been influenced by evaporation. These soil and xylem water samples frequently plot along linear trends in dual-isotope space. These trend lines are often termed "evaporation lines" and their intersection with the LMWL is often interpreted as the isotopic composition of the precipitation source water. Here we use numerical experiments based on established isotope fractionation theory to show that these trend lines are often by-products of the seasonality in evaporative fractionation and in the isotopic composition of precipitation. Thus, they are often not true evaporation lines, and, if interpreted as such, can yield highly biased estimates of the isotopic composition of the source water.

1 Introduction

Stable water isotopes (^{18}O and ^{2}H) are widely used in ecohydrology as tracers of the transport, storage, and mixing of water, from its origin as precipitation, through the soil, and ultimately to groundwater and streamflow (Kendall and McDonnell, 1998) or to plant uptake and transpiration (Dawson and Ehleringer, 1998). Water isotopes also reflect evaporation losses, through the progressive enrichment of ^{18}O and ^{2}H in the remaining liquid. Past applications of stable water isotopes in soil hydrology studies have focused on identifying evaporation fronts in the unsaturated zone (e.g., Allison and Barnes, 1983; Dawson and Ehleringer, 1998; Rothfuss et al., 2015; Sprenger et al., 2017a), quantifying groundwater recharge rates and mechanisms (e.g., Healy and Scanlon, 2010; McGuire and McDonnell, 2006; Sakakibara et al., 2017), quantifying root water uptake (e.g., Ehleringer and Dawson, 1992; Dawson et al., 2002; Volkmann et al., 2016; Rothfuss and Javaux, 2017), and identifying plant water sources (e.g., Brooks et al., 2010; Dawson and Ehleringer, 1991; Dawson and Simonin, 2011; Goldsmith et al., 2012; Evaristo et al., 2015; Hervé-Fernández et al., 2016; McCutcheon et al., 2017). A recent review by Sprenger et al. (2016) provides an extensive overview of isotope-based studies in the unsaturated zone. When expressed in the conventional δ notation and displayed together in a so-called dual-

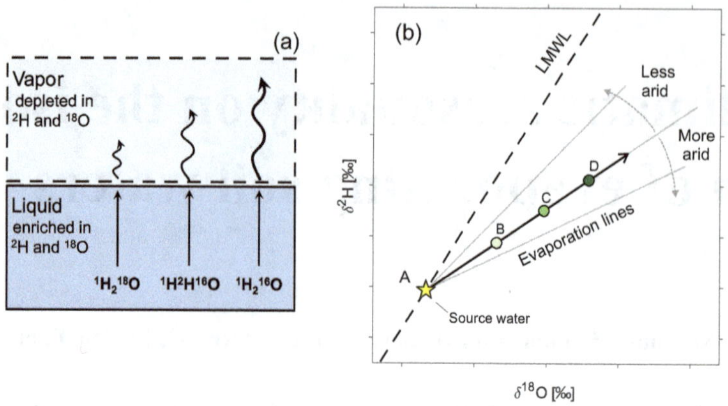

Figure 1. Fractionation effects during evaporation from an open water body. **(a)** Heavier water molecules ($^1H^2H^{16}O$, $^1H_2^{18}O$) break their bonds and evaporate less readily than lighter water molecules ($^1H_2^{16}O$), and thus have lower saturation vapor pressures. Heavier molecules also diffuse away from the evaporating surface less rapidly. As a consequence, during evaporation, lighter water molecules vaporize faster than heavier water molecules. The ratios between the evaporation rates of the different water isotopologues (net of any condensation) determine the slope of the evaporation line describing the progressive isotopic enrichment of the liquid water that is left behind. **(b)** Progressive enrichment (dots B–D) of a water source (yellow star A). The evaporation line typically lies below the local meteoric water line (LMWL), at an angle that depends on the aridity and the isotopic composition of the atmosphere (and thus on the relative rates of re-condensation of each isotopologue). **(a)** was adapted from Leibundgut et al. (2009).

isotope plot (e.g., Fig. 1), variations in $\delta^{18}O$ and δ^2H in precipitation at any given location will typically exhibit a strong linear correlation (Craig, 1961; Dansgaard, 1964) termed the local meteoric water line (LMWL). As a water parcel evaporates, its isotopic composition will evolve along an evaporation line where the slope is determined by the relative evaporation rates of the different water isotopologues (Fig. 1). This evaporation line will generally have a shallower slope than the LMWL.

Soil and xylem water samples also often lie at an angle to the LMWL, and are frequently well described by linear fits (e.g., Brooks et al., 2010; Dawson and Simonin, 2011; Goldsmith et al., 2012; Evaristo et al., 2015). If these lines are evaporation lines, then extrapolating them to their intersection with the LMWL should yield the original composition of the pre-evaporation source water. Exactly this strategy has been used to infer source water compositions for soil water and xylem water (e.g., Evaristo et al., 2015; Hervé-Fernández et al., 2016; Javaux et al., 2016), as well as groundwater (e.g., Dogramaci et al., 2012) and stream water (e.g., Telmer and Veizer, 2000). Information on the source water composition is then typically used to draw conclusions about water cycling processes in the study systems. This inference should be valid if the evaporated samples all originate from a single source water.

But what if they don't? Is a linear trend, alone, sufficient evidence that the trend is actually an evaporation line? To date, no benchmark experiment has tested whether, and under what conditions, the trend line passing through fractionated soil water samples correctly identifies their source water.

Here we use simple numerical experiments, based on established isotope fractionation theory, to model the isotopic evolution of seasonally varying precipitation inputs, under the influence of seasonally varying evaporation processes. These simulations show that the resulting evaporated samples often fall along well-defined linear trends that are markedly different from evaporation lines, and therefore do not point to any meaningful source water composition.

2 Materials and methods

We simulated the isotopic composition of evaporating soil waters using equations based on the simple and widely used linear resistance model of Craig and Gordon (1965). We then introduced the effect of climatic seasonality by applying these equations to seasonally varying isotopic sources and atmospheric conditions.

2.1 Evaporative fractionation in open water and soils

The Craig and Gordon (1965) model estimates the joint effect of equilibrium and kinetic isotopic fractionation during the phase transition from liquid water to vapor. When resistance to transport in the liquid phase is neglected, the isotopic composition of the water vapor flux can be expressed as follows:

$$\delta_E = \frac{(\delta_L - \varepsilon^+)/\alpha^+ - h \cdot \delta_A - \varepsilon_k}{1 - h + 10^{-3} \cdot \varepsilon_k}, \tag{1}$$

where δ_L and δ_A indicate the isotopic compositions of the evaporating surface and the atmosphere, h is the relative humidity of the atmosphere, α^+ and ε^+ are equilibrium fractionation factors, and ε_k is a kinetic fractionation factor. Here (and elsewhere in this paper), δ values and fractionation factors may refer to either hydrogen or oxygen isotopes unless otherwise noted. The δ notation expresses water isotope ratios as deviations, in parts per thousand, from Vienna Standard Mean Ocean Water (Kendall and Caldwell, 1998).

The equilibrium fractionation factor α^+ (–) describes differences between the isotopic compositions of liquid and vapor phases at isotopic equilibrium, and is expressed here as the super-ratio of liquid to vapor isotope ratios. Its value is slightly larger than one, reflecting the fact that lighter molecules break their bonds more readily and thus are more abundant in the vapor phase. The values of α^+ can be computed as a function of temperature T (K) using the well-established experimental results by Horita and Wesolowski (1994):

$$10^3 \ln\left[\alpha^+\left({}^2\mathrm{H}\right)\right] = 1158.8\left(T^3/10^9\right) - 1620.1\left(T^2/10^6\right) + 794.84\left(T/10^3\right) - 161.04 + 2.9992\left(10^9/T^3\right) \quad (2)$$

$$10^3 \ln\left[\alpha^+\left({}^{18}\mathrm{O}\right)\right] = -7.685 + 6.7123\left(10^3/T\right) - 1.6664\left(10^6/T^2\right) + 0.3504\left(10^9/T^3\right). \quad (3)$$

The equilibrium isotopic separation between liquid and vapor is then computed as $\varepsilon^+ = (\alpha^+ - 1)\,10^3$ (‰).

The kinetic fractionation factor ε_k quantifies isotopic effects during net evaporation associated with the higher diffusivities of isotopically lighter molecules. Variations in ε_k are generally dominated by the relative humidity (h) of the air overlying the evaporating surface. Several expressions have been derived specifically for ε_k in soils (see Mathieu and Bariac, 1996; Soderberg et al., 2012). Here we use a simplified expression given as (Gat, 1996; Horita et al., 2008)

$$\varepsilon_k = \theta n (1-h)(1 - D_i/D)\,10^3 \text{ (‰)}. \quad (4)$$

The weighting term θ (–) accounts for the possible influence of the evaporation flux on the ambient moisture, and is usually assumed to equal one for small water bodies (Gat, 1996). The term D_i/D is the ratio between the diffusivities of the heavy and light isotopes. Commonly accepted values are provided by Merlivat (1978): $D_i/D(^2\mathrm{H}) = 0.9755$ and $D_i/D(^{18}\mathrm{O}) = 0.9723$. The term n (–) accounts for the aerodynamic regime above the evaporating liquid–vapor interface. It ranges from $n = 0.5$ (fully turbulent transport that reduces kinetic fractionation, appropriate for lakes or saturated soil conditions) to $n = 1$ (fully diffusive transport, appropriate for very dry soil conditions). According to Eq. (4), in a dry atmosphere ($h = 0$), the kinetic fractionation factor is roughly 12.2–24.5 ‰ for $\varepsilon_k(^2\mathrm{H})$ and 13.8–27.7 ‰ for $\varepsilon_k(^{18}\mathrm{O})$.

We now consider the case of an isolated volume of water with initial isotopic composition δ_0 that evaporates into the atmosphere. As evaporation is the only flux, the remaining liquid volume decreases over time (a case sometimes referred to as a "desiccating" water body). We use x (–) to represent the fraction of the initial volume that has evaporated. The fraction remaining as liquid thus equals $1 - x$. Assuming that the fractionation factors do not change during the evaporation process, the equation describing the isotopic composition of the residual liquid δ_L is (Gonfiantini, 1986)

$$\delta_L = \left(\delta_0 - \delta^*\right)(1-x)^m + \delta^*, \quad (5)$$

where δ^* [(‰) represents the limiting isotopic composition (i.e., the composition that the desiccating water volume would approach upon drying up) and the term m (–) is referred to as "temporal enrichment slope" (Gibson et al., 2016). These two terms can be computed as follows:

$$\delta^* = \left(h\delta_A + \varepsilon_k + \varepsilon^+/\alpha^+\right) / \left(h - 10^{-3}\cdot\left(\varepsilon_k + \varepsilon^+/\alpha^+\right)\right) \quad (6)$$

and

$$m = \left(h - 10^{-3}\cdot\left(\varepsilon_k + \varepsilon^+/\alpha^+\right)\right) / \left(1 - h + 10^{-3}\cdot\varepsilon_k\right). \quad (7)$$

Equation (5) can represent an isolated volume of precipitation with initial isotopic composition δ_P that progressively evaporates into an atmosphere with an isotopic composition of δ_A. If the isotopic composition of the atmospheric vapor is unknown, it is common to assume that it is in equilibrium with precipitation (Gibson et al., 2008):

$$\delta_A = \left(\delta_P - \varepsilon^+\right)/\alpha^+. \quad (8)$$

As an introductory example, we modeled the isotopic evolution of an individual water volume by implementing Eqs. (1)–(8) with parameters $T = 20\,°\mathrm{(C)}$, $h = 0.75$ and $n = 1$. Figure 2 illustrates the resulting increase in the hydrogen and oxygen δ_L of the residual water during the evaporation process. The dual-isotope plot (Fig. 2c) shows the simultaneous behavior of the hydrogen and oxygen isotope ratios. As more of the water evaporates, the composition of the residual liquid gradually departs from the LMWL following a nearly linear trajectory. This trajectory is termed the evaporation line. Depending on the atmospheric parameters used in Eqs. (1)–(8), the slopes of evaporation lines will typically range from 2.5 to 5, markedly shallower than typical meteoric water lines, which usually have slopes of roughly 8 (Kendall and Caldwell, 1998).

2.2 Accounting for the seasonality of atmospheric variables

The degree of evaporative fractionation will vary seasonally, reflecting seasonal changes in temperature and relative humidity. The isotopic composition of precipitation will

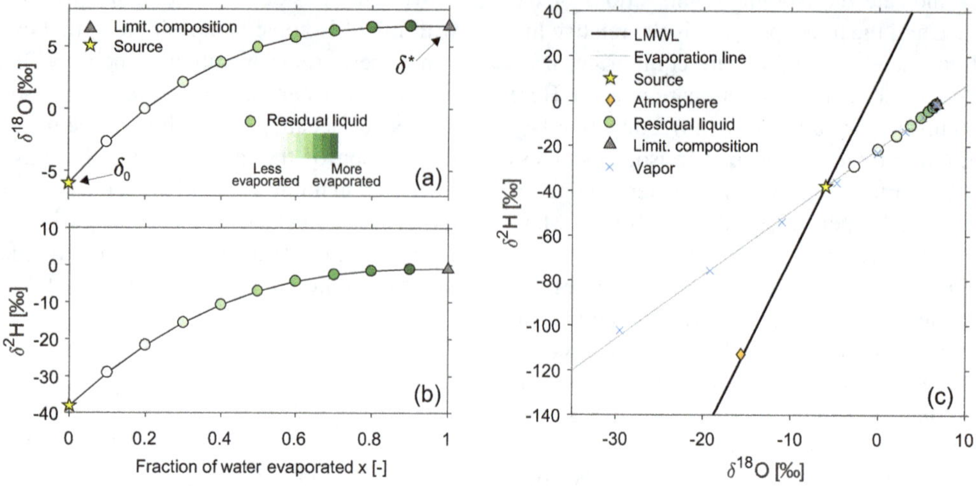

Figure 2. Introductory example showing the evolution of the isotopic composition of residual water δ_L for the case of an isolated volume of precipitation that evaporates into the atmosphere. The initial composition (source water) $\delta_0 = \delta_P$ is -6 ‰ (for δ^{18}O) and -38 ‰ (for δ^2H). **(a)** and **(b)** show the oxygen and hydrogen isotopic composition for increasing fractions of evaporation (decreasing fraction of residual liquid) as they approach the limiting composition, while **(c)** shows the same isotope effects in a dual isotope plot.

also vary seasonally, reflecting seasonal shifts in moisture sources, air mass trajectories, and cloud processes (Craig and Gordon, 1965; Rozanski et al., 1993). With this in mind, we explore how these two seasonal patterns jointly shape the isotopic composition of the residual liquid remaining after rainfall partly evaporates from a soil.

We consider a 12 month period and for each month we use the mean isotopic composition of precipitation as source water for the model outlined above. Each month's precipitation then undergoes a seasonally varying amount of evaporation, and the isotopic composition of the residual water is determined separately for each month using Eqs. (2)–(8), along with that month's average temperature and relative humidity. In this approach, the isotopic composition of monthly residual water depends only on precipitation and evaporation during the same month. This simplified approach does not explicitly account for in-soil mixing processes, whose effects are discussed in Sect. 3.

We apply this approach to real-world weather and precipitation data from the Vienna Hohe Warte station, Austria. The full isotopic dataset is freely available, along with temperature and vapor pressure data, from the Global Network of Isotopes in Precipitation (GNIP Database), provided by IAEA/WMO and accessible at: https://nucleus.iaea.org/wiser. From GNIP we obtained long-term mean monthly values of precipitation δ^{18}O at Hohe Warte, along with mean monthly air temperatures and vapor pressures, from which we calculated mean monthly relative humidities. Rather than using the precipitation δ^2H values from GNIP, we instead computed them from δ^{18}O using the equation for the LMWL at Hohe Warte: $\delta^2\text{H} = 2.12 + 7.45\delta^{18}\text{O}$. This ensured that each precipitation sample plotted exactly on the LMWL, thus aiding visualization.

The long-term mean monthly time series exhibit pronounced seasonality (Fig. 3). The seasonal temperature excursion is about 20 °C, and monthly average δ^{18}O ranges from -13 ‰ in winter to -6 ‰ in summer. The relative humidity ranges from roughly 0.85 in winter to 0.65 in spring and summer.

To investigate the effect of seasonality in evaporation rates on residual liquid composition, we represented the evaporation-to-precipitation fraction (the variable x) by sinusoidal cycles with different amplitudes and timing. We did not consider transpiration fluxes, since the isotopic effects of transpiration are generally considered to be negligible. Moreover, to keep the example simple, we did not consider the seasonality of precipitation flux, although this could be easily included. The parameter n in Eq. (4) was fixed at 0.75 throughout the year. A numerical code to implement Eqs. (1)–(8) and apply them to the case of seasonal sources and climatic conditions is freely available at https://github.com/pbenettin/evaporation-lines and provided in the Supplement.

3 Results

3.1 Seasonal patterns in evaporated soil waters

The isotopic compositions of different source waters (mean monthly values of precipitation from the Vienna Hohe Warte station) and of the residual liquid water after evaporation (computed through Eqs. 2–8) are shown in dual-isotope space in Fig. 4. For this figure, we generated two hypothet-

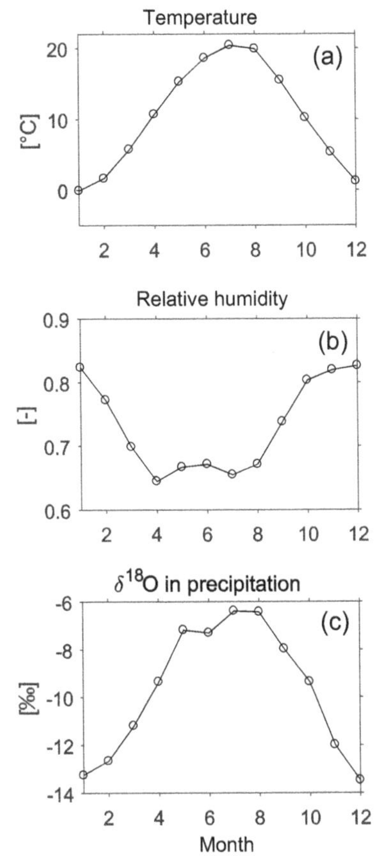

Figure 3. Long-term mean monthly air temperature **(a)**, relative humidity **(b)** and oxygen isotopic composition in precipitation **(c)** for the station Vienna Hohe Warte, Austria.

ical evaporation cycles, both peaking in July and having the same mean value $\overline{x}$ of 0.10 (–) but with different degrees of seasonality. The weakly seasonal cycle had a peak-to-peak amplitude of just 0.02, and the more strongly seasonal cycle had an amplitude of 0.16. These x values are modest, representing conditions of limited evaporation that may be found in many temperate regions.

The source waters (shown as yellow stars) vary along the LMWL reflecting the seasonal variability of atmospheric moisture sources and conditions, with isotopically lighter precipitation during colder months. The simulated residual water samples (shown as green dots) plot below the LMWL, with summer samples plotting farther from the LMWL than winter samples, reflecting their greater evaporative enrichment. The evaporation lines connecting individual source waters and residual waters are longer and shallower in summer than in winter, reflecting seasonal differences in temperature, relative humidity, and evaporated fraction x. As a result, the summer residual water samples plot farther away from the LMWL than the winter samples do, by an amount that reflects the seasonality in the evaporation process. The residual water samples follow a nearly linear trend (shown as a dashed line), which is markedly steeper than the evaporation lines for the individual source waters (shown as grey lines). The slopes of the evaporation lines range from 3.1 to 3.4; by contrast, the trend lines for the residual waters have slopes of 6.1 (Fig. 4a) and 7.1 (Fig. 4b), close to the assumed LMWL slope of 7.45. Note that whenever the residual water trend line has a slope that is close to that of the LMWL, the location of the intersection between these two lines will be highly uncertain.

Because the simulated residual water samples can be fitted easily with a simple trend line, it may seem logical to interpret this trend line as an evaporation line, and to infer an apparent source water end-member from its intersection with the LMWL. In the case of an isolated water parcel that is progressively evaporated (as in Fig. 2c), this approach could yield a reasonable estimate of the original source water. However, when residual water samples do not come from a single source, the trend line is not an evaporation line, and the intercept of this trend with the LMWL can lie far away from the average source water (Fig. 4a); the intercept can even lie far outside the range of all the source waters (Fig. 4b).

Figure 5 illustrates how different degrees of seasonality in evaporation patterns may yield different trend lines in residual water samples, with different intercepts with the LMWL. The individual source waters and evaporation lines are the same as in Fig. 4. The five trend lines in Fig. 5 are associated with different seasonal evaporation cycles, which feature similar low evaporation fractions in winter (roughly $x = 0.04$), but different evaporation fractions in summer (roughly $x = 0.15$ to $x = 0.60$). The evaporation cycles with higher summer peaks correspond to trend lines with shallower slopes and less negative intersections with the LMWL. All of the intersections lie far from the true mean source water; indeed none of them lie within the range of the individual monthly source waters.

If the seasonal cycle of evaporative fractionation is not in phase with the seasonal cycle in source water composition (that is, if the most strongly fractionated sample is not also the one with the heaviest initial isotopic signature), the residual water samples will trace out a hysteresis loop. In Fig. 6, the source waters are the same as those in Fig. 4, but the seasonal evaporation cycle has been shifted by two months. The width of the resulting hysteresis loop depends on the amplitude of the seasonal cycle in evaporation, and how far out of phase it is with the seasonal cycle in precipitation isotopes. Even where such hysteresis loops exist in nature, they may be difficult to detect due to measurement uncertainties and environmental noise.

3.2 Mixtures of evaporated soil waters

In Figs. 4–6, each residual water sample is derived from a discrete monthly precipitation source water sample. Real-world soil waters, by contrast, can be expected to contain mixtures of waters with different ages, and thus different

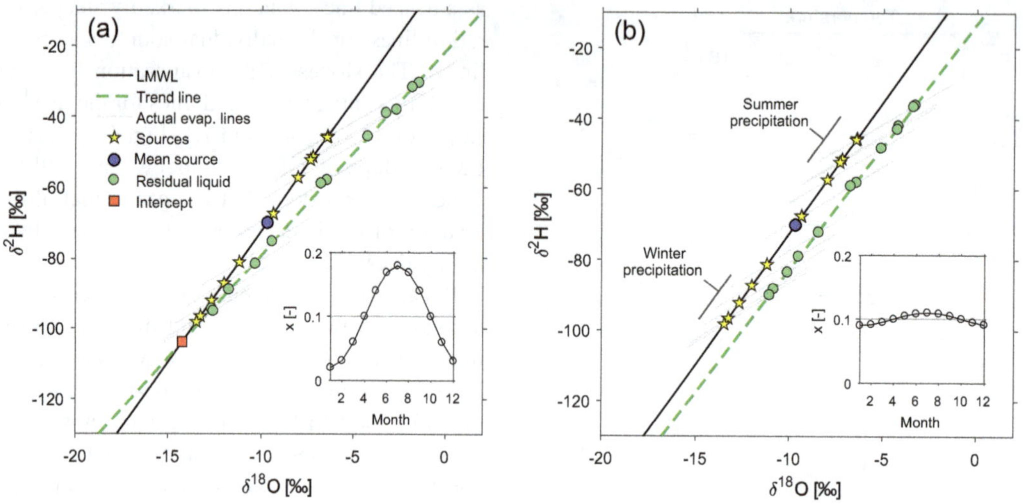

Figure 4. Effect of atmospheric seasonality on the isotopic composition of residual water from seasonally varying precipitation. The evaporatively fractionated residual water samples (green dots) cluster around a trend line (dashed line) which is much steeper than the individual evaporation lines (grey lines). The effects of strong and weak seasonality in evaporation rates (represented by x, the fraction of the initial volume that has evaporated) are shown in **(a)** and **(b)**, respectively. The insets show the assumed annual cycles in evaporated fractions x.

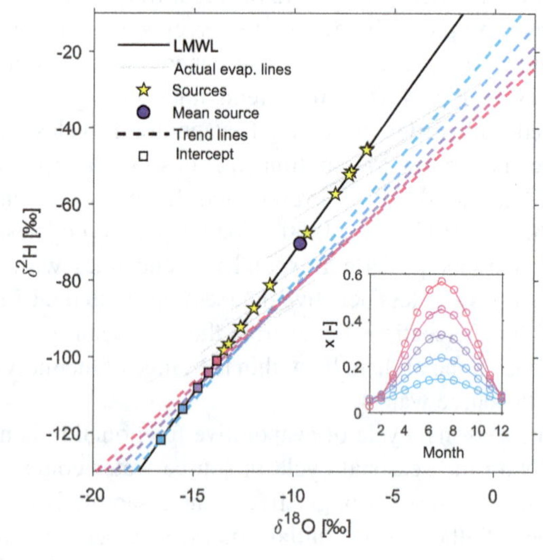

Figure 5. Examples of trend lines and intercepts arising from various seasonal evaporation patterns (inset).

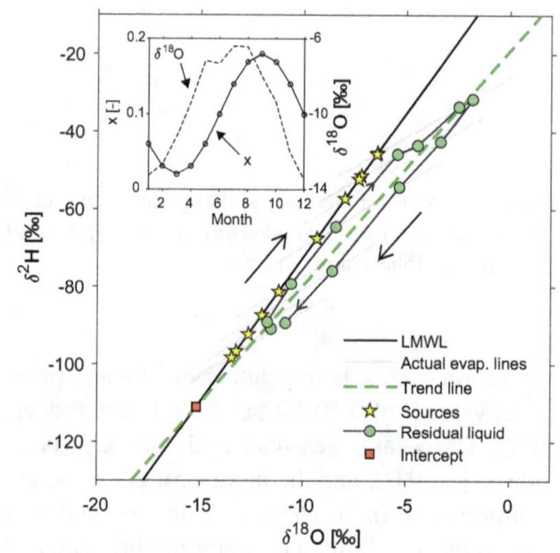

Figure 6. Hysteretic pattern arising in the computed residual liquid when the seasonality of evaporation rates is shifted relative to the seasonality of the source water composition (see inset).

source water signatures and evaporative fractionation trajectories. For simplicity, we simulated the soil as a well-mixed reservoir that integrates each month's residual waters. Mathematically this means that the composition of the soil pool is an exponentially weighted running average of the residual water samples shown in Fig. 4a. For purposes of illustration we used a time constant of six months, such that the same-month contribution to each sample is roughly 15 % and the contribution from the previous 12 months is roughly 86 % of the total. The results are shown in Fig. 7a.

The same procedure was used to create Fig. 7b, except we considered each of the approximately 600 individual monthly δ^{18}O and δ^2H values available at Hohe Warte (the cloud of grey dots) as meteoric source waters. These source waters were not constrained to lie along the LMWL, in contrast to the analyses presented above. These source waters were individually evaporated and fractionated, by amounts that depended on the individual monthly temperature and relative humidity (and the same seasonal cycle in the evaporated fraction x that was assumed in Figs. 4a, 6, and 7a). We then

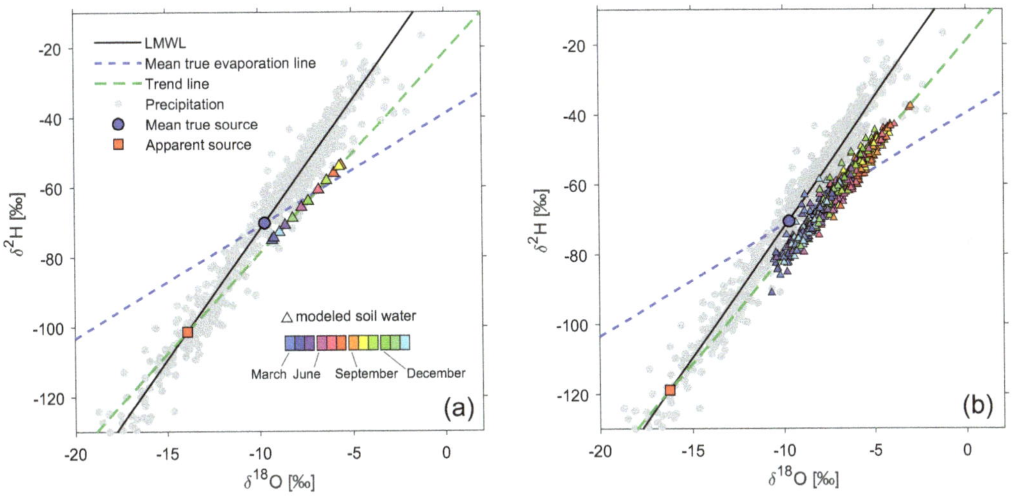

Figure 7. Isotopic composition of modeled soil waters (triangles) obtained by mixing evaporated source waters through an exponential function. Soil water isotopic compositions were obtained starting from **(a)** long-term monthly sources (as in Figs. 4–6), and **(b)** individual monthly isotopic sources, recorded at Vienna Hohe Warte station between 1961 and 2015 (grey dots).

applied the same running weighted time averaging used in Fig. 7a, with the resulting cloud of residual water samples shown in Fig. 7b. The more that the residual water samples are time-averaged, the more their scatter will be compressed and the smaller the portion of the dual-isotope plot they will occupy, but their trend line will remain almost the same. The exponentially weighted averaging used here also introduces a time lag of roughly 3–4 months between the seasonal cycle in the source water and the seasonal cycle in the time-averaged soil water. For this reason, the isotopically heaviest soil water samples are found in October even though the isotopically heaviest precipitation falls in the summer. (Different time constants in the weighted averaging would yield different lag intervals.) Similar lag periods of several months are often found in experimental studies (e.g., Sprenger et al., 2017b).

Due to the scatter among the source water samples in Fig. 7b, the evaporatively fractionated residual water samples are less collinear than in Fig. 7a. Nonetheless, in both cases the trend lines intersect the LMWL far from the true mean source water. Because the intersection point lies within the range of the individual winter precipitation samples, however, there is a risk that one could incorrectly infer that it represented a winter-precipitation source water for the evaporated soil samples (when in fact the winter precipitation in these simulations has hardly been evaporated at all).

4 Discussion and concluding remarks

The analyses presented above serve as a reminder that isotope samples need to be understood as combining the effects of source variation, mixing, and fractionation. Indeed, in our examples, all three of these effects jointly determine the isotopic patterns in the evaporated soil samples.

All else equal, the greater the isotopic variability in precipitation (and thus the larger the range of source waters), the closer the slope of the evaporated samples will lie to the LMWL (Fig. 8a). (Conversely, in the absence of any variability in precipitation, the evaporated samples would trace out evaporation lines instead.) All else equal, the greater the seasonality in evaporative fractionation, the more the slope of the evaporated samples will deviate from the LMWL (Fig. 8b). The intercept of the trend line with the LMWL is driven purely by these geometric considerations, and has no significance in its own right.

The seasonality of evaporative fractionation combines two factors: the variation in the slope of the evaporation line, and the variation in the amount of water lost to evaporation (as quantified by the evaporation fraction x), which determines how far out on the evaporation line the evaporated samples are found. In most real-world situations, the second of these factors is likely to have a greater influence on the trend line of the evaporated samples (and thus on its intersection with the LMWL).

The more mixing the evaporated samples undergo, the more their variability will be compressed. In addition, the more closely that the variations in precipitation isotopes and evaporation rates are synchronized, the more the evaporated samples will follow a trend line; conversely, if they are out of phase, they will form a hysteresis loop.

Our analysis of the effects of variability in source signatures and evaporative fractionation has been couched in terms of seasonal patterns, but similar considerations apply to variations at other timescales as well. For example, under more arid conditions the evaporation line will have a flatter slope

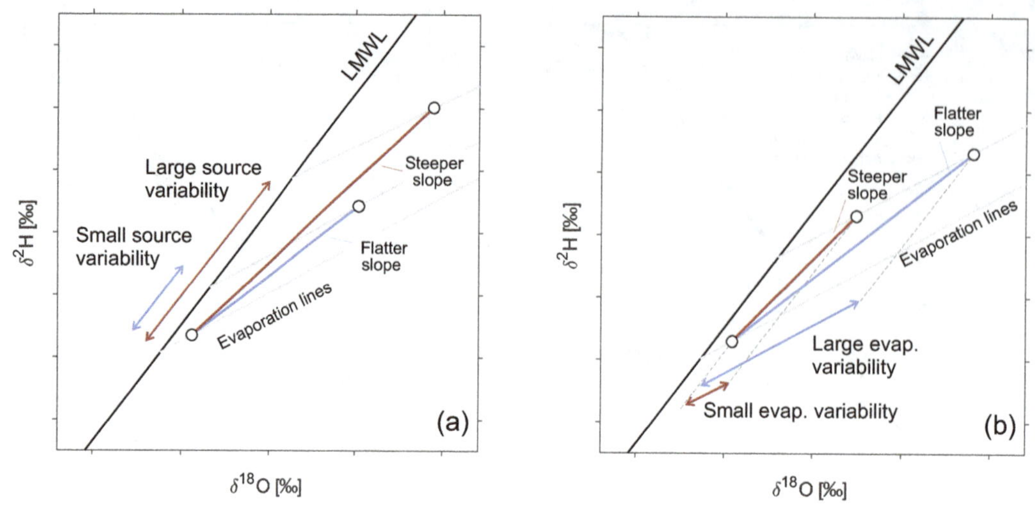

Figure 8. Effects of precipitation source variability **(a)** and evaporation variability **(b)** on the trend line that interpolates evaporated soil water samples (dots). The trend line is much steeper than the evaporation lines, unless there is no variability in the source water isotopic composition.

and evaporative losses will be greater, both factors that will push evaporated samples farther from the LMWL. If those atmospheric conditions are also correlated with isotopically heavier source waters, the resulting residual water trend line will be similar to those we have simulated here (with a slope much steeper than a true evaporation line, and typically intersecting the LMWL far from the average source water).

The intersection between this trend line and the LMWL gives a heavily biased estimate of average source water, but what could give a better one? One can see from Fig. 7 that a reasonable estimate of the average source water could be obtained by translating the individual evaporated samples back to the LMWL along assumed evaporation lines, yielding estimates of their pre-evaporation compositions which are then averaged. Ideally the slope of each evaporation line would be determined from atmospheric conditions that are specific to each evaporated sample. But even where these are unknown, any reasonable estimate of the evaporation slope will yield much better results than the slope of the trend line through the evaporated samples.

We have chosen a relatively simple model to simulate the evaporative fractionation of the residual water samples. More sophisticated models of evaporative fractionation in soil water have been proposed (see, e.g., Mathieu and Bariac, 1996; Soderberg et al., 2012; Dubbert et al., 2013; Good et al., 2014). Results from these models may provide more accurate estimates of the kinetic fractionation factors and thus of the slope of the evaporation line (and its variability). However, use of these models is unlikely to yield qualitatively different results from those in Figs. 4–7, because other reasonable estimates of the fractionation factors (and thus of the slopes of the evaporation lines) will make little difference to the slope of the trend line running through the evaporated samples. Our analysis also invokes the simplifying assumption that (for example) July's rainfall only evaporates under July conditions. But if some of July's rainfall is stored until August, September, October, etc. then some of it should also evaporate under those conditions. Arguably our analysis could be superseded by a detailed process model that simulates the time-dependent storage and release of water in soils. However, such a model would complicate the analysis considerably and we have no reason to believe that it would yield substantially different results.

The data and equations presented in this paper are not novel, and many readers will not be surprised by our conclusion that trend lines through evaporated samples can differ widely from true evaporation lines. Nonetheless, our analysis shows how residual water trend lines can result from the interplay of seasonally varying isotopic inputs and evaporation rates, and shows that their intersection with the LMWL will generally be a highly unreliable guide to the average source water composition. Analyses that have used these trend lines to identify the compositions of source waters may be substantially in error, and therefore should be re-examined. Because plant uptake is generally not strongly fractionating, isotopic variations in soil water are likely to be transferred to plant xylem, and thus we expect that our conclusions will also apply to xylem water as well. Whether on seasonal or synoptic timescales, the regional and global energy dynamics that drive variations in source water composition and evaporative fractionation are likely to be widespread. Thus, although results for individual sites and time periods may differ in quantitative details from those presented here, we expect the qualitative patterns to be general.

Competing interests. The authors declare that they have no conflict of interest.

Acknowledgements. This article was inspired by discussions at the workshop on "Isotope-based studies of water partitioning and plant-soil interactions in forested and agricultural environments", which was held in September 2017 in San Casciano in Val di Pesa, Florence, Italy. Paolo Benettin thanks the ENAC school at EPFL for financial support. Till H. M. Volkmann gratefully acknowledges support by the National Science Foundation (grant 1334452). The authors thank Nils Michelsen for catching a small error in a previous draft, and the two anonymous referees for their comments on the manuscript.

Edited by: Markus Hrachowitz

References

Allison, G. B. and Barnes, C. J.: Estimation of evaporation from non-vegetated surfaces using natural deuterium, Nature, 301, 143–145, https://doi.org/10.1038/301143a0, 1983.

Brooks, J. R., Barnard, H. R., Coulombe, R., and McDonnell, J. J.: Ecohydrologic separation of water between trees and streams in a Mediterranean climate, Nat. Geosci., 3, 100–104, https://doi.org/10.1038/ngeo722, 2010.

Craig, H.: Isotopic Variations in Meteoric Waters, Science, 133, 1702–1703, https://doi.org/10.1126/science.133.3465.1702, 1961.

Craig, H. and Gordon, L. I.: Deuterium and oxygen 18 variations in the ocean and marine atmosphere, in: Stable Isotopes in Oceanographic Studies and Paleotemperatures, Spoleto, Italy, edited by: Tongiorgi, E., Consiglio nazionale delle ricerche, Laboratorio di geologia nucleare, Spoleto, Italy, 9–130, 1965.

Dansgaard, W.: Stable isotopes in precipitation, Tellus, 16, 436–468, https://doi.org/10.1111/j.2153-3490.1964.tb00181.x, 1964.

Dawson, T. E. and Ehleringer, J. R.: Streamside trees that do not use stream water, Nature, 350, 335–337, https://doi.org/10.1038/350335a0, 1991.

Dawson, T. E. and Ehleringer, J. R.: Chapter 6 – Plants, Isotopes and Water Use: A Catchment-Scale Perspective, in: Isotope Tracers in Catchment Hydrology, edited by: Kendall, C. and McDonnell, J. J., Elsevier, Amsterdam, 165–202, https://doi.org/10.1016/B978-0-444-81546-0.50013-6, 1998.

Dawson, T. E. and Simonin, K. A.: The Roles of Stable Isotopes in Forest Hydrology and Biogeochemistry, in: Forest Hydrology and Biogeochemistry: Synthesis of Past Research and Future Directions, vol. 216 of Ecological Studies, edited by: Levia, D. F., Carlyle-Moses, D., and Tanaka, T., Springer Netherlands, Dordrecht, 137–161, https://doi.org/10.1007/978-94-007-1363-5, 2011.

Dawson, T. E., Mambelli, S., Plamboeck, A. H., Templer, P. H., and Tu, K. P.: Stable Isotopes in Plant Ecology, Annu. Rev. Ecol. Syst., 33, 507–559, https://doi.org/10.1146/annurev.ecolsys.33.020602.095451, 2002.

Dogramaci, S., Skrzypek, G., Dodson, W., and Grierson, P. F.: Stable isotope and hydrochemical evolution of groundwater in the semi-arid Hamersley Basin of subtropical northwest Australia, J. Hydrol., 475, 281–293, https://doi.org/10.1016/j.jhydrol.2012.10.004, 2012.

Dubbert, M., Cuntz, M., Piayda, A., Maguás, C., and Werner, C.: Partitioning evapotranspiration – Testing the Craig and Gordon model with field measurements of oxygen isotope ratios of evaporative fluxes, J. Hydrol., 496, 142–153, https://doi.org/10.1016/j.jhydrol.2013.05.033, 2013.

Ehleringer, J. R. and Dawson, T. E.: Water uptake by plants: perspectives from stable isotope composition, Plant Cell Environ., 15, 1073–1082, https://doi.org/10.1111/j.1365-3040.1992.tb01657.x, 1992.

Evaristo, J., Jasechko, S., and McDonnell, J. J.: Global separation of plant transpiration from groundwater and streamflow, Nature, 525, 91–94, https://doi.org/10.1038/nature14983, 2015.

Gat, J. R.: Oxygen and hydrogen isotopes in the hydrologic cycle, Annu. Rev. Earth Planet. Sci., 24, 225–262, https://doi.org/10.1146/annurev.earth.24.1.225, 1996.

Gibson, J. J., Birks, S. J., and Edwards, T. W. D.: Global prediction of δ_A and δ^2H-δ^{18}O evaporation slopes for lakes and soil water accounting for seasonality, Global Biogeochem. Cy., 22, https://doi.org/10.1029/2007GB002997, 2008.

Gibson, J. J., Birks, S., and Yi, Y.: Stable isotope mass balance of lakes: a contemporary perspective, Quaternary Sci. Rev., 131, 316–328, https://doi.org/10.1016/j.quascirev.2015.04.013, 2016.

Goldsmith, G. R., Muñoz-Villers, L. E., Holwerda, F., McDonnell, J. J., Asbjornsen, H., and Dawson, T. E.: Stable isotopes reveal linkages among ecohydrological processes in a seasonally dry tropical montane cloud forest, Ecohydrology, 5, 779–790, https://doi.org/10.1002/eco.268, 2012.

Gonfiantini, R.: Environmental isotopes in lake studies, in: The Terrestrial Environment, B, Handbook of Environmental Isotope Geochemistry, edited by: Fritz, P. and Fontes, J., Elsevier, Amsterdam, 113–168, https://doi.org/10.1016/B978-0-444-42225-5.50008-5, 1986.

Good, S. P., Soderberg, K., Guan, K., King, E. G., Scanlon, T. M., and Caylor, K. K.: δ^2H isotopic flux partitioning of evapotranspiration over a grass field following a water pulse and subsequent dry down, Water Resour. Res., 50, 1410–1432, https://doi.org/10.1002/2013WR014333, 2014.

Healy, R. W. and Scanlon, B. R.: Estimating groundwater recharge, Cambridge University Press, Cambridge, New York, 2010.

Hervé-Fernández, P., Oyarzún, C., Brumbt, C., Huygens, D., Bodé, S., Verhoest, N. E. C., and Boeckx, P.: Assessing the "two water worlds" hypothesis and water sources for native and exotic evergreen species in south-central Chile, Hydrol. Process., 30, 4227–4241, https://doi.org/10.1002/hyp.10984, 2016.

Horita, J. and Wesolowski, D.: Liquid-Vapor Fractionation of Oxygen and Hydrogen Isotopes of Water from the Freezing to the Critical-Temperature, Geochim. Cosmochim. Ac., 58, 3425–3437, https://doi.org/10.1016/0016-7037(94)90096-5, 1994.

Horita, J., Rozanski, K., and Cohen, S.: Isotope effects in the evaporation of water: a status report of the Craig–Gordon model, Isotop. Environ. Health Stud., 44, 23–49, https://doi.org/10.1080/10256010801887174, 2008.

IAEA/WMO: Global Network of Isotopes in Precipitation. The GNIP Database, available at: https://nucleus.iaea.org/wiser, 2018.

Javaux, M., Rothfuss, Y., Vanderborght, J., Vereecken, H., and Brüggemann, N.: Isotopic composition of plant water sources, Nature, 536, E1–E3, https://doi.org/10.1038/nature18946, 2016.

Kendall, C. and Caldwell, E. A.: Chapter 2 – Fundamentals of Isotope Geochemistry, in: Isotope Tracers in Catchment Hydrology, edited by: Kendall, C. and McDonnell, J. J., Elsevier, Amsterdam, 51–86, https://doi.org/10.1016/B978-0-444-81546-0.50009-4, 1998.

Kendall, C. and McDonnell, J. J.: Isotope tracers in catchment hydrology, Elsevier, Amsterdam, https://doi.org/10.1016/B978-0-444-81546-0.50001-X, 1998.

Leibundgut, C., Maloszewski, P., and Külls, C.: Tracers in hydrology, John Wiley & Sons, Ltd, Oxford, https://doi.org/10.1002/9780470747148, 2009.

Mathieu, R. and Bariac, T.: A numerical model for the simulation of stable isotope profiles in drying soils, J. Geophys. Res.-Atmos., 101, 12685–12696, https://doi.org/10.1029/96JD00223, 1996.

McCutcheon, R. J., McNamara, J. P., Kohn, M. J., and Evans, S. L.: An evaluation of the ecohydrological separation hypothesis in a semiarid catchment, Hydrol. Process., 31, 783–799, https://doi.org/10.1002/hyp.11052, 2017.

McGuire, K. J. and McDonnell, J. J.: A review and evaluation of catchment transit time modeling, J. Hydrol., 330, 543–563, https://doi.org/10.1016/j.jhydrol.2006.04.020, 2006.

Merlivat, L.: Molecular diffusivities of $H_2^{16}O$, $HD^{16}O$, and $H_2^{18}O$ in gases, J. Chem. Phys., 69, 2864–2871, https://doi.org/10.1063/1.436884, 1978.

Rothfuss, Y. and Javaux, M.: Reviews and syntheses: Isotopic approaches to quantify root water uptake: a review and comparison of methods, Biogeosciences, 14, 2199–2224, https://doi.org/10.5194/bg-14-2199-2017, 2017.

Rothfuss, Y., Merz, S., Vanderborght, J., Hermes, N., Weuthen, A., Pohlmeier, A., Vereecken, H., and Brüggemann, N.: Long-term and high-frequency non-destructive monitoring of water stable isotope profiles in an evaporating soil column, Hydrol. Earth Syst. Sci., 19, 4067–4080, https://doi.org/10.5194/hess-19-4067-2015, 2015.

Rozanski, K., Araguás-Araguás, L., and Gonfiantini, R.: Isotopic Patterns in Modern Global Precipitation, American Geophysical Union, Washington D.C., 1–36, https://doi.org/10.1029/GM078p0001, 1993.

Sakakibara, K., Tsujimura, M., Song, X., and Zhang, J.: Spatiotemporal variation of the surface water effect on the groundwater recharge in a low-precipitation region: Application of the multi-tracer approach to the Taihang Mountains, North China, J. Hydrol., 545, 132–144, https://doi.org/10.1016/j.jhydrol.2016.12.030, 2017.

Soderberg, K., Good, S. P., Wang, L., and Caylor, K.: Stable Isotopes of Water Vapor in the Vadose Zone: A Review of Measurement and Modeling Techniques, Vadose Zone J., 11, https://doi.org/10.2136/vzj2011.0165, 2012.

Sprenger, M., Leistert, H., Gimbel, K., and Weiler, M.: Illuminating hydrological processes at the soil–vegetation–atmosphere interface with water stable isotopes, Rev. Geophys., 54, 674–704, https://doi.org/10.1002/2015RG000515, 2016.

Sprenger, M., Tetzlaff, D., Buttle, J. M., Laudon, H., Leistert, H., Mitchell, C., Snelgrove, J., Weiler, M., and Soulsby, C.: Measuring and modelling stable isotopes of mobile and bulk soil water, Vadose Zone J., 17, https://doi.org/10.2136/vzj2017.08.0149, 2017a.

Sprenger, M., Tetzlaff, D., and Soulsby, C.: Soil water stable isotopes reveal evaporation dynamics at the soil–plant–atmosphere interface of the critical zone, Hydrol. Earth Syst. Sci., 21, 3839–3858, https://doi.org/10.5194/hess-21-3839-2017, 2017b.

Telmer, K. and Veizer, J.: Isotopic constraints on the transpiration, evaporation, energy, and gross primary production budgets of a large boreal watershed: Ottawa River Basin, Canada, Global Biogeochem. Cy., 14, 149–165, https://doi.org/10.1029/1999GB900078, 2000.

Volkmann, T. H. M., Haberer, K., Gessler, A., and Weiler, M.: High-resolution isotope measurements resolve rapid ecohydrological dynamics at the soil–plant interface, New Phytol., 210, 839–849, https://doi.org/10.1111/nph.13868, 2016.

11

Every apple has a voice: using stable isotopes to teach about food sourcing and the water cycle

Erik Oerter[1,2,4], **Molly Malone**[3], **Annie Putman**[1,2], **Dina Drits-Esser**[3], **Louisa Stark**[3], **and Gabriel Bowen**[1,2]

[1]Department of Geology and Geophysics, University of Utah, 115 South 1460 East, Salt Lake City, UT 84112, USA
[2]Global Change and Sustainability Center, University of Utah, 257 South 1400 East, Salt Lake City, UT 84112, USA
[3]Genetic Science Learning Center, University of Utah, 515 100 South, Salt Lake City, UT 84112, USA
[4]Current address: Lawrence Livermore National Laboratory, 7000 East Avenue, Livermore, CA 94550, USA

Correspondence to: Erik Oerter (erikjoerter@gmail.com)

Abstract. Agricultural crops such as fruits take up irrigation and meteoric water and incorporate it into their tissue ("fruit water") during growth, and the geographic origin of a fruit may be traced by comparing the H and O stable isotope composition (δ^2H and δ^{18}O values) of fruit water to the global geospatial distribution of H and O stable isotopes in precipitation. This connection between common fruits and the global water cycle provides an access point to connect with a variety of demographic groups to educate about isotope hydrology and the water cycle. Within the context of a 1-day outreach activity designed for a wide spectrum of participants (high school students, undergraduate students, high school science teachers) we developed introductory lecture materials, in-class participatory demonstrations of fruit water isotopic measurement in real time, and a computer lab exercise to couple actual fruit water isotope data with open-source online geospatial analysis software. We assessed learning outcomes with pre- and post-tests tied to learning objectives, as well as participant feedback surveys. Results indicate that this outreach activity provided effective lessons on the basics of stable isotope hydrology and the water cycle. However, the computer lab exercise needs to be more specifically tailored to the abilities of each participant group. This pilot study provides a foundation for further development of outreach materials that can effectively engage a range of participant groups in learning about the water cycle and the ways in which humans modify the water cycle through agricultural activity.

1 Introduction

Teaching of environmental science and enhancement of student engagement with associated disciplines is a key focus of science education efforts in the United States and worldwide (Quinn et al., 2012). Because human activity and climate change combine to create new pressures on water resources (Pachauri et al., 2014), a comprehensive understanding of how water moves between reservoirs in Earth's subsurface, surface, and atmosphere ("the water cycle"; e.g., Oki and Kanae, 2006) is critical knowledge for students in any field of study. The economic and ecological implications of changes to the water cycle provide motivation for increased focus on environmental sciences and improvement in the teaching of these sciences in K–12 (from kindergarten to grade 12) and undergraduate science education curricula in the United States and internationally (Osborne and Dillon, 2008; Quinn et al., 2012; Ruddell and Wagener, 2013; Wagener et al., 2007). However, the invisibility of some components of the water cycle, such as inter-reservoir fluxes (e.g., atmospheric water vapor transport) is a fundamental challenge in teaching the concepts of the global water cycle (Bar, 1989; Rappaport, 2009). Anthropogenic forcings, such as transfers between watersheds in the form of water diversions for irrigation and drinking water (Good et al., 2014), further complicate the already-complex natural system.

Students' most direct connections to the water cycle are the water they drink and the food they eat. These existing connections may be a novel entry point for education. Existing studies of food provenance have been forensic studies of counterfeiting and ingredient substitutions (Cerling et al., 2016). This present work relies on the fact that agricultural crops such as fruits and vegetables take up precipitation and irrigation water during growth and incorporate it into their tissue ("fruit water"; Kramer and Boyer, 1995), which may have a specific geochemical signature in the form of its hydrogen and oxygen isotopic composition (detailed below). Therefore, if the fruit water can be linked to its source water, the geographic origin of the fruit may be determined, and the complex relationship between agriculture, transportation of crops to market, and consumption becomes more apparent.

The distance between food source and consumption represents the movement of water out of source watersheds via trucks and trains, rather than inside pipes, a concept known as a "virtual water footprint", which is another nearly invisible water transport mechanism (Hoekstra and Mekonnen, 2012; Chen and Chen, 2013). For example, in California in 2015, 2.3×10^{10} kg of vegetable crops were harvested and transported to market (CDFA, 2016), which represents 2.17×10^{10} L of water (average vegetable water content is 95 % by weight; Spungen, 2005). While this vegetable water is 2 orders of magnitude smaller than the volume of water transported for urban uses in California in 2015 (7×10^{12} L; Mount and Hanak, 2016), the vegetable crop water is an enormous amount of water being moved around the landscape that many people may not have considered.

We designed a 1-day teaching activity within the context of a week-long outreach workshop that was centered on helping aspiring science, technology, engineering, and math (STEM) students and high school science teachers gain insight into the practice of science and to engage the participants in hands-on hydrological research. For this teaching activity, we developed introductory lecture materials, in-class participatory demonstrations of fruit water isotopic measurement in real time, and a computer lab exercise to couple actual fruit water isotope data with open-source online geospatial analysis software. The goals of the teaching activities were to introduce and establish key water cycle concepts, reveal human modification of the natural water cycle, introduce core concepts of stable isotope hydrology, and link these concepts together through scientific data analysis and synthesis in the context of a food sourcing study. Additionally, because the week-long workshop was conducted as outreach from large teaching and research universities towards ambitious students and teachers, an over-arching goal of the workshop activities (and this paper) was to inspire and nurture scientific interest in the participants by exposing them to a variety of environmental science activities.

In this paper, we describe how we (1) developed teaching modules about isotope hydrology and the water cycle, (2) developed novel laboratory and computer demonstrations and exercises to link water cycle components to the investigation of food sourcing, and (3) evaluated the efficacy of these teaching activities and materials. To do so, we first discuss the scientific background for this teaching activity, and provide a dataset of fruit source water H and O stable isotope values for use in the teaching activities. We then provide and discuss teaching materials in the form of lecture materials, as well as laboratory exercises and datasets, and provide results and discussion on the evaluation of the lesson's efficacy in achieving learning outcomes. The teaching materials are publicly available to the teaching community for nonprofit use, and are included in the accompanying Supplement.

2 Scientific background and research methods

2.1 Isotope hydrology

Isotopes are atoms of the same chemical element (defined by the number of protons in the nucleus) that differ in the number of neutrons. In the case of the water molecule (H_2O), hydrogen has two stable isotopes (termed "stable" because they do not undergo radioactive decay), ^1H and ^2H (where the superscript numeral refers to the number of protons plus neutrons), while oxygen has three stable isotopes: ^{16}O, ^{17}O, and ^{18}O, though ^{17}O is present in very low abundance in nature and is not included in this work. The heavier isotopes (^2H and ^{18}O) are more rare in natural waters than their light counterparts (^1H and ^{16}O), with ^2H comprising 0.0156 % of all hydrogen atoms and ^{18}O comprising 0.1995 % of all oxygen atoms in water. Because of the low abundance of heavy isotopes, the stable isotopes in a sample are expressed as the ratio of the rare (heavy) isotope to the common (light) isotope in delta (δ) notation as follows: $\delta = (R_{\text{sample}} / R_{\text{standard}} - 1)$, where R_{sample} and R_{standard} are the $^2H / ^1H$ or $^{18}O / ^{16}O$ ratios for the sample and standard, respectively, and values are reported in per mill (‰). The standard for H_2O δ^2H and $\delta^{18}O$ values is Vienna Standard Mean Ocean Water (VSMOW; Coplen, 1994).

The δ^2H and $\delta^{18}O$ values of precipitation (meteoric water) exhibit a linear relationship, termed the global meteoric water line (GMWL), as shown in Fig. 1: $\delta^2H = 8\,\delta^{18}O + 10$ (Craig, 1961). Precipitation during cold seasons, in continental interiors, or in cold regions like high latitudes and elevations tend to exhibit low δ^2H and $\delta^{18}O$ values, whereas precipitation in warm seasons, at low latitudes and altitudes, and near ocean coasts tend to have higher δ^2H and $\delta^{18}O$ values (Fig. 1). This results in a global geospatial distribution of δ^2H and $\delta^{18}O$ values of precipitation that correlates well with mean annual air temperature (Bowen and Revenaugh, 2003). A detailed discussion of the mechanisms behind these worldwide stable isotope patterns can be found in Bowen (2010).

Because plants take up soil water during growth, the H and O stable isotope composition (δ^2H and $\delta^{18}O$ values) of the water in the plant represents the soil water. Therefore, the δ^2H and $\delta^{18}O$ values of the plant water can be used to identify plant water sources within the soil (Dawson et al., 2002). By comparing the δ^2H and $\delta^{18}O$ values of water measured in fruit to the increasingly well-constrained global geospa-

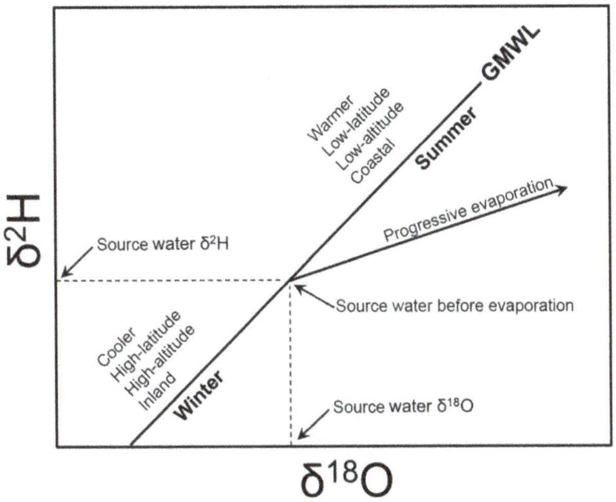

Figure 1. The generalized relationship between δ^2H and δ^{18}O values measured in precipitation worldwide. The spatial and seasonal trends along the global meteoric water line (GMWL) are described. The evaporation line is denoted by the line labeled "progressive evaporation", and samples that have undergone more evaporation will move away from the GMWL along this line. The original δ^2H and δ^{18}O values of the evaporated water's source can be determined by finding the intersection of the evaporation line with the GMWL. Base figure redrawn from sahra.arizona.edu.

tial distribution of H and O stable isotopes in precipitation (Bowen and Revenaugh, 2003), the likely growing region of a fruit may be estimated.

A potential caveat of directly linking fruit water to a source region using precipitation is that evaporation may occur from the soil water before plant uptake, after plant uptake and during fruit growth, or after harvest. Water that has undergone evaporation will have higher δ^{18}O and δ^2H values than its source water, and the change in these values proceeds along a linear trend in a plot of δ^2H and δ^{18}O, termed an "evaporation line", with a slope smaller than that of the GMWL (Fig. 1). Water that has undergone minimal evaporation will plot along the evaporation line but near the GMWL, while progressive evaporation will drive the sample further along the evaporation line and away from the intersection of the evaporation line with the GMWL (Craig, 1961). This model of progressive enrichment through evaporation allows the effects of fruit water evaporation to be detected and corrected for. The simplest approach to this correction involves fitting a regression line through measured δ^2H and δ^{18}O values from multiple samples having the same source water but somewhat different degrees of evaporation (e.g., several pieces of fruit from the same lot but of different size) and back-projecting the regression line to where it intersects with the GMWL (Fig. 1). This intersection of the evaporation line and the GMWL are the source water δ^2H and δ^{18}O values.

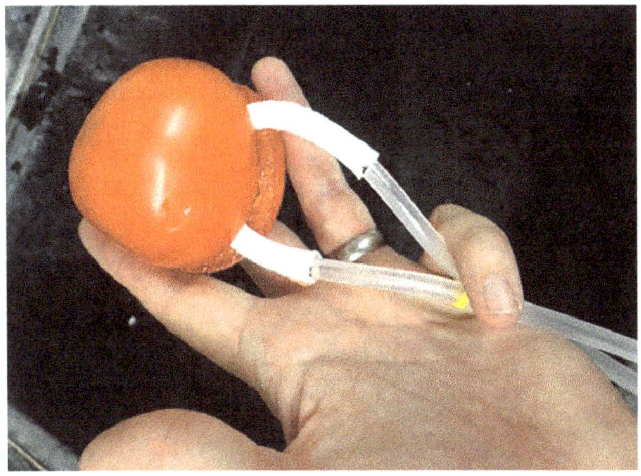

Figure 2. Photograph of membrane-inlet probe (opaque white tubing) partially inserted into a tomato.

2.2 Fruit water isotopic analysis

Water δ^2H and δ^{18}O values were measured in the fruits and vegetables listed in Table 1. Hereafter the measurements will be referred to as "fruit water" for brevity, though "vegetable water" is equivalent. Measurements were performed using a membrane-inlet water-vapor isotope analyzer system originally developed to sample soil water (Oerter and Bowen, 2017; Oerter et al., 2017), and fruit water measurement methods mimic those detailed for soil water in Oerter et al. (2017). To our knowledge, this is the first application of such a system to measure in situ fruit water δ^2H and δ^{18}O values, as well as the use of such a system as a teaching tool.

Briefly, the analytical system consists of a microporous membrane probe attached to a laser spectrometer instrument (L2130-i, Picarro, Inc., USA) that simultaneously measures water vapor concentration ([H_2O]), δ^2H, and δ^{18}O values in water vapor (δ^2H_{vap} and $\delta^{18}O_{vap}$) at 1 Hz frequency. A measurement is performed by halving or excavating a small cavity in the fruit item, placing the fruit item in a resealable plastic bag, inserting the membrane probe into the fruit so that it is completely enveloped by the fruit (Fig. 2), evacuating the bag of air by squeezing or suction, and closing the bag around the probe lines. The fruit water vapor from the probe is carried to the analyzer on dry N_2 gas. A measurement duration of 5 minutes is typically sufficient to achieve >2 min of stable [H_2O], δ^2H_{vap}, and $\delta^{18}O_{vap}$ measurements, after which the probe is removed from the fruit and replaced with a clean and dry probe for the next analysis. During each isotopic measurement, a thermometer is inserted into the fruit adjacent to the probe, and fruit temperature is recorded for the duration of isotopic measurement. Because the δ^2H_{vap} and $\delta^{18}O_{vap}$ values are measurements of water vapor and not liquid water, they were converted into liquid water δ^2H and δ^{18}O values (δ^2H_{liq} and $\delta^{18}O_{liq}$) by assuming the vapor and liquid were in isotopic equilibrium at the measured sample temperature,

Table 1. Labeled origins, average measured δ^2H_{vap} and $\delta^{18}O_{vap}$ values, and calculated δ^2H_{liq}, $\delta^{18}O_{liq}$, δ^2H_{source}, and $\delta^{18}O_{source}$ values (±1 SD) and regression relationships of fruits and vegetables measured in this study and used in the accompanying teaching materials.

Type	Varietal	Labeled origin	*n*	$\delta^{18}O_{vapor}$	δ^2H_{vapor}	$\delta^{18}O_{liquid}$	δ^2H_{liquid}	Regression	$\delta^{18}O_{source}$	δ^2H_{source}
Apple	Koru	New Zealand	4	−11.28 (±0.31)	−112.9 (±1.7)	−1.53 (±0.31)	−30.8 (±1.7)	$y = 5.1118x - 22.955$ $R^2 = 0.82$	−11.44	−81.5
Apple	Pink Lady	Washington, USA	4	−17.16 (±2.32)	−164.4 (±11.2)	−7.36 (±2.32)	−81.7 (±11.2)	$y = 4.7481x - 46.764$ $R^2 = 0.96$	−17.45	−129.7
Apricot	–	Perry, Utah, USA	4	−5.02 (±1.41)	−137.3 (±4.3)	4.78 (±1.41)	−54.6 (±4.3)	$y = 2.8924x - 68.42$ $R^2 = 0.86$	−15.35	−112.8
Banana	–	Peru	4	−5.30 (±0.54)	−92.6 (±2.1)	4.50 (±0.54)	−9.9 (±2.1)	$y = 3.7021x - 26.542$ $R^2 = 0.87$	−8.5	−58.0
Orange	Navel	California, USA	3	3.34 (±1.62)	−85.7 (±5.8)	12.94 (±1.62)	−5.6 (±5.8)	$y = 3.5646x - 51.771$ $R^2 = 0.99$	−13.93	−101.4
Tomato	Beefsteak	Colorado, USA	4	−17.55 (±1.44)	−166.1 (±5.5)	−7.85 (±1.44)	−84.5 (±5.5)	$y = 3.7579x - 54.966$ $R^2 = 0.98$	−15.32	−112.5
Tomato	Hot House	Mexico	4	−13.46 (± 0.49)	−129.7 (± 1.6)	−3.86 (± 0.49)	−49.6 (± 1.6)	$y = 2.9682x - 38.181$ $R^2 = 0.79$	−9.58	−66.6
Tomato	Roma	Mexico	3	−5.42 (±1.48)	−109.6 (±3.8)	4.28 (±1.5)	−28.0 (±3.8)	$y = 2.5133x - 38.73$ $R^2 = 0.95$	−8.94	−61.1
Tomato	On The Vine (vine no. 1)	USA	3	−9.75 (±0.83)	−107.0 (±0.4)	−0.15 (±0.83)	−26.9 (±0.4)	$y = 0.4902x - 26.857$ $R^2 = 0.97$	−3.82	−28.2
Tomato	On The Vine (vine no. 2)	USA	3	−10.52 (±0.33)	−108.0 (±0.9)	−0.92 (±0.33)	−27.9 (±0.9)	$y = 2.0799x - 25.976$ $R^2 = 0.64$	−6.08	−38.6
Tomato	On The Vine (vine no. 3)	USA	3	−7.95 (±2.87)	−101.6 (±4.6)	1.65 (±2.87)	−21.5 (±4.6)	$y = 1.6085x - 24.202$ $R^2 = 0.99$	−5.35	−32.8
Tomato	On The Vine (all vines)	USA	9	−9.41 (±1.89)	−105.6 (±3.8)	0.19 (±1.89)	−25.5 (±3.8)	$y = 1.9068x - 25.824$ $R^2 = 0.90$	−5.88	−37.0
Zucchini	–	Mexico	4	−12.50 (±1.51)	−138.9 (±4.0)	−2.60 (±1.51)	−54.0 (±4.0)	$y = 2.5938x - 47.229$ $R^2 = 0.96$	−10.59	−74.7
Zucchini	–	Colorado, USA	4	−15.23 (±1.18)	−149.8 (±3,2)	−5.23 (±1.18)	−64.6 (±3.6)	$y = 2.7881x - 50.061$ $R^2 = 0.83$	−11.52	−82.2

using the temperature-dependent equilibrium $H_2O_{liquid\text{-}water}$ fractionation factors for H and O of Majoube (1971).

It was necessary to measure several (≥ 3) individual fruits of each type in order to characterize the variation in δ^2H_{liq} and $\delta^{18}O_{liq}$ values of that fruit type's population. These fruit water δ^2H_{liq} and $\delta^{18}O_{liq}$ values were then plotted in δ^2H and $\delta^{18}O$ space, and a least-squares regression line was fitted through the fruit water δ^2H_{liq} and $\delta^{18}O_{liq}$ values (Fig. 3). Each fruit type's source water δ^2H and $\delta^{18}O$ values (δ^2H_{source} and $\delta^{18}O_{source}$) were then determined by finding the fruit water regression line's intersection with the GMWL. This can be done by either simply drawing the regression line back to the GMWL intersection (example shown in Fig. 4) or by mathematically solving the system of equations consisting of the fruit water regression line and the GMWL. An example of this calculation for use in the accompanying laboratory exercise is included in the Supplement (Supplement Material File 2).

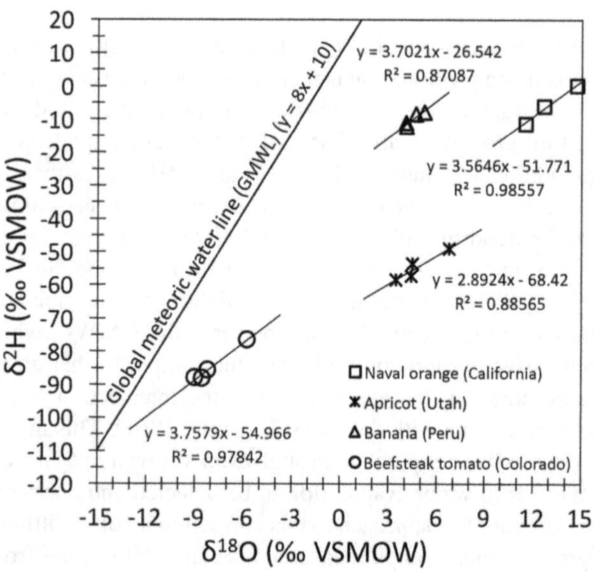

Figure 3. Values of δ^2H_{liq} and $\delta^{18}O_{liq}$ (symbols) calculated from measured δ^2H_{vap} and $\delta^{18}O_{vap}$ values (not shown) for various fruits and their associated regression lines compared to the global meteoric water line (GMWL).

3 Teaching lesson components

3.1 Overview of teaching activities

This teaching and outreach activity took place in the context of a week-long outreach workshop for nine high school students with an interest in pursuing environmental science, nine undergraduate students who are actively studying environmental science and water issues, and nine high school teachers who teach physical science subjects such as chemistry and earth science. Day one of the workshop consisted of a field-based environmental sampling session where the participants collected water samples from a small creek and documented their sampling activities. Day two involved the isotope hydrology activity discussed herein. Day three consisted of collecting responses to a short survey of questions on water-related issues in an urban setting. On day four, the students self-selected into three equally sized groups divided amongst the three days' concept areas, with the goal of producing a poster exhibit communicating the scientific results of their activities. Each group was guided by a PhD-level scientist who specialized in that area of research (and who had guided that day's lesson and sampling activities). Day five consisted of a lunchtime poster exhibition where each group presented their research to a scientific audience. The workshop was designed holistically to expose students to several types of water-related research, but each component of the workshop could be a standalone activity.

The teaching activities described below consist of three components: (1) an introductory lecture on the water cycle and isotope hydrology general concepts with a participatory demonstration of fruit water isotopic measurements in real time, (2) a computer-based laboratory exercise using the fruit water H and O stable isotope dataset, and (3) learning effectiveness evaluations consisting of pre- and post-tests (Supplement Material File 4) aligned with learning objectives for activities 1 and 2 (Table 2) . The learning objectives are tied to the three dimensions of the Next Generation Science Standards (NGSS Lead States, 2013), which are science curriculum guidelines for grades K–12 in the United States. The NGSS dimensions and the teaching activity associated with each component are as follows: (A) Disciplinary Core Ideas: The Water Cycle and stable isotope hydrology, (B) Science and Engineering Practices: data collection and analysis, (C) Crosscutting Concepts: using data and geospatial relationships to evaluate food sourcing.

A live demonstration of the use of the isotope analyzer system to make measurements of fruit water isotopic composition as described above and utilization of student participation could be repeated in situations where the instrumentation is available, but is not required. The dataset shown in Table 1 can provide the basis for implementing the following teaching activities and materials in any teaching setting.

3.2 Water cycle and isotope hydrology introductory lesson

The first activity consists of a lecture introducing basic water cycle and stable isotope concepts, and linking them with stable isotope hydrology. The content of the introductory lesson is focused on teaching to learning objectives 1–4 (Table 2). The presentation slides used in the teaching session are available for public, not-for-profit use and are provided in the Supplement. The following discussion in Sect. 3.2 and 3.3 can be utilized as a teaching guide to accompany the introductory lesson materials. Supplement Material File numbers referenced in the following discussion refer to slide numbers in the introductory lesson presentation slides (Supplement Material File 1).

The introductory lecture begins with a presentation of the water cycle (Supplement Material File 1.2). The oceans, atmosphere, and subsurface water are all introduced as "reservoirs" where water resides for some length of time. The processes such as evaporation and runoff associated with movement between reservoirs are introduced as "fluxes". The ways in which human activity and built infrastructure can alter the size of water cycle components are illustrated in Supplement Material File 1.3.

The chemical formula of the water molecule is introduced in Supplement Material File 1.4, and Supplement Material Files 1.4–1.7 illustrate how atoms are composed of protons in the nucleus with orbiting electrons, using hydrogen as an example. Isotopes are introduced and illustrated in Supplement Material Files 1.8–1.14, again using hydrogen as the example. The important concept for isotopes is that they are atoms of the same element because they have the same number of protons, but differ in the number of neutrons in the nucleus, with ^{1}H and ^{2}H as the examples. The central concept for Supplement Material Files 1.15–1.20 is that the differences in how stable isotopes of the same element "behave" in chemical and physical reactions are largely due to differences in mass. In this section, the effects for a water molecule are addressed. Supplement Material File 1.20 points out the observation that heavier molecules will evaporate less easily than lighter molecules, and this contrasting evaporation behavior is an important concept for the remainder of the learning activities.

The concept of stable isotope notation is introduced in Supplement Material Files 1.21–1.23, and which can be an obstacle for comprehension and could potentially divert a student's learning progression because the notation is represented by a mathematical equation, the use of a Greek symbol (δ), and the resulting values are all negative numbers (e.g., Orton and Frobisher, 2004). This is an important juncture to emphasize the simple concept that the notation refers to the ratio of the isotope abundances, and thus lower δ values mean more of the lighter isotopes, and higher δ values mean more of the heavier isotopes, even if all values are negative. More discussion of stable isotope notation as *troublesome knowledge* (i.e., Meyer and Land, 2003, 2005) follows in Sect. 4.3.

Supplement Material Files 1.24–1.25 show that $\delta^{2}H$ and $\delta^{18}O$ values in precipitation exhibit substantial geospatial structure and spatial autocorrelation. Supplement Material Files 1.26–1.27 introduce the idea that $\delta^{2}H$ values in precipitation are related to $\delta^{18}O$ values in a predictable

Table 2. Learning objectives with pre- and post-test assessment results. Asterisks indicate level of statistical significance (p values) as listed below results.

	Pre-test	Post-test
Concept area	% correct responses	% correct responses
(1) Students should be able to describe how water moves through the water cycle from precipitation to soil to incorporation in plants (and fruits and vegetables).	0.84 (±0.37)	0.92 (±0.28)
(2) Students should be able to define what a stable isotope is.	0.68 (±0.48)	0.96* (±0.20)
(3) Students should be able to identify the main factor that controls how stable isotopes of water are distributed around the globe.	0.48 (±0.51)	0.96** (±0.20)
(4) Students should be able to explain why and how the stable isotopes in fruit and vegetable water can yield information on the geographic origin of the food itself.	0.6 (±0.50)	0.88 (±0.38)
(5) Concepts as a whole	0.65 (±0.25)	0.93*** (±0.13)

($n = 25$. * $p = 0.005$. ** $p < 0.001$. *** $p < 0.0001$)

way that is correlated with regional air temperature patterns. Departures from the global meteoric water line are often due to the effects of evaporation, as shown by the diverging arrow in Supplement Material File 1.27. Reinforcing the relationship between average air temperature and the geographic pattern of δ^2H and δ^{18}O values in river water (which ultimately comes from precipitation) are Supplement Material Files 1.28 and 1.29, while Supplement Material Files 1.30–1.32 go into more detail about the effects of temperature. Supplement Material File 1.33 again shows the worldwide distribution of δ^2H and δ^{18}O values in precipitation, and the idea that distance from the coast also influences isotope values is introduced and expanded upon in Supplement Material Files 1.34–1.37. The combined effects of temperature and distance from the coast are well illustrated by Supplement Material File 1.38.

In Supplement Material File 1.39 the concept of the water cycle is returned to, and includes the routing of water through plants as they use soil water to form new plant tissue, including fruits and vegetables. The table in Supplement Material File 1.40 emphasizes the amount of water in common fruits and vegetables. This is a good time to point out the fact that by harvesting fruits and vegetables and bringing them to market to be sold and later consumed, we are transporting water. The example of a tomato truck is useful in this context: a tomato truck is effectively a water truck. This is also a good opportunity to introduce the idea that irrigation is necessary to grow food crops in regions where there is not enough precipitation. The map in Supplement Material File 1.41 shows that in the arid western United States, much more water is used for irrigation than falls as rain or snow. This implies that ground water or river water must be transported from a wetter region to the agricultural region. This is an explicit anthropogenic alteration to the water cycle through exporting or importing water.

Finally, the concepts of the water cycle, stable isotope hydrology, and water in fruits and vegetables can be applied to the overarching research question (Supplement Material File 1.42): Can we use the stable isotopes of hydrogen and oxygen in water as "fingerprints" to determine the origin of fruits and vegetables? The general approach to using fruit and vegetable water to identify source regions is shown in Supplement Material File 1.43.

3.3 Computer-based laboratory exercise

The computer-based laboratory exercise was designed to reinforce the concepts covered in the introductory lecture and promote hands-on learning using fruit water stable isotope data. The laboratory exercise materials are made available as Supplement Material File 2 for public, nonprofit use. The laboratory exercise makes use of the IsoMAP online isotope mapping software and resulting data products that are publicly available under a noncommercial reuse and attribution license (http://isomap.rcac.purdue.edu:8080/gridsphere). The laboratory exercise and associated data were conceptually driven by the research question: can we demonstrate how to use the stable isotopes of hydrogen and oxygen in water as "fingerprints" to determine where fruits and vegetables come from?

The students are provided with detailed handouts giving step-by-step instructions for completing the laboratory exercise, which may be worked individually or in small groups. The exercise begins with a brief review of concepts covered in the introductory lecture, and the material is made available again as a reference to use in completing the exercises. The sequence of measurement activities that yields fruit water isotope data is reviewed, and the method for converting the measured δ^2H_{vap} and $\delta^{18}O_{vap}$ values into δ^2H_{liq} and $\delta^{18}O_{liq}$ values is explained in the printed laboratory exercise document. The concept behind correcting the fruit water δ^2H_{liq}

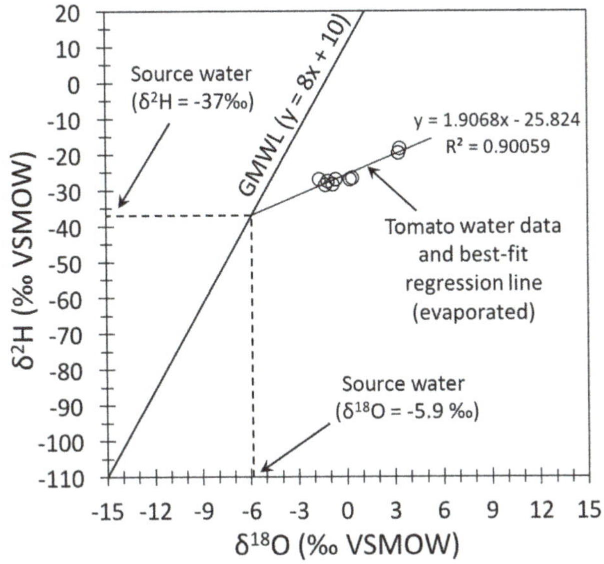

Figure 4. Graphical method of fruit source water determination by back-projecting regression line through δ^2H_{liq} and $\delta^{18}O_{liq}$ values of water inside tomatoes (open symbols) to find the source water δ^2H_{liq} and $\delta^{18}O_{liq}$ values.

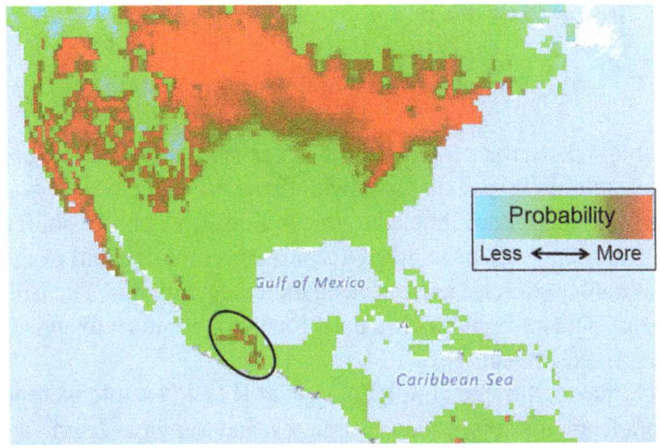

Figure 5. Example of an assignment map created with IsoMAP software. Warm colors (red to orange) indicate higher probability that the fruit water was sourced from that location, while cooler colors (blue, exclusive of ocean area) indicate lower probability. Fruit source water $\delta^{18}O$ values used are from the tomatoes shown in Fig. 4 and Table 1. Black outline over central Mexico highlights the probable source location for the tomatoes. The most immediate observation from Fig. 5 is that large regions are indicated with high probability of being the tomato source area, and that no single location or region provides a unique solution (see Sect. 4.2 for discussion).

and $\delta^{18}O_{liq}$ values for evaporation effects is explained and illustrated with examples of using the graphical and mathematical solution methods. The students are provided with paper plots of the GMWL with measured δ^2H_{liq} and $\delta^{18}O_{liq}$ values from different kinds of fruit already plotted. They use these teaching tools to graphically estimate the source water δ^2H_{liq} and $\delta^{18}O_{liq}$ values. The student-generated estimated source water values are compiled into a master list and can be compared across estimation methods and used in the following mapping activities.

The students are directed to a website that hosts downloadable html files (included as Supplement Material Files 3.1–3.4) that are interactive maps allowing the student to use their mouse pointer to hover over a spot on the map and the isotope values of that spot appear. These interactive maps provide a good opportunity for the students to synthesize what they have learned about what controls the spatial distribution of stable isotopes in precipitation (e.g., temperature and geographic location) as they explore the maps. The students can also use the interactive maps to quickly survey a geographic area and identify potential growing regions for each fruit.

Next, the students are asked to use the online mapping software at http://isomap.rcac.purdue.edu to create maps of precipitation δ^2H and $\delta^{18}O$ values for areas of the world that they find interesting, and to make maps for summer and winter seasons. This exercise illustrates and emphasizes the data and data sources that go into creating a data-rich map, such as the interactive maps the students used in the previous exercise. Then they are asked to make maps that combine the precipitation isotope maps they created with the fruit source water isotope data. These predictive maps represent the likelihood or probability that a particular pixel on the map could be the origin of the fruit water, and are displayed as color-coded "heat maps" (Fig. 5).

3.4 Assessment of learning effectiveness

We assessed learning outcomes with evaluation pre- and post-tests (Supplement Material File 4) tied to learning objectives (Table 2), as well as participant feedback surveys. The pre- and post-tests were developed by preliminary testing beforehand on an audience with similar demographics to that of the anticipated students. Problematic questions and terminology in terms of "guessability" and embedded contextual clues on the pre- and post-tests were corrected through iterative rephrasing and testing prior to administering the evaluation tests to the participant group. The pre- and post-tests consisted of identical multiple-choice questions with four candidate answers. The tests were anonymized by having each participant label their tests with a unique four-digit number (the last four digits of their phone number). With this system, there could be a possibility to track individual learning gains, though we did not do so. Instead we assessed the learning effectiveness of the group as a whole. Calculation of statistical results was done with IBM SPSS version 23 software.

4 Results and discussion

4.1 Isotopic analyses and food sourcing results

Table 1 lists the range of fruit water δ^2H and δ^{18}O values measured in this study along with their calculated source waters and their store-labeled source regions (the labeled source regions were not provided to the students until the end of the laboratory exercise as a check on their calculations). The fruit water dataset in Table 1 is used for the laboratory exercises described in Sect. 3.3.

Figure 3 shows example fruit water H and O stable isotope data from apricots, bananas, oranges, and tomatoes from varied source regions in North and South America. A primary observation is that the fruit waters are distinct from each other when plotted in δ^2H and δ^{18}O space. Given the labeled origin for each of the fruits, the organization of the different fruit values on the isotope plot roughly parallels that expected based on known controls of isotopic variation in precipitation water (Fig. 1). That is, bananas sourced from low-latitude Peru have higher δ^2H and δ^{18}O values and plot near to the warm end of the GMWL, whereas tomatoes from mid-latitude and high-elevation Colorado plot near to the cold end of the GMWL.

Another important observation from Fig. 3 is that all samples within a fruit type have generally similar δ^2H and δ^{18}O values but are separated from each other along a trend line with a lower slope than that of the GMWL. Slopes of these trend lines differ somewhat for each fruit type. These trends lines are calculated by least-squares linear regression (shown as solid lines in Fig. 3) and can be interpreted as an evaporation line characteristic of each fruit type. When these evaporation lines are projected back to their intersections with the GMWL, each fruit type's source water can be read from the plot axes (i.e., via the graphical method shown in Fig. 1). Note that the apricots, oranges, and tomatoes all plot in distinct places in Fig. 3, but their evaporation lines intersect the GMWL in a similar place, yielding similar source water δ^2H and δ^{18}O values for these fruits near the cold end of the GMWL (Fig. 1). This similarity in source water isotope values may indicate that these fruits were grown in areas with similar climates *or* that their irrigation water was sourced from areas with similar climates.

4.2 Synthesis of concepts

When the known source regions (from store labeling) of the apricots, oranges, and tomatoes are combined with the source water information derived from isotope measurements discussed above, several key water cycle concepts can be synthesized to provide the following teaching example. The low δ^2H and δ^{18}O values of the fruits indicate a cold region source for the water used to grow these crops, but none of these crops are typically considered to be sourced from cold regions. For example, the state of Florida in the USA is a well-known orange-producing region, and many students will be familiar with the Florida Orange Juice trademark that is well advertised in North America. The likely solution to the conundrum of warm region crops that appear to be grown with cold region waters is that irrigation diversions bring snowmelt-sourced water from high mountain elevations down to low valley elevations where the crops are grown. This is an example of how water can be moved across the landscape in pipes or canals for human uses rather than naturally as in rivers. In a more general sense, it is an example of how humans influence the water cycle on a large scale.

Having students utilize either their own pre-existing knowledge, or knowledge gained during the teaching activity through online or literature research, can enable better source water interpretations of the isotope data. For example, a map of the likelihood of water source locations in North America for the tomato water data shown in Fig. 4 is shown in Fig. 5 (data in Fig. 4 and Table 1). The red regions on the map indicate a higher probability that the calculated fruit source water δ^{18}O values match that observed in summer precipitation (June, July, and August), while green to blue colors indicate lower probability. The most immediate observation from Fig. 5 is that large regions are indicated with high probability of being the tomato source area, and that no single location or region provides a unique solution.

This lack of a clear-cut geographical source solution requires the students to synthesize some of the concepts introduced in the water cycle portion of the introductory lecture and include outside information in their evaluation of the map results. For example, the prediction map in Fig. 5 shows high probabilities for tomato source water in regions of high-latitude Canada, the Laurentian Great Lakes, mountainous areas of western North America and Appalachia, as well as parts of California and Mexico. This example of a scientific analysis not yielding an unambiguous result is a useful aspect to the teaching activities. In this instance, the students must assimilate external information to determine if these are likely tomato-growing regions. In practice, similar situations are common in science, when the investigator must ask themselves if their results are reasonable, and if not, what other evidence should be collected to answer the question at hand.

4.3 Communication of learning gains

The student group was reconvened after the three 1-day activity modules were completed (stream sampling for water quality, this stable isotope hydrology activity, and social science activity with surveys) and the group was asked to partition themselves into equally sized groups according to individual subject matter interest. Each group was tasked with creating a poster presentation summarizing that day's scientific activities and findings. Mentoring of the stable isotope hydrology group by one of the authors (E. Oerter) during the poster creation yielded the following qualitative observations. The participants were generally in agreement on which aspects of the water cycle and stable isotope hydrology concepts to include in the introduction portion of the poster. The participants all seemed to understand how the water cycle works and how the stable isotope signature of H and O in the fruit water could give information on where the fruit was grown. There was also general agreement about what parts of the methods were necessary to include and that photographs and conceptual plots similar to Figs. 1 and 2 would communicate what was done during the laboratory exercise. There was less agreement on what the conclusions should consist of, with roughly half of the participants feeling that a straightforward presentation of confirming the labeled origin of each food was a sufficient conclusion, while the balance of the group felt that there needed to be a stronger tie between the food sourcing aspect and general water cycle themes. Ultimately, the poster conclusions were focused largely on the food sourcing results with minor mention of the water cycle.

The teaching activities as a whole can be viewed through the lens of experiential learning theory, which is based on an iterative cycle of grasping a new concept (concrete experience and abstract conceptualization) followed by transformation of the new concept (active experimentation and reflective observation; Kolb et al., 2001; Kolb, 2014). The grasping phase consisted of the introductory lecture and fruit water isotopic measurement demonstration, followed by the transformation phase where students worked with data and explored results in the laboratory exercise. The poster creation session provided an extension to this iterative cycle, where the students were forced to articulate their new knowledge, and in so doing confront where the gaps were in their understanding and work with their peers and the instructor to fill those gaps.

The posters were presented to the entire workshop participant group with a 20 min talk where each member of the poster group focused on a particular aspect with which they felt comfortable. The presentations were followed by a brief question-and-answer period with the audience. The presentations largely confirmed observations of learning outcomes made during the small group poster creation session, with the group again divided about how far to extend the conclusions of their findings beyond simple issues of food sourcing. The question-and-answer period revealed individuals' nuanced strengths and weaknesses in specific subject areas. Audience questions that targeted tangential concepts from, or extensions of, the presented topic sometimes received answers that combined newly learned vocabulary terms in nonsensical ways. This type of response was common for questions that were at the edge of newly acquired knowledge areas, and was not limited to any particular participant demographic group. These responses suggest that the students, regardless of their age, scientific background, or public speaking experience, had not yet reached a threshold of enough exposure and usage of the new and unfamiliar knowledge of stable isotope hydrology. Threshold concept theory (Meyer and Land, 2003, 2005) posits that unfamiliar concepts without an established analogue for the student (e.g., ranges of negative numbers, such as for stable isotope values, in which the relative magnitude of each are inverse compared to that of positive numbers) can be troublesome knowledge and require sustained exposure and usage to surpass the threshold of familiarity. This outcome implies that the instructor or discussion leader should focus on core concepts over details or implications, and that more exercises in which the students work with the unfamiliar concepts are necessary to realize more learning gains.

However, presenting the activities described here in a 1 day format may be fundamentally limiting to surpassing the learning thresholds that would allow each student to transform the troublesome knowledge into tractable and readily useable concepts. A way this limitation could be overcome in a 1-day setting would be for the instructor and the course content to focus on strengthening the participants' understanding of core concepts framed in the context of pre-existing analogues. For example, stable isotope values and their relative differences could be discussed in terms of which values are "higher" and which are "lower", instead of numerical (negative) values. Additionally, the instructor should train the students to rely on their solid understanding of the basic concepts to answer questions, and to understand and acknowledge the limits of their new knowledge.

4.4 Evaluation of learning effectiveness

The evaluation pre- and post-tests (Supplement Material File 4) consisted of four questions with multiple choice answers tied to the learning objectives in Table 2. Tests were scored by tallying the number of correct responses out of the total possible (e.g., $x/4$, expressed hereafter as a decimal quotient). A paired-samples t test was conducted to evaluate whether participants increased their understanding of the content covered in the course. Results from 25 participants ($n = 25$) indicated highly significant improvement in participant scores on the concept areas as a whole (measured by total test score) from pre-test (mean (M) = 2.6, standard deviation (SD) = 1.0) to post-test ($M = 3.7$, SD = 0.5), $t(24) = -5.07$, $p < 0.0001$) (Table 2). Because of the small sample size, a Wilcoxon nonparametric test was also performed. Results confirmed the significant t test results, $z = -3.67$, $p < 0.0001$. These results indicate that this learning activity as a whole provided effective lessons on the basics of stable isotope hydrology and the water cycle.

A paired-samples t test was also conducted to evaluate increases in participant scores by test question. Bonferroni corrections, established at an alpha level of 0.0125, were used in order to decrease the possibility of finding significant differences by chance (type I error). Results from 25 participants indicated statistically significant growth in two of the items (stable isotope definitions and distribution patterns in natural systems: Table 2, items 2 and 3): item 2 pre-test ($M = 0.68$, $SD = 0.48$) to post-test ($M = 0.96$, $SD = 0.2$), $t(24) = -3.06$, $p = 0.005$) and item 3 pre-test ($M = 0.48$, $SD = 0.51$) to post-test ($M = 0.96$, $SD = 0.2$), $t(24) = -4.71$, $p < 0.001$). Increases in scores were also found in the other two items (water cycle concepts and fruit water sourcing: Table 2, items 1 and 4); however, the increase was not statistically significant: item 1 pre-test ($M = 0.84$, $SD = 0.37$) and post-test ($M = 0.92$, $SD = 0.28$) and item 4 pre-test ($M = 0.60$, $SD = 0.50$) and post-test ($M = 0.88$, $SD = 0.33$).

Because of the small sample size, nonparametric analyses, using the related-sample McNemar test, were also performed. Results confirmed the significant t test results for stable isotope distribution patterns (item 3, $p = 0.0001$). For stable isotope definitions (item 2), results were 0.0035 above significance, which is close enough to confirm the significant increase of this item from pre-test to post-test ($p = 0.016$). These nonparametric analyses also confirmed the nonsignificant t test results for water cycle concepts (item 1) and fruit water sourcing (item 4).

The evaluation pre- and post-tests were anonymized for participant identity, and we did not follow individual participant's learning progress through the exercises. In the context of a 1-day course, tracking individuals would probably be of limited value, as individualized instruction would need several sessions to implement. However, if the lessons and laboratory exercises were conducted over the course of several sessions, as would be the case in a conventional course setting, individual participant progress tracking would be beneficial and could provide improved learning outcomes for each participant.

4.5 Participant feedback and pedagogic tailoring

Informal, qualitative participant feedback was solicited from participants after the conclusion of the teaching activities in the form of post-participation surveys and was combined with informal observation during the teaching activities by one of the authors (E. Oerter). These results indicated that the audience demographic groups (high school students, high school teachers, and undergraduate students) responded to the teaching approaches differently.

Interestingly, high school teachers *and* high school students responded to teaching approaches in similar ways, even though the high school teachers had at least an undergraduate-level background in science education, and thus were more similar in background scientific education to the undergraduate students. For example, none of the high school students *or* teachers were interested in using the algebraic simultaneous solution approach to determining source water values, and some of each group even communicated their disinterest vocally. In contrast, while some undergraduates were satisfied with the graphical method for finding fruit source water H and O isotope values, others wanted to calculate more precise results using the simultaneous solution method.

Contrasting learning styles extended to the structure of the lessons, with commonalities again between high school teachers and students, but not between teachers and undergraduate students. One high school teacher suggested that the teaching model, "I do, we do, you do" (i.e., the instructor first demonstrates, then the instructor and students repeat the activity together, and finally the students perform the activity unguided) would be more effective. Undergraduate students were able to follow the lab exercise instructions without guidance, but asked clarifying questions. Perhaps these results are not surprising or unexpected as the teachers have become subconsciously accustomed to how they themselves construct learning activities, while the undergraduates are currently immersed in classes that utilize the same teaching approach as the "lecture then laboratory" model used here. These disparities in pedagogic receptiveness suggest that lesson plans and approaches need to be expressly tailored to each demographic group, and that age or education level is not a good grouping denominator.

It is also worthwhile to consider potential applications for extending the teaching activities beyond that described here as a way to increase student motivation and interest. Possibilities could include the investigation of whether fruits found at local farmer's markets are sourced as claimed by a vendor, as well as applications making use of the geographical information provided by the sourcing analysis. For example, food transportation to market may involve fossil fuel resources, and therefore the carbon cycle could be tied to the water cycle. In this way, the teaching materials discussed here provide a versatile foundation to extend or build upon.

5 Conclusions

We developed, described here, and provided teaching materials in the Supplement for an education and engagement activity that introduces hydrologic concepts using H and O stable isotope measurements in fruit water. The fruit water isotope dataset provided in this paper is a useful dataset for teaching data analysis and interpretation. The core scientific disciplinary areas of hydrology, isotope geochemistry, and predictive mapping were introduced to the students through the teaching activities and laboratory exercises. We found that students gained effective learning outcomes as measured by assessment metrics, though feedback indicated that pedagogic tailoring may be a way to increase teaching effectiveness.

This pilot study provides an example of ways that real scientific data resulting from research activities can be incorporated into teaching activities. The study described here can be used as a foundation for further development of outreach materials that may effectively engage a range of participant groups in learning about the water cycle and the ways in which humans modify the water cycle through agricultural activity.

Competing interests. The authors declare that they have no conflict of interest.

Acknowledgements. We appreciate the classroom assistance of Griffin Siebert. This research was supported by NSF EPSCoR grant IIA 1208732, awarded to Utah State University as part of the State of Utah Research Infrastructure Improvement Award. Erik Oerter's preparation of this paper was performed under the auspices of the US Department of Energy by Lawrence Livermore National Laboratory under contract DE-AC52-07NA27344, release number LLNL-JRNL-725565. Constructive comments by Sally Male and an anonymous reviewer helped improve this paper.

Edited by: Anas Ghadouani

References

Bar, V.: Children's views about the water cycle, Sci. Educ., 73, 481–500, 1989.

Bowen, G. J.: Isoscapes: spatial pattern in isotopic biogeochemistry, Annu. Rev. Earth Planet. Sc., 38, 161–187, 2010.

Bowen, G. J. and Revenaugh, J.: Interpolating the isotopic composition of modern meteoric precipitation, Water Resour. Res., 39, 1299–1312, https://doi.org/10.1029/2003WR002086, 2003.

CDFA: California Agricultural Statistics Review: 2015–2016, California Department of Food and Agriculture, Sacramento, CA, 126, 2016.

Cerling, T. E., Barnette, J. E., Bowen, G. J., Chesson, L. A., Ehleringer, J. R., Remien, C. H., Shea, P., Tipple, B. J., and West, J. B.: Forensic Stable Isotope Biogeochemistry, Annu. Rev. Earth Planet. Sc., 44, 175–206, 2016.

Chen, Z.-M. and Chen, G.: Virtual water accounting for the globalized world economy: national water footprint and international virtual water trade, Ecol. Indic., 28, 142–149, 2013.

Coplen, T. B.: Reporting of stable hydrogen, carbon, and oxygen isotopic abundances (technical report), Pure Appl. Chem., 66, 273–276, 1994.

Craig, H.: Isotopic variations in meteoric waters, Science, 133, 1702–1703, 1961.

Dawson, T. E., Mambelli, S., Plamboeck, A. H., Templer, P. H., and Tu, K. P.: Stable isotopes in plant ecology, Annu. Rev. Ecol. Syst., 33, 507–559, 2002.

Good, S. P., Kennedy, C. D., Stalker, J. C., Chesson, L. A., Valenzuela, L. O., Beasley, M. M., Ehleringer, J. R., and Bowen, G.: Patterns of local and nonlocal water resource use across the western US determined via stable isotope intercomparisons, Water Resour. Res., 50, 8034–8049, 2014.

Hoekstra, A. Y. and Mekonnen, M. M.: The water footprint of humanity, P. Natl. Acad. Sci. USA, 109, 3232–3237, 2012.

Kolb, D. A.: Experiential learning: Experience as the source of learning and development, FT Press, Upper Saddle River, New Jersey, 2014.

Kolb, D. A., Boyatzis, R. E., and Mainemelis, C.: Experiential learning theory: Previous research and new directions, Perspectives on Thinking, Learning, and Cognitive Styles, 1, 227–247, 2001.

Kramer, P. J. and Boyer, J. S.: Water relations of plants and soils, Academic Press, San Diego, California, 1995.

Majoube, M.: Fractionnement en oxygene-18 et en deuterium entre l'eau et sa vapeur, J. Chem. Phys., 68, 1423–1436, 1971.

Meyer, J. and Land, R.: Threshold concepts and troublesome knowledge: Linkages to ways of thinking and practicing within the disciplines, in: Improving Student Learning: Improving Student Learning Theory and Practice – Ten Years On, edited by: Rust, C., Oxford Centre for Staff and Learning Development, Oxford, 412–424, 2003.

Meyer, J. H. and Land, R.: Threshold concepts and troublesome knowledge (2): Epistemological considerations and a conceptual framework for teaching and learning, Higher Educ., 49, 373–388, 2005.

Mount, J. and Hanak, E.: Water use in California, Public Policy Institute of California, San Francisco, California, 2, 2016.

NGSS Lead States: The Next Generation Science Standards: For States By States, Washington D.C., The National Academies Press, Washington, D.C., 2013.

Oerter, E. J. and Bowen, G.: In situ monitoring of H and O stable isotopes in soil water reveals ecohydrologic dynamics in managed soil systems, Ecohydrology, 10, e1841, https://doi.org/10.1002/eco.1841, 2017.

Oerter, E. J., Perelet, A., Pardyjak, E., and Bowen, G.: Membrane inlet laser spectroscopy to measure H and O stable isotope compositions of soil and sediment pore water with high sample throughput, Rapid Commun. Mass Sp., 31, 75–84, 2017.

Oki, T. and Kanae, S.: Global hydrological cycles and world water resources, Science, 313, 1068–1072, 2006.

Orton, A. and Frobisher, L.: Insights Into Teaching Mathematics, A&C Black, London, 2004.

Osborne, J. and Dillon, J.: Science education in Europe: Critical reflections, The Nuffield Foundation, London, 2008.

Pachauri, R. K., Allen, M. R., Barros, V. R., Broome, J., Cramer, W., Christ, R., Church, J. A., Clarke, L., Dahe, Q., and Dasgupta, P.: Climate change 2014: synthesis report. Contribution of Working Groups I, II and III to the fifth assessment report of the Intergovernmental Panel on Climate Change, IPCC, 2014.

Quinn, H., Schweingruber, H., and Keller, T.: A framework for K-12 science education: Practices, crosscutting concepts, and core ideas, National Academies Press, Washington, D.C., 2012.

Rappaport, E. D.: What undergraduates think about clouds and fog, Journal of Geoscience Education, 57, 145–151, 2009.

Ruddell, B. L. and Wagener, T.: Grand challenges for hydrology education in the 21st century, J. Hydrol. Eng., 20, A4014001, https://doi.org/10.1061/(ASCE)HE.1943-5584.0000956, 2013.

12

Managed aquifer recharge with reverse-osmosis desalinated seawater: modeling the spreading in groundwater using stable water isotopes

Yonatan Ganot[1,2,a], **Ran Holtzman**[2], **Noam Weisbrod**[3], **Anat Bernstein**[3], **Hagar Siebner**[3], **Yoram Katz**[4], **and Daniel Kurtzman**[1]

[1]Institute of Soil, Water and Environmental Sciences, The Volcani Center, Agricultural Research Organization, Rishon LeZiyyon, 7505101, Israel
[2]Department of Soil and Water Sciences, The Hebrew University of Jerusalem, Rehovot, 7610001, Israel
[3]Department of Environmental Hydrology & Microbiology, Zuckerberg Institute for Water Research, Jacob Blaustein Institutes for Desert Research, Ben-Gurion University of the Negev, Midreshet Ben-Gurion, 8499000, Israel
[4]Mekorot, Water Company Ltd., Tel Aviv, 6713402, Israel
[a]now at: Department of Land, Air, and Water Resources, University of California, Davis, 95616, USA

Correspondence: Yonatan Ganot (yonatan.ganot@mail.huji.ac.il)

Abstract. The spreading of reverse-osmosis desalinated seawater (DSW) in the Israeli coastal aquifer was studied using groundwater modeling and stable water isotopes as tracers. The DSW produced at the Hadera seawater reverse-osmosis (SWRO) desalination plant is recharged into the aquifer through an infiltration pond at the managed aquifer recharge (MAR) site of Menashe, Israel. The distinct difference in isotope composition between DSW ($\delta^{18}O = 1.41$ ‰; $\delta^2H = 11.34$ ‰) and the natural groundwater ($\delta^{18}O = -4.48$ ‰ to -5.43 ‰; $\delta^2H = -18.41$ ‰ to -22.68 ‰) makes the water isotopes preferable for use as a tracer compared to widely used chemical tracers, such as chloride. Moreover, this distinct difference can be used to simplify the system to a binary mixture of two end-members: desalinated seawater and groundwater. This approach is validated through a sensitivity analysis, and it is especially robust when spatial data of stable water isotopes in the aquifer are scarce. A calibrated groundwater flow and transport model was used to predict the DSW plume distribution in the aquifer after 50 years of MAR with DSW. The results suggest that after 50 years, 94 % of the recharged DSW was recovered by the production wells at the Menashe MAR site. The presented methodology is useful for predicting the distribution of reverse-osmosis desalinated seawater in various downstream groundwater systems.

1 Introduction

Desalinated seawater global production is projected to double by 2040 while extending its geographical extent (Hanasaki et al., 2016). In some regions, desalinated seawater (DSW) is already the main source for fresh water (Dawoud, 2005). In Israel, for example, DSW reached 66 % of the domestic and industrial fresh water supply in 2017 (Israel Water Authority, 2018). This growing use of DSW affects downstream water systems such as reservoirs (Ronen-Eliraz et al., 2017; Negev et al., 2017; Stuyfzand et al., 2017; Ganot et al., 2017, 2018), wastewater treatment plants (Lahav et al., 2010; Negev et al., 2017) and agricultural irrigation (Lahav et al., 2010; Yermiyahu et al., 2007). One direct way by which DSW use affects the water budget is managed aquifer recharge (MAR). MAR using different water sources has been practiced for over 5 decades as part of the integrated water resource management of Israel (Dreizin et al., 2008; Gvirtzman, 2002) and is becoming a major component of water management in many Mediterranean countries (Rodríguez-Escales et al., 2018). Excess DSW produced in Israel due to operational constraints made it an attractive alternative source for MAR, raising the need to understand its effect. While the relatively rapid hydrological and geochemical processes (timescales of hours to weeks) of this new

MAR activity were recently monitored and modeled (Ganot et al., 2017, 2018; Ronen-Eliraz et al., 2017), the potential long-term (months to decades) impact of this process on the natural aquifer is still unknown, lacking observations and quantitative studies.

Stable water isotopes ^{18}O and ^{2}H are excellent tracers for water generated by seawater reverse-osmosis (SWRO) desalination. The lack of fractionation during the reverse-osmosis process, in contrast with various isotope-fractionation processes occurring in natural, fresh water (Al-Basheer et al., 2017; Gat, 1996; Kloppmann et al., 2008a, b), is the cause of the distinct difference in isotope composition between reverse-osmosis DSW and groundwater (GW) originating from natural, fresh water (Ganot et al., 2018; Kloppmann et al., 2018; Negev et al., 2017). For example, the advantage of using ^{18}O and ^{2}H as a quantitative tool for tracing treated wastewater (originating from DSW) that mixes with GW was recently demonstrated by comparing the mixing ratios of chloride, carbamazepine and water isotopes in the soil aquifer treatment (SAT) site at the Shafdan MAR system, Israel (Negev et al., 2017).

Here, we use stable water isotopes to trace the spreading of DSW in the aquifer and the production wells within the MAR site of Menashe, Israel. The DSW is produced at the Hadera SWRO desalination plant, which has operated since 2010 with an annual production capacity of 130 million cubic meters (MCM). It is one of five large SWRO desalination plants (production capacity ≥ 90 MCM, per year per plant) that were built along the Mediterranean coast of Israel during 2005–2015 (Stanhill et al., 2015). The DSW is regularly supplied directly to consumers through the centralized national water system. Periodically, operational constraints, such as maintenance of the national system, prohibit distribution of the DSW; limited reservoir capacity makes storage of this expensive surplus of DSW in the aquifer through MAR operations the only feasible solution (Ganot et al., 2017, 2018; Ronen-Eliraz et al., 2017).

There are currently only a few places that are practicing MAR with DSW, but this practice is expected to grow due to the increasing use of DSW globally. Practically, most of the known case studies of MAR with DSW involve brackish-water aquifers (mainly in the Persian Gulf countries) and not necessarily reverse-osmosis DSW. In this work we present a unique case study that explores the spreading of a reverse-osmosis DSW plume in a fresh-water aquifer. The use of two isotope-distinguished end-members, in this case, reverse-osmosis DSW and natural, fresh GW, is a prerequisite to implement the analysis presented in this paper.

Predicting the long-term DSW distribution in the aquifer and the production wells is the main objective of this study. We incorporate water isotope data of ^{18}O and ^{2}H in a regional GW flow and transport model (e.g., Boronina et al., 2005; Krabbenhoft et al., 1990; Liu et al., 2014; Reynolds and Marimuthu, 2007; Stichler et al., 2008) in order to predict the DSW distribution in the aquifer. While the methodology for measuring the present mixing of DSW and GW was reported previously (Negev et al., 2017), in the current study our GW modeling approach allows us to predict future mixing trends in the production wells of the Menashe MAR site. Predicting the DSW distribution in the aquifer is of main interest from water quantity (estimating the recovery potential of DSW originating from MAR) and quality perspectives (e.g., Birnhack et al., 2011; Ganot et al., 2018 and references therein).

2 Methods

2.1 Study area

The Menashe MAR site is located on sand dunes 28 m above sea level in the northern part of the Israeli coastal aquifer, an unconfined sandy aquifer stretching over an area of 2000 km^2 along the Mediterranean coast (Fig. 1a). The local climate is Mediterranean, with an annual average temperature of 20.2 °C, and annual mean precipitation of 566 mm yr^{-1} (Israel Meteorological Service, 2014). The aquifer thickness varies from 100 m on the coastline (to the west of the Menashe site) to a few meters in the east. It is composed of Pleistocene calcareous sandstone interleaved with discontinuous marine and continental silt and clay lenses. Thick Neogene clay (Saqiye Group), which is highly impermeable, underlies the aquifer (Kurtzman et al., 2012). The regional groundwater level is ~ 3 m a.m.s.l. (September 2014, Israel Water Authority, 2014), and the characteristic aquifer properties are a hydraulic conductivity of 10 m d^{-1}, storativity of 0.25 and porosity of 0.4 (Shavit and Furman, 2001).

The Menashe MAR site diverts the natural ephemeral flows from the Menashe Heights streams into a settling pond, and from there to three infiltration ponds. Production wells that encircle the site recover the recharged water from the aquifer (Sellinger and Aberbach, 1973). In the last few years, the southern infiltration pond is dedicated for the infiltration of a surplus of DSW from the Hadera SWRO desalination plant, located 4 km to the west, on the coastline (Fig. 1b).

2.2 Water sampling

Groundwater from 14 wells at the Menashe MAR site were sampled biannually during 2015 to 2017 ($n = 42$). In addition, water was sampled from the infiltration pond DSW inlet pipe ($n = 3$), a few locations inside the pond during MAR events ($n = 4$), shallow observation wells (OA and OB; $n = 11$) and runoff canals ($n = 1$). Stable water isotopes (expressed as $\delta^{18}O$ and $\delta^{2}H$ in ‰ vs. the VSMOW – Vienna Standard Mean Ocean Water) were measured by a cavity ring-down spectroscopy (CRDS) analyzer (L2130-i, Picarro).

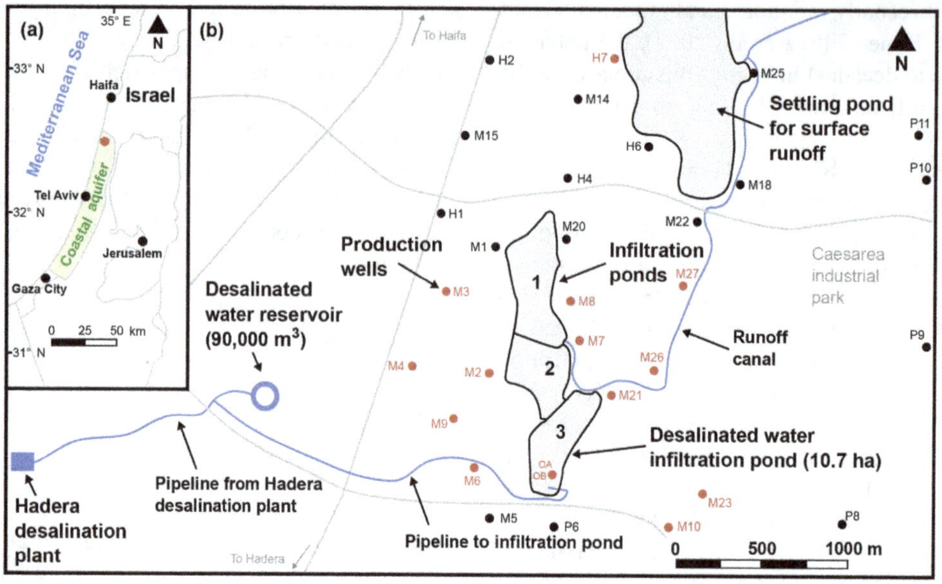

Figure 1. Map of the study area. **(a)** Location of the Israeli coastal aquifer and the Menashe MAR site (red circle). **(b)** The Menashe MAR site. A surplus of desalinated seawater is delivered from the Hadera SWRO desalination plant (lower left) to the southern infiltration basin (pond 3). The red dots represent wells that were sampled for water isotope analysis.

2.3 Groundwater flow and transport model

A detailed three-dimensional transient water flow and solute transport model was set up in order to estimate DSW spreading in the aquifer at the Menashe site area. The model covers an area of 65 km^2, including a western offshore strip of 9 km^2 (Fig. 2a). The geological data processed from well logs, geological maps and structural maps served as the basis for the conceptual model, constructed via the GMS software package (version 10.3; https://www.aquaveo.com, last access: 22 November 2018). The variety of rock types was grouped into four hydrogeological units, each characterized by a set of hydrological properties (Table 1). Over 100 well logs were analyzed using the T-PROGS software (Carle, 1999) and provided the spatial distribution of the hydrogeological units. This geostatistically generated unit array, conditioned to the borehole logs, was combined with structural map data of the major marine clay lenses present in the aquifer. The resulting model hence reflects the hydrogeological units' proportions and transition trends as well as the division into sub-aquifers by marine clay within the western part of the aquifer (Fig. 2b, c).

The model domain was discretized horizontally into 70×70 m mesh cells. The vertical section of the aquifer, with thickness ranging 50–100 m from east to west, was divided into 24 layers with a vertical spatial resolution of 5 m or smaller. The model's bottom boundary was defined by the impermeable Saqiye Group underlying the aquifer. The model's top boundary was defined by the water table representing an unconfined aquifer. Boundary conditions along the northern, eastern and southern model boundaries were set to be of a transient head type, based on periodic water level measurements. The western boundary was set to a constant head boundary dictated by the sea level. Initial conditions were based on static heads measured at several dozens of production and observation wells included in the model. Sources and sinks in the flow model include recharge by precipitation, MAR (both runoff and DSW recharge) and production wells. Natural recharge from precipitation was based on adjacent rain gauge measurements (Gan Shmuel), using an average recharge coefficient of 0.4 (which is representative of sands). Recharge flux of DSW by MAR activity was calculated by a variably saturated model of the upper 30 m of the sediment under the southern infiltration pond (Ganot et al., 2017). Pumping activity of the production wells was based on a database from the national water company of Israel, Mekorot.

The transport model considers the stable water isotopes ^{18}O and ^{2}H as conservative tracers, i.e., neglecting isotope fractionation (there is strong evidence that local groundwater tends to be isotopically uniform, see for example Krabbenhoft et al., 1990 and references therein). We normalize the tracer concentration as $C = (\delta - \delta_{min})/(\delta_{max} - \delta_{min})$, where δ is the isotope composition of $\delta^{18}O$ or $\delta^{2}H$ in the aquifer, and δ_{min} and δ_{max} are the minimum and maximum isotope composition, respectively. Since practically $\delta_{max} = \delta_{DSW}$, the normalized concentration of DSW is $C_{DSW} = 1$, whereas that of GW ranges from $C_{GW} = 0$ ($\delta^{18}O = -5.43$ ‰, $\delta^{2}H = -22.68$ ‰) to $C_{GW} = 0.13$ ($\delta^{18}O = -4.48$ ‰, $\delta^{2}H = -18.41$ ‰). Boundary conditions of the transport model are of specified mass flux (qC,

Table 1. Major rock types in the study area grouped into four hydrogeological units.

Hydrogeological unit	1	2	3	4
Rock types	Gravel, beach rock, kurkar with shells and gravel	Calcareous sandstone (kurkar), sand	Loam, sandy loam, loamy sand, marine silty sand	Clay and silt of marine or terrestrial origin
Hydraulic conductivity (K)	High	Medium	Low	Very low
Unit proportions (%)	4	59	23.5	13.5

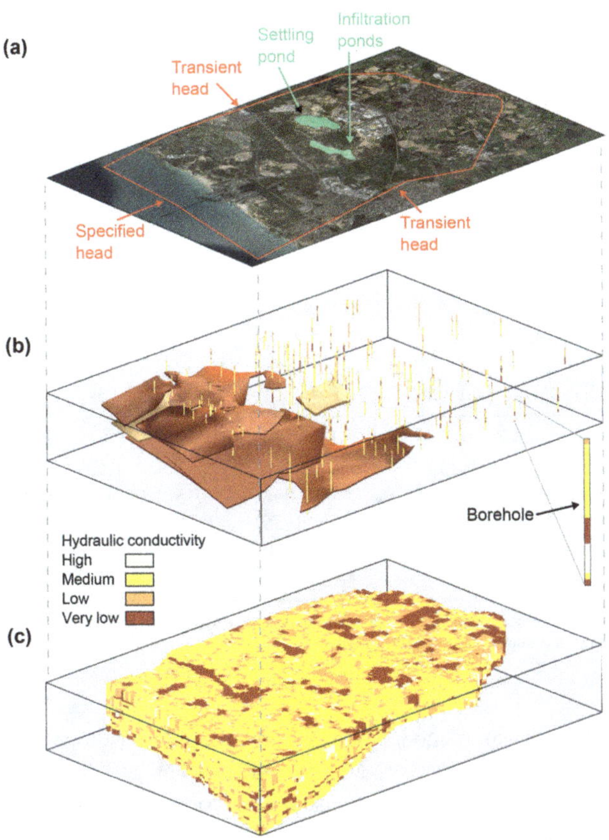

Figure 2. The model used in simulations of water flow and solute transport. **(a)** The modeled area and boundary conditions. **(b)** The major continuous marine clay lenses and the borehole log. **(c)** The combined deterministic and geostatistically generated material array representing the aquifer in the model.

where q is the specific discharge), with zero flux at the bottom boundary (considered impermeable) as well as zero flux at the northern, eastern and southern boundaries, and also with the precipitation and the runoff-pond source terms, due to their GW isotope compositions ($C_{GW} = 0$). Mass flux with the DSW isotope composition ($C_{DSW} = 1$) is given at the western boundary (sea) and the DSW infiltration pond source term. The validity of the use of a single value ($C_{GW} = 0$) for the GW mass-flux boundaries, in light of the range of isotope compositions in the aquifer prior to the MAR of DSW ($\delta^{18}O = -4.48$ ‰ to -5.43 ‰ and $\delta^2H = -18.41$ ‰ to -22.68 ‰), is discussed in Sect. 3.2.3. Initial conditions were set by interpolating the water isotope data from several production wells.

The MODFLOW (Harbaugh et al., 2000) and MT3DMS (Zheng and Wang, 1999) codes were used through the GMS user interface to numerically solve the flow and transport models, respectively. Both codes, which use a finite difference scheme, are considered reliable and are therefore widely used for regional aquifer modeling (Zhou and Li, 2011). The flow and transport model was calibrated using a dataset from 2015 to 2017. During 2015, 2016 and 2017 a volume of 2.6, 1.3 and 0.6 MCM of DSW were recharged, respectively, at the Menashe MAR site. In these years the MAR events consisted of a noncontinuous discharge of DSW to pond 3 (Fig. 1b) during January and/or February (Ganot et al., 2017, 2018). In addition, a volume of 3.2 and 1.6 MCM of runoff water were discharged to the settling pond during 2015 and 2017, respectively.

3 Results and discussion

3.1 Water isotopes

The distinct difference between the water isotopes of the production wells and DSW is shown in a δ^2H vs. $\delta^{18}O$ diagram for the period of 2015 to 2017 (Fig. 3a and Table S1 in the Supplement). During 2016, and more prominently in 2017, few wells showed a progressive change in composition towards higher isotope values – a transition from GW towards DSW on the mixing line (Fig. 3a) indicates mixing with DSW – while most wells retain a constant isotope composition. Note that for all samples in Fig. 3a, there is a strong linear correlation between $\delta^{18}O$ and δ^2H ($R^2 = 0.9991$); thus, hereafter we only report δ^2H as a tracer.

The isotope composition of δ^2H and the concentration of chloride are shown for comparison in nine wells during the years 2010 to 2018 (Fig. 3b). The chloride concentration of DSW at the Menashe MAR site is always lower than 10 mg L^{-1} (Ganot et al., 2018), while in the local GW it is found in a wider range of 40 to 140 mg L^{-1}. The large chloride concentration variability in the different wells prior to MAR with DSW (before 2015) suggests that various water sources feed the aquifer (as there is no extensive soluble salt

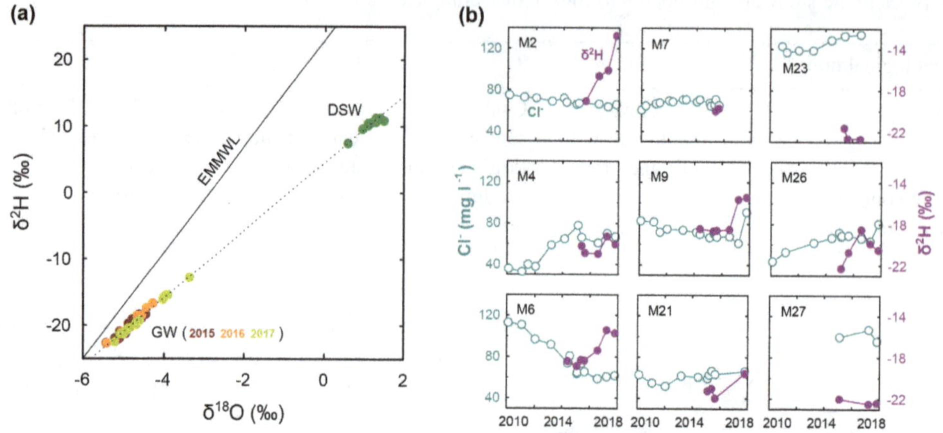

Figure 3. **(a)** Water isotopic composition of the production wells (GW) and reverse-osmosis desalinated seawater (DSW); the eastern Mediterranean meteoric water line (EMMWL) is shown for comparison (Gat and Dansgaard, 1972). **(b)** Chloride (Cl^-) and δ^2H sampled in nine production wells at the Menashe MAR site.

layer in the aquifer according to the recent available geological data). Moreover, the breakthrough of DSW in wells M2, M6 and M9, captured by an increase in δ^2H, is not reflected in the chloride concentration (expected to decrease). This implies that chloride – in general a widely used conservative tracer – is less sensitive to reverse-osmosis DSW in natural, fresh GW systems and therefore less useful for its detection.

Finally, we note that the very different DSW signature, in terms of δ^2H from the other water sources in the Menashe site, reduces the problem of mixing various water sources to a binary system: (i) DSW and (ii) all other natural sources. This is because the signatures of runoff water ($\delta^{18}O = -4.77$ ‰ and $\delta^2H = -19.5$ ‰) and rainwater ($\delta^{18}O = -5.8$ ‰ and $\delta^2H = -19.9$ ‰; Gat and Dansgaard, 1972; Goldsmith et al., 2017) are very similar to that of the local GW. Therefore, the binary system approach used in this study, which is based on conservative water isotope tracers is superior to both conservative and nonconservative chemical tracers.

3.2 Model

3.2.1 Calibration

The flow model was calibrated against head data from 13 wells (Fig. 4a). We mainly used continuous head data measured at two production wells, M5 and M8 (Fig. 4b). Well M5, situated 400 m SE of pond 3, exploiting aquifer layers bounded between −16 and −54 m MSL, was inactive during 2015–2017, making it ideal for head monitoring. Well M8, situated 1 km north of pond 3, exploiting aquifer layers bounded between −14 and −48 m MSL, was used for production during some of the study period, and thus only selected head data (representing quasi-static heads) were used for calibration.

Table 2. Calibrated parameters used for the different hydrogeological units.

Hydrogeological unit	1	2	3	4
Horizontal K ($m\,d^{-1}$)	50	12	6	0.01
Vertical K ($m\,d^{-1}$)	12.5	3	1.5	0.01
Specific storage (m^{-1})	0.002	0.0015	0.001	0.001
Specific yield	0.35	0.12	0.12	0.1
Longitudinal dispersivity* (m)	20	20	20	20
Porosity	0.35	0.19	0.17	0.1

* Transverse horizontal and vertical dispersivities are 0.1 and 0.01, respectively, of the longitudinal dispersivity (Burnett and Frind, 1987).

The transport model was calibrated against isotope data from 12 wells (M2–4, M6–10, M21, M23 and M26–27; Fig. 4c). Specifically, we used data corresponding to the breakthrough of DSW in the down-gradient (western) production wells near the DSW infiltration pond (M2, M6, and M9) since other wells showed smaller δ^2H variations (Fig. 4d). The simulated groundwater heads and δ^2H for the calibration period are generally in good agreement with the observations. The calibrated hydrological parameters are specified in Table 2.

3.2.2 DSW spreading in the aquifer

Our simulations show that at the end of 2017 the DSW plume was spreading westwards (in the direction of the natural hydraulic gradient, as expected), approaching the closest western production wells (M2, M6, M9; Fig. 5a). Note that the production wells to the east (up-gradient) show a constant δ^2H, indicating no interaction with the DSW recharge. Variability of δ^2H along the production well screens (in the vertical direction), implies that the measured δ^2H is a mixture of several aquifer layers (Fig. 5b).

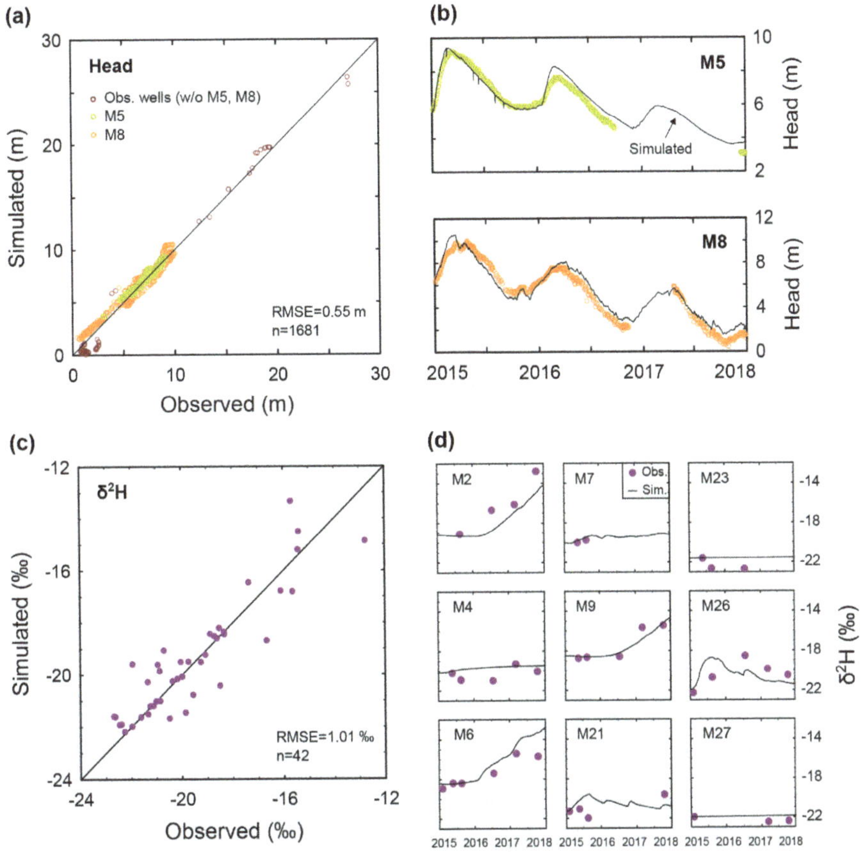

Figure 4. Model calibration. **(a)** Comparison of the simulated and observed hydraulic head. **(b)** Temporal variations of the simulated and observed hydraulic head in wells M5 and M8. **(c)** Comparison of the simulated and observed δ^2H. **(d)** Temporal variations of the simulated and observed δ^2H in nine selected wells.

The δ^2H variations shown in Fig. 5a reflect the DSW spreading in the aquifer. The highest δ^2H that was measured in the aquifer (prior to MAR of DSW) was $\delta^2\text{H} = 18.41$ ‰, and therefore any value above it indicates mixing with DSW. However, because the initial measured δ^2H values in the aquifer are in the range of $\delta^2\text{H} = -18.41$ ‰ to -22.68 ‰, the extent of DSW mixing in each well is relative to its specific initial δ^2H. This can be calculated by a mixing ratio (MR) approach with $\text{MR} = (\delta_\text{w} - \delta_\text{i}) / (\delta_\text{DSW} - \delta_\text{i})$, where δ_w is the δ^2H in the well, δ_i is the initial (background) δ^2H in the well and δ_DSW is the δ^2H of DSW. The MR value of 0 and 1 implies original aquifer water and pure DSW, respectively. Figure 5c shows the MR (expressed in %DSW) of three down-gradient wells (M2, M4 and M6), two up-gradient wells (M23 and M26) and an observation well (OA) inside the DSW pond. Wells M2 and M6 have up to a 20 % DSW portion, while M23 and OA retain original aquifer water and almost pure DSW, respectively. At the end of 2017, about 7 % of the recharged DSW was recovered by the production wells.

Knowing the water composition of the aquifer and DSW, and assuming a conservative transport of all the major ions, one can estimate the water composition in a specific well based on the calculated mixing ratio, $[X]_\text{w} = \text{MR} \times [X]_\text{DSW} + (1 - \text{MR}) \times [X]_\text{i}$. Here $[X]_\text{w}$ is the (calculated) ion concentration in the well, $[X]_\text{DSW}$ is the ion concentration in the DSW and $[X]_\text{i}$ is the initial ion concentration (background) in the well. Diversion of the observed concentration from the calculated concentration can give insight to the sediment–water reaction (e.g., Ganot et al., 2018; Ronen-Eliraz et al., 2017; Stuyfzand et al., 2017).

3.2.3 Binary system assumption

The model was based on the assumption that all water types in this system can be described by two end-members sorted by their isotope composition: (1) the "heavy" DSW ($\delta^2\text{H} = 11.34$ ‰), and (2) the "light" natural water ($\delta^2\text{H} = -22.68$ ‰) which includes all other water types (rain, runoff and GW). As pointed out before, while DSW isotope composition is constant, that of the local natural water is more variable. To examine the validity of the assumption of binary δ^2H values, we ran the simulation again for the same period of 2015 to 2017, but this time with the

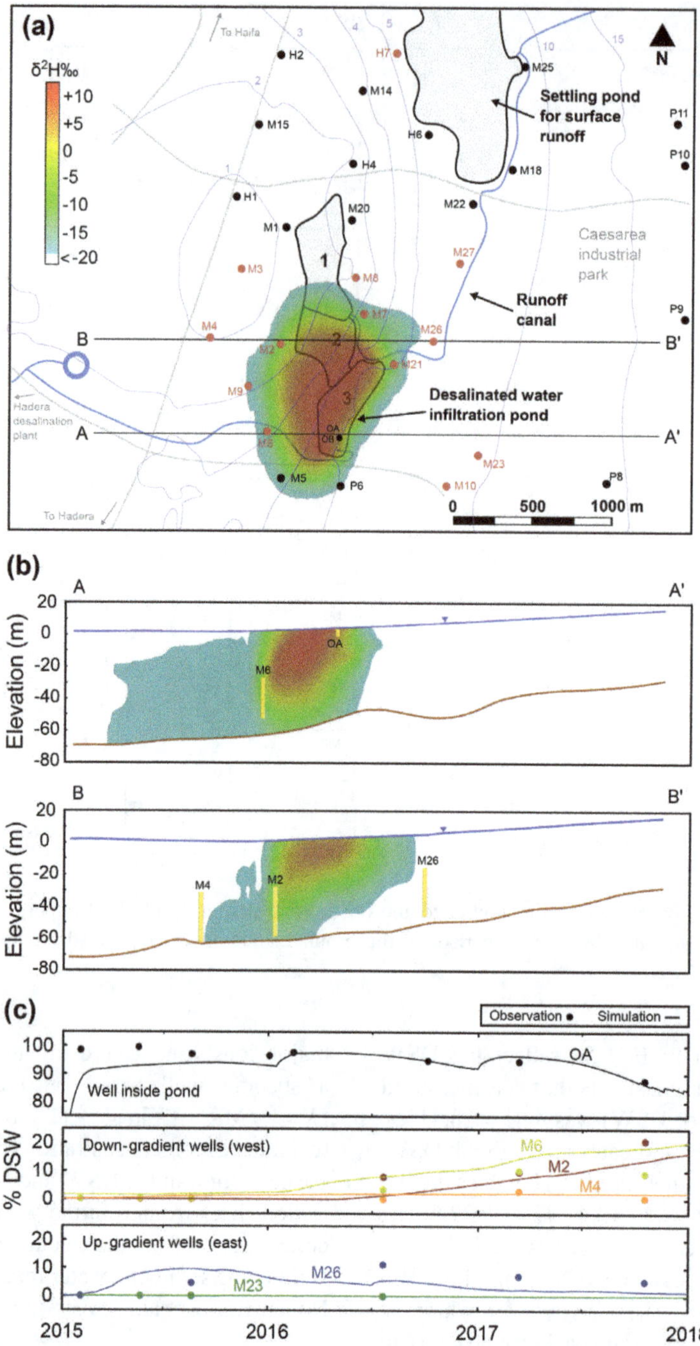

Figure 5. Simulation results showing DSW spreading at the end of 2017. **(a)** Plan view (water table); the colored area shows the DSW plume, the white area indicates natural GW (δ^2H<-20‰) and the blue contours are the GW head. **(b)** East–west cross sections through wells OA and M6 (A-A′) and wells M4, M2 and M26 (B-B′); well screens are shown in yellow. **(c)** Observed and simulated DSW fraction (%) in selected wells along the cross sections A-A′ and B-B′.

maximum value of GW δ^2H$=-18.41$‰ (in all GW boundaries and also as a rain and runoff source) in order to check the model sensitivity to the natural GW isotope variability. We subtracted the isotope composition results of the two simulations in all model cells to produce an error map (Fig. 6a) of δ^2H differences (Δ‰). In terms of δ^2H composition in the production wells (Fig. 6b), the results of both simulations were similar (Δ‰<1), while some differences (up to Δ‰$=4.3$) were found in the domain boundaries and at the upper layer that was affected by rain and runoff recharge.

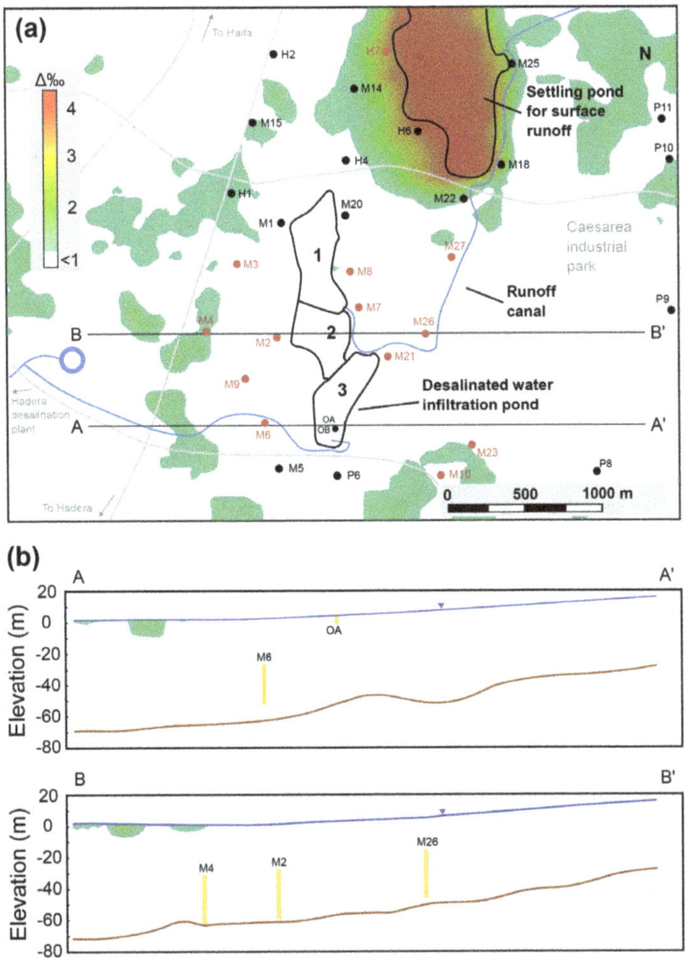

Figure 6. Examination of the validity of the assumption of binary isotopic mixing. **(a)** Plan view (water table) of δ^2H difference (Δ‰) between simulation results (2015–2017) with $\delta^2H_{max} = -18.41$ ‰ ($C_{GW} = 0.13$) and $\delta^2H_{min} = -22.68$ ‰ ($C_{GW} = 0$) at the end of 2017; the white area indicates Δ‰ < 1. **(b)** East–west cross sections through wells OA and M6 (A-A′) and wells M4, M2 and M26 (B-B′); well screens are shown in yellow.

Specifically, a notable difference is seen in the runoff settling pond, which is a source of natural water recharge. Nevertheless, for the area surrounding the DSW infiltration basin (pond 3), the binary assumption is valid due to the following conditions: (1) the distinct difference between the isotope composition of DSW and GW; (2) the model boundaries are relatively far (> 2 km) from the source of the MAR with DSW; and (3) the screens of the production wells are relatively deep (depth > 50 m). Hence, in this case we can conclude that the initial variability of isotope composition in the aquifer has a negligible impact on the simulation results. Practically, it implies that interpolation efforts of the aquifer isotope composition (prior to MAR with DSW) are unnecessary, as one can use an average isotope value to normalize the tracer concentration in the aquifer. In addition, a major advantage of the binary assumption is that it allows one to estimate the mixing when the spatial data of water isotope is limited. This was exploited in the current study, where isotope data of the model boundaries were unavailable.

The results of the error analysis also support the model assumption that isotope fractionation is negligible during GW flow (i.e., isotope composition is conservative) as the isotope composition variability in the aquifer (which originates from fractionation processes) does not impact the simulation results (Fig. 6). Moreover, our measurements at the Menashe MAR site show a similar isotope composition between the DSW source water at the surface, in the variably saturated zone and at the shallow GW (Ganot et al., 2018). Therefore, even if isotope fractionation exists in the aquifer to some extent as a slow process, it should be considered negligible compared to the distinct difference between the isotope end-members.

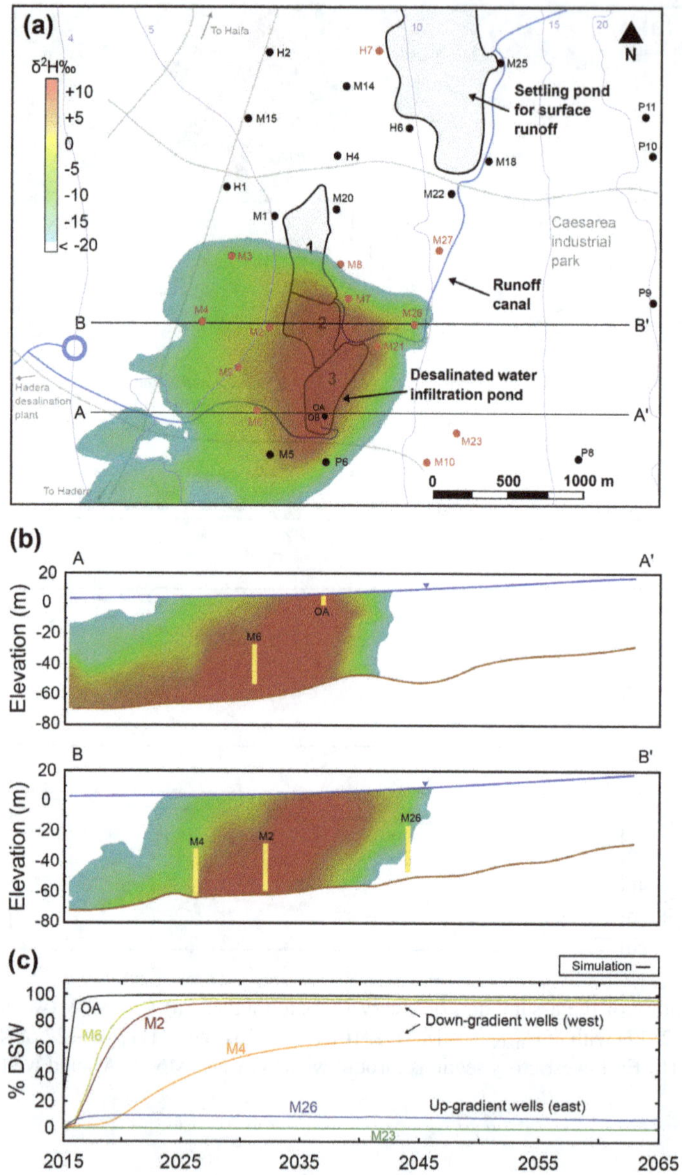

Figure 7. Long-term simulations of DSW spreading at the end of 2065 after 50 years of MAR. **(a)** Plan view (water table); the colored area shows the DSW plume, the white area indicates natural GW ($\delta^2H < -20$ ‰) and the blue contours are the GW head. **(b)** East–west cross sections through wells OA and M6 (A-A′) and wells M4, M2 and M26 (B-B′); well screens are shown in yellow. **(c)** Simulated DSW fraction (%) in selected wells along the cross sections A-A′ and B-B′.

3.2.4 Predicting long-term DSW spreading in the aquifer (2015–2065)

We test the extent of DSW spreading in the aquifer by performing long-term (50 years) simulation of MAR with DSW, considering 50 repeated annual cycles of the hydraulic conditions recorded in 2015, with a MAR event of 2.6 MCM (Fig. 7a). According to the simulation results, the water in the down-gradient (westwards) wells closest to the DSW pond, M2 and M6, will be fully exchanged with DSW after 10 years of MAR, while the up-gradient wells show little (M26) or no (M23) mixing with DSW (Fig. 7b, c). Interestingly, well M4, located further to the west, reaches a steady DSW mixing of almost 70 % after about 35 years of MAR without being fully exchanged with DSW, while the DSW plume continues to progress further west. By the end of 2065, the total DSW volume of 130 MCM recharged at the infiltration pond will be distributed as follows: 114 MCM (88 %) is recovered by the western pumping wells (M2–9, P6), 8.4 MCM (6 %) by the eastern pumping wells (M21, M26) and only 7.5 MCM (6 %) remains in the aquifer.

4 Conclusions

We track the fate of reverse-osmosis DSW that was introduced to groundwater by MAR using stable water isotopes. The use of the water isotopes of ^{18}O and ^{2}H is advantageous in this system for two reasons: (1) there is a distinct difference between the isotope composition of DSW and natural, fresh water; and (2) the water isotope composition of all natural water sources – groundwater, rain and runoff – is very similar. The former makes water stable isotopes a more sensitive tracer (compared to other natural conservative tracers such as chloride), whereas the latter reduces the problem to a binary mixture of two end-members: reverse-osmosis DSW and natural GW. We formulate a detailed three-dimensional GW flow and transport model, exploiting these advantages. The model, calibrated using field data (measured during 2015–2017), is used to predict the spreading of DSW in the aquifer during 50 years of MAR with DSW. Our simulation results suggest that most of the recharged DSW (94 %) is recovered by the production wells, indicating the efficacy of the Menashe MAR site.

The advantage of using stable water isotopes for tracing reverse-osmosis DSW in various downstream water systems is already known from previous studies. In this study we used this advantage in a modeling framework to predict future mixing and spreading trends of DSW in an aquifer. Hence, this modeling approach can be used in other MAR sites (e.g., Mazariegos et al., 2017; Negev et al., 2017; Stuyfzand et al., 2017) to predict reverse-osmosis DSW distribution in aquifers. As the production of DSW using reverse-osmosis is projected to increase and the use of MAR systems expands, we believe that the methodology presented in this paper will be highly relevant for more MAR hydrologists.

Competing interests. The authors declare that they have no conflict of interest.

Acknowledgements. The research leading to these results received funding from the joint German–Israeli water technology research program BMBF–MOST, project WT1401. We thank Amos Russak and Raz Studny (Ben-Gurion University) for sampling and analysis assistance.

Edited by: Bill X. Hu

References

Al-Basheer, W., Al-Jalal, A., and Gasmi, K.: Variations in isotopic composition of desalinated water, Water Environ. J., 31, 209–214, https://doi.org/10.1111/wej.12232, 2017.

Birnhack, L., Voutchkov, N., and Lahav, O.: Fundamental chemistry and engineering aspects of post-treatment processes for desalinated water – A review, Desalination, 273, 6–22, https://doi.org/10.1016/j.desal.2010.11.011, 2011.

Boronina, A., Balderer, W., Renard, P., and Stichler, W.: Study of stable isotopes in the Kouris catchment (Cyprus) for the description of the regional groundwater flow, J. Hydrol., 308, 214–226, https://doi.org/10.1016/j.jhydrol.2004.11.001, 2005.

Burnett R. D. and Frind E. O.: Simulation of contaminant transport in three dimensions: 2. Dimensionality effects, Water Resour. Res., 23, 695–705, https://doi.org/10.1029/WR023i004p00695, 1987.

Carle, S. F.: T-PROGS: Transition probability geostatistical software, University of California, Davis, CA, 84, available at: http://gmsdocs.aquaveo.com/t-progs.pdf (last access: 15 August 2017), 1999.

Dawoud, M. A.: The role of desalination in augmentation of water supply in GCC countries, Desalination, 186, 187–198, https://doi.org/10.1016/j.desal.2005.03.094, 2005.

Dreizin, Y., Tenne, A., and Hoffman, D.: Integrating large scale seawater desalination plants within Israel's water supply system, Desalination, 220, 132–149, https://doi.org/10.1016/j.desal.2007.01.028, 2008.

Ganot, Y., Holtzman, R., Weisbrod, N., Nitzan, I., Katz, Y., and Kurtzman, D.: Monitoring and modeling infiltration–recharge dynamics of managed aquifer recharge with desalinated seawater, Hydrol. Earth Syst. Sci., 21, 4479–4493, https://doi.org/10.5194/hess-21-4479-2017, 2017.

Ganot, Y., Holtzman, R., Weisbrod, N., Russak, A., Katz, Y., and Kurtzman, D.: Geochemical Processes During Managed Aquifer Recharge With Desalinated Seawater, Water Resour. Res., 54, 978–994, https://doi.org/10.1002/2017WR021798, 2018.

Gat, J. R.: Oxygen and hydrogen isotopes in the hydrologic cycle, Annu. Rev. Earth Pl. Sc., 24, 225–262, 1996.

Gat, J. R. and Dansgaard, W.: Stable isotope survey of the fresh water occurrences in Israel and the Northern Jordan Rift Valley, J. Hydrol., 16, 177–211, https://doi.org/10.1016/0022-1694(72)90052-2, 1972.

Goldsmith, Y., Polissar, P. J., Ayalon, A., Bar-Matthews, M., deMenocal, P. B., and Broecker, W. S.: The modern and Last Glacial Maximum hydrological cycles of the Eastern Mediterranean and the Levant from a water isotope perspective, Earth Planet. Sc. Lett., 457, 302–312, https://doi.org/10.1016/j.epsl.2016.10.017, 2017.

Gvirtzman, H.: Israel Water Resources, Chapters in Hydrology and Environmental Sciences, Yad Ben-Zvi Press, Jerusalem, 2002.

Hanasaki, N., Yoshikawa, S., Kakinuma, K., and Kanae, S.: A seawater desalination scheme for global hydrological models, Hydrol. Earth Syst. Sci., 20, 4143–4157, https://doi.org/10.5194/hess-20-4143-2016, 2016.

Harbaugh, A. W., Banta, E. R., Hill, M. C., and McDonald, M. G.: MODFLOW-2000, The US Geological Survey Modular Ground-Water Model – User Guide to Modularization Concepts and the Ground-Water Flow Process, Report, available at: http://pubs.er.usgs.gov/publication/ofr200092 (last access: 30 November 2018), 2000.

Israel Meteorological Service: Climate atlas of Israel, http://www.ims.gov.il/IMS/CLIMATE/ClimaticAtlas/ (last access: 4 June 2018), 2014.

Israel Water Authority: National groundwater levels report, http://www.water.gov.il/Hebrew/ProfessionalInfoAndData/Data-Hidrologeime/DocLib2/hydrological-report-sep14.pdf (last access: 4 June 2018), 2014.

Israel Water Authority: Water consumption report, http://www.water.gov.il/Hebrew/ProfessionalInfoAndData/Allocation-Consumption-and-production/20173/intro.pdf, last access: 19 November 2018.

Kloppmann, W., Vengosh, A., Guerrot, C., Millot, R., and Pankratov, I.: Isotope and Ion Selectivity in Reverse Osmosis Desalination: Geochemical Tracers for Man-made Freshwater, Environ. Sci. Technol., 42, 4723–4731, https://doi.org/10.1021/es7028894, 2008a.

Kloppmann, W., Van Houtte, E., Picot, G., Vandenbohede, A., Lebbe, L., Guerrot, C., Millot, R., Gaus, I., and Wintgens, T.: Monitoring Reverse Osmosis Treated Wastewater Recharge into a Coastal Aquifer by Environmental Isotopes (B, Li, O, H), Environ. Sci. Technol., 42, 8759–8765, https://doi.org/10.1021/es8011222, 2008b.

Kloppmann, W., Negev, I., Guttman, J., Goren, O., Gavrieli, I., Guerrot, C., Flehoc, C., Pettenati, M., and Burg, A.: Massive arrival of desalinated seawater in a regional urban water cycle: A multi-isotope study (B, S, O, H), Sci. Total Environ., 619/620, 272–280, https://doi.org/10.1016/j.scitotenv.2017.10.181, 2018.

Krabbenhoft, D. P., Anderson, M. P., and Bowser, C. J.: Estimating groundwater exchange with lakes: 2. Calibration of a three-dimensional, solute transport model to a stable isotope plume, Water Resour. Res., 26, 2455–2462, https://doi.org/10.1029/WR026i010p02455, 1990.

Kurtzman, D., Netzer, L., Weisbrod, N., Nasser, A., Graber, E. R., and Ronen, D.: Characterization of deep aquifer dynamics using principal component analysis of sequential multilevel data, Hydrol. Earth Syst. Sci., 16, 761–771, https://doi.org/10.5194/hess-16-761-2012, 2012.

Lahav, O., Kochva, M., and Tarchitzky, J.: Potential drawbacks associated with agricultural irrigation with treated wastewaters from desalinated water origin and possible remedies, Water Sci. Technol., 61, 2451, https://doi.org/10.2166/wst.2010.157, 2010.

Liu, Y., Yamanaka, T., Zhou, X., Tian, F. and Ma, W.: Combined use of tracer approach and numerical simulation to estimate groundwater recharge in an alluvial aquifer system: A case study of Nasunogahara area, central Japan, J. Hydrol., 519, 833–847, https://doi.org/10.1016/j.jhydrol.2014.08.017, 2014.

Mazariegos, J. G., Walker, J. C., Xu, X., and Czimczik, C. I.: Tracing Artificially Recharged Groundwater using Water and Carbon Isotopes, Radiocarb. Tucson, 59, 407–421, https://doi.org/10.1017/RDC.2016.51, 2017.

Negev, I., Guttman, J. and Kloppmann, W.: The Use of Stable Water Isotopes as Tracers in Soil Aquifer Treatment (SAT) and in Regional Water Systems, Water, 9, 73, https://doi.org/10.3390/w9020073, 2017.

Reynolds, D. A. and Marimuthu, S.: Deuterium composition and flow path analysis as additional calibration targets to calibrate groundwater flow simulation in a coastal wetlands system, Hydrogeol. J., 15, 515–535, https://doi.org/10.1007/s10040-006-0113-5, 2007.

Rodríguez-Escales, P., Canelles, A., Sanchez-Vila, X., Folch, A., Kurtzman, D., Rossetto, R., Fernández-Escalante, E., Lobo-Ferreira, J.-P., Sapiano, M., San-Sebastián, J., and Schüth, C.: A risk assessment methodology to evaluate the risk failure of managed aquifer recharge in the Mediterranean Basin, Hydrol. Earth Syst. Sci., 22, 3213–3227, https://doi.org/10.5194/hess-22-3213-2018, 2018.

Ronen-Eliraz, G., Russak, A., Nitzan, I., Guttman, J., and Kurtzman, D.: Investigating geochemical aspects of managed aquifer recharge by column experiments with alternating desalinated water and groundwater, Sci. Total Environ., 574, 1174–1181, https://doi.org/10.1016/j.scitotenv.2016.09.075, 2017.

Sellinger, A. and Aberbach, S.: Artificial recharge of coastal-plain aquifer in Israel, Underground Waste Management and Artificial Recharge, 2, 701–714, 1973.

Shavit, U. and Furman, A.: The location of deep salinity sources in the Israeli Coastal aquifer, J. Hydrol., 250, 63–77, https://doi.org/10.1016/S0022-1694(01)00406-1, 2001.

Stanhill, G., Kurtzman, D., and Rosa, R.: Estimating desalination requirements in semi-arid climates: A Mediterranean case study, Desalination, 355, 118–123, https://doi.org/10.1016/j.desal.2014.10.035, 2015.

Stichler, W., Maloszewski, P., Bertleff, B., and Watzel, R.: Use of environmental isotopes to define the capture zone of a drinking water supply situated near a dredge lake, J. Hydrol., 362, 220–233, https://doi.org/10.1016/j.jhydrol.2008.08.024, 2008.

Stuyfzand, P., Smidt, E., Zuurbier, K., Hartog, N., and Dawoud, M.: Observations and Prediction of Recovered Quality of Desalinated Seawater in the Strategic ASR Project in Liwa, Abu Dhabi, Water, 9, 177, https://doi.org/10.3390/w9030177, 2017.

Yermiyahu, U., Tal, A., Ben-Gal, A., Bar-Tal, A., Tarchitzky, J., and Lahav, O.: Rethinking Desalinated Water Quality and Agriculture, Science, 318, 920–921, https://doi.org/10.1126/science.1146339, 2007.

Zheng, C. and Wang, P. P.: MT3DMS: A modular three-dimensional multi-species transport model for simulation of advection, dispersion, and chemical reactions of contaminants in ground-water systems. Documentation and user's guide, in Contract Report SERDP-99-1, US Army Engineer Research and Development, 1999.

Zhou, Y. and Li, W.: A review of regional groundwater flow modeling, Geosci. Front., 2, 205–214, https://doi.org/10.1016/j.gsf.2011.03.003, 2011.

13

Delineating multiple salinization processes in a coastal plain aquifer, northern China: hydrochemical and isotopic evidence

Dongmei Han[1,2] **and Matthew J. Currell**[3]

[1]Key Laboratory of Water Cycle & Related Land Surface Processes, Institute of Geographic Sciences and Natural Resources Research, Chinese Academy of Sciences, Beijing, 100101, China
[2]College of Resources and Environment, University of Chinese Academy of Sciences, Beijing 100049, China
[3]School of Engineering, RMIT University, Melbourne VIC 3000, Australia

Correspondence: Dongmei Han (handm@igsnrr.ac.cn)

Abstract. Groundwater is an important water resource for agricultural irrigation and urban and industrial utilization in the coastal regions of northern China. In the past 5 decades, coastal groundwater salinization in the Yang–Dai river plain has become increasingly serious under the influence of anthropogenic activities and climatic change. It is pivotal for the scientific management of coastal water resources to accurately understand groundwater salinization processes and their causative factors. Hydrochemical (major ion and trace element) and stable isotopic ($\delta^{18}O$ and $\delta^{2}H$) analysis of different water bodies (surface water, groundwater, geothermal water and seawater) were conducted to improve understanding of groundwater salinization processes in the plain's Quaternary aquifer. Saltwater intrusion due to intensive groundwater pumping is a major process, either by vertical infiltration along riverbeds which convey saline surface water inland, and/or direct subsurface lateral inflow. Trends in salinity with depth indicate that the former may be more important than previously assumed. The proportion of seawater in groundwater is estimated to have reached up to 13 % in shallow groundwater of a local well field. End-member mixing calculations also indicate that the geothermal water with high total dissolved solids (up to $10.6\,g\,L^{-1}$) with depleted stable isotope compositions and elevated strontium concentrations ($> 10\,mg\,L^{-1}$) also mixes locally with water in the overlying Quaternary aquifers. This is particularly evident in samples with elevated Sr / Cl ratios (> 0.005 mass ratio). Deterioration of groundwater quality by salinization is also clearly exacerbated by anthropogenic pollution. Nitrate contamination via intrusion of heavily polluted marine water is evident locally (e.g., in the Zaoyuan well field); however, more widespread nitrate contamination due to other local sources such as fertilizers and/or domestic wastewater is evident on the basis of NO_3 / Cl ratios. This study provides an example of how multiple geochemical indicators can delineate different salinization processes and guide future water management practices in a densely populated water-stressed coastal region.

1 Introduction

Coastal regions are key areas for the world's social and economic development. Approximately 40 % of the world's population lives within 100 km of the coast (UN Atlas, 2010). Worldwide, these areas have become increasingly urbanized, with 14 of the world's 17 largest cities located along coasts (Creel, 2003). China has 18 000 km of continental coastline, and around 164 million people (approximately 12 % of the total population) living in 14 coastal provinces; nearly 80 % of these people inhabit three coastal "economic zones", namely Beijing–Tianjin–Hebei, the Yangtze River delta and the Pearl River delta (Shi, 2012). The rapid economic development and growing population in these regions have greatly increased demand for fresh water. Meanwhile, they are also confronted with increased sewage and other wastewater discharge into coastal environments.

Groundwater resources play crucial roles in the social, economic and ecologic function of global coastal systems (IPCC, 2007). Coastal aquifers connect with the ocean and with continental hydroecological systems (Moore, 1996; Ferguson and Gleeson, 2012). As an important freshwater resource, groundwater may be overextracted during periods of high demand, which are often periods of low recharge and/or surface water availability (Post, 2005). Overexploitation of groundwater can therefore result in seawater intrusion, as well as related environmental issues such as land subsidence. Seawater intrusion has become a global issue and related studies can be found from coastal aquifers around the world, including Israel (Sivan et al., 2005; Yechieli et al., 2009; Mazi et al., 2014), Spain (Price and Herman, 1991; Pulido-Leboeuf, 2004; Garing et al., 2013), France (Barbecot et al., 2000; de Montety et al., 2008), Italy (Giambastiani et al., 2007; Ghiglieri et al., 2012), Morocco (Bouchaou et al., 2008; El Yaouti et al., 2009), the USA (Gingerich and Voss, 2002; Masterson, 2004; Langevin et al., 2010), Australia (Zhang et al., 2004; Narayan et al., 2007; Werner, 2010), China (Xue et al., 2000; Han et al., 2011, 2015), Vietnam (An et al., 2014), Indonesia (Rahmawati et al., 2013), India (Radhakrishna, 2001; Bobba, 2002) and Brazil (Montenegro et al., 2006; Cary et al., 2015). Werner et al. (2013) provides a comprehensive review of seawater intrusion processes, investigation and management.

Seawater, or saltwater, intrusion is a complicated hydrogeological process, due to the impact of aquifer properties, anthropogenic activities (e.g., intensive groundwater pumping, irrigation practices), recharge rates, variable density flow, tidal activity and effects relating to global climate change, such as sea level rise (Ghassemi et al., 1993; Robinson et al., 1998; Smith and Turner, 2001; Simpson and Clement, 2004; Narayan et al., 2007; Werner and Simmons, 2009; Wang et al., 2015). Understanding the complex interactions between groundwater, surface water and seawater is thus essential for effective management of coastal water resources (Mondal et al., 2010). Very different salinization patterns may arise as a result of diverse interactions in coastal settings (Sherif and Singh, 1999; Bobba, 2002; Westbrook et al., 2005). Modeling has shown that generally, seawater intrusion is more sensitive to groundwater pumping and recharge rates in comparison to tidal fluctuation and sea level rise (Narayan et al., 2007; Ferguson and Gleeson, 2012). However, most models of seawater intrusion require simplification of the coastal interface zone. Relatively few studies have focused on delineating complex interactions among the surface–ground–seawater continuum in estuarine environments, including the effects of vertical infiltration of seawater into aquifers through river channels, compared to subsurface lateral landward migration of the freshwater–saltwater interface. Recent data indicate that such processes may be more important in causing historical salinization of coastal groundwater than previously appreciated (e.g., Cary et al., 2015; Lee et al., 2016; Larsen et al., 2017).

Additionally, groundwater in coastal aquifers may be affected by other salinization processes, such as input of anthropogenic contaminants or induced mixing with saline water from deeper or adjacent formations, which may include the high total dissolved solids (TDSs) in geothermal water or brines emplaced in the coastal zone over geologic history. The data from China's marine environment bulletin released on March 2015 by the State Oceanic Administration showed that the major bays, including Bohai Bay, Liaodong Bay and Hangzhou Bay, are seriously polluted, with inorganic nitrogen and active phosphate being the major pollutants (SOA, 2015). Seawater intrusion in China is most serious in the circum-Bohai Sea region (Han et al., 2011, 2016a); and due to the heavy marine pollution, the impacts of anthropogenic activities on groundwater quality in future may not simply be a case of simple saltwater intrusion. This region is also characterized by deep brines and geothermal waters (e.g., Han et al., 2014), which may migrate and mix with fresher groundwater due to intensive water extraction. Depending on the specific processes involved, additional contaminants may mix with fresh groundwater resources in parallel with seawater intrusion, and it is thus likely to be more difficult to mitigate and remediate groundwater pollution.

A variety of approaches can be used to investigate and differentiate seawater intrusion and other salinization processes, including time-series water level and salinity measurements, geophysical methods, conceptual and mathematical modeling as well as geochemical methods (see reviews by Jones et al., 1999; Werner et al., 2013). Geochemical techniques are particularly valuable in areas where the dynamics of saline intrusion are complicated and may involve long-term processes predating accurate water level records, or where multiple salinization processes may be occurring simultaneously. These techniques typically employ the use of major ion ratios such as Cl / Br and Cl / Na, which are indicative of solute origins (Edmunds, 1996; Jones et al., 1999). Other ionic ratios, involving Mg, Ca, Na, HCO_3 and SO_4, and characterization of water "types" can also be useful in determining the geochemical evolution of coastal groundwater, for example, indicating freshening or salinization, due to commonly associated ion exchange and redox reactions (Anderson et al., 2005; Walraevens, 2007). Trace elements such as strontium, lithium and boron can provide additional valuable information about sources of salinity and mixing between various end-members, as particular waters can have distinctive concentrations (and/or isotopic compositions) of these elements (e.g., Vengosh et al., 1999). Stable isotopes of water ($\delta^{18}O$ and δ^2H) are also commonly used in such studies, as they are sensitive indicators of water and salinity sources, allowing seawater to be distinguished from other salt sources (e.g., Currell et al., 2015).

This study examines the Yang–Dai coastal river plain in Qinhuangdao City, Hebei province, northern China, specifically focusing on salinization of fresh groundwater caused by groundwater exploitation in the Zaoyuan well field and surrounding areas. The study investigates groundwater salinization processes and interactions among surface water, seawater and geothermal groundwater in a dynamic environment, with significant pressure on water resources. Qinhuangdao is an important port and tourist city of northern China. In the past 30 years, many studies have investigated seawater intrusion and its influencing factors in the region using hydrochemical analysis (Xu, 1986; Yang et al., 1994, 2008; Chen and Ma, 2002; Sun and Yang, 2007; Zhang, 2012) and numerical simulations (Han, 1990; Bao, 2005; Zuo, 2009). However, these studies have yet to provide clear resolution of the different mechanisms contributing to salinization, and they have typically ignored the role of anthropogenic pollution and groundwater–surface water interaction. This study is thus a continuation of previous investigations of the region, using a range of hydrochemical and stable isotopic data to delineate the major processes responsible for increasing groundwater salinity, including lateral subsurface seawater intrusion, vertical leakage of marine-influenced surface water, induced mixing of saline geothermal water and anthropogenic pollution. The goal is to obtain a more robust conceptual model of the interconnections between the various water sources under the impact of groundwater exploitation. The results provide significant new information to assist water resources management in the coastal plain of Bohai Bay, and other similar coastal areas globally.

2 Study area

The Yang–Dai coastal river plain (Fig. 1) covers approximately 200 km^2 of the west side of Beidaihe District of Qinhuangdao City, northeastern Hebei Province. It is surrounded by the Yanshan Mountains to the north and west, and the southern boundary of the study area is the Bohai Sea. The plain declines in topographic elevation (with an average slope of 0.008) from approximately 390 m a.s.l. (meters above sea level) in the northwest to 1–25 m in the southeast, forming a fan-shaped distribution of incised piedmont alluvial plain sediments. Zaoyuan well field, located in the southern edge of the alluvial fan, approximately 4.3 km from the Yang River estuary, was built in 1959 (Xu, 1986) as a major water supply for the region (Fig. 1).

2.1 Climate and hydrology

The study area is in a warm and semihumid monsoon climate. On the basis of a 56-year record in Qinhuangdao, the mean annual rainfall is approximately 640 mm, the average annual temperature is approximately 11 °C, and mean potential evaporation is 1469 mm. A total of 75 % of the total annual rainfall falls in July–September (Zuo, 2006), during the East Asian Summer Monsoon. The average annual tide level is 0.86 m (meters above Yellow Sea base level), while the high and low tides are approximately 2.48 and −1.43 m.

The Yanghe River and Daihe River, originating from the Yanshan Mountains, are the major surface water bodies in the area, flowing southward into the Bohai Sea (Fig. 1). The Yang River is approximately 100 km long with a catchment area of 1029 km^2 and average annual runoff of $1.11 \times 10^8 m^3 a^{-1}$ (Han, 1988). Dai River has a length of 35 km and catchment area of 290 km^2, with annual runoff of $0.27 \times 10^8 m^3 a^{-1}$. The rivers become full during intense rain events, and revert to minimal flow during the dry season – in part this is related to impoundment of flow in upstream reservoirs.

2.2 Geological and hydrogeological setting

Groundwater in the area includes water in Quaternary porous sediment as well as fractured bedrock in the northern platform area. Fractured rock groundwater volume mainly depends on the degree of weathering and the nature and regularity of fault zones (Fig. 1). The strata outcropping in the west, north and eastern edge of the plain include Archean, Proterozoic and Jurassic metamorphic and igneous rocks, which also underlie the Quaternary sediments of the plain (from which most samples in this study were collected). The basement faults under Quaternary cover are mainly NE-trending and NW-trending (Fig. 1); these structures control the development and thickness of the overlying sediments, as well as the distribution of hot springs and geothermal anomalies. Fault zones are thought to be the main channel for transport of thermal water from deeper to shallower depths.

The Quaternary sediments are widely distributed in the area, with the thickness ranging from approximately 5–80 m (mostly 20–40 m) up to more than 100 m immediately adjacent to the coastline. The bottom of the Holocene (Q_4) unit in most areas consists of clay, making the groundwater in the coastal zone confined or semiconfined, although there are no regional, continuous aquitards between several layers of aquifer-forming sediments (Fig. 1b). The aquifer is mainly composed of medium sand, coarse sand and gravel layers with a water table depth of 1–4 m in the phreatic aquifer, and deeper semiconfined groundwater (where present and hydraulically separated from the phreatic aquifer) hosted in similar deposits with a potentiometric surface 1–5 m below topographic elevation (Zuo, 2006).

The general flow direction of groundwater is from northwest to south, according to the topography. The main sources of recharge are from infiltration of rainfall, river water and irrigation return flow, as well as lateral subsurface inflow from the piedmont area. Naturally, groundwater discharges into the rivers and the Bohai Sea. Apart from phreatic water evaporation, groundwater pumping for agricultural, industrial and domestic usage (including seasonal tourism) are currently the main pathways of groundwater discharge.

Geothermal water discharges into shallow Quaternary sediments near the fault zones, evident as geothermal anomalies (Hui, 2009). The temperature of thermal water ranges from 27 to 57 °C in this low-to-medium temperature geothermal field (Zeng, 1991). Deeper thermal water is stored in the Archaeozoic granite and metamorphic rocks; major fracture

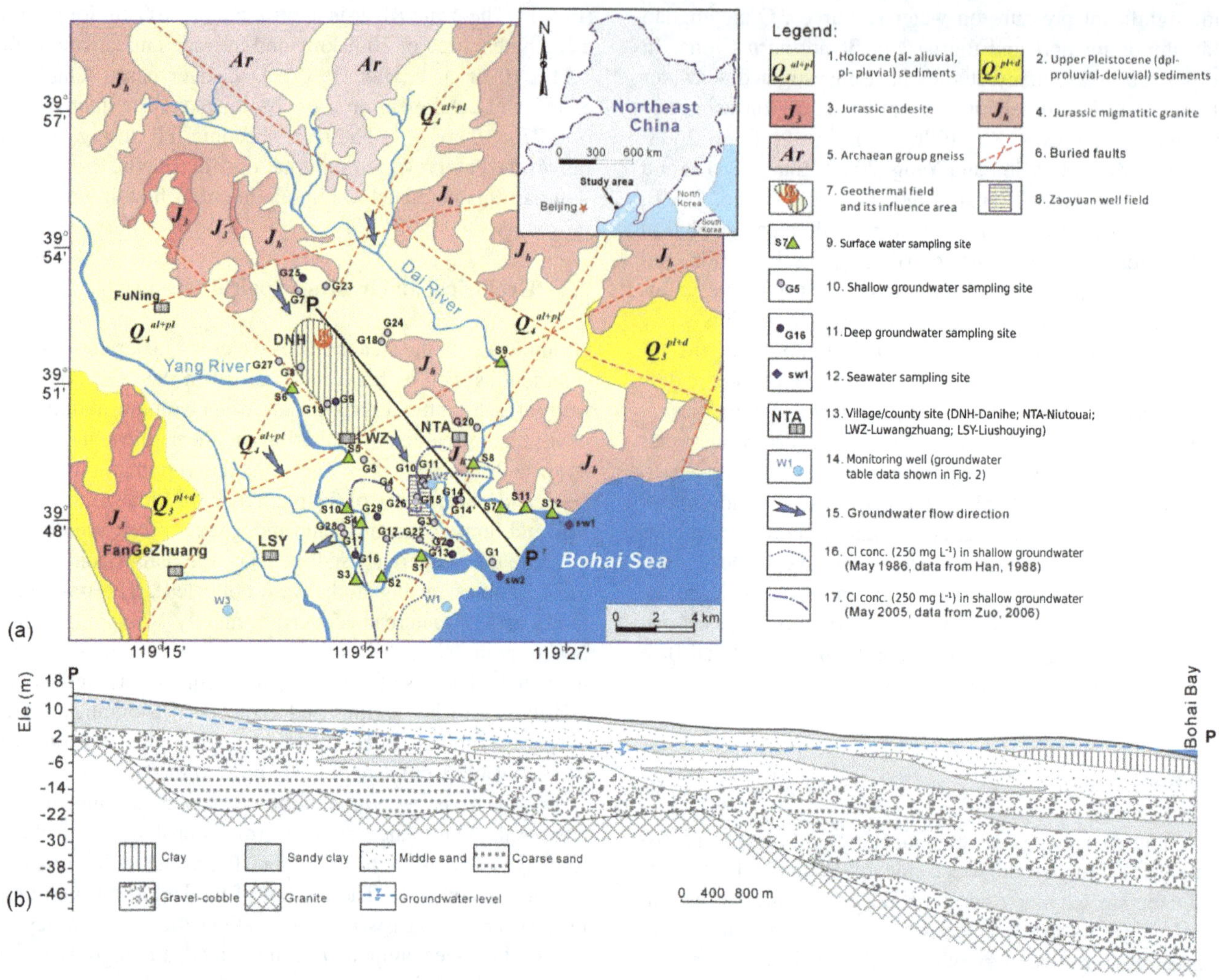

Figure 1. Location map **(a)** for showing geological background and water sampling sites in the study area, and **(b)** hydrogeological cross section of Yang–Dai river Plain ($P - P'$ in **a**) (modified from Han, 1988). The surface area covered from the $> 250\,\mathrm{mg\,Cl\,L^{-1}}$ contour line to the coastline refers the seawater zones.

zones provide pathways into the overlying Quaternary sediments (Pan, 1990; Shen et al., 1993; Yang, 2011).

2.3 Groundwater usage and seawater intrusion history

Shallow groundwater pumped from the Quaternary aquifer occupies 94 % of total groundwater exploitation, and is used for agricultural irrigation (52 % of total groundwater use), industrial (32 %) and domestic water (16 %) (Meng, 2004). Many large and medium-sized reservoirs were built in the 1960s and 1970s, meaning that the surface water was intercepted and downstream runoff dropped sharply, even causing rivers to dry up in drought years. With the intensification of human socioeconomic activities and growing urbanization, coupled with extended drought years (severe drought during 1976–1989 in north China) (Wilhite,1993; Han et al., 2015), increased groundwater exploitation to meet the ever-growing fresh water demand resulted in groundwater table declines and seawater intrusion (SWI) in the aquifers.

The pumping rate in the Zaoyuan well field gradually increased from 1.25 million $\mathrm{m^3\,a^{-1}}$ in the early 1960s to 3.5 million $\mathrm{m^3\,a^{-1}}$ in the late 1970s, and beyond 10 million $\mathrm{m^3\,a^{-1}}$ in the 1980s. During 1966–1989, the planting of paddy fields became common, resulting in significant agricultural water consumption. This caused formation of a cone of depression in the Quaternary aquifer system. Groundwater pumping in this region mainly occurs in spring and early summer, with typical pumping rates of 70 000–80 000 $\mathrm{m^3\,d^{-1}}$. Pumping from the Zaoyuan well field occurs in wells approximately 15 to 20 m deep. Groundwater tables decline sharply and reach their lowest level during May, before the summer rains begin, and recover to their yearly high in January–February (Fig. 2). In May 1986, the groundwater table in the depression center, which is located in

Zaoyuan–Jiangying (Fig. S1 in the Supplement), decreased to −2 m a.s.l. and the area with groundwater tables below sea level covered 28.2 km^2. The local government commenced reduction in groundwater exploitation in this area after 1992, and groundwater tables began to decrease more slowly after 1995, even showing recovery in some wells. However, during an extreme drought year (1999), increased water demand resulted in renewed groundwater table declines in the region (Fig. 2). Since 2000, the groundwater tables have responded seasonally to water demand peaks and recharge (Figs. 2 and S1).

From 1990, the rapid development of township enterprises (mainly paper mills), also began to cause groundwater overexploitation in the western area of the plain. The groundwater pumping rate for paper mills reached 55 000 $m^3\,d^{-1}$ in 2002, resulting in groundwater table depressions around Liushouying and Fangezhuang (Fig. 1). The groundwater table in the western depression associated with this pumping reached −11.6 in 1991 and −17.4 m a.s.l. in 2002. After the implementation of "Transferring Qing River water to Qinhuangdao" project in 1992, the intensity of groundwater pumping generally reduced, and the groundwater table in the depression center recovered to −4.3 m a.s.l. in July 2006.

Overall, the depression area (groundwater tables below mean sea level) was recorded as 132.3 km^2 in May 2004 and the shape of the depression has generally been elliptical with the major axis aligned east–west. In addition to groundwater overexploitation, climate change-induced recharge reduction has also likely contributed to groundwater table declines and hence seawater intrusion (Fig. S2). The annual average rainfall declined from 639.7 between 1954 and 1979 to 594.2 mm between 1980 and 2010, a significant decrease over the last 30 years (Zhang, 2012). As indicated in Fig. S2, the severity of seawater intrusion (indicated by changes in Cl concentration, and the total area impacted by SWI, as defined by the 250 $mg\,L^{-1}$ Cl contour) correlates with periods of below-average rainfall – indicated by monthly cumulative rainfall departure (CRD, Weber and Stewart, 2004).

Groundwater quality of the area gradually became more saline from the early 1980s, with chloride concentrations increasing year by year. As early as 1979, seawater intrusion was recorded in the Zaoyuan well field. The intrusion area with groundwater chloride concentration greater than 250 $mg\,L^{-1}$ was 21.8 in 1984, 32.4 in 1991, 52.6 in 2004 and 57.3 km^2 in 2007 (Zuo, 2006; Zang et al., 2010). The chloride concentration of groundwater pumped from a monitored well field well (depth of 18 m, G10 in Fig. 1) changed from 90 in 1963 to 218 $mg\,L^{-1}$ in 1978, 567 in 1986, 459 in 1995 and 1367 $mg\,L^{-1}$ in 2002 (Zuo, 2006), reducing to 812 $mg\,L^{-1}$ in July 2007. The distance of estimated seawater intrusion into the inland area from the coastline had reached 6.5 in 1991 and 8.75 km in 2008 (Zang et al., 2010). In the early 1990s, 16 of 21 pumping wells in the well field were abandoned due to the saline water quality (Liang et al., 2010). Additionally, 370 of 520 pumping wells were abandoned in the wider Yang–Dai coastal river plain during 1982–1991 (Zuo, 2006).

3 Methods

In total, 80 water samples were collected from the Yang–Dai coastal river plain, including 58 groundwater samples, 19 river water samples (from 12 sites) and 3 seawater samples, during 3 sampling campaigns (June 2008, September 2009 and August 2010). Groundwater samples were pumped from 28 production wells with depths between 6 and 110 m, including 7 deep wells with depths greater than 60 m (Fig. 1). While sampling for geochemical parameters would ideally be conducted on monitoring wells, due to an absence of these, production wells were utilized. In most cases, the screened interval of these wells encompasses aquifer thicknesses of approximately 5 to 15 m above the depths indicated in Table 1.

In this study, sampling focused predominantly on low temperature groundwater; however, geothermal water from around Danihe was also considered a potentially important ongoing source of groundwater salinity. As such, while geothermal water samples were not accessible during our sampling campaigns (as the area is now protected), data reported by Zeng (1991) were compiled and analyzed in conjunction with the sampled wells.

Measurements of physicochemical parameters (pH, temperature and electrical conductivity (EC)) were conducted in situ using a portable meter (WTW Multi 3500i). All water samples were filtered with 0.45 µm membrane filters before analysis of hydrochemical composition. Two aliquots in polyethylene 100 mL bottles at each site were collected for major cation and anion analysis, respectively. Samples for cation analysis (Na^+, K^+, Mg^{2+} and Ca^{2+}) were treated with 6N HNO_3 to prevent precipitation. Water samples were sealed and stored at 4 °C until analysis. Bicarbonate was determined by titration within 12 hr of sampling. Concentrations of cations and some trace elements (B, Sr, Li) were analyzed by inductively coupled plasma-optical emission spectrometry (ICP-OES) in the chemical laboratory of the Institute of Geographic Sciences and Natural Resources Research (IGSNRR), Chinese Academy Sciences (CAS). Only the Sr data are reported here, as the other trace elements were not relevant to the interpretations discussed (Table 1). The detection limits for analysis of Na^+, K^+, Mg^{2+} and Ca^{2+} are 0.03, 0.05, 0.009 and 0.02 $mg\,L^{-1}$, respectively. Concentrations of major anions (i.e., Cl^-, SO_4^{2-}, NO_3^- and F^-) were analyzed using a high-performance ion chromatograph (SHIMADZU, LC-10ADvp) at the IGSNRR, CAS. The detection limits for analysis of Cl^-, SO_4^{2-}, NO_3^- and F^- are 0.007, 0.018, 0.016 and 0.006 $mg\,L^{-1}$, respectively. The testing precision of the cation and anion analysis is 0.1–5.0 %. Charge balance errors were less than 8 %. Stable isotopes ($\delta^{18}O$ and δ^2H) of water samples were measured using a Finnigan MAT 253 mass

Table 1. Physical, hydrochemical and isotopic data of the water samples collected from the Yang–Dai coastal river plain.

Water type	ID	Sampling time	Ele. m	Well depth m	Water table depth (m)	EC $\mu s\,cm^{-1}$	pH	T °C	ORP mV	DO $mg\,L^{-1}$	Cl^- $mg\,L^{-1}$	NO_3^- $mg\,L^{-1}$	SO_4^{2-} $mg\,L^{-1}$	HCO_3^- $mg\,L^{-1}$	Ca^{2+} $mg\,L^{-1}$	Na^+ $mg\,L^{-1}$	K^+ $mg\,L^{-1}$	Mg^{2+} $mg\,L^{-1}$	Sr $mg\,L^{-1}$	δ^2H ‰	$\delta^{18}O$ ‰
Shallow groundwater samples:																					
	G4	Aug 2010	5	15	4.2	741	6.5	20.5	6	3.9	124.3	92.6	89.7	136.9	70.3	69.2	2.3	21.8	0.67	−50	−6.9
	G27	Aug 2010	3	12		1014	6.4	14.8	16	3.8	177.5	163.0	121.0	128.0	122.0	50.6	2.6	33.4	1.22	−50	−6.4
	G12	Aug 2010	8	10		1152	7.2	25.0	9	5.9	276.9	31.9	69.8	148.8	15.8	201.9	11.1	16.2	0.25	−51	−6.8
	G17	Aug 2010	5	25	5.7	624	7.1	19.1	13	3.3	88.8	39.5	58.3	223.3	98.4	41.4	4.3	7.5	0.33	−76	−11.0
	G18	Aug 2010	5	23	1.7	934	7.4	14.9	12	7.1	124.3	137.8	98.7	235.2	133.3	48.3	2.2	23.2	0.71	−55	−8.2
	G1	Sep 2009	6	8	2.4	1673	8.3	19.4			220.1	2.0	148.3	207.9	84.3	119.6	17.1	49.9	0.66	−51	−7.1
	G4	Sep 2009	6	15	4.6	1295	7.9	13.9			124.3	178.5	108.7	92.4	104.4	64.4	2.7	35.9	0.85	−44	−6.9
	G5	Sep 2009	11	16	6.5	1544	7.8	13.3			303.4	162.3	108.2	38.5	124.5	103.7	1.1	39.1	1.14	−47	−6.5
	G23	Sep 2009	15	30	3.0	556	7.9	14.9			88.8	163.4	54.0	51.9	97.4	34.5	0.5	15.6	0.55	−48	−6.6
	G7	Sep 2009	17	8	1.7	901	8.4	14.9			81.7	41.2	74.2	69.3	64.0	33.2	1.0	16.5	0.41	−47	−6.4
	G8	Sep 2009	4	13	4.3	1621	7.8	14.0			334.6	85.4	90.1	69.3	132.5	91.9	1.4	38.5	1.19	−44	−5.3
	G11	Sep 2009	4	13		1237	8.6	18.4			222.9	74.5	98.7	30.8	87.4	80.1	1.8	29.0	0.84	−43	−5.5
Fresh groundwater	G12	Sep 2009	8	10		1114	7.9	15.0			174.0	46.0	76.4	107.8	86.2	73.6	4.3	27.2	0.53	−48	−6.0
	G28	Sep 2009	6	23	6.8	1748	8.2	14.6			127.8	22.5	38.4	107.8	88.2	33.3	1.3	14.0	0.58		
	G18	Sep 2009	9	23	5.5	2850	7.8	13.7			110.1	152.9	76.7	53.9	120.3	23.1	1.1	20.5	0.65	−54	−7.4
	G20	Sep 2009	12	30		1602	7.7	19.5			246.9	107.1	135.9	107.8	158.2	82.8	1.7	26.3	3.94	−50	−6.7
	G17	Sep 2009	6	33	7.9	819	8.1	14.3			243.5	126.4	141.7	161.7	176.3	105.5	1.8	24.6	0.71	−53	−6.9
	G3	Jun 2018	4	20		1573	7.3	14.5			315.0		184.4	54.9	67.3	152.1	3.8	33.2	0.46	−43	−4.2
	G4	Jun 2018	6	6		688	7.5	14.1			74.6	75.8	83.2	60.4	58.2	45.2	1.2	20.6	0.47	−48	−4.8
	G5	Jun 2018	5	11		1455	7.4	14.7			310.9	119.5	92.7	52.2	121.3	95.4	0.9	29.9	1.43	−54	−6.3
	G8	Jun 2018	4	30		1402	7.3	14.7			349.7	55.9	81.5	85.1	118.0	88.1	1.2	36.9	1.08	−40	−4.7
	G12	Jun 2018	8	10		1285	7.6	19.3			281.0	29.8	56.8	74.1	9.7	205.7	11.0	13.7	0.19	−50	−6.0
	G17	Jun 2018	6	33		1462	8.1	15.0			227.9	92.6	123.8	175.7	179.2	72.1	1.1	15.9	0.57	−56	−6.1
	G18	Jun 2018	9	23		1125	7.9	18.8			115.5	144.2	44.4	90.6	103.7	31.8	1.4	13.7	0.41	−53	−6.8
	G20	Jun 2018	12	30		1210	9.1	28.3			234.3	18.2	90.6	230.6	122.0	81.4	2.6	22.4	2.10	−39	−4.4
	G1	Aug 2010	6	11		2750	7.4	17.3	143	1.5	312.4	120.0	337.1	205.4	112.0	294.4	29.9	41.3	0.54	−56	−7.7
	G3	Aug 2010	4	8		2490	5.9	14.7	74	1.4	654.3	106.0	263.7	86.3	128.2	283.6	6.6	78.6	0.97	−47	−5.8
	G15	Aug 2010	4	30		2290	6.6	15.4	3	3.3	435.7	181.3	263.7	244.1	164.3	242.8	6.6	60.8	1.60	−50	−6.3
	G26	Aug 2010	4	18	1.8	4060	6.4	15.2	145	1.3	784.6	339.4	341.4	485.2	192.3	538.2	36.0	72.1	1.20	−46	−6.8
	G11	Aug 2010	6	13	2.3	3490	6.5	15.0	47	1.3	646.1	414.1	402.7	402.6	204.8	441.8	34.9	76.3	1.51	−52	−6.8
	G20	Aug 2010	12	30		1513	6.7	18.8	4	4.3	596.4	177.9	245.0	184.6	268.6	225.4	2.0	29.2	3.95	−51	−7.0
	G5	Aug 2010	5	16	5.4	1500	6.2	15.6	7	5.9	447.3	253.3	304.7	79.3	204.1	188.6	0.8	47.5	0.78	−54	−7.7
	G8	Aug 2010	4	30	2.1	1563	6.3	15.7	120	4.9	596.4	141.4	188.1	104.2	196.3	220.8	3.2	38.7	1.87	−44	−5.3
Brackish groundwater	G7	Aug 2010	5	8	0.9	1438	6.8	19.3	24	6.1	299.6	338.3	192.7	128.0	164.2	151.8	0.3	50.0	0.66	−60	−8.9
	G19	Aug 2010	8	11	2.8	2380	6.5	16.1	42	3.2	600.0	211.9	192.1	119.1	241.4	199.5	6.6	42.4	7.10	−52	−6.8
	G24	Aug 2010	4	21		4400	6.8	15.5	4	5.9	646.1	952.1	813.3	223.3	484.1	349.6	3.4	117.4	2.44	−52	−8.1
	G22	Sep 2009	2	20	2.0	2170	8.0	17.8			408.3	2.0	277.5	130.9	109.2	226.8	21.3	54.2	0.70	−48	−6.5
	G10	Sep 2009	3	18		9770	7.5	15.1			2563.1	27.7	25.3	870.2	181.4	1340.0	98.1	123.0	1.92	−44	−5.7
	G14	Sep 2009	2	8.5		1886	8.5	16.6			291.1	109.8	216.4	92.4	117.8	164.7	19.3	46.5	0.82	−55	−7.9
	G24	Sep 2009	11	21	5.3	1003	7.8	13.6			454.4	441.5	128.1	115.5	252.3	165.6	0.6	36.2	0.93		
	G19	Sep 2009	12	11	4.7	2560	8.1	14.6			622.0	91.5	115.6	69.3	214.9	180.7	6.0	49.4	4.87	−48	−6.2
	G1	Jun 2018	6	8		3150	8.5	15.0			717.1	87.7	265.6	98.8	9.1	527.4	26.7	26.6	0.17	−54	−7.2
	G11	Jun 2018	6	13		2370	7.2	12.2			451.2		211.3	162.0	145.1	201.2	22.0	52.6	1.00	−49	−6.0
	G14	Jun 2018	3	8.5		2050	7.4	18.6			372.8	267.3	206.1	49.4	136.4	171.3	12.7	49.5	0.93	−48	−6.5
	G15	Jun 2018	4	18		5990	7.5	15.6			1675.6	146.8	110.1	474.9	246.5	764.7	20.2	104.9	2.17	−45	−5.1

Table 1. Continued.

Water type	ID	Sampling time	Ele. m	Well depth m	Water table depth (m)	EC $\mu s\,cm^{-1}$	pH	T °C	ORP mV	DO $mg\,L^{-1}$	Cl^- $mg\,L^{-1}$	NO_3^- $mg\,L^{-1}$	SO_4^{2-} $mg\,L^{-1}$	HCO_3^- $mg\,L^{-1}$	Ca^{2+} $mg\,L^{-1}$	Na^+ $mg\,L^{-1}$	K^+ $mg\,L^{-1}$	Mg^{2+} $mg\,L^{-1}$	Sr $mg\,L^{-1}$	δ^2H ‰	$\delta^{18}O$ ‰
Deep groundwater samples:																					
	G25	Aug 2010	4	95		444	6.6	23.5	41	4.4	68.1	71.0	48.2	89.3	52.5	35.9	1.1	10.4	0.25	−56	−7.5
Fresh groundwater	G16	Sep 2009	3	110		1214	7.9	19.9			214.4		66.1	100.1	76.3	97.8	2.8	19.5	2.68	−58	−7.7
	G29	Sep 2009	6	60		1291	8.1	20.9			255.6	26.3	57.5	69.3	8.0	205.9	12.3	14.6	0.20	−51	−6.6
	G29	Aug 2010	6	60	6.2	3220	7.2	24.1	12	5.6	803.4	143.0	264.7	178.6	386.8	189.3	4.3	77.6	6.95	−50	−6.5
	G16	Aug 2010	3	110		1733	7.3	21.5	16	4.8	766.8	5.5	67.1	205.4	246.2	197.8	4.5	37.9	4.00	−56	−7.8
	G9	Jun 2018	9	104		3110	7.8	15.0			823.6	47.4	110.5	85.1	255.1	222.2	5.1	36.7	7.40	−47	−5.3
Brackish groundwater	G9	Sep 2009	9	104		3190	8.5	20.4			917.4	65.8	116.6	61.6	279.8	245.1	5.7	40.6	8.02	−51	−6.2
	G9	Aug 2010	9	104		4600	6.3	24.3	18	4.0	1228.3	337.4	296.3	92.3	392.2	455.4	5.5	45.3	11.59	−48	−6.6
	G14'	Aug 2010	5	110	2.1	2850	6.1	22.9	120	1.8	553.8	433.7	491.4	134.0	186.0	312.8	52.3	96.6	1.06	−53	−6.8
	G13	Aug 2010	3	90		3230	8.9	19.2	36	2.4	908.8	210.6	231.8	92.3	92.0	413.5	29.0	115.7	0.26	−45	−5.2
	G13	Jun 2018	3	90		3180	7.9	13.5			945.4		122.2	52.2	51.4	351.9	25.0	105.4	0.55	−45	−5.1
	G13	Sep 2009	3	90	2.6	3070	8.1	15.3			882.0		91.9	38.5	36.5	362.9	26.0	105.3	0.35	−43	−5.2
	G2	Jun 2018	2	60		3780	7.7	18.5			1093.4		200.3	96.1	67.3	484.1	25.8	103.1	1.03	−46	−5.1
River water samples:																					
Fresh water samples																					
Dai River	S9	Aug 2010				511	7.2	22.2	22	5.5	80.3	65.2	107.9	86.3	72.1	29.9	3.7	16.7	0.34	−66	−9.4
Dai River	S12	Aug 2010				485	7.5	25.8	18	7.3	71.3	41.9	83.7	83.4	54.3	27.6	4.0	15.1	0.30	−69	−9.7
Dai River	S8	Aug 2010				495	7.3	22.1	24	5.8	68.6	54.9	96.8	89.3	62.3	27.2	4.1	15.9	0.32	−67	−9.6
Yang River	S6	Aug 2010				507	7.0	23.3	17	4.5	66.0	51.9	80.5	107.2	61.6	32.2	3.8	16.8	0.30	−65	−9.2
Yang River	S2	Aug 2010				435	7.3	7.3	6	4.9	63.2	40.2	92.2	101.2	59.3	29.9	4.3	14.0	0.26	−71	−10.1
Yang River	S5	Sep 2009				718	8.4	24.4			85.2	12.9	62.0	107.8	46.1	52.0	6.4	17.4	0.36	−45	−5.8
Yang River	S4	Sep 2009				2630	8.1	24.7			733.1	6.6	142.2	115.5						−42	−5.3
Yang River	S6	Sep 2009				718	8.3	25.6			99.4	6.6	57.5	107.8	44.3	59.8	5.0	16.4	0.30	−43	−7.2
Dai River	S9	Sep 2009			2.0	560	8.0	23.6			88.8	6.8	60.5	92.4	57.2	33.2	3.3	16.7	0.36	−46	−5.8
Dai River	S8	Sep 2009				1013	8.2	23.6			174.0	6.4	82.2	92.4	52.1	98.1	6.0	23.6	0.55	−43	−5.4
Yang River	S6	Jun 2018				1166	7.5	12.6			81.7	12.6	71.3	112.5	53.6	47.5	4.9	18.0	0.31	−49	−5.5
Dai River	S10	Jun 2018				1255	9.1	28.0			208.9	23.1	73.7	175.7	60.7	123.2	10.8	24.2	0.57	−44	−3.3
Dai River	S9	Jun 2018				1163	7.8	11.8			92.3	7.5	52.2	90.6	54.7	31.1	3.6	19.5	0.35	−40	−3.9
Brackish and saltwater samples																					
Yang River	S3	Sep 2009				34800	7.8	15.5			11289.4		1684.7	115.5	251.0	5658.7	231.3	690.1	4.29	−21	−2.4
Dai River	S12	Sep 2009				47100	7.5	23.7			16766.3		2416.5	77.0	412.1	10074.3	398.2	1306.0	7.64	−12	−1.2
Dai River	S11	Sep 2009				52500	8.5	23.4			1601.5	2.8	258.7	84.7	79.7	801.4	32.9	102.7	0.93	−41	−5.2
Yang River	S1	Jun 2018				39800	8.7	24.2			14953.5		2035.9	134.5	313.0	7496.0	270.5	928.3	5.55	−15	−1.1
Yang River	S2	Jun 2018				20200	8.8	28.5			8328.3		912.1	189.4	233.4	4094.2	147.9	495.9	3.37	−28	−2.5
Dai River	S7	Jun 2018				49500	8.4	11.5			16677.1		2261.3	113.9	349.0	8730.0	326.4	1084.0	6.36	−11	−0.6
	SW1	Aug 2010				45600	7.8	25.5	83	4.90	14768.3	810.1	4047.0	148.8	312.7	8326.4	293.6	1007.0	5.79	3.8	1.1
Seawater	SW1	Sep 2009				47700	7.8	24.2			16568.0		2394.3	92.4	352.2	8922.0	322.2	1107.0	6.45	−10	−1.1
	SW2	Sep 2009				39500	7.3	23.2			14484.8		1926.6	107.8	313.2	7214.0	267.9	916.9	5.43	−17	−2.0

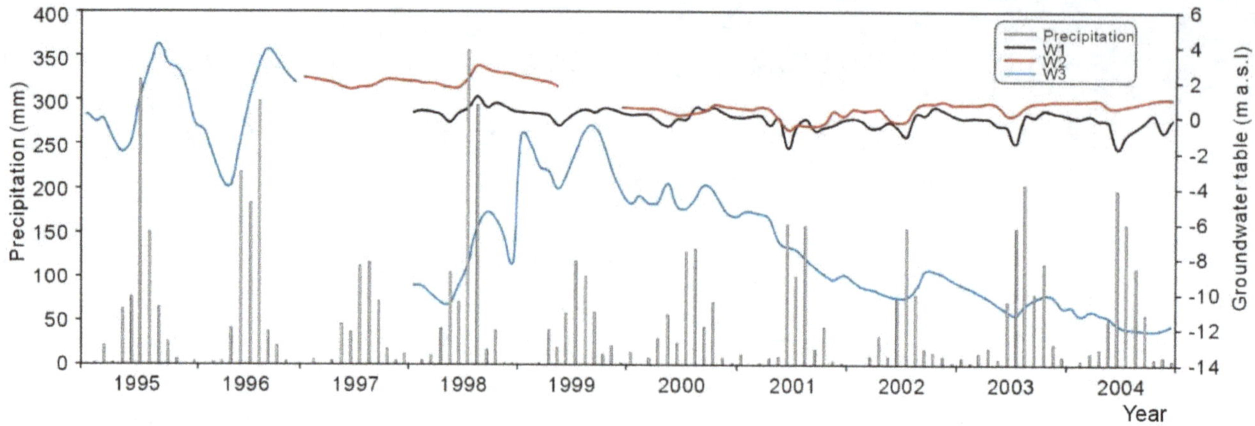

Figure 2. Distribution of precipitation and dynamics of the water table in the study area. Locations of the monitoring wells (W1, W2 and W3) can be seen from Fig. 1.

spectrometer after on-line pyrolysis with a Thermo Finnigan TC/EA in the Stable Isotope Laboratory of the IGSNRR, CAS. The results are expressed in per mill (‰) relative to international standards (V-SMOW, Vienna Standard Mean Ocean Water) and resulting δ^{18}O and δ^2H values are shown in Table 1. The analytical precision for δ^2H is ±2 ‰ and for δ^{18}O is ±0.5 ‰ . All hydrochemical, physicochemical and isotope data are reported in Table 1.

Mixing calculations were also conducted on the basis of Cl^- concentrations of the samples under a conservative freshwater–seawater mixing system (Fidelibus et al., 1993; Appelo, 1994). The seawater contribution for each sample is expressed as a fraction of seawater (f_{sw}), using the formula according to Appelo and Postma (2005):

$$f_{sw} = \frac{C_{Cl,sam} - C_{Cl,f}}{C_{Cl,sw} - C_{Cl,f}}, \quad (1)$$

where $C_{Cl,sam}$, $C_{Cl,f}$ and $C_{Cl,sw}$ refer to the Cl^- concentration in the sample, freshwater and seawater, respectively.

4 Results

4.1 Water stable isotopes (δ^2H and δ^{18}O)

The local meteoric water line (LMWL, $\delta^2H = 6.6\delta^{18}O + 0.3$, $n = 64$, $r^2 = 0.88$) is based on δ^2H and δ^{18}O mean monthly rainfall values between 1985 and 2003 from Tianjin station some 120 km southwest of Qinhuangdao City (IAEA/WMO, 2006). Due to similar climate and position relative to the coast, this can be regarded as representative of the study area. Surface water samples collected from Yang River and Dai River ($n = 19$) have δ^{18}O and δ^2H values ranging from −10.1 to −0.6 ‰ (mean = −5.4 ‰) and −71 to −11 ‰ (mean = −43 ‰), respectively. Stable isotope compositions for surface water appear to exhibit significant seasonal variation (Fig. S3): for Yang River samples from the relatively dry season (June 2008, $n = 3$) had mean δ^{18}O and δ^2H values of −3.0 and −31 ‰, respectively; samples from the wet season (August 2009 and September 2010, $n = 6$) had mean δ^{18}O and δ^2H values of −6.6 and −48 ‰, respectively. Dai River samples showed similar results: the dry season mean δ^{18}O and δ^2H values ($n = 3$) were −2.6 and −32 ‰, respectively; wet season samples ($n = 7$) had mean δ^{18}O and δ^2H values of −6.6 and −49 ‰, respectively (Fig. 3).

The 56 groundwater samples were characterized by δ^{18}O and δ^2H values ranging from −11.0 to −4.2 ‰ (mean = −6.5 ‰) and −76 to −39 ‰ (mean = −50 ‰), respectively. Among these, shallow and deep groundwater samples showed similar mean values, although deep groundwater samples ($n = 13$) showed relatively narrow overall ranges (−7.8 to −5.1 ‰ and mean = −6.3 ‰ for δ^{18}O; −58 to −43 ‰ and mean = −50 ‰ for δ^2H; Fig. 3). Slight seasonal variation was evident in the groundwater isotope compositions; shallow groundwater from the dry season ($n = 12$) showed δ^{18}O and δ^2H values from −7.2 to −4.2 ‰ (mean = −5.7 ‰) and δ^2H values from −56 to −39 ‰ (mean = −48 ‰); while during the wet season ($n = 31$) δ^{18}O and δ^2H values ranged from −11.0 to −5.3 ‰ (mean = −6.9 ‰) and −76 to −43 ‰ (mean = −51 ‰), respectively. Some variability was also evident in deep groundwater compositions, although only three deep samples were collected during the dry season.

From Fig. 3, it can be seen that surface water exhibits a much wider range of δ^{18}O and δ^2H values relative to groundwater, with shallow groundwater in turn more spatially variable than deep groundwater. Water samples collected in the wet season showed wider ranges of δ^{18}O and δ^2H values relative to the dry season. Most water samples of all types plot to the right of (below) the LMWL, with some surface water samples showing similar compositions to the local seawater (Fig. 3). The local seawater plots below (more neg-

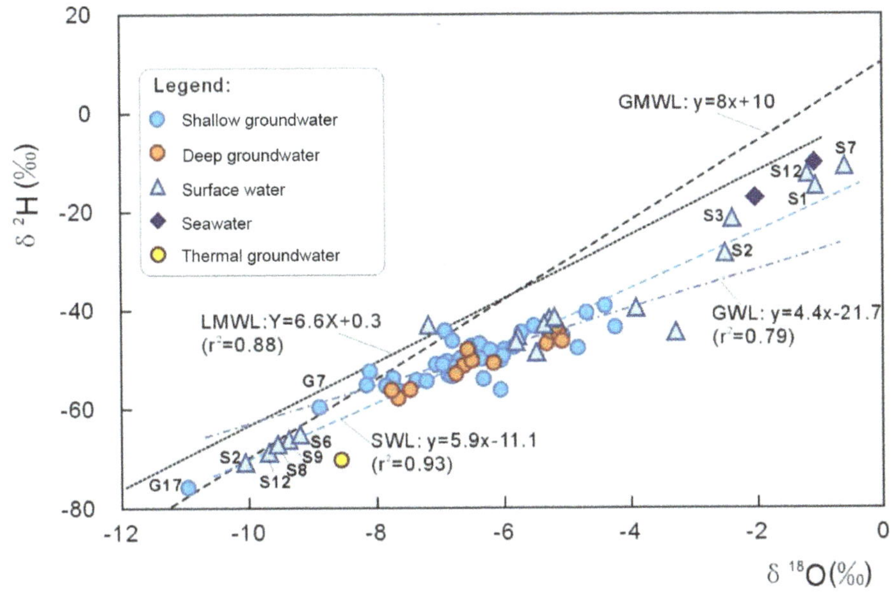

Figure 3. Stable isotope compositions of different water samples collected from the study area. LMWL: local meteoric water line; GMWL: global meteoric water line (Craig, 1961); GWL: groundwater line; SWL: surface (river) water line.

ative) than typically assumed values (e.g., VSMOW = 0 ‰) for both δ^2H and δ^{18}O, and this water appears to represent an end-member involved in mixing with meteoric-derived waters in both ground and surface water (Fig. 3).

4.2 Water salinity and dissolved ions

TDS concentrations of surface water samples from Dai River range from 0.3 to 31.4 g L^{-1}, with Na^+ and Ca^{2+} comprising 22–78 and 4–56 % of total cations and Cl^- comprising 36–91 % of total anions. The composition changes from Ca•Na•Mg-Cl•HCO_3 to Na-Cl water type from upstream to downstream locations along with increasing salinity; Cl^- concentrations vary from approximately 70 upstream to 16 700 mg L^{-1} near the coastline, due to marine influence. Similar variation occurs along the Yang River, where samples had TDS concentrations between 0.3 and 26.1 g L^{-1} with increasing concentrations and proportions of Cl^- (63.2–14 953.5 mg L^{-1}) from upstream to downstream locations. Nitrate concentrations also range from 2.8 to 65.2 mg L^{-1} in the surface water samples, increasing downstream.

In the early 1960s, groundwater pumped from the Zaoyuan well field exhibited Ca-HCO_3 water type and chloride concentrations of 90–130 mg L^{-1}; this was followed by rapid salinization since the 1980s (see Sect. 2.3). In this study, shallow groundwater is characterized by TDS concentrations of 0.4–4.8 g L^{-1}, with Cl^- (34–77 %), Na^+ (12–85 %) and Ca^{2+} (5–69 %) being the predominant major anion and cations, respectively. Groundwater hydrochemical types vary from Ca-HCO_3•Cl, Ca•Na-Cl, Na•Ca-Cl to Na-Cl (Fig. 4). Deep groundwater has TDS concentrations between 0.3 and 2.8 g L^{-1}, dominated by Ca (up to 77 % of major cations) in the upstream area and Na (up to 85 % or major cations) near the coast, with water type evolving from Ca-Cl•HCO_3 to Ca•Na-Cl and Na•Mg-Cl (Fig. 4). At present, the TDS of groundwater from the well field reaches 3.31 g L^{-1} with Na-Cl water type (see well G15). The highest observed mixing proportions of seawater occur in shallow well G10 and deep well G2, respectively, with calculated f_{sw} values (according to Eq. 1) of 12.95 and 5.35 %, respectively.

Hydrochemical features of thermal water from the Danihe–Luwangzhuang area (Fig. 1a) are distinct from the normal-to-low temperature groundwater. Previous work by Zeng (1991) and Hui (2009) identified geothermal water with high TDS in the fractures of deep metamorphic rock. The geothermal water was characterized by TDS values between 6.2 and 10.6 g L^{-1} and Ca•Na-Cl water type, while Cl^- concentrations ranged from 5.4 to 6.5 g L^{-1} and Sr concentrations from 6.73 to 89.8 mg L^{-1}. Some normal-to-low temperature groundwater samples collected in this study from wells G8, G19 and G9 featured by Ca•Na-Cl water type with relative high TDS ranges (0.8–1.4, 1.3–1.6 and 1.5–2.8 g L^{-1}, respectively) and strontium concentrations (1.1–1.9, 4.9–7.1 and 7.3–11.6 mg L^{-1}, respectively), showing similarity with the geothermal system. Low temperature groundwater sampled in this study had Sr / Cl mass ratios ranging from 2.4×10^{-4} to 1.6×10^{-2}, with higher ratios in deep groundwater (range: 9.4×10^{-4} to 1.3×10^{-2}, median: 3.7×10^{-3}) compared to shallow groundwater (median: 3.1×10^{-3}), and groundwater generally higher than saline surface water (range: 3.7×10^{-4} to 5.8×10^{-4}, median: 3.9×10^{-4}; Table S1).

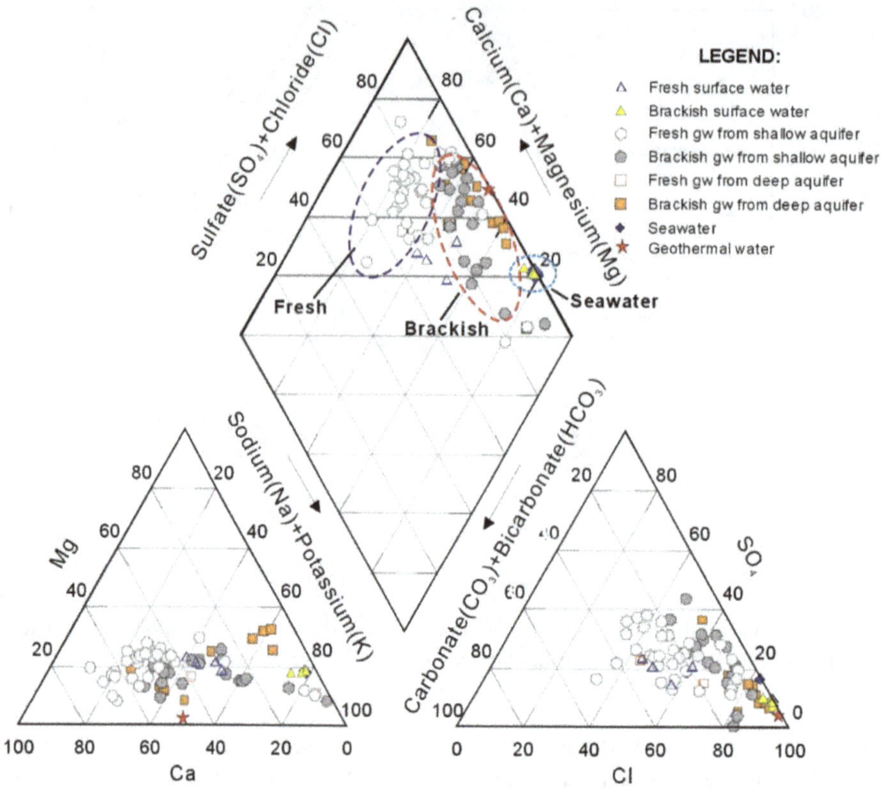

Figure 4. Piper plot of different water samples. Groundwater: gw.

Nitrate concentrations in groundwater range from 2.0 to 178.5 mg L^{-1} (mean 90.1 mg L^{-1}) for shallow groundwater, and 2.0–952.1 mg L^{-1} (mean 232.1 mg L^{-1}) for the deep groundwater, with most samples exceeding the WHO drinking water standard (50 mg L^{-1}).

5 Discussion

5.1 Groundwater isotopes and hydrochemistry as indicators of mixing processes

The Quaternary groundwater system in the Yang–Dai coastal river plain may be recharged by precipitation, irrigation return flow, river infiltration and lateral subsurface runoff (e.g., from mountain-front regions). Groundwater geochemical characteristics are then controlled by hydrogeological conditions and mixing processes, including mixing induced by extensive groundwater pumping, as well as natural mixing and water–rock interaction. It is evident from the geochemistry that mixing has occurred between groundwater and seawater in the coastal areas, as well as between normal/low temperature groundwater and geothermal water in the inland areas (e.g., near the Danihe geothermal field). Different sources of water are generally characterized by somewhat distinctive stable isotopic and hydrochemical compositions, allowing mixing calculations to aid understanding of the groundwater salinization and mixing processes, as discussed below.

Stable isotopes of O and H in groundwater and surface water fall on a best-fit regression line (dashed line in Fig. 3) with slope of $\delta^2H = 4.4 \times \delta^{18}O - 21.7$, significantly lower than either the local or global meteoric water lines. Three processes are likely to be responsible for the measured range of isotopic compositions: (1) mixing between saline surface water (e.g., seawater or saline river water affected by tidal ingress) and fresher, meteoric-derived groundwater or surface water; (2) mixing between fresh meteoric-derived groundwater and saline thermal water; (3) evaporative enrichment of surface water and/or irrigation return flow, which may infiltrate groundwater in some areas. A subgroup of surface water samples (e.g., S1 to S3, S7 and S12; termed "brackish surface water") show marine-like stable isotopic compositions and major ion compositions (Figs. 3 and 5). The "fresh" surface water samples (e.g., EC values $< 1500\,\mu S\,cm^{-1}$) exhibit meteoric-like stable isotope compositions, with some samples (such as S9 and S10) showing clear evidence of evaporative enrichment in the form of higher δ^2H and particularly, $\delta^{18}O$ values (Fig. 3).

Fresh groundwater has depleted $\delta^{18}O$ and δ^2H values relative to seawater and show a clear meteoric origin, albeit with modification due to mixing. Theoretically, the mix-

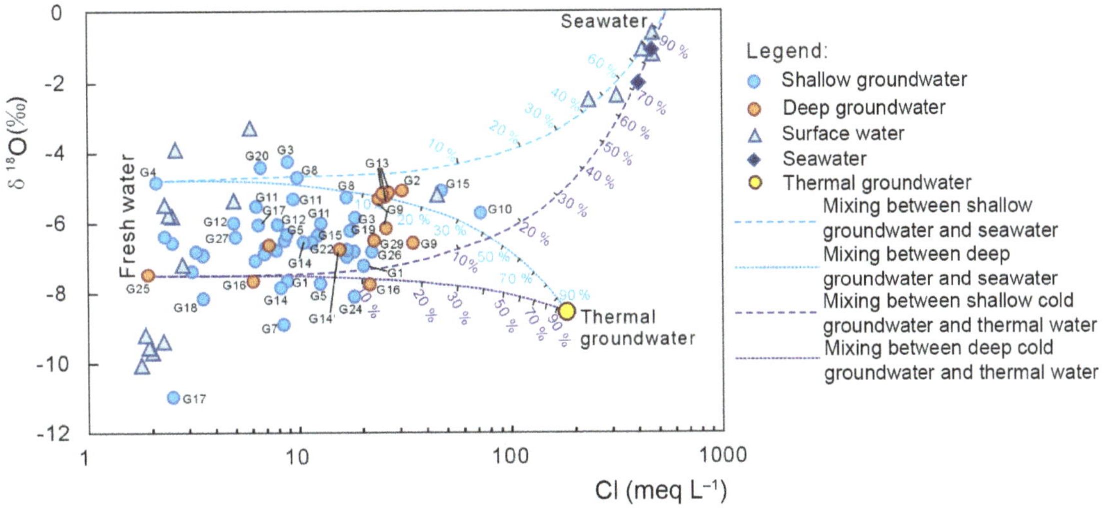

Figure 5. Relationship between chloride content and isotopic signature of different water samples as a means to differentiate mixing processes in the area. The data of thermal water are from Zeng (1991).

ing of meteoric-derived fresh groundwater and marine water should result in a straight mixing line connecting the two end-members; however, this is also complicated in the study area by the possible mixing with geothermal water. The thermal groundwater has distinctive stable isotopic and major ion composition (Han, 1988; Zeng, 1991), allowing these mixing processes to be partly delineated. Stable isotopes of thermal groundwater are more depleted than low-temperature groundwater (e.g., $\delta^{18}O$ values of approximately -8 ‰, Fig. 8), indicating this likely originates from the mountainous areas to the north; Zeng (1991) estimated the elevation of the recharge area for the geothermal field to be from 1200 to 1500 m a.s.l. Based on a bivariate plot of $\delta^{18}O$ vs. Cl^- with mixing lines and defined fresh and saline end-members, Fig. 5 shows the estimated degree of mixing between fresh groundwater, including shallow (G4) and deep (G25) groundwater end-members, and saline water, including seawater and geothermal end-members.

The two fresh end-members were selected to represent a range of different groundwater compositions and/or recharge sources, from shallow water that is impacted by infiltration of partially evaporated recharge (fresh but with enriched $\delta^{18}O$) to deeper groundwater unaffected by such enrichment (fresh and with relatively depleted $\delta^{18}O$). The narrower range and relatively enriched stable isotopes in shallow groundwater samples collected during the dry season compared with the wet season indicate some influence of seasonal recharge by either rainfall (fresh, with relatively depleted stable isotopes) or irrigation water subject to evaporative enrichment (more saline, with enriched stable isotopes and high nitrate concentrations; Currell et al., 2010) and/or surface water leakage. While there is overlap in the isotopic and hydrochemical compositions of shallow and deep groundwater (Figs. 3 and 4), this effect appears to only affect the shallow aquifer.

Based on Fig. 5, the shallow groundwater samples (e.g., G15, G10, G11, G14) collected from or around the Zaoyuan well field appear to be characterized by mixing between fresh meteoric water and seawater (plotting in the upper part of Fig. 5); while some deeper groundwater samples (e.g., G13, G2, G16, G14) collected from the coastal zone also appear to indicate mixing with seawater. Groundwater sampled relatively close to the geothermal field (e.g., G9, G19) shows compositions consistent with mixing between low-temperature fresh water and saline thermal water (lower part of Fig. 5). This is more evident in deep groundwater than shallow groundwater, which is consistent with mixing from below, as expected for the deep-source geothermal water. Other samples impacted by salinization show more ambiguous compositions between the various mixing lines, which may arise due to mixing with either seawater, geothermal water or a combination of both (e.g., G29).

The estimated mixing fraction (f_{sw}) of marine water for the shallow brackish groundwater ranges from 1.2 to 13.0 % and 2.6 to 6.0 % for the deep brackish groundwater. The highest fraction of 13 % was recorded in G10, located in the northern part of the Zaoyuan well field, which is located near a tidally impacted tributary of the Yang River (Fig. 1). Relatively higher fractions of marine water in relatively shallow samples (including those from the well field) compared to deeper samples may indicate a more "top down" salinization process, related to leakage of saline surface water through the riverbed, rather than "classic" lateral sea water intrusion, which typically causes salinization at deeper levels due to migration of a saltwater "wedge" (e.g., Werner et al., 2013); this is consistent with results of resistivity surveys conducted in the region (Fig. 6). The profile of chloride concentrations vs. depth indicates that salinization affects shallow and deep

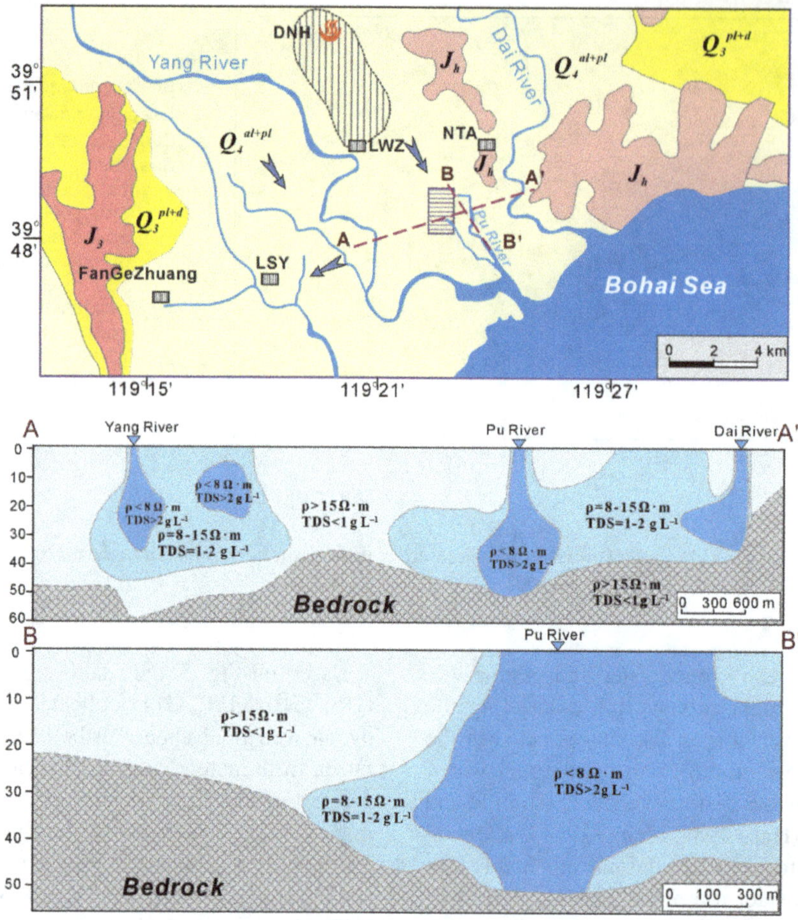

Figure 6. Cross sections showing results obtained from application of geophysical resistivity method (employed in May 2004, data and methods described in Zuo, 2006).

samples alike, with the most saline samples being relatively shallow wells in the Zaoyuan well field (Fig. 7).

In general, brackish and fresh groundwater samples show distinctive major ion compositions, with the more saline water typically showing higher proportions of Na and Cl (Fig. 4). This contrasts with historic data collected from the Zaoyuan well field, which showed Ca-HCO_3 type water with Cl concentrations ranging from 130 to 170 mg L^{-1}. This provides additional evidence that the salinization in this area is largely due to marine water mixing. More Ca-dominated compositions are evident in the region near the geothermal well field further in-land (e.g., G5, G8, G19, G29 and G24), consistent with a component of salinization that is unrelated to marine water intrusion. Plots of ionic ratios of Na / Cl and Mg / Ca vs. Cl also reveal a subset of relatively saline deep groundwater samples which appear to evolve towards the geothermal-type signatures with increasing salinity (Fig. 8).

Stronger evidence of mixing of the geothermal water in the Quaternary aquifers (particularly deep groundwater) is provided by examining strontium concentrations in conjunction with chloride (Fig. 9). The geothermal water from Danihe geothermal field has much higher Sr concentrations (up to 89.8 mg L^{-1}) than seawater (5.4–6.5 mg L^{-1} in this study), due to Sr-bearing minerals (i.e., celestite, strontianite) with Sr contents of 300–2000 mg kg^{-1} present in the bedrock (Hebei Geology Survey, 1987). Groundwater sampled from near the geothermal field in this study has the highest Sr concentrations, e.g., G9 with Sr concentrations ranging from 7.4 to 11.6 mg L^{-1} and G19 from 4.9 to 7.1 mg L^{-1}.

The plot of chloride vs. strontium concentrations (Fig. 9) shows that these samples and others (e.g., G16, G20, G27, G29) plot close to a mixing line between fresh low-temperature and saline thermal groundwater. Mass ratios of Sr / Cl in these samples are also elevated relative to seawater by an order of magnitude or more (e.g., Sr / Cl > 5.0×10^{-3}, compared to 3.9×10^{-4} in seawater, Table S1). Other samples from closer to the coast (e.g., G4) also approach the thermal-to-low temperature mixing line, indicating probable input of thermal water. Samples collected from the Zaoyuan well field generally plot closer to the Sr / Cl seawater mixing line (consistent with salinization largely due to marine water – Fig. 9); however, samples mostly plot slightly above

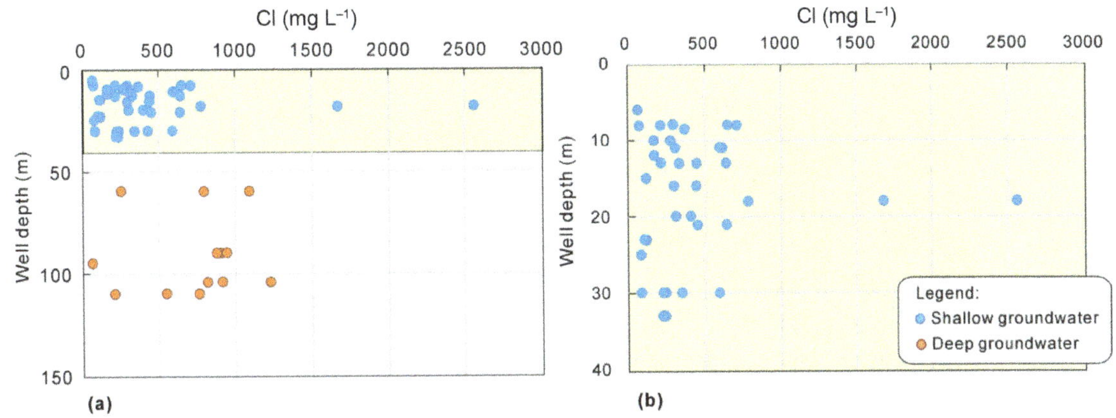

Figure 7. Chloride concentration vs. well depth for groundwater samples.

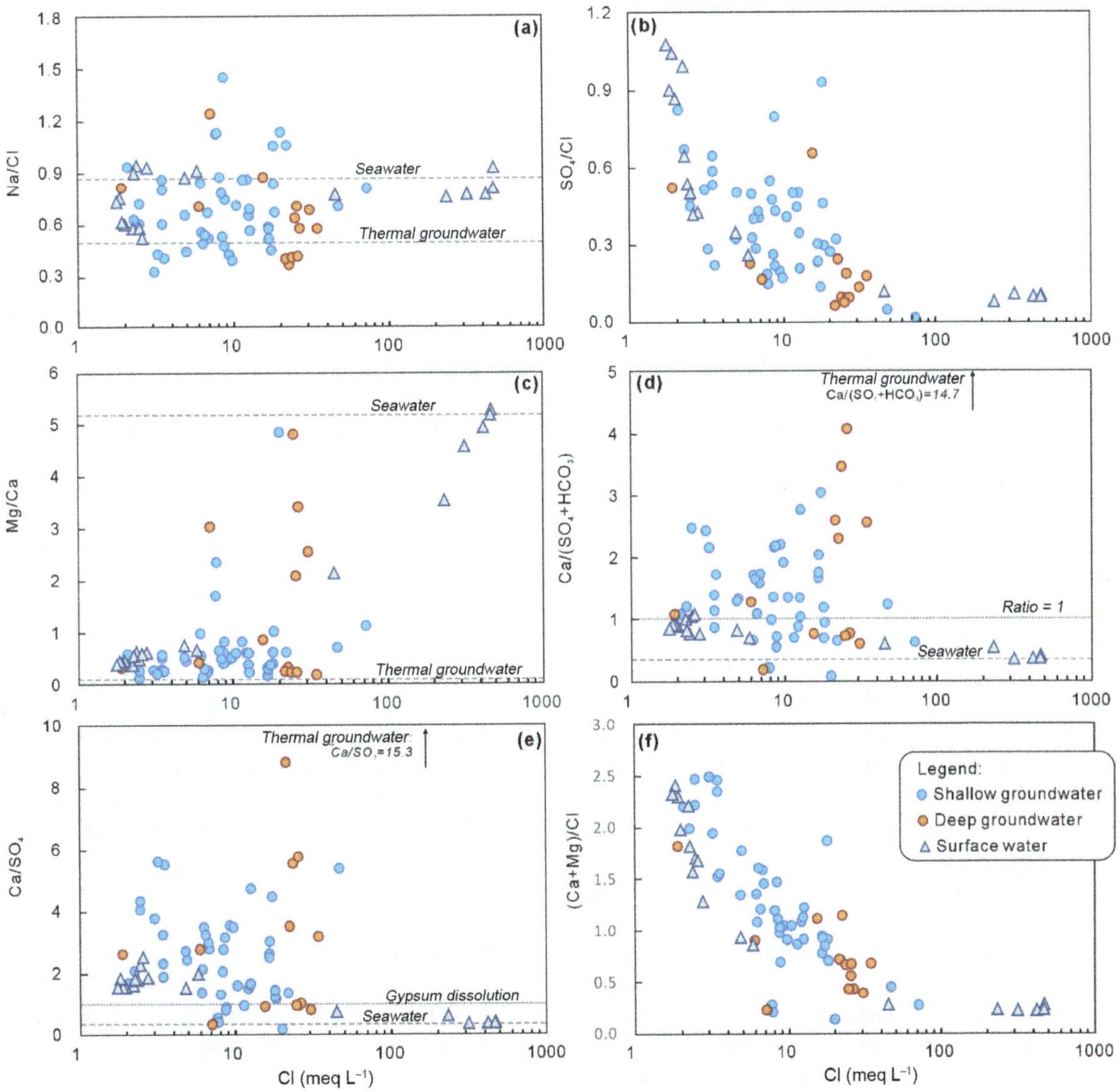

Figure 8. Molar ratios of major ions vs. chloride concentrations for different water samples from the study area.

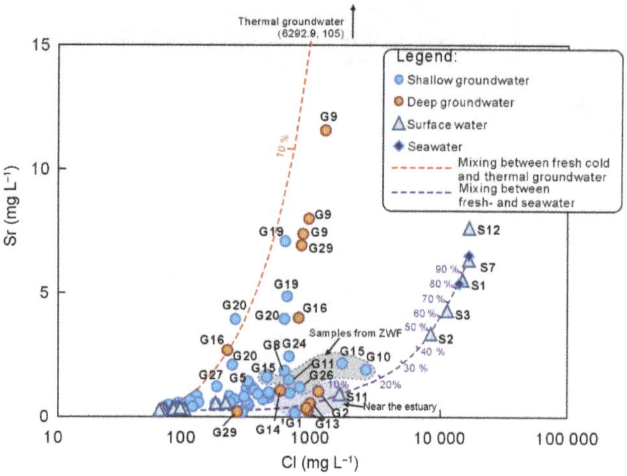

Figure 9. Chloride vs. strontium concentrations of different water samples. ZWF: Zaoyuan well field.

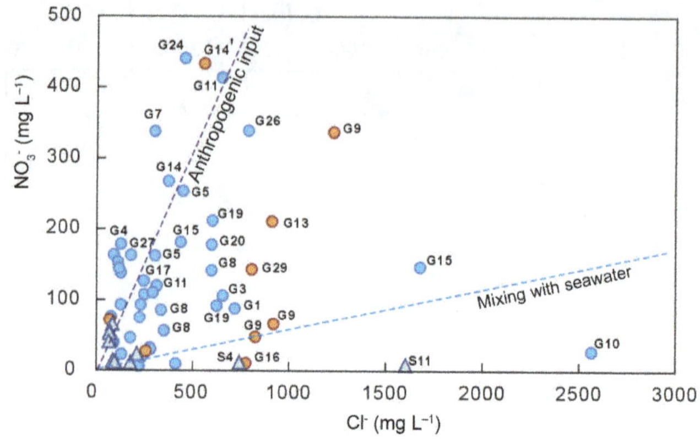

Figure 10. Plot of chloride vs. nitrate concentrations in groundwater, with seawater and anthropogenic pollution mixing trajectories.

the mixing line with additional Sr, which may indicate more widespread (but volumetrically minor) mixing with the thermal water in addition to seawater.

5.2 Anthropogenic pollution of groundwater

The occurrence of high nitrate (and possibly also sulfate) concentrations in groundwater in both coastal and in-land areas also indicates that anthropogenic pollution is an important process impacting groundwater quality and salinity (Fig. 10, Table 1). Seawater from Bohai Sea is heavily affected by nutrient contamination, showing NO_3^- concentrations of $810\,mg\,L^{-1}$ in this study, and up to $1092\,mg\,L^{-1}$ in seawater further north of the bay near Dalian (Han et al., 2015), primarily due to wastewater discharge into the sea. The historic sampled NO_3^- concentration of groundwater in the well field increased from 5.4 in May 1985 to 146.8–339.4 $mg\,L^{-1}$ in August 2010, while the concentration in seawater changed from 57.4 in May 1985 to 810.1 $mg\,L^{-1}$ in August 2010. A bivariate plot of Cl^- vs. NO_3^- concentrations in groundwater (Fig. 10) can thus be used to identify nitrate sources and mixing trends, including infiltration with contaminated seawater, and other on-land anthropogenic NO_3 sources (e.g., domestic or industrial wastewater discharge and/or NO_3^--bearing fertilizer input through irrigation return flow).

From this plot (Fig. 10) it appears that the major source of NO_3^- in groundwater is on-land anthropogenic inputs rather than mixing with seawater, which would result in relatively large increases in Cl along with NO_3. Samples G10 and G15 (from the well field) are exceptions to this trend, showing clear mixing with nitrate-contaminated seawater. Deep groundwater (e.g., G9, G14) is also extensively contaminated with high NO_3^- concentrations; this is likely associated with leakage from the surface via poorly constructed or abandoned wells – a problem of growing significance in China (see Han et al., 2016b; Currell and Han, 2017). According to one investigation by Zang et al. (2010), 14 of 21 pumping wells in the Zaoyuan well field have been abandoned due to poor water quality, and 307 pumping irrigation wells (occupied two-thirds of total pumping wells for irrigation) in the region have also been abandoned. Local authorities have, however, not implemented measures to deal with abandoned wells, meaning they are a future legacy contamination risk – e.g., by allowing surface runoff impacted by nitrate contamination to infiltrate down well annuli.

5.3 Hydrochemical evolution during salinization

A hydrogeochemical facies evolution diagram (HFE-D) proposed by Giménez-Forcada (2010) was used to analyze the geochemical evolution of groundwater during seawater intrusion and/or freshening phases (Fig. 11). In the coastal zone, the river water shows an obvious mixing trend between fresh and saline end-members. Some shallow groundwaters (e.g., G2, G4, G10, G13, G15) are also close to the mixing line between the surface-water end-members on this figure, indicating mixing with seawater without significant additional modification by typical water–rock interaction processes (e.g., ion exchange). Most brackish groundwaters (e.g., G11, G16, G17, G20, G25, G28, G29) have evolved in the series $Ca\text{-}HCO_3 \rightarrow Ca\text{-}Cl \rightarrow Na\text{-}Cl$, according to classic seawater intrusion. A relative depletion in Na (shown in lower than marine Na / Cl ratios) and enrichment in Ca (shown as enriched Ca / SO_4 ratios) is evident in groundwater with intermediate salinities (e.g., Figs. 8 and 11), indicating classic base exchange between Na and Ca during salinization (Appelo and Postma, 2005). Locally, certain brackish water samples (e.g., G1, G12, G26) appear to plot in the "freshening" part of the HFE diagram (potentially indicating slowing or reversal of salinization due to reduce in groundwater use), although these do not follow a conclusive trajectory. Water samples from the geothermal field (G5, G8, G9, and G19) plot in a

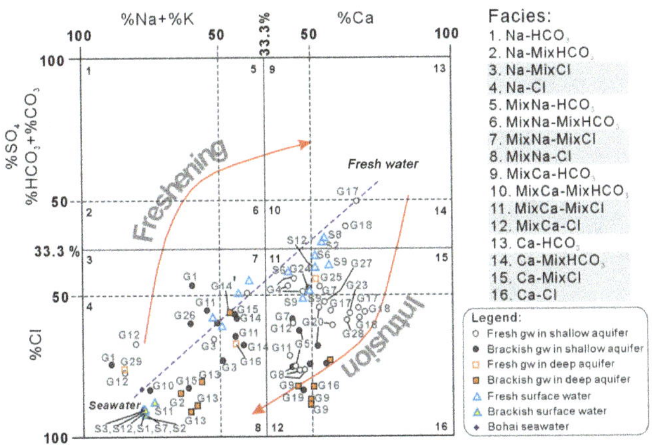

Figure 11. Hydrogeochemical facies evolution diagram (HFE-D) for the collected water samples. Groundwater: gw.

particular corner of the HFE diagram away from other samples (being particularly Ca-rich); a result of their distinctive geochemical evolution during deep transport through the basement rocks at high temperatures.

5.4 Conceptual model of salinization and management implications

Coastal zones encompass the complex interaction among different water bodies (i.e., river water, seawater and groundwater). The interactions between surface water and groundwater in the Yang–Dai coastal river plain have generally been ignored in previous studies. However, the surface water chemistry data show that the distribution of saltwater has historically reached more than 10 km inland along the estuary of the Yang River, and approximately 4 km inland in the Dai River (Han, 1988). The relatively higher proportion of seawater-intrusion-derived salinity in shallow samples in this study, along with the evidence from resistivity surveys (Fig. 6, Zuo, 2006) indicate that intrusion by vertical leakage from these estuaries is therefore an important process. The hazard associated with this pathway in recent times has been reduced by the construction of a tidal dam, which now restricts seawater ingress along the Yang estuary to within 4 km of the coastline. This may alleviate salinization to an extent in future in the shallow aquifer by removing one of the salinization pathways; however, as described, there are multiple other salinization processes impacting the groundwater in the Quaternary aquifers of the region.

A conceptual model of the groundwater flow system in the Yang–Dai coastal river plain is summarized in Fig. 12. This model presents an advance on the previous understanding of the study area, by delineating four major processes responsible for groundwater salinization in this area. These are as follows: (1) seawater intrusion by lateral subsurface flow; (2) interaction between saline surface water and groundwater (e.g., vertical leakage of saline water from the river estuaries); (3) mixing between low-temperature groundwater and deep geothermal water; and, (4) irrigation return flow and associated anthropogenic contamination. Both the lateral and vertical intrusion of saline water are driven by the long-term overpumping of groundwater from fresh aquifers in the region. The irrigation return flow from local agriculture results from overirrigation of crops and is responsible for extensive nitrate pollution (up to 340 mg L^{-1} NO_3^- in groundwater of this area), probably due to dissolution of fertilizers during infiltration. The somewhat enriched stable isotopes in shallow groundwater (more pronounced in the dry season) also indicate that such return flow may recharge water impacted by evaporative salinization into the aquifer. The geothermal water, with distinctive chemical composition (e.g., depleted stable isotopes, high TDS, Ca and Sr concentrations), is also demonstrated in this study to be a significant contributor to groundwater salinization, via upward mixing. The study area is therefore in a situation of unusual vulnerability, in the sense that it faces salinization threats simultaneously from lateral, downward and upward migration of saline water bodies.

According to drinking water standards and guidelines from the China Environmental Protection Authority (GB 5749-2006) and/or US EPA and WHO, chloride concentration in drinking water should not exceed 250 mg L^{-1}. At the salinity levels observed in this study – many samples impacted by salinization contain > 500 mg L^{-1} of chloride (Table 1) – a large amount of groundwater is now or will soon be unsuitable for domestic usage, as well as irrigation or industrial utilization. So far, this has enhanced the scarcity of fresh water resources in this region, leading to a cycle of groundwater table decline → seawater intrusion → loss of available freshwater → increased pumping of remaining fresh water. If this cycle continues, it is likely to further degrade groundwater quality and restrict its usage in the future. Such a situation is typical of the coastal water resources "squeeze" highlighted by Michael et al. (2017). Alternative management strategies, such as restricting water usage in particular high-use sectors, such as agriculture, industry or tourism, that are based on a comprehensive assessment of the social, economic and environmental benefits and costs of these activities, warrants urgent and careful consideration.

6 Conclusions

Groundwater in the Quaternary aquifers of the Yang–Dai coastal river plain is an important water resource for agricultural irrigation, domestic use (including for tourism) and industrial activity. Extensive groundwater utilization has made the problem of groundwater salinization in this area increasingly prominent, resulting in the closure of wells in the area. Based on the analysis of hydrochemical and stable isotopic compositions of different water bodies, we delineated the key groundwater salinization processes. Seawater intrusion

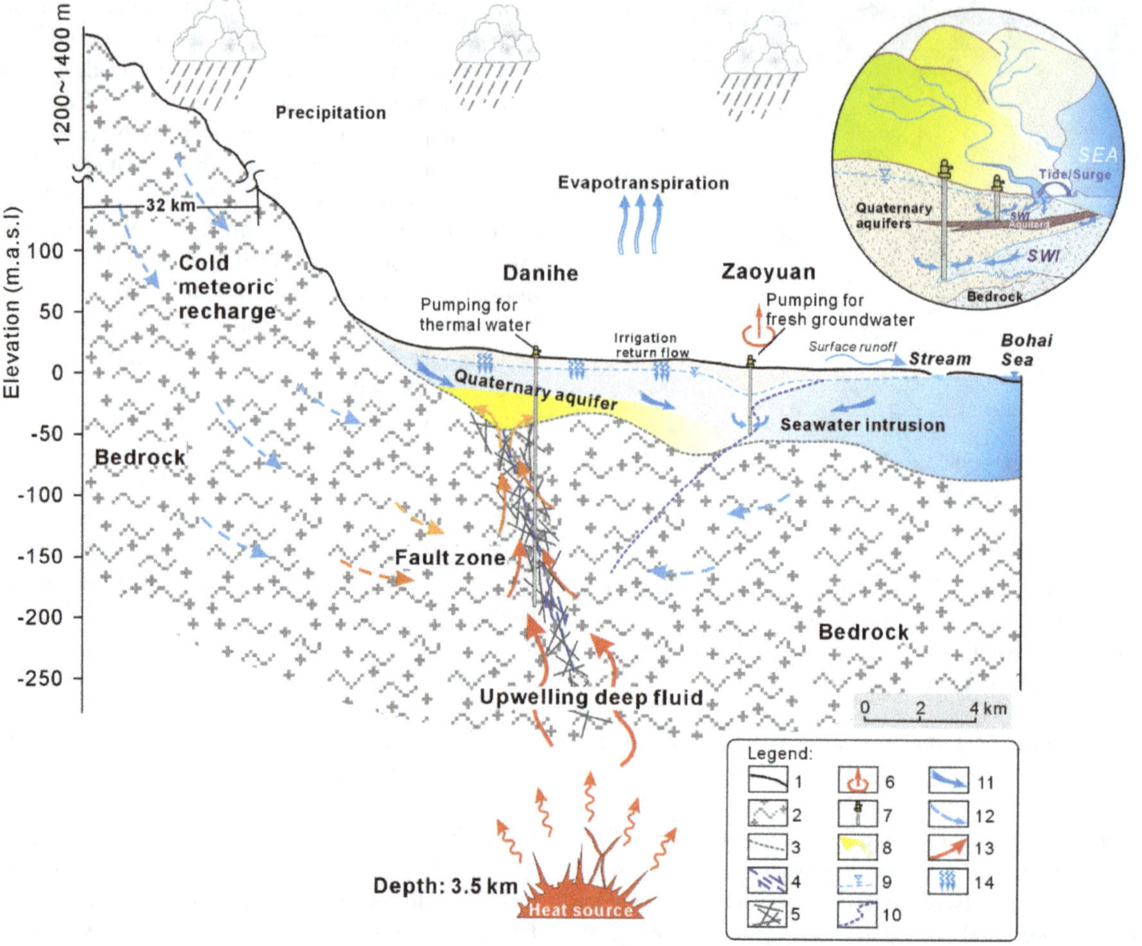

Figure 12. Conceptual model of groundwater flow system in the Yang–Dai coastal river plain. Explanation: (1) land surface; (2) bedrock; (3) boundary between quaternary sediments and bedrock; (4) fault; (5) permeable fracture zone; (6) concentrated groundwater pumping zone; (7) pumping wells; (8) zone affected by upflow of geothermal fluids; (9) water table; (10) potential interface between fresh- and saltwater; (11) groundwater flow direction in Quaternary aquifers; (12) groundwater flow in bedrocks; (13) geothermal groundwater flow direction; (14) irrigation return flow.

is the main process responsible for salinization in the coastal zone; however, this likely includes vertical saltwater infiltration along the riverbed into aquifers as well as lateral seawater intrusion caused by pumping for fresh groundwater at the Zaoyuan well field. The upward mixing of high TDS-geothermal water into the Quaternary aquifers is also evident, particularly through the use of stable isotope, chloride and strontium end-member mixing analysis. Additionally, significant nitrate pollution from the anthropogenic activities (e.g., agricultural irrigation return flow with dissolution of fertilizers) and local intrusion of heavily polluted seawater are also evident.

Groundwater salinization has become a prominent water environment problem in the coastal areas of northern China (Han et al., 2014, 2015, 2016a) and threatens to create further paucity of fresh water resources, which may prove a significant impediment to further social and economic development in these regions. Since the 1990s, the local government has begun to pay attention to the problem of seawater intrusion, and irrational exploitation of groundwater has been restricted in some cases. The Zaoyuan well field ceased to pump groundwater in 2007, while an anti-tide dam (designed to protect against tidal surge events) established in the Yang River estuary may also reduce saline intrusion in future. However, due to the significant lag time associated with groundwater systems, a response in terms of water quality may take time to emerge, and in the meantime the other salinization and pollution impacts documented here may continue to threaten water quality. In this regard, we recommend continued monitoring of groundwater quality and levels, and active programs to reduce input of anthropogenic contaminants such as nitrate from fertilizers, as well as appropriate well construction and decommissioning protocols to prevent contamination through preferential pathways.

Competing interests. The authors declare that they have no conflict of interest.

Acknowledgements. This research was supported by Zhu Kezhen Outstanding Young Scholars Program in the Institute of Geographic Sciences and Natural Resources Research, Chinese Academy of Sciences (CAS) (grant number 2015RC102), and Outstanding member program of the Youth Innovation Promotion Association, CAS (grant number 2012040). The authors appreciate the helpful field work and data collection carried out by Yang Jilong from the Tianjin Institute of Geology and Mineral Resources, and Wang Peng and Liu Xin from Chinese Academy of Sciences.

Edited by: Bill X. Hu

References

An, T. D., Tsujimura, M., Phu, V. L., Kawachi, A., and Ha, D. T.: Chemical Characteristics of Surface Water and Groundwater in Coastal Watershed, Mekong Delta, Vietnam, The 4th International Conference on Sustainable Future for Human Security, SustaiN 2013, Procedia Environ. Sci., 20, 712–721, 2014.

Andersen, M. S., Jakobsen, V. N. R., and Postma, D.: Geochemical processes and solute transport at the seawater/freshwater interface of a sandy aquifer, Geochim. Cosmochim. Ac., 69, 3979–3994, 2005.

Appelo, C. A. J.: Cation and proton exchange, pH variations, and carbonate reactions in a freshening aquifer, Water Resour. Res., 30, 2793–2805, 1994.

Appelo, C. A. J. and Postma, D.: Geochemistry, Groundwater and Pollution, 2 Edn., A. A. Balkema Publishers, Amsterdam, the Netherlands, 2005.

Bao, J.: Two-dimensional numerical modeling of seawater intrusion in Qinhuangdao Region, MA thesis, Tongji University, Shanghai, China, 2005 (in Chinese with English abstract).

Barbecot, F., Marlin, C., Gibert, E., and Dever, L.: Hydrochemical and isotopic characterisation of the Bathonian and Bajocian coastal aquifer of the Caen area (northern France), Appl. Geochem., 15, 791–805, 2000.

Bobba, A. G.: Numerical modelling of salt-water intrusion due to human activities and sea-level change in the Godavari Delta, India, Hydrological Sciences, 47, S67–S80, 2002.

Bouchaou, L., Michelot, J. L., Vengosh, A., Hsissou, Y., Qurtobi, M., Gaye, C. B., Bullen, T. D., and Zuppi, G. M.: Application of multiple isotopic and geochemical tracers for investigation of recharge, salinization, and residence time of water in the Souss–Massa aquifer, southwest of Morocco, J. Hydrol., 352, 267–287, 2008.

Cary, L., Petelet-Giraud, E., Bertrand, G., Kloppmann, W., Aquilina, L., Martins, V., Hirata, R., Montenegro, S., Pauwels, H., Chatton, E., Franzen, M., Aurouet, A., Lasseur, E., Picot, G., Guerrot, C., Fléhoc, C., Labasque, T., Santos, J. G., Paiva, A., Braibant, G., and Pierre, D.: Origins and processes of groundwater salinization in the urban coastal aquifers of Recife (Pernambuco, Brazil): A multi-isotope approach, Sci. Total. Environ., 530–531, 411–429, 2015.

Chen, M. and Ma, F.: Groundwater resources and the environment in China, Seismological Press, Beijing, 255–281, 2002 (in Chinese).

Craig, H.: Standard for reporting concentration of deuterium and oxygen-18 in natural water, Science, 133, 1833–1834, 1961.

Creel, L.: Ripple effects: Population and coastal regions, Population Reference Bureau, 1–7, available at: http://www.prb.org/Publications/Reports/2003/RippleEffectsPopulationandCoastalRegions.aspx (last access: 25 June 2016), 2003.

Currell, M. J. and Han, D.: The Global Drain: Why China's water pollution problems should matter to the rest of the world, Environment: Science and Policy for Sustainable Development, 59, 16–29, 2017.

Currell, M. J., Cartwright, I., Bradley, D. C., and Han, D. M.: Recharge history and controls on groundwater quality in the Yuncheng Basin, north China, J. Hydrol. 385, 216–229, 2010.

Currell, M. J., Dahlhaus, P. D., and Ii, H: Stable isotopes as indicators of water and salinity sources in a southeast Australian coastal wetland: identifying relict marine water, and implications for future change, Hydrogeol. J., 23, 235–248, 2015.

de Montety, V., Radakovitch, O., Vallet-Coulomb, C., Blavoux, B., Hermitte, D., and Valles, V.: . Origin of groundwater salinity and hydrogeochemical processes in a confined coastal aquifer: Case of the Rhône delta (Southern France), Appl. Geochem., 23, 2337–2349, 2008.

Edmunds, W. M.: Bromine geochemistry of British groundwaters, Mineral. Mag., 60, 275–284, 1996.

El Yaouti, F., El Mandour, A., Khattach, D., Benavente, J., and Kaufmann, O.: Salinization processes in the unconfined aquifer of Bou-Areg (NE Morocco): A geostatistical, geochemical, and tomographic study, Appl. Geochem., 24, 16–31, 2009.

Ferguson, G. and Gleeson, T.: Vulnerability of coastal aquifers to groundwater use and climate change, Nat. Clim. Change., 2, 342–345, 2012.

Fidelibus, M. D., Giménez, E., Morell, I., and Tulipano, L.: Salinization processes in the Castellon Plain aquifer (Spain), in: Study and Modelling of Saltwater Intrusion into Aquifers, Centro Internacional de Métodos Numéricos en Ingeniería, Barcelona, edited by: Custodio, E. and Galofré, A., 267–283, 1993.

Garing, C., Luquot, L., Pezard, P. A., and Gouze, P.: Geochemical investigations of saltwater intrusion into the coastal carbonate aquifer of Mallorca, Spain, Appl. Geochem., 39, 1–10, 2013.

Ghassemi, F., Chen, T. H., Jakeman, A. J., and Jacobson, G.: Two and three-dimensional simulation of seawater intrusion: performances of the "SUTRA" and "HST3D" models, AGSO J. Aust. Geol. Geophys., 14, 219–226, 1993.

Ghiglieri, G., Carletti, A., and Pittalis, D.: Analysis of salinization processes in the coastal carbonate aquifer of Porto Torres (NW Sardinia, Italy), J. Hydrol., 432–433, 43–51, 2012.

Giambastiani, B. M. S., Antonellini, M., Oude Essink, G. H. P., and Stuurman, R. J.: Saltwater intrusion in the unconfined coastal aquifer of Ravenna (Italy): A numerical model, J. Hydrol., 340, 91–104, 2007.

Giménez-Forcada, E.: Dynamic of sea water interface using hydrochemical facies evolution diagram, Ground Water, 48, 212–216, 2010.

Gingerich, S. and Voss, C.: Three-dimensional variable-density flow simulation of a coastal aquifer in southern Oahu, Hawaii, USA, in: Proceedings SWIM17 Delft 2002, edited by: Boekelman, R., Delft Univ. of Technol., Delft, the Netherlands, 93–103, 2002.

Han, D. M., Kohfahl, C., Song, X. F., Xiao, G. Q., and Yang, J. L.: Geochemical and isotopic evidence for palaeo-seawater intrusion into the south coast aquifer of Laizhou Bay, China, Appl. Geochem., 26, 863–883, 2011.

Han, D. M., Song, X. F., Currell, M. J., Yang, J. L., and Xiao, G. Q.: Chemical and isotopic constraints on the evolution of groundwater salinization in the coastal plain aquifer of Laizhou Bay, China, J. Hydrol., 508, 12–27, 2014.

Han, D., Post, V. E. A., and Song, X.: Groundwater salinization processes and reversibility of seawater intrusion in coastal carbonate aquifers, J. Hydrol., 531, 1067–1080, 2015.

Han, D., Song, X., and Currell, M. J.: Identification of anthropogenic and natural inputs of sulfate into a karstic coastal groundwater system in northeast China: evidence from major ions, $\delta^{13}C_{DIC}$ and $\delta^{34}S_{SO_4}$, Hydrol. Earth Syst. Sci., 20, 1983–1999, https://doi.org/10.5194/hess-20-1983-2016, 2016a.

Han, D., Currell, M. J., and Cao, G.: Deep challenges for China's war on water pollution, Environ. Pollut., 218, 1222–1233, 2016b.

Han, Z.: Seawater intrusion into coastal porous aquifer- a case study of the alluvial plain of Yang River and Dai River in Qinhuangdao City, China University of Geosciences, Beijing, China, 1988 (in Chinese with English abstract).

Han, Z.: Controlling and harnessing of the seawater intrusion the alluvial plain of Yang River and Dai River in Qinhuangdao, Geoscience, Journal of Graduate School, China University of Geosciences, 4, 140–151, 1990 (in Chinese with English abstract).

Hebei Geology Survey: Report of regional geological and tetonic background in Hebei Province, Internal materials, Geological Publishing House, Beijing, 1987.

Hui, G.: Characteristics and formation mechanism of hydrochemistry of geothermal water in Danihe, Funing District of Qinhuangdao region, Science & Technology Information, 9, 762–763, 2009 (in Chinese).

IAEA/WMO: Global Network of Isotopes in Precipitation, The GNIP Database, Vienna, available at: http://www-naweb.iaea.org/napc/ih/IHS_resources_gnip.html (last access: 18 November 2013), 2006.

IPCC, Climate Change: The Physical Science Basis, Contribution of Working Group I to the Fourth Assessment Report of the Intergovernmental Panel on Climate Change, edited by: Solomon, S., Qin, D., and Manning, M., Cambridge Univ. Press, Cambridge, 2007.

Jones, B. F., Vengosh, A., Rosenthal, E., and Yechieli, Y.: Geochemical investigations, in: Seawater Intrusion in Coastal Aquifers – Concepts, Methods and Practices, edited by: Bear, J., Cheng A. H.-D., Sorek, S., Ouazar, D., and Herrera, I., Spinger Science+Business Media, B.V., Dordrecht, the Netherlands, 51-72, 1999.

Langevin, C. D., Zygnerski, M. R., White, J. T., and Hughes, J. D.: Effect of sea-level rise on future coastal groundwater resources in southern Florida, USA, SWIM21-21st Salt Water Intrusion Meeting, Azores, Portugal, 21–26 June 2010, 125–128, 2010.

Larsen, F., Tran, L. V., Hoang, H. V., Tran, L. T., Chistiansen, A. V., and Pham, N. Q.: Groundwater salinity influenced by Holocene seawater trapped in the incised valleys in the Red River delta plain, Nat. Geosci., 10, 376–382, 2017.

Lee, S., Currell, M., and Cendon, D. I.: Marine water from mid-Holocene sea level highstand trapped in a coastal aquifer: Evidence from groundwater isotopes, and environmental significance, Sci. Total. Environ., 544, 995–1007, 2016.

Liang, X. J., Liu, H. B., and Xing H. Y.: Seawater intrusion and countermeasures in the coastal area of south Funing County, Hebei Province, Water Resources Planning and Design, 5, 30–33, 2010 (in Chinese).

Masterson, J. P.: Simulated interaction between freshwater and saltwater and effects of ground-water pumping and sea-level change, Lower Cape Cod aquifer system, Massachusetts: U.S. Geological Survey Scientific Investigations Report, 2004–5014, p. 72, 2004.

Mazi, K., Koussis, A. D., and Destouni, G.: Intensively exploited Mediterranean aquifers: resilience to seawater intrusion and proximity to critical thresholds, Hydrol. Earth Syst. Sci., 18, 1663–1677, https://doi.org/10.5194/hess-18-1663-2014, 2014.

Meng, F.: Rational development of groundwater resources on the coastal zone of Qinhuangdao area, Marine Geology Letters, 20, 22–25, 2004 (in Chinese with English abstract).

Michael, H. A., Post, V. E. A., Wilson, A. M., and Werner, A. D.: Science, society and the coastal groundwater squeeze, Water Resour. Res., 53, 2610–2617, 2017.

Mondal, N. C., Singh, V. P., Singh, V. S., and Saxena, V. K.: Determining the interaction between groundwater and saline water through groundwater major ions chemistry, J. Hydrol., 388, 100–111, 2010.

Montenegro, S. M. G., Montenegro, A. A. A., Cabral, J. J. S. P., and Cavalcanti, G.: Intensive exploitation and groundwater salinity in Recife coastal plain (Brazil): monitoring and management perspectives, Proceedings 1st SWIM-SWICA Joint Saltwater Intrusion Conference, Cagliari-Chia Laguna, Italy, 24–29 September 2006, 79–85, 2006.

Moore, W. S.: Large groundwater inputs to coastal waters revealed by 226Ra enrichments, Nature, 380 612–214, 1996.

Narayan, K. A., Schleeberger, C., and Bristow, K. L.: Modelling seawater intrusion in the Burdekin Delta Irrigation Area, North Queensland, Australia, Agr. Water Manage., 89, 217–228, 2007.

Pan, G., Yang Y., and Zhang, L.:Survey report of geothermal water in Qinhuangdao city of Hebei Province, Team of mineral hydrological and engineering geology from the Bureau of Geology and mineral Resources of Hebei Province, China, 1990 (in Chinese).

Post, V. E. A.: Fresh and saline groundwater interaction in coastal aquifers: Is our technology ready for the problems ahead? Hydrogeol. J., 13, 120–123, 2005.

Price, R. M. and Herman, J. S.: Geochemical investigation of saltwater intrusion into a coastal carbonate aquifer: Mallorca, Spain, Geol. Soc. Am. Bull., 103, 1270–1279, 1991.

Pulido-Leboeuf, P.: Seawater intrusion and associated processes in a small coastal complex aquifer (Castell de Ferro, Spain), Appl. Geochem., 19, 1517–1527, 2004.

Radhakrishna, I.: Saline fresh water interface structure in Mahanadi delta region, Orissa, India, Environ. Geol., 40, 369–380, 2001.

Rahmawati, N., Vuillaume, J., and Purnama, I. L. S.: Salt intrusion in Coastal and Lowland areas of Semarang City, J. Hydrol., 494, 146–159, 2013.

Robinson, M. A., Gallagher, D. L., and Reay, W. G.: Field observations of tidal and seasonal variations in ground water discharge to estuarine surface waters, Ground Water Monit. R., 18, 83–92, 1998.

Shen, Z., Zhu, Y., and Zhong, Z.: Theoretical basis of hydrogeochemistry, Geological Publishing House, Beijing, China, 1993 (in Chinese).

Sherif, M. M. and Singh, V. P.: Effect of climate change on sea water intrusion in coastal aquifers, Hydrol. Process., 13, 1277–1287, 1999.

Shi, M. Q.: Spatial distribution of population in the low elevation coastal zone and assessment on vulnerability of natural disaster in the coastal area of China, MA thesis, Shanghai Normal University, 24–32, 2012 (in Chinese with English abstract).

Simpson, M. J. and Clement, T. P.: Improving the worthiness of the Henry problem as a benchmark for density-dependent groundwater flow models, Water Resour. Res., 40, W01504, https://doi.org/10.1029/2003WR002199, 2004.

Sivan, O., Yechieli, Y., Herut, B., and Lazar, B.: Geochemical evolution and timescale of seawater intrusion into the coastal aquifer of Israel, Geochim. Cosmochim. Ac., 69, 579–592, 2005.

Smith, A. J. and Turner, J. V.: Density-dependent surface water-groundwater interaction and nutrient discharge in the Swan-Canning Estuary, Hydrolological Processes, 15, 2595–2616, 2001.

SOA (State Oceanic Administration of the People's Republic of China): China's Marine Environment Bulletin, 2014, released 11 March, 2015 (in Chinese).

Sun, J. and Yang, Y.: Seawater intrusion characteristics in Qinhuangdao, Journal of Environmental Management College of China, 17, 51–54, 2007 (in Chinese with English abstract).

UN Atlas: UN Atlas: 44 Percent of us Live in Coastal Areas, available at: http://coastalchallenges.com/2010/01/31/un-atlas-60-of-us-live-in-the-coastal-areas/ (last access: 10 August 2013), 2010.

Vengosh, A., Spivack, A. J., Artzi, Y., and Ayalon, A.: Geochemical and boron, strontium, and oxygen isotopic constraints on the origin of the salinity in groundwater from the Mediterranean coast of Isreal, Water Resour. Res., 35, 1877–1894, 1999.

Walraevens, K., Cardenal-Escarcena, J., and Van Camp, M.: Reaction transport modelling of a freshening aquifer (Tertiary Ledo-Paniselian Aquifer, Flanders-Belgium), Appl. Geochem., 22, 289–305, 2007.

Wang, J., Yao, P., Bianchi, T. S., Li, D., Zhao, B., Cui, X., Pan, H., Zhang, T., and Yu, Z.: The effect of particle density on the sources, distribution, and degradation of sedimentary organic carbon in the Changjiang Estuary and adjacent shelf, Chem. Geol., 402, 52–67, 2015.

Weber, K. and Stewart, M., A critical analysis of the cumulative rainfall departure concept, Ground Water, 42, 935–938, 2004.

Werner, A. D.: A review of seawater intrusion and its management in Australia, Hydrogeol. J., 18, 281–285, 2010.

Werner, A. D. and Simmons, C. T.: Impact of sea-level rise on seawater intrusion in coastal aquifers, Ground Water, 47, 197–204, 2009.

Werner, A. D., Bakker, M., Post, V. E. A., Vandenbohede, A., Lu, C., Ataie-Ashtiani, B., Simmons, C. T., and Barry, D. A.: Seawater intrusion processes, investigation and management: Recent advances and future challenges, Adv. Water Resour., 51, 3–26, 2013.

Westbrook, S. J., Rayner, J. L., Davis, G. B., Clement, T. P., Bjerg, P. L., and Fisher S. J.: Interaction between shallow groundwater, saline surface water and contaminant discharge at a seasonally and tidally forced estuarine boundary, J. Hydrol., 302, 255–269, 2005.

Wilhite, D. A.: Drought Assessment, Management, and Planning: Theory and Case Studies, Springer, USA, p. 628, 1993.

Xu, G.: Analysis of seawater intruding into aquifer system in Beidaihe Region, Hydrogeology & Engineering Geology, 2, 7–10, 1986 (in Chinese).

Xue, Y. Q., Wu, J. C., Ye, S. J., and Zhang, Y. X.: Hydrogeological and hydrogeochemical studies for salt water intrusion on the South Coast of Laizhou Bay, China, Ground Water, 38, 38–45, 2000.

Yang, L.: Formation mechanism of bedrock fracture type-geothermal water in Qinhuangdao area, West-china Exploration Engineering, Urumchi, China, 10, 151–152, 2011 (in Chinese).

Yang, Y., He, Q., Xie, Y., and Cao, C.: Grey model prediction of seawater intrusion of Qinhuangdao, The Chinese Journal of Geological Hazard and Control, 5, 181–183, 1994 (in Chinese with English abstract).

Yang, Y., Gao, S., and Xie, Y.: Assessment and control countermeasures of seawater intrusion hazard on Qinhuangdao Region, The Chinese Journal of Geological Hazard and Control, 19, 139–143, 2008 (in Chinese with English abstract).

Yechieli, Y., Kafri, U., and Sivan, O.: The inter-relationship between coastal sub-aquifers and the Mediterranean Sea, deduced from radioactive isotopes analysis, Hydrogeol. J., 17, 265–274, 2009.

Zang, W., Liu, W., Guo, J., and Zhang, X.: Geological hazards of seawater intrusion and its control measures in Qinhuangdao City, Hebei Province, The Chinese Journal of Geological Hazard and Control, 21, 120–125, 2010 (Chinese with English abstract).

Zeng, J.: Geochemistry of geothermal water in Qinhuangdao area, Hebei Province, Bull. Institute of Hydrogeology and Engineering Geology, Chinese Academy of Geological Sciences, 7, 111–127, 1991.

Zhang, B.: Mechanism of Seawater Intrusion Using Hydrochemistry and Environmental Isotopes in Qinhuangdao Yang Dai River Plain, MA thesis, Xiamen University, Fujian, China, 2012 (in Chinese with English abstract).

Zhang, Q., Volker, R. E., and Lockington, D. A.: Numerical investigation of seawater intrusion at Gooburrum, Bundaberg, Queensland, Australia, Hydrogeol. J., 12, 674–687, 2004.

Zuo, W.: Survey and Research on Seawater Intrusion in the Yandaihe Plain of Qinhuangdao City, PhD thesis, China University of Geosciences, Beijing, China, 2006 (in Chinese with English abstract).

Zuo, W., Yang, Y., Dong, Y., and Liang, M.: The numerical study for seawater intrusion in Yanghe and Daihe coastal plain of Qinhuangdao City, Journal of Natural Resources, 24, 2087–2095, 2009 (in Chinese with English abstract).

14

ENSO–cave drip water hydrochemical relationship: a 7-year dataset from south-eastern Australia

Carol V. Tadros[1,2], **Pauline C. Treble**[1,2], **Andy Baker**[2], **Ian Fairchild**[3,4], **Stuart Hankin**[1], **Regina Roach**[5], **Monika Markowska**[1,2], **and Janece McDonald**[6]

[1]Australian Nuclear Science and Technology Organisation, Locked Bag 2001, Kirrawee DC, NSW 2232, Australia
[2]Connected Waters Initiative Research Centre, UNSW Australia, Kensington NSW 2052, Australia
[3]School of Geography, Earth and Environmental Sciences, University of Birmingham, Edgbaston, Birmingham, UK
[4]Birmingham Institute for Forest Research, University of Birmingham, Edgbaston, Birmingham, UK
[5]NSW National Parks and Wildlife Service, Sydney, NSW, Australia
[6]Environmental and Climate Change Research Group, School of Environmental and Life Sciences, University of Newcastle, Callaghan, NSW 2308, Australia

Correspondence to: Carol V. Tadros (carol.tadros@ansto.gov.au)

Abstract. Speleothems (cave deposits), used for palaeoenvironmental reconstructions, are deposited from cave drip water. Differentiating climate and karst processes within a drip-water signal is fundamental for the correct identification of palaeoenvironmental proxies and ultimately their interpretation within speleothem records. We investigate the potential use of trace element and stable oxygen-isotope ($\delta^{18}O$) variations in cave drip water as palaeorainfall proxies in an Australian alpine karst site. This paper presents the first extensive hydrochemical and $\delta^{18}O$ dataset from Harrie Wood Cave, in the Snowy Mountains, south-eastern (SE) Australia. Using a 7-year long rainfall $\delta^{18}O$ and drip-water Ca, Cl, Mg / Ca, Sr / Ca and $\delta^{18}O$ datasets from three drip sites, we determined that the processes of mixing, dilution, flow path change, carbonate mineral dissolution and prior calcite precipitation (PCP) accounted for the observed variations in the drip-water geochemical composition. We identify that the three monitored drip sites are fed by fracture flow from a well-mixed epikarst storage reservoir, supplied by variable concentrations of dissolved ions from soil and bedrock dissolution. We constrained the influence of multiple processes and controls on drip-water composition in a region dominated by El Niño–Southern Oscillation (ENSO). During the El Niño and dry periods, enhanced PCP, a flow path change and dissolution due to increased soil CO_2 production occurred in response to warmer than average temperatures in contrast to the La Niña phase, where dilution dominated and reduced PCP were observed. We present a conceptual model, illustrating the key processes impacting the drip-water chemistry. We identified a robust relationship between ENSO and drip-water trace element concentrations and propose that variations in speleothem Mg / Ca and Sr / Ca ratios may be interpreted to reflect palaeorainfall conditions. These findings inform palaeorainfall reconstruction from speleothems regionally and provide a basis for palaeoclimate studies globally, in regions where there is intermittent recharge variability.

1 Introduction

The El Niño–Southern Oscillation (ENSO) is the leading mode of rainfall variation in south-eastern (SE) Australia (Dai et al., 1997), where extreme events of rainfall variability, such as droughts, floods, bush fires and cyclones associated with ENSO, are prominent (Risbey et al., 2009). Severe drought between 2001 and 2008 and enhanced El Niño conditions resulted in record low inflows from the alpine headwaters of the Murray River (Murphy and Timbrel, 2008; Cai and Cowan, 2008; Nicholls, 2010), strongly impacting the water resource availability of the Murray–Darling basin and agricultural production, affecting the livelihood of urban and rural Australians (Barros and Bowden, 2008; McGowan et al., 2009). Reconstructing past ENSO variability from speleothems (calcium carbonate cave deposits) located within the Yarrangobilly Caves system in the Snowy Mountains alpine region will provide a basis for understanding future regional impacts, therefore assisting with water resource management policy making and the global impacts that ENSO-driven climate variability has on the environment, agricultural production, water resources, ecosystems as well as on human life, emergency management and disease (Power and Smith, 2007).

Studies have shown that trace element time series constructed from the central growth axis of a speleothem provide potential proxy evidence of palaeorainfall conditions (Roberts et al., 1998; Lauritzen et al., 1999; Fairchild et al., 2001; Johnson et al., 2006; Cruz et al., 2007; Jo et al., 2010). The concentration of trace elements in drip water is dependent on the evolution of the drip-water geochemistry, which is influenced by site-specific characteristics (Spötl et al., 2005) and a range of surface and karst processes (Baldini et al., 2006). Elements may be atmospherically derived from meteoric precipitation (P), dust supply (Goede et al., 1998; Dredge et al., 2013), marine aerosols (Baker et al., 2000; Fairchild et al., 2000), volcanic eruption activity (Frisia et al., 2005, 2008), atmospheric pollutants (Spötl et al., 2005; Wynn et al., 2008), the host rock and soil (Tooth and Fairchild, 2003). Surface processes: deforestation (Borsato et al., 2007) and fire (Coleborn et al., 2016; Nagra et al., 2016); soil processes: water–sediment and water–rock interaction (Fairchild et al., 2000), colloid, particle or solute mobilization (Hartland et al., 2012), and temperature, water availability and CO_2 changes (Cuthbert et al., 2014; Rutlidge et al., 2014; Treble et al., 2016); and karst hydrological processes: hydrological flow routes, mixing and dilution effects, degassing and calcite precipitation, differential and incongruent dissolution, and selective leaching (Fairchild et al., 2000; Tooth and Fairchild, 2003) potentially modulate the concentration of elements in the drip water. Constraining these processes and understanding potential climatic signals in the hydrochemistry of drip water is quintessential in successfully using trace elements as a palaeoclimate proxy (Fairchild et al., 2006).

As such, long-term datasets of stable oxygen-isotope ($\delta^{18}O$) and geochemistry (trace element concentrations and ratios) of rainfall and cave drip water provide an empirical basis for identifying factors influencing trace element variability ultimately recorded in speleothems (Baldini et al., 2002; Treble et al., 2003; Riechelmann et al., 2011; Oster et al., 2012; Frappier, 2013; Partin et al., 2013). For example, McDonald et al. (2004) demonstrated Mg / Ca and Sr / Ca ratios in the drip water doubled in response to an El Niño event that occurred during a 2.5-year baseline monitoring study at Wombeyan Caves, SE Australia. This was an important finding that raised the potential for using speleothem records to reconstruct past ENSO variability for this region. The Wombeyan Caves site lies in the Sydney catchment. The study site used here lies ~85 km away in the headwaters of the Murray–Darling basin and as such provides an opportunity to further examine the ENSO signal in cave drip water at a second site from this region, with a longer dataset. Additionally, the interpretation of this new hydrochemical dataset is conducted within an established framework. The dominant controls on precipitation stable isotope variability in this alpine region have been examined. Callow et al. (2014) conducted event-based $\delta^{18}O$ precipitation (rainfall and snow) sampling across 18 sites ($n = 70$; from February 2010 to March 2012) from a transect in the Snowy Mountains; they determined that the origin of moisture, pathway and terrain effects were the dominant controls on precipitation stable isotope variability in this alpine region. Furthermore, the unsaturated zone hydrology of our studied cave has been investigated. Markowska et al. (2015) presented rainfall, soil moisture saturation and drip discharge data at 14 sites within the same cave studied here, between October 2011 and January 2013. Markowska et al. (2015) applied a statistical approach to classify the drip types and identified five flow regimes, which were represented using a combined conceptual flow and box hydrological model.

The emphasis of this study is to understand the relationship between modern climatic and environmental controls on the cave drip water in a region strongly influenced by ENSO, to aid in the interpretation of speleothem-based palaeoenvironmental records and ultimately to develop climate proxy records from suitable speleothems (Tooth and Fairchild, 2003). Here we present the first comprehensive climate and drip water monitoring study, which commenced at three drip sites in 2006, in Harrie Wood Cave (Yarrangobilly, New South Wales, Australia). This record encapsulated the last 3 years of the "Millennium Drought" (1997–2009), which had a large impact across SE Australia (CSIRO and BoM, 2015) and the 2010/2011 La Niña event, which produced widespread flooding across SE Australia (CSIRO and BoM, 2015). Within this framework, spanning 7 years, we employed the results to ascertain the key hydrological processes that control the drip hydrochemistry during La Niña and El Niño events and categorized the flow regime. Our findings form the basis for palaeoclimate interpretation of speleothem trace element and $\delta^{18}O$ records located within the Yarrangobilly Caves system and are pertinent for speleothem palaeoclimate research in other ENSO-dominated regions globally.

2 Study area

2.1 Study site and climate

Harrie Wood Cave (35°44′ S, 148°30′ E) is located in a limestone belt ca. 14 km long and 1.5 km wide along the Yarrangobilly River in the north of Kosciuszko National Park, New South Wales, Australia (Fig. 1a). The Yarrangobilly

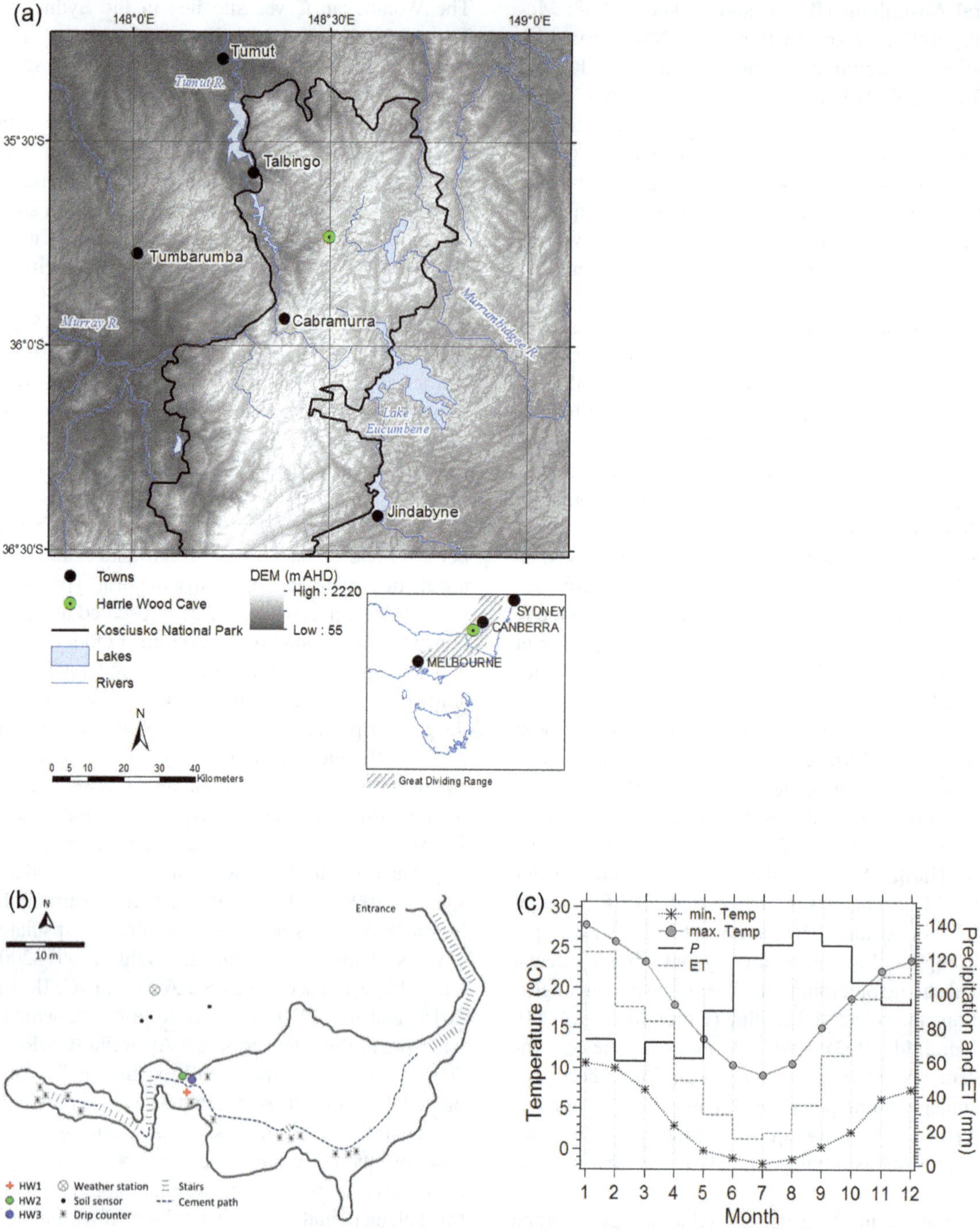

Figure 1. **(a)** Location of Harrie Wood Cave, Snowy Mountains, Australia. **(b)** Survey map of the Harrie Wood Cave system (adapted from Nicholl, 1974) showing location of the three drip-water sites (HW1–3), which have been monitored since 2006, the location of automatic drip counters underneath 14 drip sites reported by Markowska et al. (2015), and overlayed is the location of the weather station and Stevens Hydra Probe® soil sensors. **(c)** Site monthly mean air temperature (°C) (2006–2013), precipitation (mm) (1983–2013) and ET; the sum of transpiration and soil evaporation (1967–1990; parameter F_{WE} (m day^{-1}) compiled by the WaterDyn model, from the Australian Water Availability Project (AWAP) database; Raupach et al., 2009, 2011).

Limestone formed in the upper Silurian period from a coral reef (Worboys, 1982). Harrie Wood Cave is located within the Yarrangobilly Caves system, which includes over 250 independent limestone caves that began to develop in the Pleistocene (Worboys, 1982). Harrie Wood Cave is hosted within a highly fractured hard massive limestone and the drip sites are in close proximity to a fracture contact zone. The host limestone bedrock contains red/brown palaeokarst features and little to no dolomite. The cave entrance is ca. 965 m above sea level (a.s.l.) on a north-dipping steep rocky gorge. Harrie Wood is a restricted access, medium-sized south-dipping cave. The cave chamber is 80 ± 2 m in length and 34 ± 1 m deep (Nicholl, 1974).

Surrounding vegetation consists of open snow gum (*Eucalyptus pauciflora* subsp. *Pauciflora*) and black sallee (*E. stelullata*) woodland dominated by a snow grass (*Poa sieberi*) understorey (Aplin et al., 2010). Above the cave, vegetation coverage is sparsely developed on shallow rocky soil that lacks clearly defined horizons, and are dominated by angular clasts (typically 2–10 cm size) indicating mechanical weathering processes. The surface over the cave was burnt by an intense wildfire in 2003 and the shrubby vegetation that is present over the cave shows evidence of regeneration post-fire (Coleborn, 2016). There is no evidence of a distinct zone of infiltration on the surface directly above and upslope of Harrie Wood Cave. There is also no evidence of surface runoff following rain events. We interpret from our field observations that infiltration is via pervasive cracks and fissures.

The Yarrangobilly Caves system is part of the Australian Alps bioregion and is dominated by a montane climate (Stern et al., 2000), being characterized by mild dry summers and cold wet winters (Fig. 1c). The median annual rainfall at the Bureau of Meteorology (BoM) weather station at Yarrangobilly Caves (BoM station 72141), calculated using the climatological median 1985–2013, is 1178 ± 29 mm with a winter maximum of 349 ± 15 mm. The two dominant synoptic weather categories delivering most of this cool-season rainfall to SE Australia are cut-off low-pressure systems out of the westerlies and systems from the mid-latitudes including mid-latitude storms and fronts (Chubb et al., 2011; Pook et al., 2014; Callow et al., 2014). Modelled total evapotranspiration (ET) is maximum in summer and minimum in winter, and mean annual ET is 838 ± 40 mm over the period 1961–1990. The site has a mean annual temperature of 10.5 °C, with mean maximum temperatures in January and mean minimum temperatures in July of 27.8 and −1.8 °C, respectively. The growing season is limited by cold winter temperatures.

2.2 Drip site setting

The three drip-water monitoring sites (HW1–3) in this study are measured from active stalactites (Table 1) located centrally within the cave (Fig. 1b) at a depth of 38 m to the surface. HW1 and 2 were feeding actively forming stalagmites ca. 0.5 m apart on either side of the main path. These stalagmites were removed for palaeoclimate studies in 2006. In order to remove HW2, a small adjacent calcite column that had formed from a fused stalactite–stalagmite pair ca. 10–15 cm from HW2 (see Supplement S1 for photograph), had to be removed also. This resulted in re-invigoration of the drip point that had formed the column, which we included in our sampling program and refer to here as HW3.

Table 1. Summary of the drip-point characteristics for the monitoring sites.

Site	Description
HW1	The tip of the 1 m long massive stalactite (130 mm wide) was broken, presumably when the path was constructed in ∼ 1911 CE. A 7 cm long soda-straw stalactite has formed from the base of the massive stalactite and drips onto a 14 cm tall and 8 cm wide stalagmite developed on a flowstone on pebble ground. The stalagmite was removed in 2006.
HW2	Drip emanates from a 1 m long, 110 mm wide stalactite. The 160 mm by 70 mm stalagmite was removed in 2006. HW2 is an overflow of HW3 (see Supplement S1 and Sect. 5.1 for further discussion).
HW3	Drip point formed when stalagmite from site HW2 was removed.

Unsaturated zone hydrology of Harrie Wood Cave has recently been characterized (Markowska et al., 2015) and 5 drip-water regimes were identified. All flow types are fed by a theoretical soil storage and epikarst storage reservoir by fracture/fissure drainage. The five discharge flow types are as follows. Type 1 is designated mixed-flow/storage connectivity (low flow/high flow). Water at these discharge points is drained from a bulk homogenized epikarst storage reservoir. At the high-flow sub-type, during periods of water excess, the epikarst store is bypassed and water is routed directly from the soil storage reservoir. At the low-flow sub-type, discharge is from a pocket reservoir with a variable head within the epikarst storage reservoir. With Type 2 extreme events activated drip sites, a large intense infiltration event is required to initiate flow from the epikarst store. Type 3 are overflow sites, with discharge at the drip point fed by overflow from the pocket reservoir. Type 4 are non-linear flow sites; based on intra-karst dynamics, flow is intermittent between both storage reservoirs. Type 5 are underflow sites; during high infiltration, discharge is preferential underflow where both reservoirs are bypassed, and during base-flow conditions flow is sourced from the epikarst store. The sampling points in this study, HW1–3, are situated within the transect monitored by Markowska et al. (2015) (Fig. 1b).

3 Methods

3.1 Meteorological data

Daily rainfall was measured from the BoM standard 203 mm rain gauge at 09:00 local time (LT) each morning. Provided precipitation was 2 mm or greater, an aliquot of this precipitation was collected in a 10 mL amber bottle, ensuring zero headspace, and stored for stable isotope analysis ($\delta^{18}O$, $\delta^{2}H$). Maximum and minimum air temperatures are also recorded at 09:00 LT using BoM standard procedures.

A network of two automatic weather stations were installed above Harrie Wood and Jillabenan Caves on the 14 October 2011 and 6 September 2012, respectively. Atmospheric measurements of pressure, humidity, rainfall, temperature and wind speed and direction are recorded by a Davis Vantage Pro2™. Within the soil zone above each cave system, soil sensor probes were installed and buried in holes drilled with an auger to a depth of 25–30 cm at three localities (Fig. 1b). The Stevens Hydra Probe® soil sensor measurements include temperature, soil moisture and electrical conductivity and complex dielectric permittivity (both corrected from 0 to 35 °C). All data are recorded every 15 min by a dataTaker DT80 data logger. A detailed description of the parameters measured by the various instrumentation and data available from the network are outlined in Supplement S2.

3.2 Geochemistry

Drip water analysed for this study was collected approximately fortnightly over the period July 2006 to December 2013 ($n = 468$). The drip rate was measured manually as the time interval elapsed between two drips recorded using a stopwatch and the drips emanating from the stalactites accumulated in 1 L HDPE containers. Since March 2011 and sample volume permitting, in situ field measurements of electrical conductivity (referred to 25 °C; ±1 %), temperature (±0.1 °C) and pH measurements (±0.01 pH) were made using a TPH1-MyronL TechPro II handheld meter, which was calibrated using buffer solutions that were kept at cave temperature.

At each drip site, aliquots from the bulk water sample were collected and filtered through a mixed cellulose ester 0.45 µm filter and split into (i) two clean 50 mL polypropylene bottles for cation and anion analysis and (ii) a 10 mL amber glass bottle for $\delta^{18}O$ and $\delta^{2}H$ analysis. On exiting the cave, samples were refrigerated, transported to the laboratory in an insulated container and maintained at 4 °C until analysis.

Cation and anion analysis was conducted at the Environmental Research Chemistry Laboratory at the Australian Nuclear Science and Technology Organisation (ANSTO). Cation analysis was carried out on a Varian™ Vista Pro AX ICP-AES; prior to analysis aliquots were acidified with 1 % HNO_3. Anion analysis was conducted on an un-acidified aliquot using a Dionex 600 Instrument with an auto suppressor. Analytical error on cation and anion analyses was ≤ 5 %.

Representative limestone samples ($n = 7$) were collected from bedrock exposures above the cave and within Harrie Wood Cave. Freshly cleaved samples were dried at 40 °C and powdered using a ROCKLABS® TC-40 tungsten carbide ball mill. Then, 0.2 g of sample was microwave digested at 180 °C for 15 min using aqua regia ($HCl / HNO_3 = 1 : 3$) and analysed by ICP-AES at the Environmental Research Chemistry Laboratory at ANSTO.

The isotopic composition ($\delta^{18}O$ and $\delta^{2}H$) of the rainfall and drip-water samples were performed on a LGR-24 d off-axis, integrated cavity output, cavity ringdown mass spectrometer at University of New South Wales (UNSW) Australia. $\delta^{18}O$ and $\delta^{2}H$ values are reported relative to V-SMOW2. Analytical precision for $\delta^{18}O$ is 0.17 and 0.6 ‰ for $\delta^{2}H$.

3.3 Mixing and prior calcite precipitation calculations

To determine whether the geochemical evolution of the drip water was principally due to prior calcite precipitation (PCP), the hydrochemistry was assessed based on the accepted mathematical methods after Sinclair et al. (2012) and Tremaine and Froelich (2013). First, to compare between the two methods, ln(Sr / Ca) (mmol mol^{-1}) vs. ln(Mg / Ca) (mmol mol^{-1}) ratios in dissolved host bedrock endmembers overlayed with the drip-water ratios was graphed. The Mg / Ca and Sr / Ca ratios during the step change in 2007, wet period and high dry-season Ca values were differentiated from the complete dataset to isolate processes affecting the drip-water chemical evolution during these periods. Then, at each drip site the slopes of the linear regression of the ln(Sr / Ca) vs. ln(Mg / Ca) (weight ratio) graph was calculated based on the model suggested by Sinclair et al. (2012), where a slope of 0.709–1.003 is an indicator of water–rock interactions, i.e. PCP and incongruent dissolution-driven processes.

4 Results

4.1 Bedrock composition

The Ca concentration and Mg / Ca and Sr / Ca ratios for the suite of bedrock samples are listed in Table 2. Bedrock Mg / Ca ratios range from 2.6 to 15.3 mmol mol^{-1} and Sr / Ca ratios lie between 0.03 and 0.32 mmol mol^{-1}. These low ratios indicate the host limestone rock above and within Harrie Wood Cave are a low-Mg calcite type, indicating diagenesis of the original bedrock material. We observe three groupings of samples; Ca does not vary significantly between the samples but there is an observed difference in the Mg / Ca and Sr / Ca ratios between groups. The ratios for samples (R7–R10) are similar and there is no difference between colour variations, the limestone conglomerate (R12) has the highest Mg / Ca ratio and the palaeokarst samples (R13, R14) have the lowest Mg / Ca and Sr / Ca ratios.

4.2 Precipitation and infiltration

The observed daily rainfall and monthly cumulative water balance (CWB) at the study site and the Southern Oscillation Index (SOI) are shown in Fig. 2. The CWB represents a residual mass curve to show trends in the cumulative monthly water budget trends, following the method of Hurst (1951). It is calculated as the cumulative sum of the monthly *P*–ET anomalies from the climatological mean (1961–1990). The

Table 2. Sample name, sampling location and description, Ca concentration (mol kg^{-1}), Mg / Ca (mmol mol^{-1}) and Sr / Ca (mmol mol^{-1}) ratios in bedrock samples. On the slope above Harrie Wood Cave, representative samples were collected from the soil surface. The outcrop above the cave is located 33 m west of the cave entrance along the path and at the base of the slope. Within the cave, only exposed limestone surfaces were sampled. Colours of samples are based on a visual comparison with the Munsell® rock-colour chart.

Sample name	Sampling location: description, colour	Ca (mol kg^{-1})	Mg / Ca (mmol mol^{-1})	Sr / Ca (mmol mol^{-1})
YGB_R7	Slope above Harrie Wood Cave: 21 cm weathered limestone boulder, very light grey, contains moderate orange pink patches	10.0	11.7	0.27
YGB_R8	Outcrop above Harrie Wood Cave: limestone, greyish black	10.2	11.5	0.32
YGB_R9	Harrie Wood Cave, lower chamber: limestone, moderate reddish orange	9.7	9.9	0.23
YGB_R10	Harrie Wood Cave, lower chamber: limestone, medium dark grey	9.7	8.8	0.24
YGB_R12	Slope above Harrie Wood Cave: limestone, conglomeratic – round very light grey fragments over 2 mm, cemented by moderate yellow finer material	7.5	15.3	0.18
YGB_R13	Inception horizon along bedding plane in outcrop above Harrie Wood Cave: palaeokarst, moderate reddish brown	10.3	2.6	0.08
YGB_R13 (duplicate)		10.2	2.6	0.08
YGB_R14		8.7	3.8	0.03

earlier half of the record overlaps the latter 4 years of the "Millennium Drought", including SE Australia's most persistent rainfall deficit that occurred between 1997 and 2009 (CSIRO and BoM, 2015). Hence, the site experienced below-average rainfall from 2006 to 2009, and for much of the preceding decade. For example, from May 06 to December 06 (Fig. 3a), annual rainfall totals were 52.6 % below the mean attributable to a weak El Niño event (BoM, 2015a). The years 2007–2009 were a period of generally dry conditions, where the annual total precipitation was on average 10.3 % below the 30-year average. There were exceptions of above-average monthly rainfall for February and May 2007, November 2007 to April 2008, July 2008 and in April, July and September 2009 (Fig. 3b; BoM, 2015b). However, the reduced annual totals until 2010 resulted in a decline in the calculated cumulative water balance (Fig. 2b).

By contrast, in 2010–2012, average annual rainfall at Yarrangobilly Caves (BoM station 72141) was 152, 135 and 115 % above the mean for these 3 years, due to the strong 2010–2012 La Niña event (BoM, 2015a). The wettest interval in our study occurred from July 2010 to March 2011 (Figs. 2b and 3c) during which the calculated cumulative water balance increased (Fig. 2b). The subsequent 6 months from May to October 2011 were relatively dry (Fig. 3d), followed by a 5-month period of above-average monthly rainfall from November 2011 to March 2012 (Fig. 3e).

Based on measurements of soil moisture saturation between 14 October 2011 and 9 January 2013, Markowska et al. (2015) interpreted that a daily rainfall threshold range between 13 and 31.4 mm for Harrie Wood Cave was required to initiate a discharge response at 14 monitored drip sites. During the drier 2006 to mid-2010 period of our study, there were 75 such events in total, representing 30 % of total rain days. By contrast, during the wetter mid-2010 to 2014 interval, there were 124 events, representing 37 % of rain days.

4.3 Variability of $\delta^{18}O$ in precipitation and drip water

Daily rainfall, $\delta^{18}O$, monthly precipitation-weighted mean (PWM), as well as drip-water $\delta^{18}O$, are shown in Fig. 2. The monthly PWM of rain water $\delta^{18}O$ values ranges from −13.6 to −2.1 ‰ over the sampling period, with an overall $\delta^{18}O$ PWM of −6.9 ‰ of all rainfall events. At Harrie Wood Cave, rainfall events greater than 13 mm initiate recharge (Markowska et al., 2015); the $\delta^{18}O$ PWM value from these recharge events (−6.7 ‰) is not significantly different than that of all rainfall events. The monthly $\delta^{18}O$ PWM values show a large 5–10 ‰ intra-annual variation following the general seasonal trend of depleted isotopic values in winter and isotopically enriched values in summer, although the winter isotopic depletion is much less pronounced in 2008, 2009 and 2011. During the dry period prior to mid-2010, the PWM was enriched (−6.2 ‰) compared to the 2010/2012 wet period (−6.9 ‰).

With regards to drip water, there is ca. 0.5–1 ‰ variability over the observation period, which is largely dampened com-

Figure 2. The combined time series of **(a)** SOI as a measure of ENSO, **(b)** daily rainfall (BoM), infiltration corresponding to when rainfall exceeds 13 mm, and cumulative water balance (CWB), **(c)** daily $\delta^{18}O_{rain}$ overlayed with the monthly precipitation-weighted mean (PWM), **(d)** drip rate, **(e)** $\delta^{18}O_{dripwater}$, **(f)** Ca, **(g)** Mg / Ca, **(h)** Sr / Ca and **(i)** Cl at drip site HW1–3. Gaps within a time series indicate no available data. The dotted vertical line at July 2010 signifies the onset of the wet period. Blue and red arrows indicate an increase and decrease respectively. The pink vertical bar highlights the step change due to a flow path change.

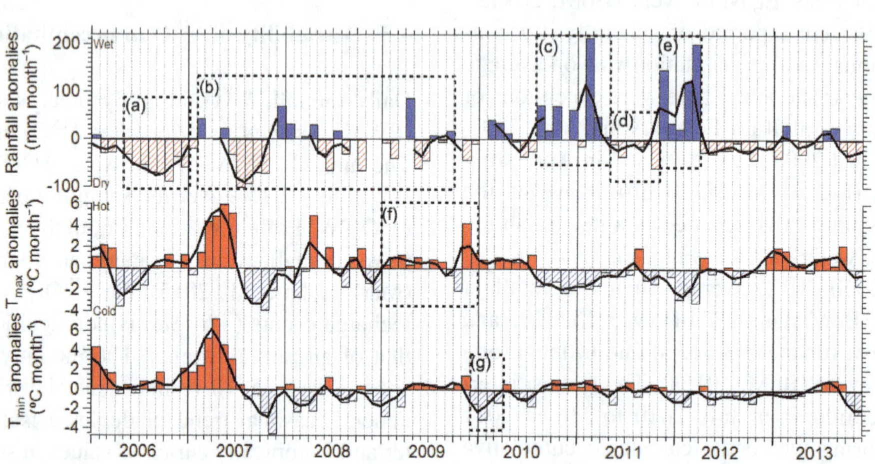

Figure 3. Monthly rainfall, minimum temperature (T_{min}) and maximum temperature (T_{max}) anomalies at Yarrangobilly Caves (BoM station 72141) during the cave drip-water monitoring study period. The solid line is the result of applying a binomial smoothing, one-pass Gaussian filter to the monthly data.

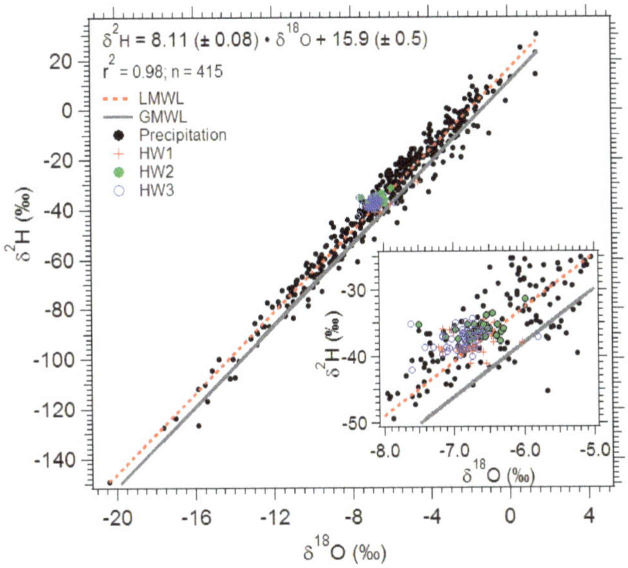

Figure 4. Linear fit representing the LMWL for the relationship between δ^2H and δ^{18}O in daily precipitation samples (dashed red line) and cave drip water (HW1–3). The solid grey line denotes the position of the GMWL.

pared to rainfall. Site HW2 had a wider range of δ^{18}O values (−5.4 – −7.7 ‰) than site HW1 (−6.0 – −7.4 ‰) and HW3 (−5.7 – −7.9 ‰). The δ^{18}O arithmetic mean values, close to the PWM (−6.9 ‰), were -6.8 ± 0.3 ‰ for HW1 and 2 and -6.9 ± 0.3 ‰ for HW3, suggesting significant mixing of infiltration events and that evaporation of infiltrating water is not significant. There is no clear seasonality in drip-water δ^{18}O, in contrast to the rainfall δ^{18}O, and there appears to be no simple isotopic response to infiltration of winter rainfall (Fig. 2). Furthermore, at all sites the mean drip-water δ^{18}O values are similar prior to mid-2010 during the dry period compared to the wettest interval between mid-2010 and March 2011 (HW1: −6.7 ‰ cf. −7.0 ‰; HW2: −6.6 ‰ cf. −6.9 ‰; and HW3: −6.9 ‰ cf. −6.9 ‰), although there is a subtle trend to lighter δ^{18}O values.

Figure 4 shows δ^2H vs. δ^{18}O calculated local meteoric water line (LMWL; $\delta^2\mathrm{H} = (8.11 \pm 0.08) \times \delta^{18}\mathrm{O} + (15.9 \pm 0.5)$, $n = 415$) compared with the global meteoric water line (GMWL; Rozanski et al., 1993). The LMWL intercept is greater than the GMWL, but is in agreement with the LMWL established by Hughes and Crawford (2013) from a 3-year precipitation dataset from Big Hill in the Southern Highlands ($\delta D = (8.10 \pm 0.12) \times \delta^{18}\mathrm{O} + (16.3 \pm 0.8)$; 188 km northeast (NE) of our site, 652 m a.s.l.). Crawford et al. (2013) attributed the high intercept value to a larger contribution of moisture recycling from the land surface to the local moisture budget.

4.4 Drip rates

Drip rates throughout the 7-year period demonstrate irregular multi-annual variations (Fig. 2d). All sites remained hydrologically active, with a mean drip rate of 2.2 ± 0.6, 2.2 ± 1.7 and $3.5 \pm 1.9 \times 10^{-3}$ mL s^{-1} at site HW1–3 respectively (Table 3). Prior to mid-2010, during the long phase of below average monthly rainfall and water deficit, a base-flow drip rate of $2.1 \pm 0.5 \times 10^{-3}$ mL s^{-1} for HW1, $1.9 \pm 1.4 \times 10^{-3}$ mL s^{-1} for HW2 and $3.1 \pm 1.3 \times 10^{-3}$ mL s^{-1} for HW3 was maintained, only increasing slightly in response to above-average monthly rainfall anomalies during this period (Sect. 4.2), therefore suggesting the drip sites receive flow from a storage reservoir. The increased rise in discharge is often only captured by one fortnightly sampling event, but is often present at all three sites. During the wettest interval between mid-2010 and March 2011, the mean drip rate increased to between 2.8 and 4.7×10^{-3} mL s^{-1} (Fig. 2d). Site HW2 is a slower dripping site compared to HW1, and HW3 is the fastest dripping stalactite. Drip rates at all three sites are mostly in phase, recording a similar pattern.

4.5 Drip-water chemistry

Concentrations of major cations and anions in the drip-water samples from the three drip sites are listed in Table 3. Drip-water samples from all three sites represent a Ca–HCO_3 dominating water type (pH 6.8–8.3) (Drever, 1982); accordingly, Ca is the main cation in the drip-water solutions. The mean Ca concentrations observed at drip sites HW1–3 are 66.4, 56.4 and 63.0 mg L^{-1} respectively, and range from 31.1 to 84.1 mg L^{-1}. Mean concentrations of other ions are similar across the three drip sites, further supporting that the drip water is sourced from the same well-mixed storage reservoir.

Time series of Ca, Mg / Ca, Sr / Ca and Cl are shown in Fig. 2. Broadly, Ca concentrations are declining from 80 to 60 mg L^{-1} at sites HW1 and 3 from the beginning of the study until early 2008 when they become more similar to one another, but still greater than HW2 (which remains lower at 50 mg L^{-1} through the same period). There is a prominent rise in Ca at all sites beginning in the summer of 2008/2009 and peaking in November 2009 at ca. 80 mg L^{-1}, before falling back to ca. 60 mg L^{-1} in autumn 2010. After this time, and throughout the following wetter interval, Ca concentrations at all sites become closer in value and less variable overall with some excursions to lower Ca at individual drip sites over periods of 0.5–2 months.

The prominent Ca peak in 2009 also occurs during a dry period when Ca is increasing rather than decreasing. In this case, this prominent peak in Ca occurs during a persistent run of above-average surface temperatures dominating the entire year (Fig. 3). Specifically, an unusually warm winter was experienced in 2009 (Fig. 3f), followed by the hottest November on record (BoM; 2015b). We propose that, in this

Table 3. Summary statistics of dominant ions in fortnightly drip-water samples analysed between 2006 and 2013.

	HW1 mean ± SD (min, max)	HW2 mean ± SD (min, max)	HW3 mean ± SD (min, max)
Drip rate ($\times 10^{-3}$ mL s^{-1})	2.2 ± 0.6 (0.8, 4.8)	2.2 ± 1.7 (0.2, 15)	3.5 ± 1.9 (0.06, 13.6)
T (°C)	11.1 ± 0.4 (10.4, 11.9)	11.1 ± 0.4 (10.4, 11.8)	11.0 ± 0.4 (10.3, 11.8)
EC (µS cm^{-1})	311 ± 16 (275, 356)	272 ± 33 (161, 330)	293 ± 30 (199, 333)
pH	7.82 ± 0.53 (7.17, 11.20)	7.86 ± 0.45 (7.37, 11.00)	7.79 ± 0.43 (7.32, 11.00)
Cation			
Ca (mg L^{-1})	66.4 ± 5.5 (55.7, 84.1)	56.5 ± 7.8 (31.1, 81.4)	63.0 ± 7.2 (41.5, 80.6)
Si (mg L^{-1})	2.3 ± 0.6 (1.4, 3.5)	2.3 ± 0.6 (1.4, 3.8)	2.3 ± 0.6 (1.3, 3.5)
*Na (mg L^{-1})	0.7 ± 0.04 (0.6, 0.8)	0.7 ± 0.1 (0.6, 0.9)	0.7 ± 0.2 (0.6, 0.8)
Mg (mg L^{-1})	0.58 ± 0.05 (0.44,0.68)	0.58 ± 0.05 (0.45, 0.73)	0.58 ± 0.05 (0.45, 0.82)
K (mg L^{-1})	0.12 ± 0.04 (0.07, 0.32)	0.13 ± 0.05 (0.08, 0.35)	0.11 ± 0.04 (0.06, 0.32)
Sr (µg L^{-1})	51.8 ± 6.7 (31.6, 61.0)	51.0 ± 6.3 (32.3, 64.0)	51.3 ± 6.4 (31.9, 76.0)
Anion			
Cl (mg L^{-1})	1.5 ± 0.4 (1.0, 2.7)	1.6 ± 0.4 (1.1, 2.5)	1.6 ± 0.4 (1.0, 3.0)
SO_4 (mg L^{-1})	0.6 ± 0.3 (0.2, 1.3)	0.6 ± 0.2 (0.2, 1.1)	0.6 ± 0.2 (0.2, 1.1)
Elemental ratio			
Mg / Ca (mmol mol^{-1})	14.3 ± 1.3 (10.5, 17.1)	17.1 ± 2.1 (13.1, 30.1)	15.3 ± 2.0 (10.7, 23.8)
Sr / Ca (mmol mol^{-1})	0.36 ± 0.05 (0.21, 0.44)	0.42 ± 0.1 (0.26, 0.71)	0.06 (0.21, 0.57)

* $n = 68$ samples from 15 March 2011 to 23 December 2013.

interval, higher than average drip-water Ca concentrations are driven by a temperature control on net soil CO_2 due to respiration. We therefore propose that the warmer than average temperatures, particularly during the winter months, induced higher soil CO_2 respiration and thereby increasing $CaCO_3$ dissolution. Drip-water Ca returned to mean levels in January 2010 (Fig. 2f), consistent with cooler temperatures (minimum temperatures were 3.1 °C lower than the long-term average) restoring soil respiration conditions to previous levels (Fig. 3g).

Chloride concentrations show consistent trends through time across the three sites. Chloride concentrations are the highest at the beginning of our study (ca. 2.3 mg L^{-1}) and decline steadily until mid-2010; at which time there is a more abrupt decrease in mean Cl at all three sites. After this transition, Cl concentrations are relatively stable at ca. 1.3 mg L^{-1}. Chloride is our most conservative of the measured ions. During the transition from the relatively drier to the relatively wetter interval in our study, the abrupt decline in Cl, coinciding with an increase in discharge is consistent with dilution by recharge at the onset of the wetter interval (Fig. 2i). The earlier downward trend in Cl (2007–mid-2010) appears initially inconsistent with this, but we propose that reduced recharge events to the epikarst storage reservoir during this interval resulted in a decreasing supply of Cl from the soil zone (discussed further in Sect. 5.1).

4.6 Drip-water mixing and PCP evolution

The observations above suggest processes include the mixing and dilution of drip water as indicated by Cl concentrations. To diagnose water–rock interactions including PCP and mixing of waters, we used a number of approaches (Sect. 3.3). Figure 5 shows a cross-plot of ln(Mg / Ca) vs. ln(Sr / Ca) drip-water ratios presented together with the bedrock ln(Mg / Ca) and ln(Sr / Ca) ratios (after Sinclair et al., 2012, and Tremaine and Froelich., 2013). Since all three drip sites are influenced by the same processes, as the same Mg(Sr) / Ca trends are observed at all sites (Supplement S3), we present drip-water data at site HW2 for visual clarity as the processes are enhanced due to the slower discharge at this site. Many of the bedrock values are similar but lower than the drip-water ln(Mg / Ca) and ln(Sr / Ca) ratios (Fig. 5). The drip-water data themselves appear to fall into two groups: the pre-October 2007 group (red squares) with relatively lower Sr / Ca ratios, and the post-November 2007 group with relatively higher Sr / Ca ratios. If the geochemical evolution of the drip water was solely due to PCP, Sinclair et al. (2012) mathematically demonstrated that the drip-water ratios would produce a straight-line correlation of known slope through the whole dataset (Sect. 3.3). The observed step change indicates a change in the initial composition of the chemistry at equilibration. We interpret this to indicate a flow path change, which we discuss further in

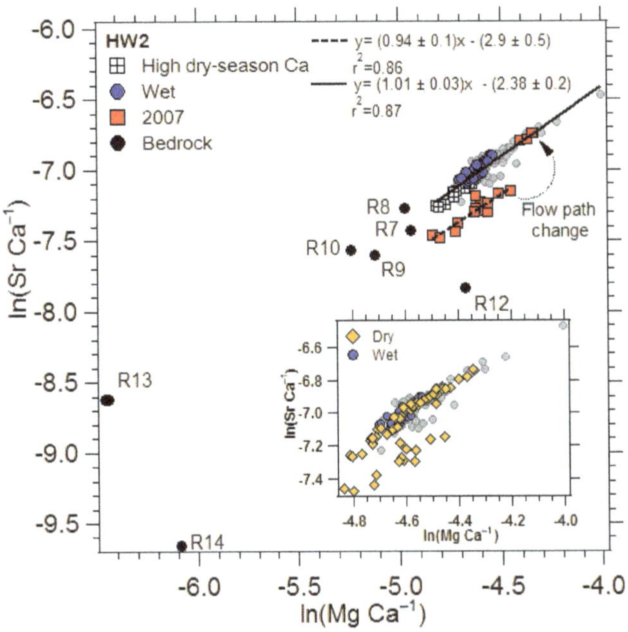

Figure 5. Drip-water ln(Sr / Ca) vs. ln(Mg / Ca) ratios at site HW2 are graphed together with the bedrock composition (see Table 2). To enhance clarity, high dry-season Ca, wet and 2007 drip-water chemistry trends are discriminated from the complete dataset (light grey circles). The step change, offsetting between the two linear trends, is indicated by a curved arrow. Inset highlights dry and wet period data.

Sect. 5.1. All HW2 drip-water values are consistent with evolution from the bedrock values, notably R7 to R10.

Linear regressions are plotted for each group of data in Fig. 5. There is some scatter of the data around these lines suggesting that more than one process may be influencing the resulting drip-water Sr / Ca and Mg / Ca ratios. During low discharge and dry rainfall conditions the drip-water composition shifts to higher Mg / Ca and Sr / Ca ratios away from the bedrock plotting in the distal outer ends, representing progressive CO_2 degassing due to increased PCP. During the transition period from dry to wet conditions in mid-2010 when discharge increased (blue filled circles in Fig. 5), drip-water Mg / Ca and Sr / Ca ratios plot closer to the bedrock region, indicating reduced PCP. During the high dry-season Ca values (November 2009, square window), Mg(Sr) / Ca ratios are close to the bedrock ratios. As drip water reached maximum Ca values, Mg(Sr) / Ca ratios evolved towards slightly lower values, due to dissolution of the host limestone, caused by increased soil CO_2 from bioproductivity. An alternative hypothesis for the rise in Ca in late 2009 could be a flow path change; however, the Sr / Ca and Mg / Ca data show no evidence to support this.

The calculated drip-water ln(Sr / Ca) vs. ln(Mg / Ca) (weight ratio) correlation slopes of the two groups of data at each drip site range between 0.66 ± 0.17 and 1.01 ± 0.07, within the range of predicted slopes suggested by Sinclair et al. (2012). Although a slight slope change is observed between the two flow paths, this suggests that cave drip water evolves under PCP.

Unlike the late 2007 step change, we observe a rise in Mg(Sr) / Ca and a drop in Ca in 2011 and 2013 (arrows; Fig. 2). These events are short-lived and are most notable at one site only (HW2); there is insufficient data to evaluate this site-specific change and nonetheless are features that would not be preserved in the speleothem.

Excluding 2007, the long-term Mg values are increasing relative to Ca and there is a drift towards higher Mg / Ca values overall but less evident in the Sr / Ca time series (Fig. 2). There is a long-term rise in Mg concentrations of 0.15 ppm from 2008 to 2013. A more complete characterization will include an investigation of aerosols and soil in a future study.

5 Discussion

Speleothem geochemical proxy records archive environmental and climate signals from the surface to the cave (Fairchild and Treble, 2009). However, informed palaeorainfall reconstruction from speleothems requires an understanding of the drip hydrological pathway and the karst hydrogeological influences, preferably based on modern cave monitoring data from similar climates. In this section we discuss the key results used to constrain the hydrological processes that characterize drip sites HW1–3 based on discharge, geochemical and stable isotope parameters in contrasting rainfall conditions, highlighting the importance of constraining the evolution of the drip water in the context of the local environment. Multiple lines of evidence indicate the drip sites are constrained by multiple processes, which we summarize in a conceptual model. We examine the relationship between the observed hydrogeochemical changes with the climate. Our findings underlie the use of geochemical tracers toward informing speleothem proxy records and we also discuss the implications of these results for palaeoenvironmental reconstruction.

5.1 Hydrogeochemical processes influencing drip-water chemistry

Based on hydrogeochemical observations presented in Sect. 4, we suggest in Harrie Wood Cave the flow paths feeding stalagmites from drip sites HW1–3 are from a Type 1, mixed-flow/storage connectivity low-flow sub-type, i.e. fracture flow from a "pocket reservoir" in the well-mixed epikarst storage reservoir (Markowska et al., 2015; see Sect. 2.2). Our data confirm the drip water that precipitates speleothem calcite drains from bulk homogenized epikarst store water, the primary karst storage reservoir at Harrie Wood Cave. The results from the oxygen-isotope data indicate that infiltrating water is well-mixed, because the range of drip-water $\delta^{18}O$ values is narrow in comparison to the rainfall, and the arithmetic mean of the drip water reflects the weighted mean of precipitation (Sect. 4.3); moreover, from the drip-water isotope time series there is a strong buffering of the extremely low $\delta^{18}O$ winter rainfall values in 2007, 2010, 2012 and 2013 (Fig. 2). Furthermore, consistency in trace element trends among the sites (Table 3) confirms the karst waters are well homogenized.

A significant proportion of flow to the drip points is drained from a storage reservoir, owing to sustained base-flow discharge levels during the weak El Niño in 2006 and the low rainfall period between 2007 and 2009 (Sect. 4.4). Furthermore, as base flow is maintained and a large infiltration event or threshold level is not required to activate the drip site; this suggests the karst plumbing system feeding the stalagmites is not as described by Markowska et al. (2015), i.e. a Type 2 extreme event activated or Type 3 overflow site (see Sect. 2.2).

Dilution of the epikarst storage reservoir was observed during the transition to the La Niña phase in 2010/2011, consistent with the observation of a clear increase in discharge followed by a decrease in Cl concentrations. This further suggests that after recharge the epikarst storage reservoir volume reached close to maximum capacity maintaining "steady state", supported by the drip-water Mg / Ca and Sr / Ca ratios reaching mean levels indicating minimum PCP. Additionally, this indicates that discharge at the drip sites are not from a Type 1 (high flow sub-type), Type 4 or Type 5 site, as during periods of water excess discharge at these sites receive flow directly from the soil storage reservoir, and therefore a rise in Cl would be anticipated due to higher ET to the system as a whole; however, the converse was observed.

During the dry period from 2007 to mid-2010, we suggest the decreasing trend from elevated Cl levels indicates limited recharge. During the dry period we observe a decline in the CWB and reduced drip rate, consistent with fewer recharge events from the soil–vadose zone to the epikarst store, and therefore decreasing drip-water Cl concentrations. Alternatively, it may also reflect higher ET in the soil–vadose zone, but this would also result in an enriched drip-water $\delta^{18}O$ signal. ET is relevant at this karst site; however, we do not favour this explanation, as the similarity between drip-water $\delta^{18}O$ values during the dry period compared to the wetter period suggests evaporation of the infiltrating drip water was not significant (Sect. 4.3).

A flow path change at October 2007 is inferred from the shift to higher Sr / Ca ratios (Sect. 4.6; Fig. 5). This process occurred during the drying conditions when the CWB was decreasing (Fig. 2b), but when base-flow levels were maintained at all drip sites during this period. We infer that a flow path re-direction through a higher Sr / Ca endmember occurred due to soil–vadose zone drying to sustain discharge. This indicates a non-linear response of the system to progressive drying. A possible explanation is that calcification of the flow path produced a threshold change causing re-routing through a higher Sr / Ca source.

The three drip-water monitoring sites (HW1–3) are located within a small area (Fig. 1b); however, there are differences in the chemistry, which is notable in 2007 during the earliest interval of the dry period. In particular a lower drip rate (Sect. 4.4) and a lower Ca concentration and higher Mg / Ca and Sr / Ca ratio (Sect. 4.5) of site HW2 compared to the neighbouring sites HW3 and 1, which are only 10–15 cm away. Equivalent Cl concentrations over this period (Sect. 4.5) suggest differences have subsequently arisen in the carbonate chemistry. A potential explanation for this is that all three drip sites are fed by the same chemistry, but the lower drip rate at site HW2 is consistent with greater in-cave PCP on the stalactite tip, inducing a lower Ca concentration and enhancing PCP (Treble et al., 2015).

Thus, we deduce that the studied drip water is from a Type 1 mixed-flow/storage connectivity low-flow sub-type, the system is open to CO_2, and a ventilated air pocket with variable height and lower pCO_2 provides the potential for degassing and calcite precipitation from the drip water (Tooth and Fairchild, 2003). Although we observed a hydrological flow path change, a distinction in drip-water Mg / Ca and Sr / Ca composition between drier and wetter periods is evident. During dry periods, the PCP mechanism was enhanced and the highest Mg / Ca and Sr / Ca ratios in this study are recorded during the El Niño and intervals of below average rainfall, as PCP is promoted due to a dewatering of spaces. In contrast, for the duration of the strong La Niña phase in 2010/11 and above-average rainfall, reduced PCP (mean Mg / Ca and Sr / Ca ratios) is noted (blue circles, Fig. 5) since a reduced reservoir head space limits degassing.

We also examine the role of calcite dissolution and soil zone CO_2 in more detail. Drip-water Ca concentrations reached maximum values ($80\,mg\,L^{-1}$) in November 2009 (see Fig. 2f), as unseasonal temperatures increased. We suggest that soil microbial production increased the CO_2 concentration of infiltrating waters driving $CaCO_3$ mineral dissolution (square window, Fig. 5). Presumably vadose zone CO_2 production also drives calcite dissolution during the dry period (Atkinson et al., 1977). During the dry period there is a long-term decline in the CWB and progressive draining of the epikarst store; therefore, more unsaturated zone is available for CO_2 production, which may also lead to increased calcite solubility.

An alternative hypothesis is the Ca trend may be a fire-driven process as the site was affected by fire 4 years before the monitoring period, which may have decreased soil CO_2 production, associated calcite dissolution and Ca concentrations (Coleborn et al., 2006). That effect is most likely in the first decade after the fire. However, a decreasing trend in Ca in the early monitoring period (the opposite of what might be expected) in combination with thin soil and sparse vegetation indicates that any fire-induced soil biogeochemical changes had relatively little impact on the drip-water signa-

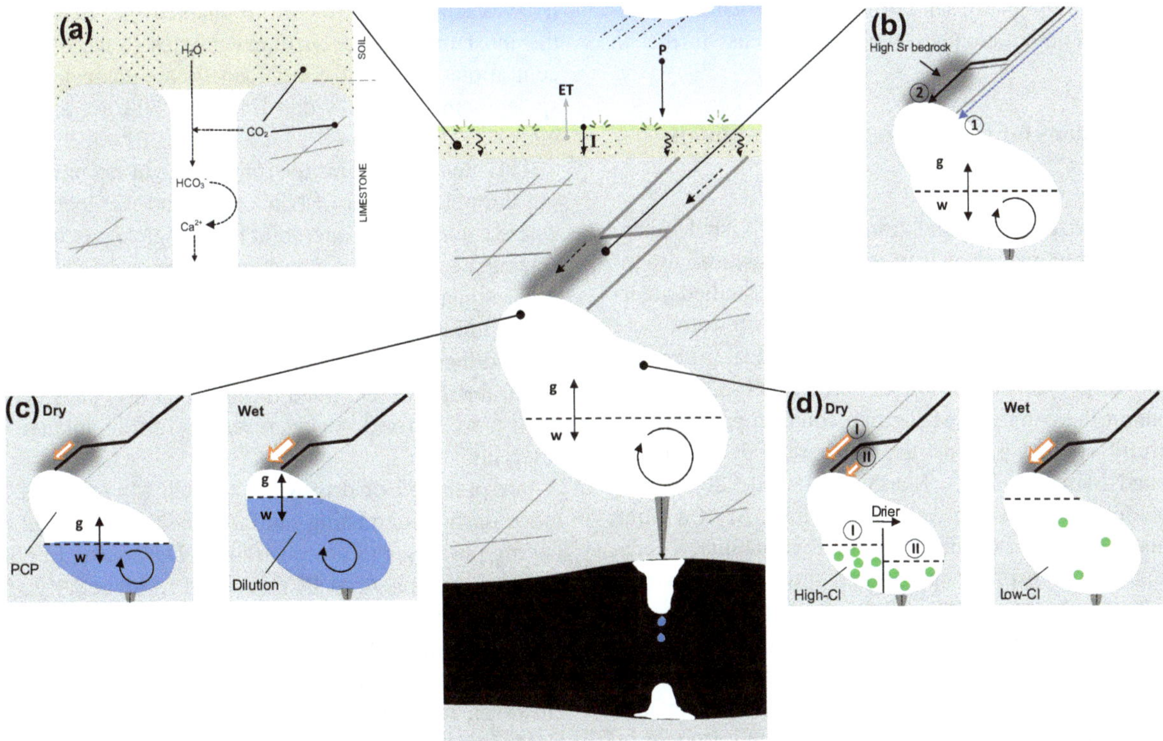

Figure 6. Conceptual model summarizing key process affecting drip-water hydrochemical variations at Harrie Wood Cave. Shown are precipitation (P), evapotranspiration (ET) and infiltration (I) infiltrations corresponding to when rainfall exceeds 13 mm. Stalactites from drip sites HW1–3 are fed by fracture flow from a ventilated well-mixed pocket reservoir with a variable head within the epikarst. **(a)** Soil and unsaturated/vadose zone CO_2 drive $CaCO_3$ dissolution, increasing drip-water Ca. **(b)** The host bedrock varies geochemically and the dark grey shading along the fracture represents bedrock of higher Sr concentrations. A flow path change from 1 to 2 (October 2007) results in drip water being enriched in Sr. The size of flow arrows in **(c)** and **(d)** qualitatively correspond to recharge to the epikarst storage reservoir. **(c)** PCP within the store is enhanced during dry periods as the water level (w) is low, fairly constant and in contact with a greater ventilated gas phase (g). Dilution occurs during wet intervals due to greater inflow and PCP decreases as a reduced reservoir head space limits degassing ($w > g$). **(d)** During dry periods, the response to reduced inflow (I to II) is a decrease in Cl concentrations within the storage reservoir and therefore drip water. In wet periods, Cl is flushed from the soil–vadose zone and then diluted within the store, resulting in a sharp decline in drip-water Cl.

ture. Also, in the short term, the concentration of elements in the soil can increase in response to a fire. Considering this, during the drying trend when there is reduced recharge and based on the observed trend of the conservative tracer Cl, the effect would be an analogous decrease in Mg and Sr concentrations in the drip water, whereas a long-term rise in Mg is noted (Sect. 4.6), as such we do not consider the 2003 fire event had an impact on the drip-water dataset.

Based on a 7-year observational study, we have unveiled a complex non-linear geochemical response, as depicted in Fig. 6. We now attempt to discern whether these long-term trends in the geochemistry are climate related. During the drying trend (2007 to mid-2010), we observed low discharge, a flow path change, decreased Cl, enhanced and increased Ca caused by increased $CaCO_3$ solubility due to higher soil CO_2 and bioproductivity. This is a response consistent with reducing recharge conditions. During the wetter period, we also observed trends consistent with increased recharge, increased discharge, dilution and reduced PCP. Therefore, we interpret these as evidence of climatic induced changes. Over the long-term, drip discharge and hydrogeochemical variations between the relatively dry and wet period are driven by variations in water availability due to ENSO and are of palaeoclimatic relevance.

We have confidently constrained the possible hydrological processes, which can occur in a karst system situated in a region greatly impacted by ENSO episodes. We attribute this to long-term monitoring that encapsulated the shift between the two extremes of ENSO, namely from dry El Niño to wet La Niña conditions. Based on our findings and with the knowledge that the impact of each ENSO episode on Australia varies, we have identified how Harrie Wood Cave responds to a changing climate signal (rainfall), and through geochemical proxies ($\delta^{18}O$, Ca, Cl, and Mg / Ca and Sr / Ca ratios) identified how this response has been transformed and is recorded in cave drip water over time. These results have important implications in informing palaeoclimate findings from speleothem archives, which we discuss further in Sect. 5.2.

5.2 Implications for speleothems as palaeorainfall recorders

Our results provide a foundation to inform speleothem palaeoclimate records where drip-water compositions are influenced by multiple processes. During our studied interval, element concentrations and ratios in the drip water were driven by climate (ENSO) and karst hydrological processes. Based on drip characterization of sites HW1–3, Mg / Ca variations in the conjugate stalagmite could in principle, be successfully applied to construct palaeorainfall conditions (Tooth and Fairchild, 2003; Markowska et al., 2015). We have demonstrated that drip-water Mg / Ca and Sr / Ca ratios are a relative measure of modern rainfall variability. Therefore, in a highly resolved speleothem time series, we anticipate displacements of $[Mg / Ca]_{calcite}$ and $[Sr / Ca]_{calcite}$ from a baseline ratio (bedrock) to higher and mean ratios will differentiate between dry and wet periods respectively. McDonald et al. (2004) also identified a relationship between drip-water Mg / Ca and Sr / Ca ratios and drought in Wombeyan Caves, New South Wales (NSW); however, their interpretation of the trace element stalagmite record varies slightly. At Wombeyan, PCP is a more dominant control on the karst hydrochemistry, although it is in the same climate region, but a warmer site with greater evapotranspiration and a longer growing season. Consequently, we would expect that the resultant drip water (and stalagmite calcite) would be enriched in Mg / Ca and Sr / Ca. As such we expect to observe a greater displacement of Mg / Ca ratios away from the bedrock composition during any given dry phase at Wombeyan Caves, in comparison to a corresponding record at Harrie Wood Cave. For wet periods, no La Niña events occurred during their study period (July 2001–January 2004); therefore, McDonald et al. (2004) could only speculate Mg / Ca calcite ratios would shift towards bedrock values due to a decreased effect of PCP. In contrast, extended monitoring through the La Niña mode of an ENSO cycle provides us with a more informed interpretation; whereby we anticipate La Niña events in the stalagmite trace element time series may be resolved from baseline ratios by a shift to mean ratios.

6 Conclusions

This research targeted the Snowy Mountains region of the Australian Alps; a key water resource region in SE Australia where rainfall variability and, therefore, future water resource availability in this climatically sensitive region is uncertain. Our results have advanced the scientific knowledge of the in-cave drip response to modern-ENSO variability, via rainfall, and therefore enabled the identification of the most reliable geochemical proxies (trace element concentrations and $\delta^{18}O$) for palaeoclimate reconstruction from speleothem archives within Harrie Wood Cave and in the region.

This study underpins the importance of extended monitoring through a modern ENSO cycle and we document a site where a number of concurrent hydrological processes are occurring, occasioning a complex non-linear geochemical drip-water signature. The local karst hydrogeological processes, which influence trace element and $\delta^{18}O$ proxies in drip water, and therefore signatures in the speleothem calcite, were constrained. We identified discharge at drip sites HW1–3 is from a well-mixed pocket reservoir in the epikarst storage reservoir.

Interpretation of drip-water Ca, Cl, Mg / Ca and Sr / Ca ratios during contrasting rainfall conditions allowed the following processes to be constrained. During the El Niño and dry periods, enhanced PCP resulted in higher Mg / Ca and Sr / Ca ratios. While during the La Niña and wet phase, Cl concentrations and discharge were used to constrain the process of dilution and reduced PCP controlled drip chemistry. However, we found a number of non-linear responses embedded in a linear drying climate trend. An interpretation of drip-water Sr / Ca vs. Mg / Ca ratios compared to different endmember bedrock ratios showed a shift to higher Sr values, suggesting a flow-path change. Decreasing drip-water Cl levels indicated reduced recharge of the epikarst storage reservoir, and an unexpected rise in the Ca time series was shown to be caused by carbonate dissolution due to an increase in soil and epikarst–vadose zone CO_2 concentrations. The data presented here highlight the complex interplay of a drip-water signal in response to the climate signal, this has been achieved only through long-term monitoring over a 7-year (2007–2013) period. A conceptual model was constructed illustrating these key processes controlling the drip-water composition.

These processes are shown to be linked to climate induced changes and propose Mg / Ca and Sr / Ca variations in the conjugate stalagmite could, in principle, be successfully applied to construct palaeorainfall conditions. This study has extended our understanding of changing climatic controls on proxy variations within the Harrie Wood Cave system and, therefore, provides a benchmark for its application regionally and globally when using speleothems for palaeoclimate reconstruction.

Author contributions. Carol V. Tadros conceptualized the research, collected data, conducted all data analysis and interpretation, generated graphs, the conceptual model and wrote the manuscript. Pauline C. Treble and Andy Bake provided guidance, reviewed and edited the manuscript in their function as my supervisors, as did Ian Fairchild, and Stuart Hankin installed and maintained the weather stations and assisted with generating Fig. 1a. Regina Roach collected the fortnightly drip-water samples for this study. Monika Markowska performed the stable isotope sample analysis. Janece McDonald set up the drip-water monitoring study in 2006.

Acknowledgements. The authors are grateful to Suzanne Hollins for supporting this research. We thank Jagoda Crawford and Darrell Tremaine for useful discussion. Henri Wong, Chris Vardanega, Robert Chisari and Barbora Gallagher are thanked for their assistance with sample analysis. George Bradford and the staff at Yarrangobilly Caves and NSW NPWS are also thanked for their dedication and on-going field support and access permission. Silvia Frisia and Andrea Borsato are thanked for their assistance in the field regarding the geology of karst in the Snowy Mountains alpine region. Peter Briggs and Alan Griffiths are acknowledged for providing the AWAP data. PCT acknowledges the support of a Land & Water Australia grant (project number ANU52) for this study. We thank Bill Hu and two anonymous reviewers for their constructive and thoughtful reviews that helped improved the manuscript.

Edited by: B. Hu

References

Aplin, K., Ford, F., and Hiscock, P.: Early Holocene human occupation and environment of the southeast Australian Alps: New evidence for the Yarrangobilly Plateau, New South Wales, in: Altered Ecologies: Fire, Climate and Human Influence on Terrestrial Landscapes, edited by: Haberle, S., Stevenson, J., Prebble, M., 187–212, 2010.

Atkinson, T. C.: Carbon dioxide in the atmosphere of the unsaturated zone: an important control of groundwater hardness in limestones, J. Hydrol., 35, 111–123, 1977.

Baker, A., Genty, D., and Fairchild, I. J.: Hydrological characterisation of stalagmite dripwaters at Grotte de Villars, Dordogne, by the analysis of inorganic species and luminescent organic matter, Hydrol. Earth Syst. Sci., 4, 439–449, doi:10.5194/hess-4-439-2000, 2000.

Baldini, J. U. L., McDermott, F., and Fairchild, I. J.: Structure of the 8200 year cold event revealed by a speleothem trace element record, Science, 296, 2203–2206, 2002.

Baldini, J. U. L., McDermott, F., and Fairchild, I. J.: Spatial variability in cave drip water hydrochemistry: Implications for stalagmite paleoclimate records, Chem. Geol., 235, 390–404, 2006.

Barros, A. P. and Bowden, G. J.: Toward long-lead operational forecasts of drought: An experimental study in the Murray-Darling River Basin, J. Hydrol., 357, 349–367, 2008.

BoM: http://www.bom.gov.au/climate/enso/ (last access: 24 December 2015), 2015a.

BoM: http://www.bom.gov.au/climate/mwr/ (last access: 24 December 2015), 2015b.

Borsato, A., Frisia, S., Fairchild, I. J., Somogyi, A., and Susini, J.: Trace element distribution in annual stalagmite laminae mapped by micrometer-resolution X-ray fluorescence: Implications for incorporation of environmentally significant species, Geochim. Cosmochim. Ac., 71, 1494–1512, 2007.

Cai, W. and Cowan, T.: Evidence of impacts from rising temperature on inflows to the Murray–Darling Basin, Geophys. Res. Lett., 35, L07701, doi:10.1029/2008GL033390, 2008.

Callow, N., McGowan, H., Warren, L., and Speirs, J.: Drivers of precipitation stable isotope variability in an alpine setting, Snowy Mountains, Australia, J. Geophys. Res.-Atmos., 119, 3016–3031, 2014.

Chubb., T. H., Siems, S. T., and Manton, M. J.: On the decline of wintertime precipitation in the snowy mountains of Southeastern Australia, J. Hydrometeorol., 12, 1483–1497, 2011.

Coleborn, K., Spate, A., Tozer, M., Anderson, M. S., Fairchild, I. J., MacKenzie, B., Treble, P. C., Meehan, S., Baker, A., and Baker, A.: Effects of wildfire on long-term soil CO_2 concentration: implications for karst processes, Environ. Earth Sci., 75, 1–12, 2016.

Crawford, J., Hughes, C. E., and Parkes, S. D.: Is the isotopic composition of event based precipitation driven by moisture source or synoptic scale weather in the Sydney Basin, Australia?, J. Hydrol., 507, 213–226, 2013.

Cruz Jr., F. W., Burns, S. J., Jercinovic, M., Karmann, I., Sharp, W. D., and Vuille, M.: Evidence of rainfall variations in Southern Brazil from trace element ratios (Mg / Ca and Sr / Ca) in a Late Pleistocene stalagmite, Geochim. Cosmochim. Ac., 71, 2250–2263, 2007.

CSIRO and Bureau of Meteorology: Climate Change in Australia Information for Australia's Natural Resource Management Regions: Technical Report, CSIRO and Bureau of Meteorology, Australia, 1–222, 2015.

Cuthbert, M. O., Baker, A., Jex, C. N., Graham, P. W., Treble, P. C., Andersen, M. S., and Acworth, R. I.: Drip water isotopes in semi-arid karst: implications for speleothem paleoclimatology, Earth Planet. Sci. Lett., 395, 194–204, 2014.

Dai, A., Fung, I. Y., and Del Genio, A. D.: Surface observed global land precipitation variations during 1900–1988, J. Climate., 10, 2943–2962, 1997.

Drever, J. I.: The Geochemistry of Natural Waters, Prentice-Hall Inc., Englewood Cliffs, NJ, 388 pp., 1982.

Dredge, J., Fairchild, I. J., Harrison, R. M., Fernandez-Cortes, A., Sanchez-Moral, S., Jurado, V., Gunn, J., Smith, A., Spötl, C., Mattey, D., Wynn, P. M., and Grassineau, N.: Cave aerosols: Distribution and contribution to speleothem geochemistry, Quat. Sci. Rev., 63, 23–41, 2013.

Fairchild, I. J. and Treble, P. C.: Trace elements in speleothems as recorders of environmental change, Quat. Sci. Rev., 28, 449–468, 2009.

Fairchild, I. J., Borsato, A., Tooth, A. F., Frisia, S., Hawkesworth, C. J., Huang, Y., McDermott, F., and Spiro, B.: Controls on trace element (Sr-Mg) compositions of carbonate cave water: implications for speleothem climatic records, Chem. Geol., 166, 255–269, 2000.

Fairchild, I. J., Baker, A., Borsato, A., Frisia,S., Hinton, R. W., McDermott, F., and Tooth, A. F.: High-resolution, multiple trace-element variation in speleothems, J. Geol. Soc. London, 158, 831–841, 2001.

Fairchild, I. J., Smith, C. L., Baker, A., Fuller, L., Spötl, C., Mattey, D., and McDermott, F.: Modification and preservation of environmental signals in speleothems, Earth Sci. Rev., 75, 105–153, 2006.

Frappier, A. B.: Masking of interannual climate proxy signals by residual tropical cyclone rainwater: Evidence and challenges for low-latitude speleothem paleoclimatology, Geochem. Geophys. Geosyst., 14, 3632–3647, doi:10.1002/ggge.20218, 2013.

Frisia, S., Borsato, A., Fairchild, I. J., and Susini, J.: Variations in atmospheric sulphate recorded in stalagmites by synchrotron micro-XRF and XANES analyses, Earth Planet. Sci. Lett., 235, 729–740, 2005.

Frisia, S., Borsato, A., and Susini, J.: Synchrotron radiation applications to past volcanism archived in speleothems: An overview, J. Volcanol. Geoth. Res., 117, 96–100, 2008.

Goede, A., McCulloch, M., McDermott, F., and Hawkesworth, C.: Aeolian contribution to strontium and strontium isotope variations in a Tasmanian speleothem, Chem. Geol., 149, 37–50, 1998.

Hartland, A., Fairchild, I. J., Lead, J. R., Borsato, A., Baker, A., Frisia, S., and Baalousha, M.: From soil to cave: Transport of trace metals by natural organic matter in karst dripwaters, Chem. Geol., 304–305, 68–82, 2012.

Hughes, C. E. and Crawford, J.: Spatial and temporal variation in precipitation isotopes in the Sydney Basin, Australia, J. Hydrol., 489, 42–55, 2013.

Hurst, H.: Methods of long-term storage in reservoirs, T. Am. Soc. Civ. Eng., 116, 519–543, 1951.

Jo, K., Kyung, S., Gi, H., Kim, S., and Suk, B.: Rainfall and hydrological controls on speleothem geochemistry during climatic events (droughts and typhoons): an example from Seopdong Cave, Republic of Korea, Earth Planet. Sci. Lett., 295, 441–450, 2010.

Johnson, K., Hu, C., Belshaw, N., and Henderson, G.: Seasonal trace-element and stable-isotope variations in a Chinese speleothem: the potential for high-resolution paleomonsoon reconstruction, Earth Planet. Sci. Lett., 244, 394–407, 2006.

Lauritzen, S. E. and Lundberg, J.: Speleothems and climate: A special issue of the Holocene, Holocene, 9, 643–647, 1999.

Markowska, M., Baker, A., Treble, P. C., Andersen, M. S., Hankin, S., Jex, C. N., Tadros, C. V., and Roach, R.: Unsaturated zone hydrology and cave drip discharge water response: Implications for speleothem paleoclimate record variability, J. Hydrol., 529, 662–675, 2015.

McDonald, J., Drysdale, R., and Hill, D.: The 2002–2003 El Niño recorded in Australian cave drip waters: Implications for reconstructing rainfall histories using stalagmites, Geophys. Res. Lett., 31, L22202, doi:10.1029/2004GL020859, 2004.

McGowan, H. A., Marx, S. K., Denholm, J., Soderholm, J., and Kamber, B. S.: Reconstructing annual inflows to the headwater catchments of the Murray River, Australia, using the Pacific Decadal Oscillation, Geophys. Res. Lett., 36, L06707, doi:10.1029/2008GL037049, 2009.

Murphy, B. F. and Timbal, B.: A review of recent climate variability and climate change in Southeastern Australia, Int. J. Climatol., 28, 859–879, 2008.

Nagra, G., Treble, P. C., Andersen, M. S., Fairchild, I. J., Coleborn, K., and Baker, A.: A post-wildfire response in cave dripwater chemistry, Hydrol. Earth Syst. Sci., 20, 2745–2758, doi:10.5194/hess-20-2745-2016, 2016.

Nicholl, O.: Harrie Wood Cave Map – Y26, Yarrangobilly Research Group and Canberra Speleological Society, Canberra, 1974.

Nicholls, N.: Local and remote causes of the southern Australian autumn–winter rainfall decline, 1958–2007, Clim. Dynam., 34, 835–845, 2010.

Oster, J. L., Montañez, I. P., and Kelley, N. P.: Response of a modern cave system to large seasonal precipitation variability, Geochim. Cosmochim. Ac., 91, 92–108, 2012.

Partin, J. W., Cobb, K. M., Adkins, J. F., Tuen, A. A., and Clark, B.: Trace metal and carbon isotopic variations in cave dripwater and stalagmite geochemistry from northern Borneo, Geochem. Geophys. Geosyst., 14, 3567–3585, doi:10.1002/ggge.20215, 2013.

Pook, M. J., Risbey, J. S., and McIntosh, P. C.: A comparative synoptic climatology of cool-season rainfall in major grain-growing regions of southern Australia, Theor. Appl. Climatol., 117, 521–533, 2014.

Power, S. B. and Smith, I. N.: Weakening of the Walker Circulation and apparent dominance of El Niño both reach record levels, but has ENSO really changed?, Geophys. Res. Lett., 34, L18702, doi:10.1029/2007GL030854, 2007.

Raupach, M. R., Briggs, P. R., Haverd, V., King, E. A., Paget, M., and Trudinger, C. M.: Australian Water Availability Project (AWAP): CSIRO Marine and Atmospheric Research Component: Final Report for Phase 3. CAWCR, Technical Report No. 013, 67 pp., 2009.

Raupach, M. R., Briggs, P. R., Haverd, V., King, E. A., Paget, M., and Trudinger, C. M.: Australian Water Availability Project. CSIRO Marine and Atmospheric Research, Canberra, Australia, http://www.csiro.au/awap (last access: 15 November 2015), 2011.

Riechelmann, D. F. C., Schroder-Ritzrau, A., Scholz, D., Fohlmeister, J., Spotl, C., Richter, D. K., and Mangini, A.: Monitoring Bunker Cave (NW Germany): A prerequisite to interpret geochemical proxy data of speleothems from this site, J. Hydrol., 409, 682–695, 2011.

Risbey, J. S., Pook, M. J., McIntosh, P. C., Wheeler, M. C., and Hendon, H. H.: On the remote drivers of rainfall variability in Australia, Mon. Weather Rev., 137, 3233–3253, 2009.

Roberts, M., Smart, P., and Baker, A.: Annual trace element variations in a Holocene speleothem, Earth Planet. Sci. Lett., 154, 237–246, 1998.

Rozanski, K., Araguas-Araguas, L., and Gonfiantini, R.: Isotopic patterns in modern global precipitation, in: Climate Change in Continental Isotopic Records, edited by: Swart, P. K., Lohmann, K. C., McKenzie, J., and Savin, S., American Geophysical Union, Geophysical Monograph No. 78, Washington, D.C., 1–36, 1993.

Rutlidge, H., Baker, A., Marjo, C. E., Andersen, M. S., Graham, P. W., Cuthbert, M. O., Rau, G. C., Roshan, H., Markowska, M., Mariethoz, G., and Jex, C. N.: Dripwater organic matter and trace element geochemistry in a semi-arid karst environment: Implications for speleothem paleoclimatology, Geochim. Cosmochim. Ac., 135, 217–230, 2014.

Sinclair, D. J., Banner, J. L., Taylor, F. W., Partin, J., Jenson, J., Mylroie, J., Goddard, E., Quinn, T., Jocson, J., and Miklavic, B.: Magnesium and strontium systematics in tropical speleothems from the Western Pacific, Chem. Geol., 294, 1–17, 2012.

Spötl, C., Fairchild, I. J., and Tooth, A. F.: Cave air control on dripwater geochemistry, Obir Caves (Austria): Implications for speleothem deposition in dynamically ventilated caves, Geochim. Cosmochim. Ac., 69, 2451–2468, 2005.

Stern, H., de Hoedt, G., and Ernst, J.: Objective classification of Australian climates, Australian Meteorological Magazine, 49, 87–96, 2000.

Tooth, A. F. and Fairchild, I. J.: Soil and karst aquifer hydrological controls on the geochemical evolution of speleothem-forming drip waters, Crag Cave, southwest Ireland, J. Hydrol., 273, 51–68, 2003.

Treble, P., Shelley, J. M. G., and Chappell, J.: Comparison of high resolution sub-annual records of trace elements in a modern (1911–1992) speleothem with instrumental climate data from southwest Australia, Earth Planet. Sci. Lett., 216, 141–153, 2003.

Treble, P. C., Fairchild, I. J., Griffiths, A., Baker, A., Meredith, K. T., Wood, A., and McGuire, E.: Impacts of cave air ventilation and in-cave prior calcite precipitation on Golgotha Cave dripwater chemistry, southwest Australia, Quaternary Sci. Rev., 127, 61–72, doi:10.1016/j.quascirev.2015.06.001, 2015.

Treble, P. C., Fairchild, I. J., Baker, A., Meredith, K. M., Andersen, M. S., Salmon, S. U., Bradley, C., Wynn, P. M., Hankin, S., Wood, A., and McGuire, E.: Roles of bioproductivity, transpiration and fire in a nine-year record of cave dripwater chemistry from southwest Australia, Geochim. Cosmochim. Ac., 184, 132–150, doi:10.1016/j.gca.2016.04.017, 2016.

Tremaine, D. M. and Froelich, P. N.: Speleothem trace element signatures: A hydrologic geochemical study of modern cave dripwaters and farmed calcite, Geochim. Cosmochim. Ac., 121, 522–545, 2013.

Worboys, G.: Kosciuosko National Park Geology and Geomorphology, National Parks and Wildlife Service, Sydney, 1–26, 1982.

Wynn, P. M., Fairchild, I. J., Baker, A., Baldini, J. U., and McDermott, F.: Isotopic archives of sulphate in speleothems, Geochim. Cosmochim. Ac., 72, 2465–2477, 2008.

PERMISSIONS

All chapters in this book were first published in HESS, by Copernicus Publications; hereby published with permission under the Creative Commons Attribution License or equivalent. Every chapter published in this book has been scrutinized by our experts. Their significance has been extensively debated. The topics covered herein carry significant findings which will fuel the growth of the discipline. They may even be implemented as practical applications or may be referred to as a beginning point for another development.

The contributors of this book come from diverse backgrounds, making this book a truly international effort. This book will bring forth new frontiers with its revolutionizing research information and detailed analysis of the nascent developments around the world.

We would like to thank all the contributing authors for lending their expertise to make the book truly unique. They have played a crucial role in the development of this book. Without their invaluable contributions this book wouldn't have been possible. They have made vital efforts to compile up to date information on the varied aspects of this subject to make this book a valuable addition to the collection of many professionals and students.

This book was conceptualized with the vision of imparting up-to-date information and advanced data in this field. To ensure the same, a matchless editorial board was set up. Every individual on the board went through rigorous rounds of assessment to prove their worth. After which they invested a large part of their time researching and compiling the most relevant data for our readers.

The editorial board has been involved in producing this book since its inception. They have spent rigorous hours researching and exploring the diverse topics which have resulted in the successful publishing of this book. They have passed on their knowledge of decades through this book. To expedite this challenging task, the publisher supported the team at every step. A small team of assistant editors was also appointed to further simplify the editing procedure and attain best results for the readers.

Apart from the editorial board, the designing team has also invested a significant amount of their time in understanding the subject and creating the most relevant covers. They scrutinized every image to scout for the most suitable representation of the subject and create an appropriate cover for the book.

The publishing team has been an ardent support to the editorial, designing and production team. Their endless efforts to recruit the best for this project, has resulted in the accomplishment of this book. They are a veteran in the field of academics and their pool of knowledge is as vast as their experience in printing. Their expertise and guidance has proved useful at every step. Their uncompromising quality standards have made this book an exceptional effort. Their encouragement from time to time has been an inspiration for everyone.

The publisher and the editorial board hope that this book will prove to be a valuable piece of knowledge for researchers, students, practitioners and scholars across the globe.

LIST OF CONTRIBUTORS

T. H. M. Volkmann and M. Weiler
Chair of Hydrology, Faculty of Environment and Natural Resources, University of Freiburg, Fahnenbergplatz, 79098 Freiburg, Germany

Jana von Freyberg and James W. Kirchner
Department of Environmental Systems Science, ETH Zurich, Zurich, Switzerland
Swiss Federal Institute for Forest, Snow and Landscape Research (WSL), Birmensdorf, Switzerland

Bjørn Studer
Department of Environmental Systems Science, ETH Zurich, Zurich, Switzerland

Michael Rinderer
Chair of Hydrology, University of Freiburg, Freiburg im Breisgau, Germany

Nguyen Le Duy
GFZ German Research Centre for Geosciences, Section 5.4 - Hydrology, Potsdam, Germany
SIWRR Southern Institute of Water Resources Research, Ho Chi Minh City, Vietnam

Ingo Heidbüchel and Heiko Apel
GFZ German Research Centre for Geosciences, Section 5.4 - Hydrology, Potsdam, Germany

Hanno Meyer
Alfred Wegener Institute for Polar and Marine Research, Research Unit Potsdam, Potsdam, Germany

Bruno Merz
GFZ German Research Centre for Geosciences, Section 5.4 - Hydrology, Potsdam, Germany
University Potsdam, Institute of Earth and Environmental Science, Potsdam, Germany

Natalie Orlowski
Global Institute for Water Security, School of Environment and Sustainability, University of Saskatchewan, Saskatoon, Canada
Institute for Landscape Ecology and Resources Management (ILR), Research Centre for BioSystems, Land Use and Nutrition (IFZ), Justus Liebig University Giessen, Giessen, Germany
Hydrology, Faculty of Environment and Natural Resources, University of Freiburg, Freiburg, Germany

Craig S. Cook
Department of Ecosystem Science and Management, Stable Isotope Facility, University of Wyoming, Laramie, Wyoming, USA

Pedro Hervé-Fernández
Isotope Bioscience Laboratory (ISOFYS), Faculty of Bioscience Engineering, Ghent University, Ghent, Belgium
Instituto de Ciencias de la Tierra, Universidad Austral de Chile, Valdivia, Chile
Laboratory of Hydrology and Water Management, Faculty of Bioscience Engineering, Ghent University, Ghent, Belgium

Christophe Hissler
Luxembourg Institute of Science and Technology (LIST), Department of Environmental Research and Innovation (ERIN), Esch-sur-Alzette, Luxembourg, Luxembourg

Paul Koeniger
German Federal Institute for Geosciences and Natural Resources (BGR), Hanover, Germany

Arnaud Legout
INRA UR1138 Biogéochimie des Ecosystèmes Forestiers, INRA Centre de Nancy, Champenoux, France

Regine Redelstein
Plant Ecology and Ecosystems Research, University of Göttingen, Göttingen, Germany

Christof Seidler
Ecophysiology of Plants, Technical University of Munich, Munich, Germany

Christine Stumpp
Institute of Groundwater Ecology, German Research Center for Environmental Health, Helmholtz Zentrum München, Neuherberg, Germany
Institute of Hydraulics and Rural Water Management (IHLW), University of Natural and Life Sciences (BOKU), Vienna, Austria

Simon Thomsen
Institute of Soil Science, University of Hamburg, Hamburg, Germany

Maren Dubbert and Christiane Werner
BayCEER, University of Bayreuth, Bayreuth, Germany
Ecosystem Physiology, University of Freiburg, Freiburg, Germany

Xin Luo and Jiu Jimmy Jiao
Department of Earth Sciences, The University of Hong Kong, Hong Kong, P. R. China
The University of Hong Kong, Shenzhen Research Institute (SRI), Shenzhen, P. R. China
The University of Hong Kong-Zhejiang Institute of Research and Innovation (HKU-ZIRI), Hangzhou, PR China

Xingxing Kuang and Hailong Li
School of Environmental Science and Engineering, Southern University of Science and Technology, 1088 Xueyuan Rd., Shenzhen, China

Sihai Liang
School of Water Resources & Environment, China University of Geosciences, 29 Xueyuan Road, Beijing, China

Rong Mao and Xiaolang Zhang
Department of Earth Sciences, The University of Hong Kong, Hong Kong, P. R. China
The University of Hong Kong, Shenzhen Research Institute (SRI), Shenzhen, P. R. China
School of Environmental Science and Engineering, Southern University of Science and Technology, 1088 Xueyuan Rd., Shenzhen, China

Matthias Sprenger, Doerthe Tetzlaff and Chris Soulsby
Northern Rivers Institute, School of Geosciences, University of Aberdeen, Aberdeen, AB24 3UF, UK

Katarina David
School of Minerals and Energy Resource Engineering, and Connected Waters Initiative, University of New South Wales, Sydney, Australia

Wendy Timms
School of Engineering, Deakin University, Waurn Ponds, Australia

Catherine E. Hughes and Jagoda Crawford
Australian Nuclear Science and Technology Organisation, Sydney, Australia

Dayna McGeeney
Australian Museum, Sydney, Australia

Martin Suchy and Gwyn Graham
Environment and Climate Change Canada, 201-401 Burrard St., Vancouver, BC, Canada

Leonard I. Wassenaar
International Atomic Energy Agency, Vienna International Centre, 1400, Vienna, Austria

Bernie Zebarth
Fredericton Research and Development Centre, Agriculture and Agri-Food Canada, 850 Lincoln Rd., NB, E3B 4Z7, Canada

Yusuf Jameel and Richard P. Fiorella
Geology and Geophysics, University of Utah, 115 S 1460 E, Salt Lake City, Utah 84112, USA

Simon Brewer
Department of Geography, University of Utah, 332 S 1400 E, Salt Lake City, Utah 84112, USA

Brett J. Tipple
Global Change and Sustainability Center, University of Utah, 115 S 1460 E, Salt Lake City, Utah 84112, USA
Department of Biology, University of Utah, 257 S 1400 E, Salt Lake City, Utah 84112, USA

Shazelle Terry
Jordan Valley Water Conservancy District, 8215 S 1300 W, West Jordan, Utah 84088, USA

Gabriel J. Bowen
Geology and Geophysics, University of Utah, 115 S 1460 E, Salt Lake City, Utah 84112, USA
Global Change and Sustainability Center, University of Utah, 115 S 1460 E, Salt Lake City, Utah 84112, USA

Paolo Benettin
Laboratory of Ecohydrology ENAC/IIE/ECHO, École Polytechinque Fédérale de Lausanne (EPFL), Lausanne, Switzerland

Till H. M. Volkmann
Biosphere 2, University of Arizona, Tucson, AZ, USA

Jana von Freyberg
Department of Environmental Systems Science, ETH Zurich, Zurich, Switzerland
Research Unit Mountain Hydrology and Mass Movements, Swiss Federal Institute for Forest, Snow and Landscape Research (WSL), Birmensdorf, Switzerland

Jay Frentress
Faculty of Science and Technology, Free University of Bolzano, Bolzano, Italy

Daniele Penna
Department of Agricultural, Food and Forestry Systems, University of Florence, Florence, Italy

Todd E. Dawson
Department of Integrative Biology, University of California, Berkeley, CA, USA

James W. Kirchner
Department of Environmental Systems Science, ETH Zurich, Zurich, Switzerland
Research Unit Mountain Hydrology and Mass Movements, Swiss Federal Institute for Forest, Snow and Landscape Research (WSL), Birmensdorf, Switzerland
Department of Earth and Planetary Science, University of California, Berkeley, CA, USA

Erik Oerter
Department of Geology and Geophysics, University of Utah, 115 South 1460 East, Salt Lake City, UT 84112, USA
Global Change and Sustainability Center, University of Utah, 257 South 1400 East, Salt Lake City, UT 84112, USA
Lawrence Livermore National Laboratory, 7000 East Avenue, Livermore, CA 94550, USA

Molly Malone, Dina Drits-Esser and Louisa Stark
Genetic Science Learning Center, University of Utah, 515 100 South, Salt Lake City, UT 84112, USA

Annie Putman and Gabriel Bowen
Department of Geology and Geophysics, University of Utah, 115 South 1460 East, Salt Lake City, UT 84112, USA
Global Change and Sustainability Center, University of Utah, 257 South 1400 East, Salt Lake City, UT 84112, USA

Yonatan Ganot
Institute of Soil, Water and Environmental Sciences, The Volcani Center, Agricultural Research Organization, Rishon LeZiyyon, 7505101, Israel
Department of Soil and Water Sciences, The Hebrew University of Jerusalem, Rehovot, 7610001, Israel
Department of Land, Air, and Water Resources, University of California, Davis, 95616, USA

Ran Holtzman
Department of Soil and Water Sciences, The Hebrew University of Jerusalem, Rehovot, 7610001, Israel

Noam Weisbrod, Anat Bernstein and Hagar Siebner
Department of Environmental Hydrology & Microbiology, Zuckerberg Institute for Water Research, Jacob Blaustein Institutes for Desert Research, Ben-Gurion University of the Negev, Midreshet Ben-Gurion, 8499000, Israel

Yoram Katz
Mekorot, Water Company Ltd., Tel Aviv, 6713402, Israel

Daniel Kurtzman
Institute of Soil, Water and Environmental Sciences, The Volcani Center, Agricultural Research Organization, Rishon LeZiyyon, 7505101, Israel

Dongmei Han
Key Laboratory of Water Cycle & Related Land Surface Processes, Institute of Geographic Sciences and Natural Resources Research, Chinese Academy of Sciences, Beijing, 100101, China
College of Resources and Environment, University of Chinese Academy of Sciences, Beijing 100049, China

Matthew J. Currell
School of Engineering, RMIT University, Melbourne VIC 3000, Australia

Index

S

T

U

W

www.ingramcontent.com/pod-product-compliance
Lightning Source LLC
LaVergne TN
LVHW060512210926
848485LV00005B/1068